COLLECTION COLLOQUES ET SÉMINAIRES

IFP EXPLORATION RESEARCH CONFERENCES
41   Thermal Phenomena in Sedimentary Basins
44   Thermal Modeling in Sedimentary Basins

The complete list of the
Collection Colloques et Séminaires
is at the end of the book

**COLLECTION
COLLOQUES ET SÉMINAIRES 44**

Edited by **Jean BURRUS**
Institut Français du Pétrole

Organized by
**INSTITUT FRANÇAIS DU PÉTROLE
CENTRE NATIONAL
DE LA RECHERCHE SCIENTIFIQUE**

THERMAL MODELING IN SEDIMENTARY BASINS

1st IFP EXPLORATION RESEARCH CONFERENCE
CARCANS, FRANCE, JUNE 3-7, 1985

1986

Distributed in the United States by
GULF PUBLISHING COMPANY
Houston, Texas

ÉDITIONS TECHNIP 27 RUE GINOUX 75737 PARIS CEDEX 15   techniP

Randall Library UNC-W

© 1986. Editions Technip - Paris

All rights reserved. No part of this publication may be reproduced or
transmitted in any form or by any means, electronic or mechanical, including
photocopy, recording, or any information storage and retrieval system,
without the prior written permission of the publisher

ISSN 0073-8360
ISBN 2-7108-0504-9

Printed in France
by CID Éditions, 44800 Saint-Herblain

WHY IFP EXPLORATION RESEARCH CONFERENCES?

The idea for organizing IFP Research Conferences came from the success encountered by a meeting organized by IFP and CNRS on "Thermal Phenomena in Sedimentary Basins" in 1983 near Bordeaux (France). Indeed it demonstrated that people from various disciplines and from both Research Institutions and Industry can meet fruitfully to discuss scientific problems of interest for Oil Exploration. Such conferences are useful because, more than ever, Oil Exploration needs to integrate recent advances in every relevant field of sciences, with the immediate target of improving the discovery ratio of exploration wells.

This immediate target does not preclude a long term goal; whatever the short term fluctuations, it will be more and more difficult to discover new hydrocarbon pools in the future.

The topics of this first Research Conference held between 3-7 June 1985 reflected an irreversible trend in Exploration towards a quantitative approach to the petroleum evaluation of sedimentary basins. One can expect that in the future, sedimentary basins will be considered by Explorationists as oil fields by Reservoir Engineers: Modeling will be a normal tool for helping decision in Exploration. But presently modeling is also an essential research tool for defining which are the domains where more progress has to be made.

IFP by its structure and its mission is particularly interested and designated for organizing such meetings. We anticipate therefore other Research Conferences in the future on various topics of geology, geochemistry and geophysics relevant to hydrocarbon exploration.

L. Montadert
Director of Exploration
Institut Français du Pétrole

FOREWORD

Two years after a first symposium held at St Médard en Jalles in 1983 and devoted to the "Thermal Phenomena in Sedimentary Basins" the Research Conference on "Thermal Modeling in Sedimentary Basins" was held from 3 to 7 June 1985 in Carcans-Maubuisson, near Bordeaux (France). Sponsored by the Institut Français du Pétrole and the Centre National de la Recherche Scientifique, the symposium was efficiently impulsed at IFP by L. Montadert, Director of Exploration, and B. Durand, Head of the Geochemistry Department, and at CNRS by M. Combarnous, Director of the Sciences for the Engineer. More than 60 research scientists from both academic and industrial institutions of ten different countries involved in such various fields as physics of the Interior of the Earth, hydrogeology, heat-transfer in porous medium, compaction and over pressuring, petroleum occurrence, organic geochemistry, etc. met with the goal to find common interest in exchanging experience, ideas and questions.

The Conference could not have been held without essential support from C. Sallé, Chief Geologist, B. Tissot, Scientific Director, J. Funck, Director of Information and Documentation, and C. Bois, Director of Geological Research, at IFP. The six oil companies who accepted to contribute to presentations on their own fields of interest played a crucial role in the success of the meeting. Having been in charge of the organisation of the Conference, I wish to thank M. Combarnous who presented the opening address, and B. Delfour from IFP and R. Beignet from ADERA, who kindly and efficiently ensured the secretariat. All the participants are indebted to Nicole Combarnous, who managed to find a convenient place between the pines of the Forêt des Landes and the sandy shore of the Atlantic, and introduced them to the pleasant "Bordeaux way of life".

Finally, J.C. Marcolin in charge of publications at IFP and J. Monicat, from Editions Technip, provided precious help in the edition of this volume, which contains most of the papers presented at Carcans.

The editor
J. BURRUS
Institut Français du Pétrole

CONTENTS

Why IFP exploration research conferences? by L. Montadert V

Foreword, by J. Burrus VII

List of contributors XV

J. BURRUS
Introduction 1

CHAPTER 1
MODELING THE GEODYNAMIC EVOLUTION

V. ČERMÁK, L. BODRI
Temperature structure of the lithosphere based on 2-D temperature modelling applied to Central and Eastern Europe 7

C. JAUPART
On the average amount and vertical distribution of radioactivity in the continental crust 33

L. ROYDEN
A simple method for analyzing subsidence and heat flow in extensional basins 49

F. LUCAZEAU
The post-rift evolution of the Massif Central (France) 75

H.J. NEUGEBAUER
On the thermal evolution of interior platform basins. Dynamic thermal processes in the lithosphere 91

CONTENTS

I. MORETTI, C. FROIDEVAUX
Physical models of extensional tectonics 107

CHAPTER 2
HEAT TRANSFER IN SEDIMENTS

PART A
POROSITY AND COMPACTION

K. MAGARA
Porosity-depth relationship during compaction in hydrostatic
and non-hydrostatic cases 129

N.C. DUTTA
Shale compaction, burial diagenesis and geopressures: a dynamic
model, solution and some results 149

B. DOLIGEZ, F. BESSIS, J. BURRUS, P. UNGERER, P.Y. CHENET
Integrated numerical simulation of the sedimentation, heat
transfer, hydrocarbon formation and fluid migration in a
sedimentary basin: the THEMIS model 173

PART B
HEAT FLOW AND WATER FLOW

W.D. GOSNOLD, D.W. FISCHER
Heat flow studies in sedimentary basins 199

M.N. LUHESHI, D. JACKSON
Conductive and convective heat transfer in sedimentary basins.. 219

P. GOBLET, E. LEDOUX, G. de MARSILY
Possibilities of abnormal flow in sedimentary basins: some
examples 235

CONTENTS

C. HERMANRUD

On the importance to the petroleum generation of heating effects from compaction derived water: an example from the northern North Sea 247

M. QUINTARD, D. BERNARD

Free convection in sediments: numerical modelling and time space scaling 271

PART C
CONDUCTIVITIES, PERMEABILITIES, RADIOACTIVE HEAT GENERATION IN SEDIMENTS

A.J. SILVA, R.H. MORIN

The sensitivity of sediment physical properties to changes in temperature, pressure and porosity 289

L. RYBACH

Amount and significance of radioactive heat sources in sediments 311

V.V. PALCIAUSKAS

Models for thermal conductivity and permeability in normally compacting basins 323

PART D
EXAMPLES OF THERMAL RECONSTRUCTIONS

F. HORVATH, A. SZALAY, P. DOVENYI, J. RUMPLER

Structural and thermal evolution of the Pannonian Basin: an overview 339

I. HUTCHISON

Numerical modelling of oceanic heat flow. A western Mediterranean case study 359

CONTENTS

F. LUCAZEAU, J.P. BARRIOT, S. LE DOUARAN
Gravity constraints on thermal models for extensional basin: example of the Provencal Basin 375

J. BURRUS, F. BESSIS
Thermal modeling in the Provencal Basin (NW-Mediterranean) 393

J.A. NUNN
Subsidence and thermal history of the Michigan Basin 417

CHAPTER 3

CONTROL BY GEOCHEMICAL METHODS

PART A
ORGANIC METHODS

B. DURAND, B. ALPERN, J.L. PITTION, B. PRADIER
Reflectance of vitrinite as a control of thermal history of sediments 441

J. ESPITALIE
Use of Tmax as a maturation index for different types of organic matter. Comparison with vitrinite reflectance 475

E. BROSSE, G. DEROO, J. ROUCACHE, T.A. BOTNEVA
Organic geochemistry as a test of validity for the results of a modelisation in the Pripiat Basin (Bielorussia) 497

E. BROSSE, A.Y. HUC
Organic parameters as indicators of thermal evolution in the Viking Graben 517

CONTENTS

P. UNGERER, J. ESPITALIE, F. MARQUIS, B. DURAND
Use of kinetic models of organic matter evolution for the reconstruction of paleotemperatures. Application to the case of the Gironville Well (France) 531

J.J. SWEENEY, A.K. BURNHAM, R.L. BRAUN
A model of hydrocarbon maturation in the Uinta Basin, Utah, USA 547

PART B
MINERAL METHODS

M. PAGEL, F. WALGENWITZ, J. DUBESSY
Fluid inclusions in oil and gas-bearing sedimentary formations 565

D.W. PHELPS, T.M. HARRISON
Application of $^{40}Ar/^{39}Ar$ thermochronology on detrital potassium feldspars to the study of sedimentary basins 585

LIST OF PARTICIPANT CONTRIBUTORS

ALVAREZ Francis
Laboratoire de Géodynamique
Université P. & M. Curie
4, place Jussieu - Tour 15
75230 Paris CEDEX 05
France

BERNARD Dominique
LEPT - ENSAM
Esplanade des Arts & Métiers
33405 Talence CEDEX
France

BROSSE Etienne
Institut Français du Pétrole
Boite Postale 311
92506 Rueil Malmaison CEDEX
France

CERMAK Vladimir
Geophysical Institute
Czechosl. Acad. Sciences
14131 Praha
Czechoslovakia

DOLIGEZ Brigitte
Institut Français du Pétrole
Boite Postale 311
92506 Rueil Malmaison CEDEX
France

DUTTA N.C.
Geophysics Research Dept.
Exploration & Production Res.
Arco Oil & Gas Co.
P.O. Box 2819
Dallas - Tex 75221
U.S.A

FROIDEVAUX Claude
Laboratoire de Géophysique
Université de Paris Sud
91405 ORSAY
France

BERTHELOT Francis
Laboratoire de Géodynamique
Université P. & M. Curie
4, place Jussieu - Tour 15
75230 Paris CEDEX 05
France

BESSIS Frédéric
Institut Français du Pétrole
Boite Postale 311
92506 Rueil Malmaison CEDEX
France

BURRUS Jean
Institut Français du Pétrole
Boite Postale 311
92506 Rueil Malmaison CEDEX
France

CHéNET Pierre-Yves
Institut Français du Pétrole
Boite Postale 311
92506 Rueil Malmaison CEDEX
France

DURAND Bernard
Institut Français du Pétrole
Boite Postale 311
92506 Rueil Malmaison CEDEX
France

ESPITALIé Jean
Institut Français du Pétrole
Boite Postale 311
92506 Rueil Malmaison CEDEX
France

GOSNOLD D.
Department of Geology
University of North Dakota
Grand Fork ND 58202
USA

LIST OF PARTICIPANT CONTRIBUTORS

HERMANRUD Christian
Statoil, Forus
P.O. Box 300
N. 4001 Stavanger
Norway

HORVATH Frank
Geophysical Department
Eotvos University
1083 Budapest
Hungary

HUTCHISON Iain
B.P. Exploration
Britannic House
Moor Lane
London EC2Y9BU
United Kingdom

JAUPART Claude
Institut de Physique du Globe
Université P. & M. Curie
4, place Jussieu
75230 PARIS CEDEX 05
France

LE DOUARAN Sylvie
SNEA(P)
Tour Elf
La Défense 6
92078 Paris La Défense CEDEX 45
France

LEDOUX Emmanuel
Ecole des Mines de Paris
Centre d'informatique géologique
35, rue Saint Honoré
77305 Fontainebleau
France

LUCAZEAU Francis
Centre Géologique et Géophysique
Univ. Sciences & Techn. Languedoc
34060 Montpellier CEDEX
France

LUHESHI M.N.
B.P. Exploration
Britannic House
Moor Lane
London EC2Y9BU
United Kingdom

MAGARA Kinji
Faculty of Earth Sciences
King Abdulaziz University
Jeddah
Saudi Arabia

MORETTI Isabelle
Institut Français du Pétrole
Boite Postale 311
92506 Rueil Malmaison CEDEX
France

NEUGEBAUER Horst J.
Institut fur Geophysik
Technische Univ.Clausthal
D. 3392 Clausthal
R.F.A

NUNN J.A
Department of Geology
Louisiana State University
Baton Rouge
Louisiana 70803 4101
U.S.A.

OXBURGH E.R.
Department of Earth Sciences
University of Cambridge
Dowing Street
Cambridge CB2 3EQ
United Kingdom

PAGEL Maurice
CREGU
Boite Postale 23
54501 Vandoeuvre les Nancy CEDEX
France

LIST OF PARTICIPANT CONTRIBUTORS

PALCIAUSKAS V.V
Chevron Oil Research Company
P.O. Box 446
La Habra - CA 90631
U.S.A.

QUINTARD Michel
LEPT - ENSAM
Esplanade des Arts & Métiers
33405 Talence CEDEX
France

RYBACH L.
Institute of Geophysics ETC
CH 8093 Zurich
Switzerland

SWEENEY J.J.
Lawrence Livermore National Lab.
P.O. Box 808
Livermore - CA
U.S.A

PHELPS David W.
Exxon Prod. Research Co
P.O. Box 2189
Houston - TEX 77096
U.S.A.

ROYDEN Leigh
Department of Earth, Atmospheric
Planetary Sciences
Massachussetts Inst. of Techn.
Cambridge - MA 02139
U.S.A.

SILVA Armand J.
University of Rhode Island
Narragansett RI 02882
U.S.A

UNGERER Philippe
Institut Français du Pétrole
Boite Postale 311
92506 Rueil Malmaison CEDEX
France

LIST OF OTHER PARTICIPANTS

ANUNZIATA Romero
Institut Mexicano del Petroleo
Eje Central Lazaro Cardenas 152
Delegacion Gustavo A. Madero
Mexico DF CP 07730
Mexico

BACCIANA Mr.
ESSO REP
213, cours Victor Hugo
33130 Begles
France

CHIARAMONTE Mario A.
AGIP s.p.a.
20097 S. Donato Milanese
Milano
Italy

COMBARNOUS Michel
LEPT-ENSAM
Esplanade des Arts & Métiers
33405 Talence CEDEX
France

DUJON Saint-Clair
Laboratoire de Géologie
Ecole Normale Supérieure
46, rue d'Ulm
75005 Paris
France

MEUNIER Mr.
CREGU
Boite Postale 23
34501 Vandoeuvre les Nancy CEDEX
France

NEGRONI P.
BEICIP
Boite Postale 213
92502 Rueil Malmaison CEDEX
France

BALLY Albert W.
Rice University
Department of geology
Wiess School of Natural Sciences
P.O. Box 1892
Houston - TX 77251
U.S.A.

BREVART Olivier
SNEA(P)
Boite Postale 65
64018 Pau CEDEX
France

CHRISTIE H.J.
Rogaland Research Institute
Stavanger
Norway

DROUILLER Y.
BEICIP
Boite Postale 213
92502 Rueil Malmaison CEDEX
France

HARVY Charlie
SHELL Development Co
P.O. Bos 481
Houston - TX 77001
U.S.A.

MONTADERT Lucien
Institut Français du Pétrole
Boite Postale 311
92506 Rueil Malmaison CEDEX
France

NOVELLI Luciano
AGIP s.p.a.
20097 S. Donato Milanese
Milano
Italy

LIST OF OTHER PARTICIPANTS

NURUSMAN Mr.
Faculté des Sciences
Laboratoire de Géophysique
Univ. de Bretagne Occidentale
6, avenue Le Gorgeu
29287 Brest CEDEX
France

OUDIN Jean-Louis
TOTAL CFP
218, avenue haut-Levêque
33605 Pessac CEDEX
France

RIIS Fridtjof
Norvegian Petroleum Directorate
P.O. Box 600
N 4001 Stavanger
Norway

SCHULZ Rudiger
N L f B
Postfach 510153
3000 Hannover 1
R.F.A

SHVEIMA Joseph S.
Phillips Petroleum C
Bartlesville
Oklahoma 74004
U.S.A.

STORZER Dieter
C.N.R.S.
Labo. de minéralogie du Muséum
61, rue Buffon
75005 Paris
France

SVEIN EGGEN
STATOIL
Forus
P.O. Box 300
4001 Stavanger
Norway

OHM Sverre Ekrene
Norvegian Petroleum Directorate
P.O. Box 600
N. 4001 Stavanger
Norway

RABILLIER Philippe
SNEA(P)
Tour Elf
La Défense 6
92078 Paris La Défense CEDEX 45
France

ROBERT Paul
SNEA(P)
Centre de Recherche de Boussens
31360 Saint Martory
France

SELO Madeleine
C.N.R.S.
Laboratoire de minéralogie
61, rue Buffon
75005 Paris
France

SPEERS Gordon C.
Norsk Hydro Research Center
P.O. Box 4313
N. 5013 Nygardstargen
Bergen
Norway

SUNDVOR Eirik
Seismologicval Observatory
University of Bergen
N. 5000 Bergen
Norway

VELDE Bruce
Laboratoire de géologie
Ecole Normale Supérieure
46, rue d'Ulm
75230 Paris
France

XIX

INTRODUCTION

Considerable efforts have been carried out the last decade in order to quantify the geological and geochemical processes pertaining to the history of sedimentary basins, in particular for extensional basins, such as rifts, passive margins, and intracratonic basins. Basic physical concepts for subsidence, heat-flow and temperature reconstruction have been widely developped, and usefully applied for practical purposes (oil occurrence, but also geothermal energy and ore-deposits). In the mean-while, various phenomena such as compaction of sediments or thermal effect of water-flow in the sediments were better understood, and tentatively quantified through geological time. The first symposium organized in 1983 by IFP proved that a wide range of scientists found themselves involved by the study of "Thermal Phenomena in Sedimentary Basins". Two years after, new advances have been carried out: an increasing number of "modeling groups" are in activity throughout the world; integrated models have been developped, which treat the various aspects of thermal reconstruction, from sedimentation to compaction, convection and conduction; besides experimental progresses in the various fields of geochemistry made it possible to improve the reliability of paleothermometers. The organizers found this was the moment for an informal exchange of ideas on these various fields.

For this reason, the Conference was focussed on the state of the art in "Thermal Modeling", i.e. on the various concepts and techniques usefull to reconstruct or constrain quantitatively the history of temperatures in extensional sedimentary basins. These are three-fold, which correspond to the three parts of this volume:

-**The Geodynamic Evolution** of the Extensional Basin is the cause of the heat input into the Basin, which is generally a function of space and time; this input has tentatively been described by considering the evolution of the whole litho-asthenospheric system. Physical processes such as Lithospheric Extension, Asthenospheric Convection, Magmatism, Radiogenic heating, Mantle diapirism have been addressed in the first part of the Conference. Of particular practical interest for the purpose of many "Modeling Geologists" among the participants was the discussion of whether the popular "Lithospheric Extension" concept, first introduced by Mc Kenzie (1978) and then transformed into the "Non-Uniform Extension" concept by Royden, Keen and others in the 1980's, was or not

INTRODUCTION

compatible with the more complex litho-asthenospherical convective model developed in the mean-while. It turned out from the exchanges that the latter concept gave a physical justification to the idea of "Non-Uniform Extension", rather than it made it irrelevant. Nevertheless, it was emphasized that the geodynamic theories have been constrained in the past by indirect observations (subsidence, heat-flow, gravity, etc.), but that direct controls provided by deep crustal seismic data (COCORP, ECORS, and other surveys) have still to be taken into account.

- **The Heat** input in the Basin **is transferred** in the sediments undergoing compaction. The conduction is generaly responsible for most of the transfer, but water-flow, either due to regional flow or to free convection might be significant. In all the cases, the description of the evolution of porosity with depth is essential for both conductive and convective reconstructions; beside normal compaction trends, non-hydrostatic evolution is very often encountered and was discussed in terms of impervious burial, effective stress, aquathermal pressuring, oil generation, smectite diagenesis. The absence of paper on carbonate compaction is a good indication that efforts are still required in this field. The important problem of water circulation was especially addressed: examples of thermal anomalies clearly due to regional discharge were given, but the difficulty of reconstructing these circulations at present time and even more in the past million years was emphasized, as such undetermined factors as paleotopographies, paleoclimates are strongly involved. If compaction driven water-flow was shown to be generally negligible in the heat budget, it is not the case for the free convection. The description of thermal effects of convective cells is theoretically possible, but was said to be practically very difficult due the great sensitivity to the geological inhomogeneities in the sediments. Nevertheless quantitative boundaries can be set to the global changes induced in the heat budget by free convection. Finally, numerical integrated models accounting for heat-flow and water-flow in compacting basin were presented, and several examples discussed, but it was emphasized that our knowledge of the parameters (conductivities, permeabilities) is still well behind our modeling ability. Examples of theoretical and experimental researches for determining conductivities, permeabilities, radioactivities were given, but in most cases, one is faced with a "cruel lack of data". As a consequence, the uncertitude on the crucial parameters should be systematically evaluated and taken into account in the precision of the outputs of the models.

- Finally, temperature reconstructions have to be controlled: **Geochemical methods** provide usefull tools for this purpose. Organic geochemical methods are by far the most commonly used. Nevertheless, it was claimed that these methods (vitrinite

INTRODUCTION

reflectance, Tmax of pyrolysis, etc.) could lead to significant errors if not very properly carried out. The problem of calibrating the kinetics of hydrocarbons generation was also addressed by several speakers. By the end, mineral geochemistry was shown to provide promising techniques such as fluid inclusions, Argon and hydrocarbon thermochronology.

The above presentation of the discussions and papers presented at the Conference appears probably artificially "deterministic": it is not true that, in real cases, the "Modeling Geologist" first reconstructs the geodynamic evolution, then the heat and water-flow in the basin, and finally controls the outputs of his model by any geochemical method. More than often, the complete chain cannot be worked out, because crucial data or essential phenomena (as geodynamic history) are not constrained. As a consequence, the geochemical data or "paleothermometers" are often considered as the main input of the reconstruction, and all other concepts (geodynamic history for instance) are only working hypothesis: despite these different point of views, it appeared clearly from the discussions that the future research in "Thermal Modeling" must conjugate three parallel approaches:

. first, focussing on elementary processes such as compaction processes, paleo-hydrogeological reconstruction, or litho-asthenospherical evolution, paleothermometer calibration, etc.

. second, integrating these elementary processes in comprehensive models, because geological phenomena are always interacting, and can often not be studied individually,

. third, constraining our knowledge of inputs: conductivities, permeabilities, radiogenic distribution by theoretical studies on porous medium and by experimental measurements.

As pointed out during the Conference, advances in Thermal Modeling will result from simultaneous efforts in these three fields, and from subsequent communication between researchers, and between researchers and people involved in the operations.

Paris,
Décembre 1985
J. BURRUS
Institut Français du Pétrole

CHAPTER 1
MODELING THE GEODYNAMIC EVOLUTION

V. ČERMÁK[1], L. BODRI[2]

TEMPERATURE STRUCTURE OF THE LITHOSPHERE BASED ON 2-D TEMPERATURE MODELLING, APPLIED TO CENTRAL AND EASTERN EUROPE

Abstract :

With the available data on crustal structure, deep temperature distribution was calculated along five geotraverses in Central and Eastern Europe that cross all major tectonic units of Precambrian European craton, and of Variscan and Alpine units in this area. A two-dimensional numerical solution of the heat conduction equation was obtained by means of finite differences method. Seismic velocity distribution was used to defining individual crustal blocks, a specific value of thermal conductivity and of heat generation being ascribed to each of them. An experimental relation between seismic velocity and radiogenic heat production was introduced, heat production decreasing exponentially with depth. Thermal conductivity is temperature dependent. Low crustal temperatures are typical of the Precambrian East European platform /Moho temperature is 350-500 °C/ with a clear minimum beneath the Ukrainian shield. Moho temperatures slightly increase beneath Variscan units /500-600 °C/ and attain 600-800 °C in the Alpine realm. The highest temperatures may exist beneath hyperthermal basins, such as the Pannonian Basin /over 800 °C/. The results clearly confirm a possible existence of the asthenosphere at a depth as shallow as 50-60 km in the areas of very high heat flow. The regional variations of the Moho heat flow may range from 15-20 to 40-60 $mW.m^{-2}$.

---

(1) *Geophysical Institute, Czechoslovakian Academy of Sciences, Prague, Czechoslovakia.*
(2) *Department of Geophysics, L. Eötvös University, Budapest, Hungary.*

TEMPERATURE STRUCTURE OF THE LITHOSPHERE

I. INTRODUCTION

Historically, the terrestrial heat flow on continents has been used as a constraint on all models of chemical and physical processes occurring within the earth. The knowledge of the surface heat flow completed with certain assumptions on the vertical distribution of thermal conductivity and heat production allows the extrapolation of the near-surface data to a greater depth and the crustal temperatures to be calculated. As temperature controls the behaviour of the physical properties of rocks, such information is vital in further interpretation of geological and geophysical data.

On the territory of Central and Eastern Europe the basic geophysical studies are organized within the Research Programme of the Commission of the Academies of Sciences for Planetary Geophysics /KAPG/. Following various KAPG-research projects the attention of geologists and geophysicists has been recently focused to studying the detailed structure along five continent-run geotraverses /Fig.1/. These East European Geotraverses /EEGT/ were proposed on the basis of numerous national and international deep seismic sounding studies performed in 1963-1978 /SOLLOGUB et al., 1978, 1979/. Three of geotraverses are oriented approximately south-north : 1 - Alpine geosyncline-Bohemian Massif-East European platform /1100 km/, 2 - Dinarides-Pannonian Basin-Ukrainian shield /1800 km/, 3 - Black Sea-Crimea-Dnieper-Donetz aulacogen-Moscow syncline-Pechora syncline /3200 km/. Two geotraverses are west-east oriented : 4 - North Germany-Poland-Ukrainian shield /2200 km/, 5 - Bohemian Massif-Pannonian Basin-East Carpathians /1500 km/.

The present paper is an attempt at applying the available data on the crustal and upper mantle structure and to calculating the deep temperature distribution up to a depth of 60-70 km along these geotraverses.

II. TECTONIC SETTING

In the study area, the dominant role is played by the ancient East European craton surrounded by younger folded units. This oldest part of Europe /∼ 3100-600 Ma/ forms the nucleus of the whole continent. It comprises the Baltic and Ukrainian shields with exposed Precambrian crust and the East European platform where Precambrian

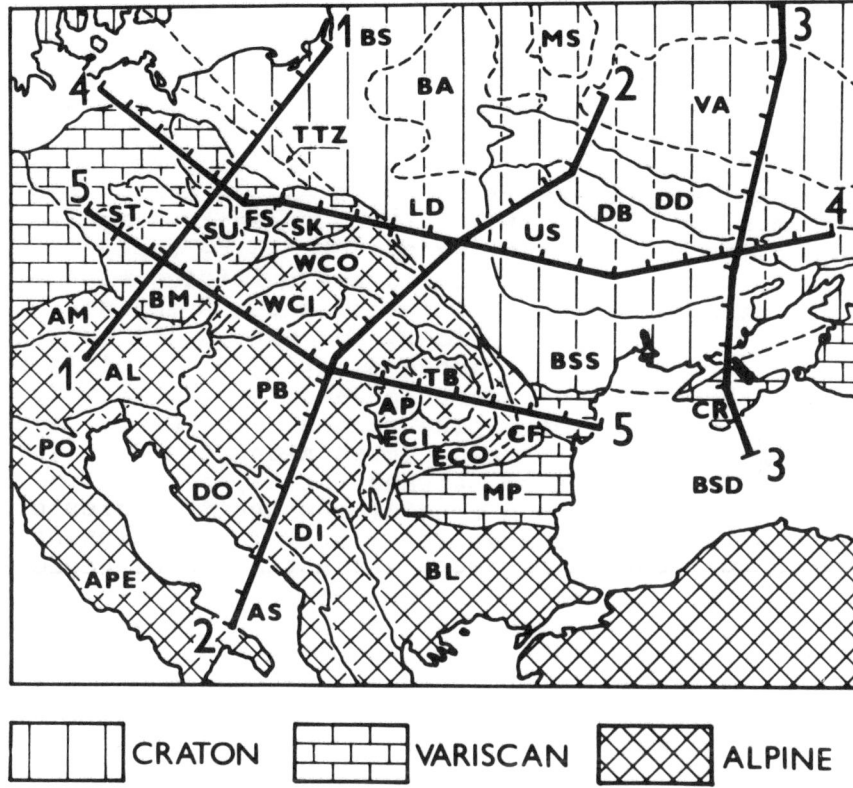

Fig.1. Simplified tectonic setting of Central and Eastern Europe together with the position of five geotraverses. EEGT 3 continues to the north and ends in the Timan-Pechora platform.

crystalline rocks occur beneath younger sediments. The ancient basement stretches to the south-west beneath the North-German-Polish Lowland and to the North Sea and to the north-east beneath the Timan-Pechora platform. The deep structure of both these areas is relatively poorly known and is usually considered to be of epi-Baikalian or, partly, epi-Caledonian age. Caledonian Europe /∼ 600-400 Ma/ encompasses predominantly the British Isles and Scandinavia and in the study area only a narrow Caledonian belt can be traced along the south-western border of the European craton. Further to the south-west, the craton is surrounded by the Variscan /Hercynian/ fold belt, an area of Late Paleozoic consolidation /∼ 400-230 Ma/. Once a wide mountain belt, it is now composed of a number of massifs of varying sizes cropping out in recently uplifted mountain units, while in the plains the Paleozoic, and locally, even older basement is burried under a Meso-Cenozoic platform cover. These units are more developed in Western Europe, in the investigated area they are only represented by the Bohemian Massif and by the Moesian Plate. In the south, the Variscan fold belt adjoins Alpine Europe, from which it is separated by a discontinuous zone of foredeeps. Alpine Europe /∼ 230-to present/ comprises the young mountainous regions of the Mediterranean area which were formed in relation to the collision of European and African plates. It includes the Apennines, Alps, Carpathians, Balkans, Dinarides, and extends to the east comprising Crimean Mts., Caucasus, etc. Of specific position is the large intramontane Pannonian Basin wedged between two branches of Alpine ranges. The Carpathians directly border the European craton along the Carpathian foredeep, but to the east between the Alpine belt and the ancient craton there is a zone of younger platforms with a Variscan and epi-Variscan basement. This zone includes the Black Sea region, Crimean plains and the Cis-Caucasus area. The epi-Variscan Donetz-Caspian system has a peculiar structure, the Black Sea and the South Caspian basins are characterized by an intermediate crust type, where the "granite" layer is absent or reduced in thickness and the lower crust is covered with an enormously thick layer of sediments. In the east, the European craton is framed by the huge Ural-Mongolian fold belt /Variscan/ separating it from the Siberian craton.

III. HEAT FLOW

With the use of more than 3000 heat flow data, a map of the distribution of the surface heat flow on the territory of Europe was prepared /ČERMÁK and HURTIG, 1979/. The regional heat flow field is clearly dominated by a general north-east to south-west increase of the geothermal activity, which is an obvious consequence of the continental tectonic evolution. A large low heat flow zone /30-45 $mW.m^{-2}$/ covers most of northern and eastern Europe and is surrounded by normal to high heat flow values spread in western, southern and south-eastern Europe.

The lowest heat flow is typical of the Baltic and Ukrainian shields /30-40 $mW.m^{-2}$/ and also of the major part of the East European platform /40-50 $mW.m^{-2}$/, i.e. of areas of the ancient Precambrian consolidated basement. Of great interest is the marked position of the graben structure of the Dnieper-Donetz aulacogen, the mean geothermal activity increases here to more than 50 mW.m^{-2}, as compared with the surrounding areas. In the east, the chain of the Ural Mts. generally closes the low heat flow zone of the East European platform.

To the south-west, the low heat flow is bordered by a line stretching from South Norway to the Black Sea /so called North Sea-Dobrudja lineament, which coincides with the Teisseyre-Tornquist zone/. The tectonic structure of Central Europe north of the Alpine-Carpathian fold belt is quite complicated and the heat flow field is dominated by an elongated zone of increased heat flow /Holland-Altmark-Sudeten/ sweeping from the east to the west, roughly following the southern margin of the Mid-European platform. In the east, the anomaly is diminishing on the northern edge of the Western Carpathians. The Bohemian Massif is typical by a closed anomaly of low heat flow, the minimum of which corresponds to the maximum crustal thickness.

The Alpine Carpathian Mts. range is generally characterized by elevated heat flow, however, there are strong local variations and the relation of the observed heat flow field to the local tectonic structure may be different in various parts of the whole system. While the range of the Western Alps is marked by an anomalously high heat flow, the heat flow pattern of the Eastern Alps is not so clear and the proper crest area of the Western Carpathians is the zone of a pronounced horizontal gradient of the heat flow field rather than an area of high heat flow. The whole Carpathian curved arc is typical by an increase of heat flow from the outer to the inner tectonic units. The extensive Pannonian Basin, situated inside the Carpathians, represents an imposing feature on the heat flow map with high to very high heat flow /80-100 mW.m-2/. The number of reliable heat flow data from the Balkan peninsula does not allow definite heat flow isolines to be constructed, but it seems that the heat flow pattern here is broken to a number of local anomalies. Low heat flow is chracteristic of Moldavian and Moesian platforms and surprisingly enough, also of the Transsylvanian depression. Local high heat flow zones are probably connected with hydrothermal features and thus difficult to be related to the deep crustal structure.

The area south of the East European platform on the territory of the USSR is characterized by highly variable heat flow, ranging from 55 to 80 mW.m^{-2}. Increased geothermal activity can be observed within most of the

Scythian Plate and quite prominent is the Stavropol uplift anomaly with a heat flow over 90 mW.m^{-2}. Local anomalies are usually west-east oriented and they follow the main trends of the Cenozoic folded structures. Also the observed heat flow values in the Crimea peninsula cover a wide range, 35-90 mW.m^{-2}, with a generally lower mean geothermal activity. Low heat flow is characteristic of all the Black Sea area, however, due to the enormous thickness of sediments, the measured data are problematic as yet, for the magnitude of the appropriate correction for sedimentation is still questionable.

IV. HEAT CONDUCTION MODEL

To describe the temperature distribution in the earth's crust and uppermost mantle, a two-dimensional numerical solution of a steady-state equation of heat conduction in an inhomogenous medium was applied :

$$\frac{\partial}{\partial x}\left(k \frac{\partial T}{\partial x}\right) + \frac{\partial}{\partial z}\left(k \frac{\partial T}{\partial z}\right) = A(x,z) ,$$

where $k(x,z,T)$ is thermal conductivity, $A(x,z)$ heat production, T temperature and x, z are coordinates.
As boundary conditions we used :
/1/ $T(x,z=0) = T_o(x)$, i.e. the knowledge of the surface temperature, which for simplicity can be taken 0 °C.
/2/ $dT/dx = 0$ at $x=0$, and at $x=L$, L is the length of the profile, i.e. the symmetry of the temperature field at vertical boundaries.
/3/ The last boundary conditions - the knowledge of temperature or heat flow at the base of the model, $T_M(x)$ or $Q_M(x)$, is problematic and as a rule this information is not available.

The solution of the above equation is thus ill posed and requires first the solution of the inverse problem, when the observed surface heat flow $Q(x,z=0) = Q_o$ can be used to evaluating the heat flow at depth, Q_M. For this purpose, the surface heat flow can be assumed to be the sum of two components : the crustal contribution due to crustal heat sources, $\bar{Q}$, and the mantle heat flow, Q_M. By repeated calculations of the surface heat flow corresponding to the model, Q_B, starting with an arbitrary value of Q_M and successively correcting the Q_M-value calculated by the difference $(Q_o - Q_B)$, it is possible to assess the conditions on the lower boundary of the model within a certain limit of accuracy.

TEMPERATURE STRUCTURE OF THE LITHOSPHERE

Several slightly different approaches to this problem were discussed by different authors and tested in concrete geological terrains. Successive overrelaxation method was applied to the Pannonian Basin /BODRI, 1981/, the regularization process to limit the maximum variation of Q_M per unit distance combined with the optimalization and redistribution of heat sources in the upper crust was studied e.g. by STROMEYER /1984/ and ŠAFANDA /1985/.

In principle it is possible to attain Q_M-distribution corresponding to any surface heat flow pattern for any required accuracy, however, at the expense of either unreal horizontal heat flow gradients or unreal distribution of additional heat sources. Therefore it is necessary to appreciate carefully the variation of Q_M per unit distance with respect to the local tectonic structure and also to assess the quality of data which serve as the criteria of the solution and to seek an acceptable compromise.
In the present paper, in which stress was laid on the evaluation of crustal temperatures along long-run geotraverses, certain simplification was necessary to reveal the universal character of the work performed. For reasons given below, the distribution of the heat sources was considered separately in the upper-ten-kilometres layer and in the lower bulk of the crust.

V. CONVERSION OF SEISMIC VELOCITIES INTO HEAT PRODUCTION

As radioactivity is supposed to be the main source of heat within the crust, empirical relation between seismic velocity, v_p, and radiogenic heat production, A, was employed to ascribe the characteristic heat production to individual crustal blocks based on their mean seismic velocities. On the basis of numerous laboratory measurements, RYBACH and BUNTERBARTH /1984/ reported relations, which are believed to be valid for rocks varying in composition from granite to ultrabasites. As older rocks usually show lower radioactivity, two formulas were proposed for Precambrian rocks, $\ln A = 12.6 - 2.17\, v_p$, and/or for Phanerozoic rocks, $\ln A = 13.7 - 2.17\, v_p$, respectively, demonstrating the limit of existing values, A being in $\mu W.m^{-3}$, v_p in km/s.

Laboratory determinations of v_p were performed at room temperature and at 100 MPa pressure, it is therefore necessary to make "in situ" and "laboratory" data comparable. For this purpose the correcting factor B was introduced:

$$v_P(20,100) = v_P(T,P)\left[1+B/v_P\right],$$

where

$$B = \frac{\partial v_P}{\partial T}\Delta T + \frac{\partial v_P}{\partial P}\Delta P.$$

Correcting factor B includes the temperature and pressure derivatives of seismic velocity, which for practical purposes were found to fit to ample experimental material compiled by KERN /1982/ and GEBRANDE /1982/, see Tables I and II.

If characteristic pressure-versus-depth and temperature-versus-depth curves are constructed for the investigated territory /Fig.2/ and completed with the typical seismic velocity-versus-depth profiles /Fig.3/, the "laboratory" formulas proposed by RYBACH and BUNTERBARTH /1984/ can be

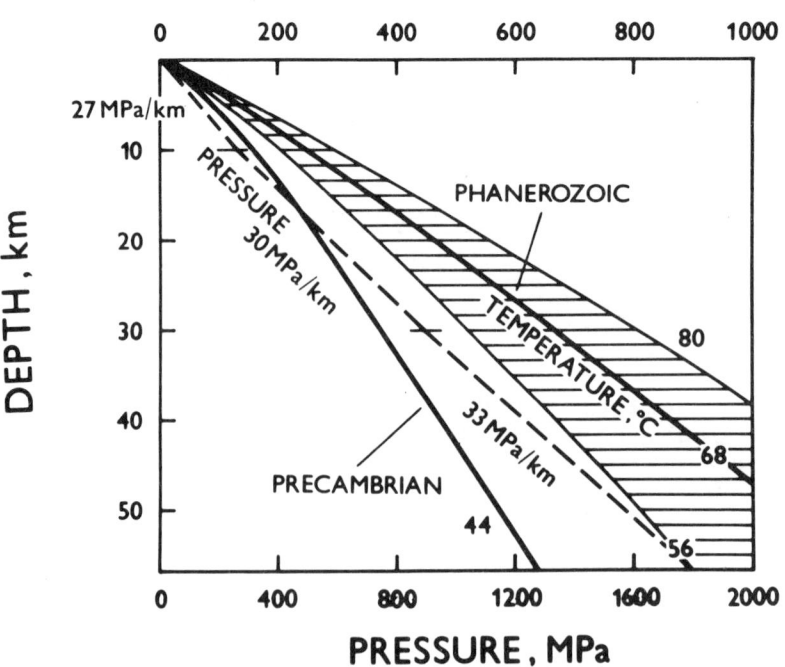

Fig.2. Characteristic curves of pressure /dashed/ and temperature /solid/ versus depth. Temperature curves are labelled by corresponding surface heat flow in $mW.m^{-2}$.

TABLE I. Temperature derivatives of seismic velocity for specific types of crustal rocks based on experimental data compiled by KERN /1982/.

Temperature range, °C	$-\frac{\partial v_P}{\partial T}$ (10^{-4} km.s^{-1}K^{-1})		
	acidic	basic	ultrabasic
0 - 100	2	2	3.5
100 - 200	2.5	2	4
200 - 300	3.5	2.5	4.5
300 - 400	5	3.5	5
400 - 500	6.5	5	5.5
500 - 600	8.5	7	6.5

TABLE II. Pressure derivatives of seismic velocity for crustal and upper mantle rocks, based on experimental data compiled by GEBRANDE /1982/.

Pressure range MPa	$\frac{\partial v_P}{\partial P}$ (10^{-4} km.s^{-1}MPa^{-1})
0 - 100	high, variable
100 - 200	10
200 - 400	6
400 - 600	3
600 - 1000	2
1000	1

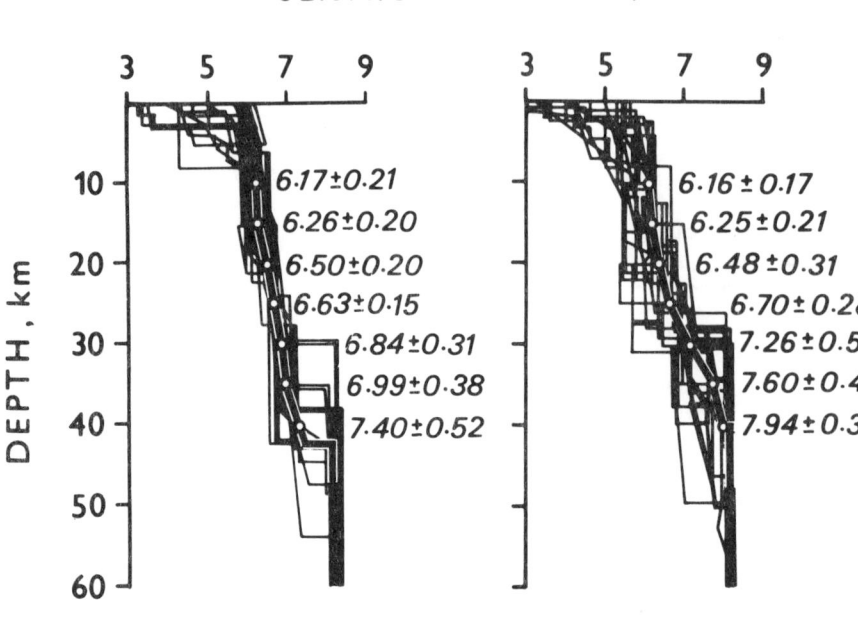

Fig.3. Characteristic distributions of seismic velocity versus depth in Precambrian and Phanerozoic units of Central and Eastern Europe.

converted to "in situ" conditions, which are valid in the respective region, see figure 4. The latter relationships were used to ascribe the heat production values to the individual crustal blocks along each geotraverse /Table III/. It was further supposed that within each block the radiogenic heat production exponentially decreases with depth :
$A(z) = A_o \exp(-z/D)$.

VI. HEAT PRODUCTION IN THE UPPER-TEN-KILOMETRES LAYER

The geological as well as the tectonic structure of the uppermost part of the crust is most complicated. The presence of microcracks dominates the physical properties of rocks at pressures up to a few kilobars, this part of the crust is thus characteristic by highly variable value of the pressure derivative of seismic velocity, which produces great uncertainty in determining the value of correcting factor B. Furthermore, according to COSTAIN /1979/

TEMPERATURE STRUCTURE OF THE LITHOSPHERE

the network of the microcracks facilitates the redistribution of U and Th by migrating underground waters and the original radioactive content of rocks may have been considerably altered. From all these reasons, the application of the A-v_p relationship may be problematic here and this area was treated separately.

There is a direct possibility of evaluating the distribution of radiogenic heat sources in the near-surface layer which can be furnished by the interpretation of the linear

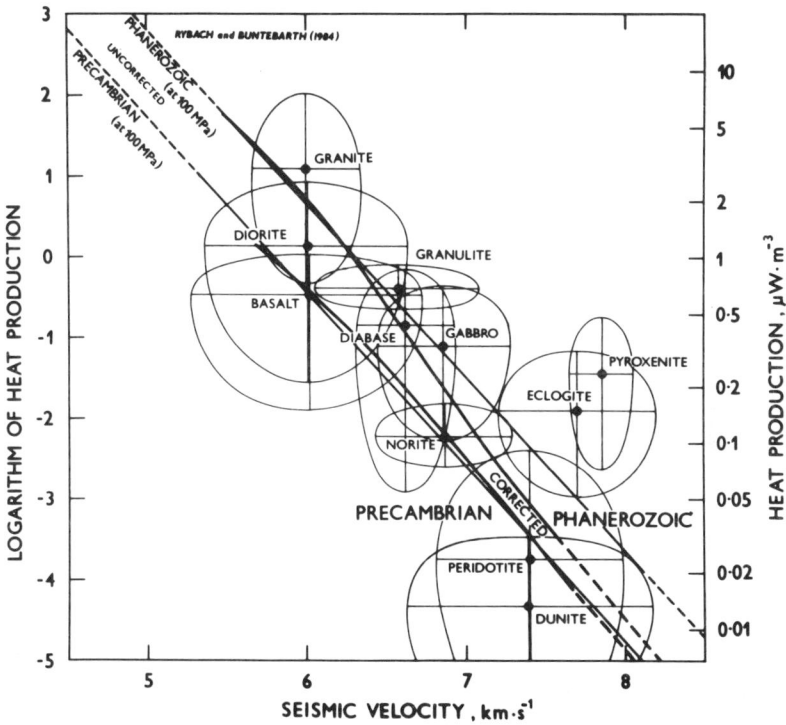

Fig.4. Heat production versus seismic velocity. Thin lines correspond to RYBACH and BUNTERBARTH's /1984/ "laboratory" formula, heavy lines correspond to above relationship after application of corrections and are valid for "in situ" conditions. Dots represent mean values for selected rocks, seismic velocities after KERN /1982/, heat productions after RYBACH and ČERMÁK /1982/, error bars bound the range of existing values.

TABLE III. Summary of heat production values, $\mu W \cdot m^{-3}$, used for two-dimensional temperature modelling and corresponding codes /pC-Precambrian, Ph-Phanerozoic/

Upper ten kilometres :

Sediments /code S/   1.2 /pC/ and/or 1.0 /Ph/

Crystalline or metamorphosed rocks /code O/

$A(z) = A_0 \exp(-z/D)$, $D = 10$ km /fixed/

a-version - minimum : $A_0 = 0.4$ Q/D
b-version - maximum : $A_0 = 0.4$ Q/D$(1-e^{-1})$

Heat production values converted from seismic velocity

Code	Seismic velocity	Heat production pC	Ph	Code	Seismic velocity	Heat production pC	Ph
A	< 6.0	0.82	2.60	M	7.0-7.1	0.08	0.13
B	6.0-6.1	0.65	2.00	N	7.1-7.2	0.06	0.10
C	6.1-6.2	0.56	1.55	P	7.2-7.3	0.05	0.08
D	6.2-6.3	0.48	1.20	R	7.3-7.5	0.035	0.06
E	6.3-6.4	0.39	0.92	T	7.5-7.7	0.02	0.035
F	6.4-6.5	0.32	0.72	U	7.7-7.9	0.015	0.02
G	6.5-6.6	0.26	0.53	V	7.9-8.1	0.01	0.015
H	6.6-6.7	0.20	0.41	W	8.1-8.3	0.01	0.015
J	6.7-6.8	0.16	0.31	X	8.3-8.7	0.005	0.005
K	6.8-6.9	0.12	0.23	Y	≥ 8.7	0.004	0.004
L	6.9-7.0	0.10	0.18	Z	Asth.	0.05	0.05

TABLE IV. Thermal conductivity values used for temperature modelling : $k = k_0/(1+CT)$

	Codes	k_0 $Wm^{-1}K^{-1}$	C K^{-1}
upper crust	A-J,O,S	3.0	0.001
lower crust	K-U	2.0	0
subcrustal lithosphere	V-Y	2.5	-0.00025
asthenosphere	Z	3.0	-0.0004

relationship between heat flow and heat generation /ROY et al., 1968, LACHENBRUCH, 1968/: $Q = q_o + D A_o$, where the intercept value /reduced heat flow/, q_o, corresponds to heat flow from below a certain depth, and the slope value, D, gives the depth scale of heat sources distribution.

This relationship was proved to be valid in both crystalline and metamorphic terrains and seems to be universally valid. Usually two crustal models are mentioned to explain it : /a/ exponential model /LACHENBRUCH, 1968/, in which D defines the logarithmic decrement of $A(z) = A_o \exp(-z/D)$, A_o being the surface heat production, and /b/ step model /ROY et al., 1968/, in which D defines the thickness of the radioactively enriched near surface layer.

The value of the D-parameter varies between 4 and 16 km, but its typical value is about 10 km. The value of 10 km was therefore taken to settle the thickness of the upper part of the crust, where the application of $A-v_p$ relationship may be unreliable, and where for the evaluation of heat sources distribution the above linear relationship was used complemented with another empirical relation between mean surface heat flow, $\hat{Q}$, and the reduced heat flow, q_o : $q_o = 0.6 \hat{Q}$ /POLLACK and CHAPMAN, 1977/.

Actually two heat production distributions in the upper ten-kilometres layer /UTK-layer/ were proposed here, believed to correspond to the extreme cases :

/a/ Let approximately $Q = \hat{Q}$. Combining both relationships and applying the exponentially decreasing crustal radioactivity, we get $A_o = 0.4 Q/D$, which in this case can be directly considered as the surface radiogenic heat production. This distribution resembles the exponential model of LACHENBRUCH /1968/.

/b/ The obtained A_o-value from the above case can be regarded as the mean heat production within the UTK-layer. Then
$$A_o = (1/D) \int_0^D A_o^- \exp(-z/D) \, dz, \text{ which gives } A_o^- = A_o/(1-e^{-1})$$
for the actual surface heat generation. This distribution somehow corresponds to the step model of ROY et al./1968/.

Two temperature profiles were calculated for each geotraverse for heat generation exponentially decreasing in the UTK-layer with a fixed value of $D = 10$ km and the surface value of heat production equal to $0.4 Q/D$ and/or to $0.4 Q/D(1-e^{-1})$, respectively, labelled further "a" or "b" and believed to correspond to the minimum and/or maximum radioactive content in the uppermost crust.

VII. COMPUTATION PROCEDURE

For a proper calculation of the temperature distribution, the solution of the heat conductivity equation /§IV/ was used. The following procedure was applied to each of five geotraverses, here being demonstrated on the example of the EEGT 5 /Bohemian Massif-Pannonian Basin-East Carpathians/.

The results of deep seismic sounding /SOLLOGUB et al., 1979/ were employed to define the gross crustal block structure /Fig.5/, and the characteristic seismic velocities of the individual blocks were converted into heat production applying RYBACH and BUNTERBARTH´s /1984/ formu-

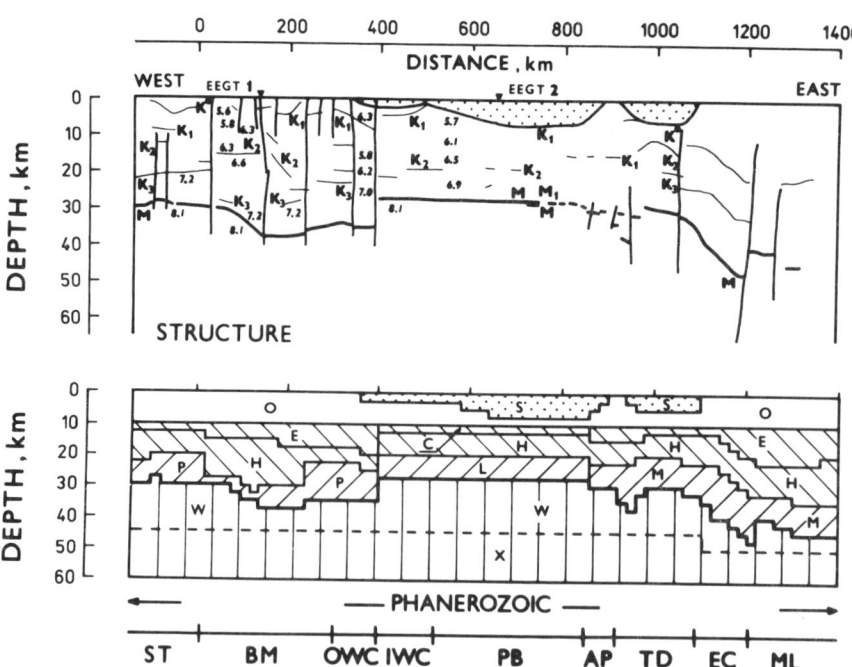

Fig.5. Top : Simplified crustal structure and seismic velocities along the EEGT 5 /after SOLLOGUB et al., 1979/. Bottom : Derived crustal block structure, individual blocks are labelled according to their heat generation, for codes see Table III.

TEMPERATURE STRUCTURE OF THE LITHOSPHERE

1a plus corresponding corrections /see §V and Tab.III./. Within each block heat production is exponentially decreasing with depth and the value of the logarithmic decrement D is automatically calculated to ensure the continuity of the A(z) - curve with depth, i.e. the bottom A(z)-value of the upper block equals the top A(z)-value of the underlying block. Within the UTK-layer the heat production was determined from the linear heat flow -- heat generation relationship /§ VI/.

The thermal conductivity is temperature dependent: $k = k_0/(1+CT)$, where k_0 is the thermal conductivity at surface conditions and C is an experimentally determined constant which controls the behaviour of the conductivity with temperature. At relatively low temperatures the

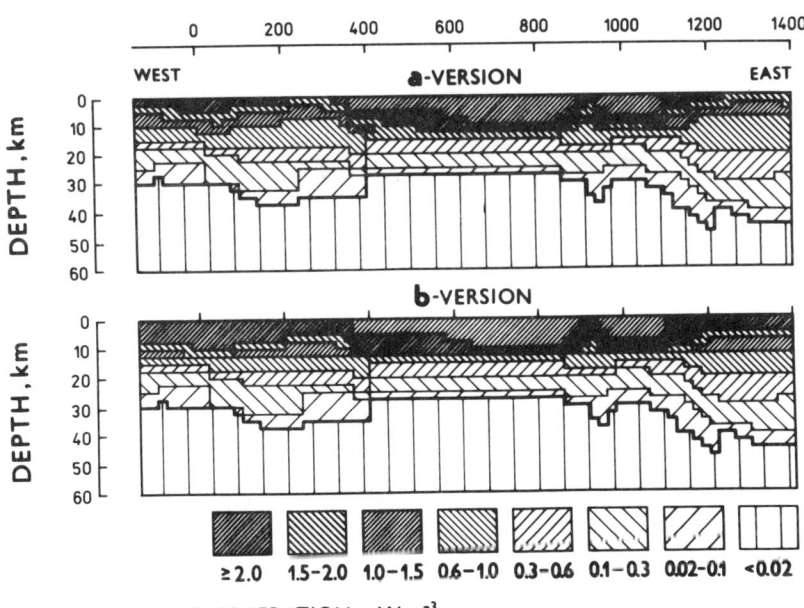

Fig.6. Heat generation distribution along the geotraverse EEGT 5.

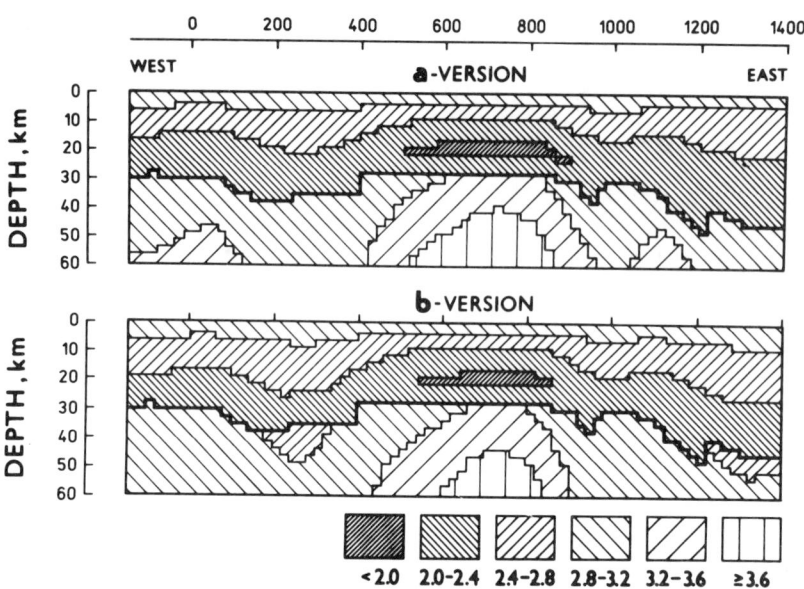

Fig.7. Thermal conductivity distribution along the geotraverse EEGT 5.

conductivity generally decreases with increasing temperature, i.e. C > 0 /ČERMÁK and RYBACH, 1982/, while at 300-500 °C the conductivity of basic rocks does not depend on temperature, C ≃ 0. At temperatures above 500 °C the radiative heat transfer begins to be more efficient and even if the lattice component of conductivity decreases further with increasing temperature, the total themal conductivity increases, i.e. C < 0. The thermal conductivities used for our calculation are summarized in Table IV.

The 2-D distributions of heat generation and of thermal conductivity along the EEGT 5 used during the temperature computation are shown in figures 6 and 7.

TEMPERATURE STRUCTURE OF THE LITHOSPHERE

The solution of the temperature distribution requires the inverse problem to be solved, i.e. to find the heat flow at the base of the model, which is approximately equal to the Moho heat flow, Q_M /see §IV/. Starting with $Q_M = 0$, and successively correcting the Q_M-value by the obtained difference between the observed and calculated heat flow on the surface corresponding to the model at the moment, Q_B, usually three to four computer runs were necessary to obtain agreement better than 2-5 mWm^{-2} between Q_O and Q_B. At the same moment the horizontal gradients of Q_O and Q_M were comparable. In regions of small scale local anomalies of surface heat flow with pronounced horizontal gradients, it is sometimes not possible to achieve the above agreement unless we get unrealistic horizontal variations of Q_M. As these anomalies are obviously of crustal origin, of certain advantage can be completing the crustal model with information on deep faults, possible conductivity contrast, distribution of zones of hydrogeological disturbances, etc., but such additional information would go far beyond the scope of the present work.

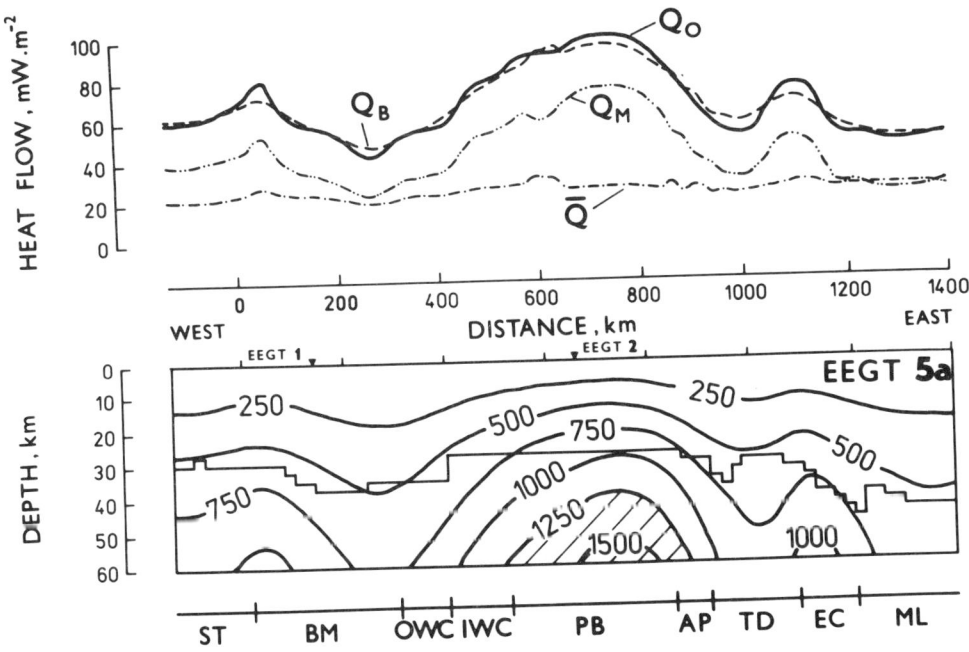

Fig.8a. Calculated 2-D temperature distribution /a-version/ along the EEGT 5 together with the observed surface heat flow, Q_O, calculated surface heat flow, Q_B, Moho heat flow, Q_M, and crustal contribution, $\bar{Q}$. For tectonic codes see Appendix.

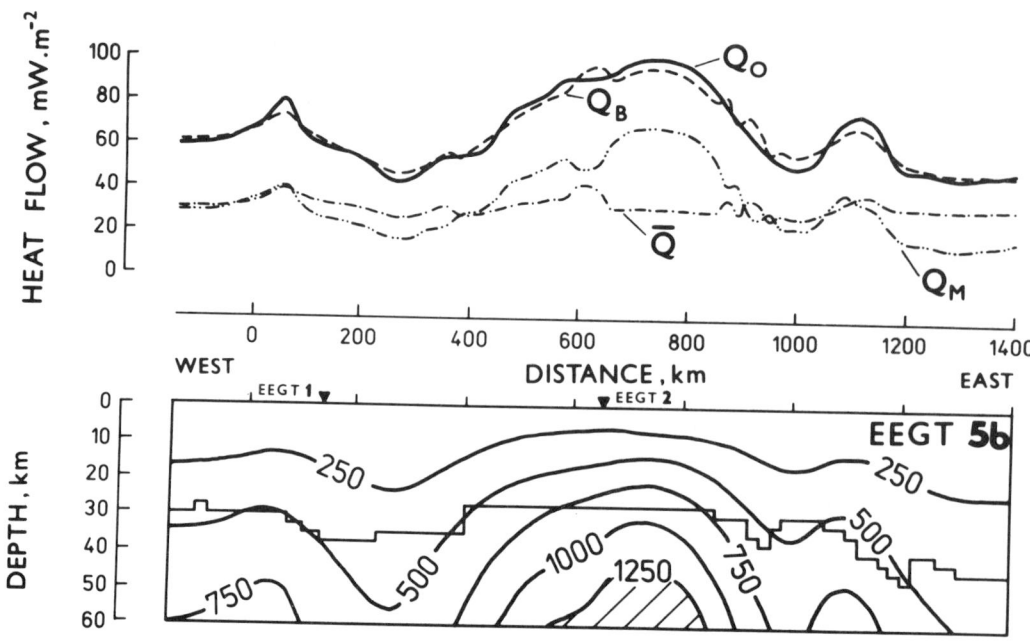

Fig.8b. The same as Figure 8a for b-version.

The relations between surface heat flow, Moho heat flow, thermal conductivity and heat generation were described by a linear algebraic system. The least squares method was applied to overcome the non-uniqueness and instability of the solution. The proper calculation of the temperature field was performed by a finite differences technique. Horizontal spacing of the grid was 20 km, vertical spacing was 2.5 km.

Figures 8a and 8b show the measured surface heat flow, Q_O, the calculated surface heat flow, Q_B, crustal contribution, $\bar{Q}$, and the calculated heat flow at the base of the model, Q_M. The calculated temperature fields are shown in the bottom parts of figures together with the position of the Mohorovičić discontinuity. The a-version of the heat production within the UTK-layer gives the minimum radioactivity, corresponding to a higher Moho heat flow and to higher crustal temperatures. The difference between a- and b-versions amount to 50-80 °C in temperature at the model base, and to 4-6 mW.m^{-2} in the Moho heat flow. To save space, only the a-version of the temperature calculations for the remaining geotraverses will be shown. The a-version is also believed to be more realistic than b-version, which gives rather low Q_M-values in some cases.

TEMPERATURE STRUCTURE OF THE LITHOSPHERE

VIII. CRUSTAL TEMPERATURES

Two-dimensional temperature distributions along all geotraverses, EEGT 1 through EEGT 5 /a-versions/ are shown in figures 9 to 12 and in figure 8.

The relatively lowest temperatures are characteristic of the Precambrian East European platform /Moho temperature is 350-500 °C/ with a clear minimum beneath the Ukrainian shield. The crustal structure of the shield itself is quite complicated, the shield is composed of several blocks, the thickness of which varies from 35 to 55 km, Moho temperature thus lies in a broad interval of 300-500 °C with a characteristic value of 450 °C. Heat

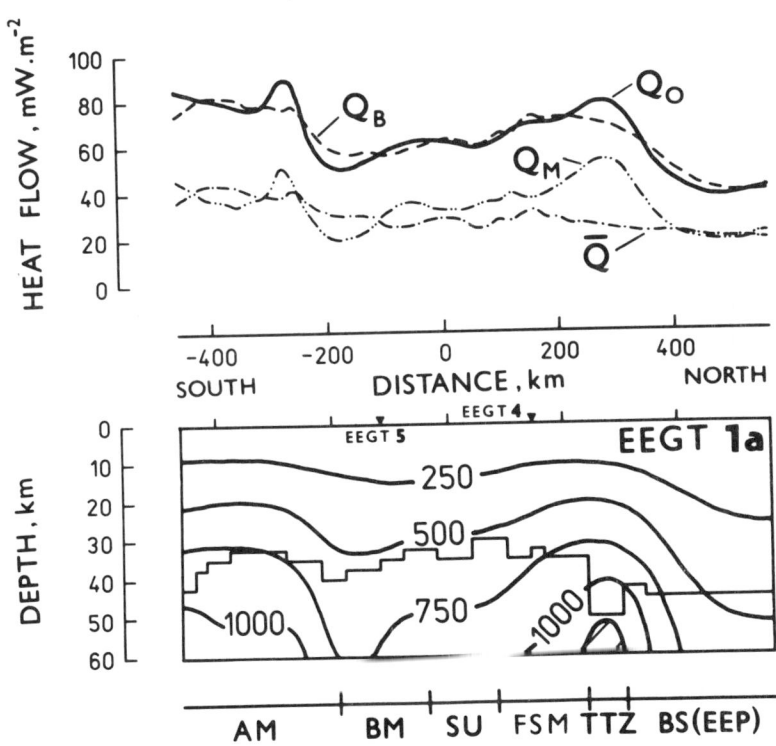

Fig.9. Temperature distribution along the geotraverse EEGT 1.

flow from the upper mantle, Q_M, attains 15-20 mW.m^{-2}. Low crustal temperatures are also characteristic of some other structures within the ancient platform : Voronezh massif and Yarinskaya and Pechora depressions /$T_M \sim$ 400-500 °C, $Q_M \sim$ 18-20 mW.m^{-2}/, which display low heat flow on the surface, $Q_O \sim$ 40 mW.m^{-2}. Relatively higher heat flow /$Q_O \sim$ 45-50 mW.m^{-2}/ was observed in aulacogens and in buried ridge structures within the paltform, such as Dnieper-Donetz, Pachelmskiy and Timan, and, partly, also in the Moscow syneclise, which are then characterized by Moho temperatures of 550-680 °C and Moho heat flow of 22-25 mW.m^{-2}.

Moho temperatures of 550-650 °C, and $Q_M \sim$ 22-30 mW.m^{-2}, correspond to Paleozoic folded units of lower than normal surface heat flow, $Q_O \sim$ 45-60 mW.m^{-2}, such as the consolidated parts of the Bohemian Massif and of Moesian plate. Crustal temperatures increase in accordance with the higher observed surface heat flow, beneath Paleozoic structures such as East-Labe massif, Saxothuringikum, Sudeten, Fore-Sudetic monocline, temperature on the Mohorovičić discontinuity may reach 600-700 °C /$Q_M \sim$ 30-45 mW.m^{-2}/, higher temperatures /$T_M \sim$ 700 °C/ are espacially typical of areas of locally weakened crust or of fractured zones, which manifest themselves by increased surface heat flow /Q_O over 70 mW.m^{-2}/. Moho heat flow may reach here 45-50 mW.m^{-2}.

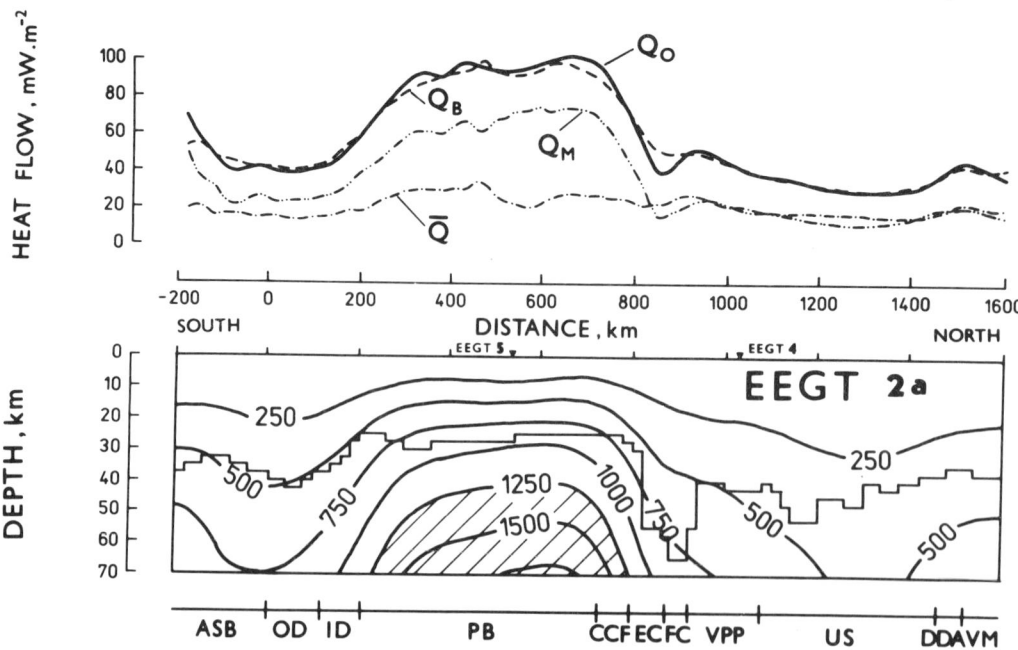

Fig.10. Temperature distribution along the geotraverse EEGT 2.

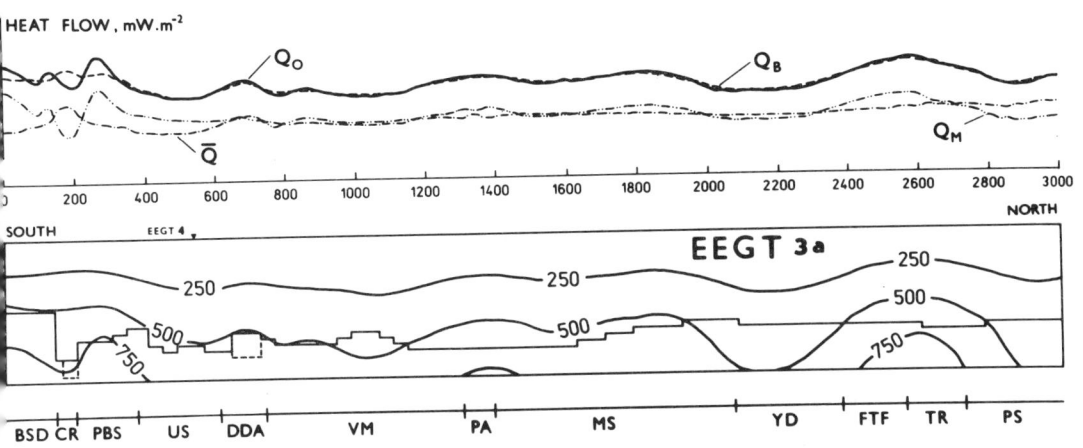

Fig.11. Temperature distribution along the geotraverse EEGT 3.

Likewise, the Alpine structures exhibit a broad interval of Moho temperatures /550-800 °C/ and of Moho heat flows /30-60 mW.m^{-2}/. The highest crustal temperatures in the study area exist beneath the Pannonian Basin /$Q_O \sim$ 80-100 mW.m^{-2}, T_M over 800 °C, and Q_M over 60 mW.m^{-2}/ even though the crustal thickness is reduced here to less than 30 km.

The agreement of the calculated temperatures in the points of intersection of the individual geotraverses is generally good, so is the agreement of the projected temperatures for the units of similar tectonic history. Certain doubts may be raised only in the case of the temperature field in the contact zone between the East European platform and Paleozoic Europe along the EEGT 1, km 200-300 /T_M over 1000 °C/, when compared with the more reasonable value of $T_M \sim$ 750-800 °C calculated for the Teisseyre-Tornquist zone in case of the EEGT 4. The former value is based on the rather problematic surface heat flow anomaly, which was not confirmed by recent observations. The crustal temperature field along the whole Teisseyre-Tornquist zone challenges an extra effort and will be studied in detail in another paper.

The calculated temperature distribution within the crust is very dependent on the magnitude of the surface heat flow and is thus vulnerable to any inaccuracy in its correct determination. The Q_O-values used for the present calculations were taken from the map, in which the interval of the isoline pattern is 10 mW.m^{-2}, this value thus

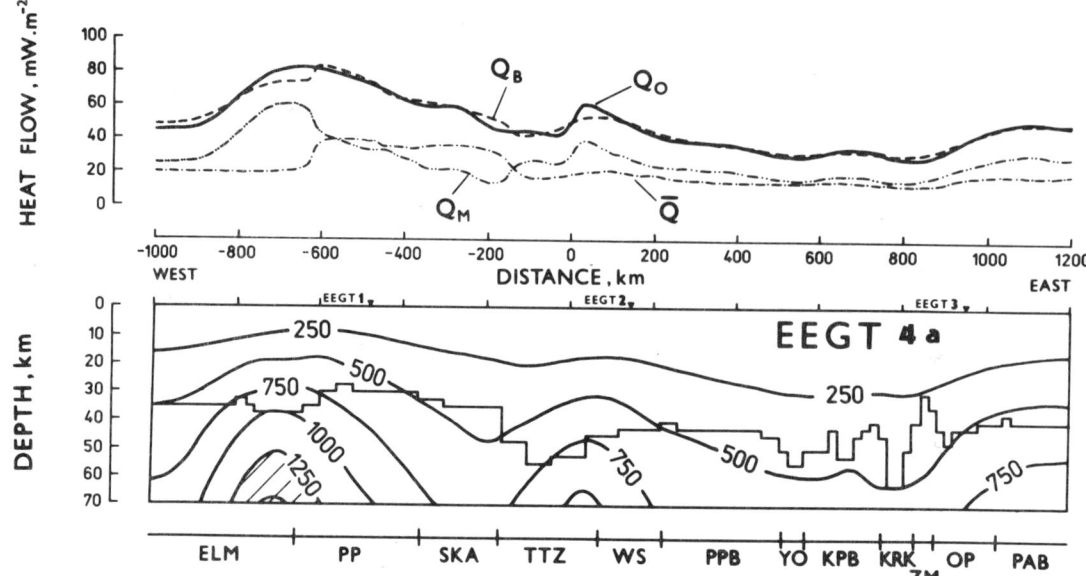

Fig.12. Temperature distribution along the geotraverse EEGT 4.

limits the validity of the calculation. As the difference of 10 mW.m^{-2} in the surface heat flow may produce the uncertainty of about 50-80 °C in the calculated crustal temperature field, the credibility of the present 2-D temperature distributions cannot be exaggerated over this limitation.

The knowledge of deep temperatures in combination with the mantle melting relations can be used to estimating the top of the seismic low velocity channel and to assessing the lithospheric thickness /POLLACK and CHAPMAN, 1977/. Lithospheric thickness decreases sharply in areas of high heat flow and beneath hyperthermal basins, such as e.g. Pannonian Basin, the lithosphere may be as thin as 50-60 km /ČERMÁK, 1982/. Even though this paper was not aimed at evaluating the temperature conditions in the upper mantle, the data obtained evidently support the idea of a very shallow asthenosphere beneath the Pannonian Basin, where deep temperatures may reach 1100-1200 °C at a depth of 50-60 km.

TEMPERATURE STRUCTURE OF THE LITHOSPHERE

Appendix :

Tectonic codes for figures 1,5,8,9,10,11, and 12

AL	- Alps	MP	- Moesian plate
AM	- Alpine molasse	MS	- Moscow syneclise
AP	- Apuseni Mts.	OD /DO/	Outer dinarides
APE	- Apennines	OP	- Orekhovo-Pavlograd- skaya geosyncline
ASB	- Adriatic Sea basin		
BA	- Belorussian anticli- norium	OWC /WCO/	Outer Western Carpathians
BL	- Balkans	PA	- Pachelmskiy aulacogen
BM	- Bohemian massif	PAB	- Pri-Azov block
BS	- Baltic syneclise	PB	- Pannonian basin
BSB	- Black Sea basin	PBS	- Pri-Black Sea depres- sion
CCF	- Cis-Carpathian fore- deep	PO	- Po basin
CHC	- Chernyshev crest	PP	- Paleozoic platform /Poland/
CR	- Crimea Mts.		
DB	- Donetz basin	PPB	- Podolsk platform block
DDA /DD/	Dnieper-Donetz aulacogen	PS	- Pechora syneclise
EC	- Eastern Carpathians	SKA /SK/	Swieto-Krzske anticlinorium
EEP	- East-European plat- form	ST	- Saxothuringikum
		SU	- Sudeten
ELM	- East Labe massif	TD /TB/	Transsylvanian depression
FCF	- Fore Carpathian foredeep	TR	- Timan ridge
FSM	- Fore Sudetic mono- cline	TTZ	- Teisseyre-Tornquist zone
FTF	- Fore Timan foredeep	US	- Ukrainian shield
ID /DI/	Inner Dinarides	VF	- Vorkuta foredeep
IWC /WCI/	Inner Wtsren Carpathians	VM /VA/	Voronezh massif
KP	- Korostenskiy pluton	VPP	- Volyno-Podolsk plate
KPB	- Kirovograd platform block	WS	- Western slopes of the Ukrainian shield
KRK	- Krivorozhsko-Kremen chutskaya geosyncline	YD	- Yarinskaya depression
		YO	- Yadlovsko-Odesskaya geosyncline
LD	- Lvov depression	ZM	- Zaporozhskiy central massif
ML	- Moldavian platform		

REFERENCES

BODRI L., 1981. Geothermal model of the earth´s crust in the Pannonian Basin. Tectonophysics, 72: 61-73.

ČERMÁK V., 1982. Regional pattern of the lithosphere thickness in Europe. In: V.ČERMÁK and R.HAENEL /Eds./, Geothermics and Geothermal Energy. E.Schweizerbart´sche Verlagsbuchhandlung, Stuttgart, pp. 1-10.

ČERMÁK V. and HURTIG E., 1979. Heat flow map of Europe, 1:5,000,000. In: V.ČERMÁK and L.RYBACH /Eds./, Terrestrial Heat Flow in Europe. Springer Verlag, Berlin, Heidelberg, New York, enclosure.

ČERMÁK V. and RYBACH L., 1982. Thermal conductivity and specific heat of minerals and rocks. In: G.ANGENHEISTER /Ed./, Landolt-Börnstein New Series, Vol.VIa, Physical Properties of Rocks, Springer Verlag, Berlin, Heidelberg, New York, pp. 305-343.

COSTAIN, J.K., 1978. A new model for the linear relationship between heat flow and heat generation. EOS, Trans. Am.Geophys.Un., 59: 392

GEBRANDE H., 1982. Elastic wave velocities and constants of elasticity of rocks at room temperature and pressures up to 1 GPa. In: G.ANGENHEISTER /Ed./, Landolt-Börnstein New Series, Vol.VIb, Physical Properties of Rocks, Springer Verlag, Berlin, Heidelberg, New York, pp. 35-99.

KERN H., 1982. Elastic wave velocities and constants of elasticity of rocks at elevated pressures and temperatures. In: G.ANGENHEISTER /Ed./, Landolt-Börnstein New Series, Vol. VIb, Physical Properties of Rocks, Springer Verlag, Berlin, Heidelberg, New York, pp.99-140.

LACHENBRUCH A.H., 1968. Preliminary geothermal model of the Sierra Nevada. J.Geophys.Res., 73: 6977-6989.

POLLACK H.N. and CHAPMAN D.S., 1977. On the regional variation of heat flow, geotherms, and lithospheric thickness. Tectonophysics, 38: 279-296.

ROY R.F., BLACKWELL D.D. and BIRCH F., 1968. Heat generation of plutonic rocks and continental heat flow provinces. Earth Planet.Sci.Lett., 5: 1-12.

RYBACH L. and BUNTEBARTH G., 1984. The variation of heat generation, density and seismic velocity with rock type in the continenetal lithosphere. In: V.ČERMÁK, L.RYBACH and D.S.CHAPMAN /Eds./, Terrestrial Heat Flow Studies and the Structure of the Lithosphere. Tectonophysics, 103: 335-344.

RYBACH L. and ČERMÁK V., 1982. Radioactive heat generation in rocks. In: G.ANGENHEISTER /Ed./, Landolt-Börnstein New Series, Vol.VIa, Physical Properties of Rocks, Springer Verlag, Berlin, Heidelberg, New York, pp.353-371.

ŠAFANDA J., 1985. Calculation of temperature distribution in the two-dimensional geothermal profile. Stud.geoph. et geod., 28: 000-000 /in press/.

SOLLOGUB V.B., GUTERCH A. and PROSEN D. /Eds./, 1978. Stroyeniye zemnoy kori i verkhney mantiyi Centralnoy i Vostochnoy Evropi. Naukova Dumka, Kiev, 272 pp. /in Russian/.

SOLLOGUB V.B., GUTERCH A., and PROSEN D. /Eds./, 1979. Struktura zemnoy kori i verkhney mantiyi po dannim geofizicheskikh issledovaniy. Naukova Dumka, Kiev, 208 pp. /in Russian/.

STROMEYER D., 1984. Downward continuation of heat flow data by means of the least squares method. In: V.ČERMÁK, L.RYBACH and D.S.CHAPMAN /Eds./, Terrestrial Heat Flow Studies and the Structure of the Lithosphere. Tectonophysics, 103: 55-66.

C. JAUPART[1]

ON THE AVERAGE AMOUNT AND VERTICAL DISTRIBUTION OF RADIOACTIVITY IN THE CONTINENTAL CRUST

Abstract.

We present a review of present knowledge on the amount and distribution of radioactivity in the continental crust. The average concentrations of uranium, thorium and potassium can be estimated independently using heat flow data and geochemical mass budgets. The range of values for the average heat production is $0.7-1.0$ $\mu W/m^3$, which translates into a range of $15-26$ mW/m^2 for the mantle heat flow beneath old cratons. The vertical distribution of radioactivity is studied using the linear heat flow relation of Birch. We compare the apparent thickness D deduced from it to the structure of the upper crust determined by independent methods. We show that this apparent thickness is significantly affected by the averaging effects of horizontal heat conduction and cannot be related to any physical dimension. The reduced heat flow deduced from the Birch relation represents an average heat flow which is free from the shallowest radioactivity variations. It integrates the heat production distribution over lateral distances as large as 300 km. To compute the mantle heat flow beneath continents, it is therefore necessary to use a large-scale average of heat production which is only available through geochemical budgets.

---

(1) *Institut de Physique du Globe de Paris, Paris, France.*

ON THE AVERAGE AMOUNT AND VERTICAL DISTRIBUTION OF RADIOACTIVITY

I. INTRODUCTION.

The thermal structure of the continental lithosphere plays a key role in the evolution of mountain belts or sedimentary basins. Yet, despite this obvious importance, much remains to be learned about it. Our constraints stem primarily from heat flow data, which are plagued with a fundamental difficulty. In steady-state conditions, the surface heat flow is the sum of two components: the heat flux supplied to the base of the lithosphere and the heat released by radioactive decay in crustal rocks. Present estimates for the latter vary from about 50% to 100% of the total. This uncertainty leads to severe limitations in models of sedimentary basins for two complementary reasons:
(1) most models are based on the plate model which was developed for ocean basins, which implies a specific value of the mantle heat flow at the base of the lithosphere.
(2) extension leads to thin the continental crust, decreasing its thickness and hence the amount of heat produced by radioactive decay.

The oceanic plate model relies on a rather high value of the mantle heat flow and, as a consequence, on a rather low value of the amount of crustal heat production. There are presently no definitive arguments to show that this model is correct for the continents. The best approach would be to start from what is directly accessible: the continental crust. However, estimating its average radioactivity is difficult for two reasons. The first is that it is extremely heterogeneous (Fountain & Salisbury, 1981). The second is that the radiogenic elements uranium and thorium are trace elements, which implies that their concentrations are not related to bulk composition and hence to other physical characteristics such as density or seismic velocity. This is exemplified by the detailed and work of Galson (1983) in the Ivrea Zone (Swiss Alps). In this region where most crustal rock types and metamorphic facies crop out, he found no meaningful relationship between density and heat production (Fig.1). The data show a fundamental fact: mafic rocks (high density) are poor in uranium and thorium, whereas felsic rocks (low density) are markedly enriched. The plot shown in Fig.1 must be interpreted as follows: little variation and small values at high density, large scatter and high mean value at lower density values. Where radioactivity is largest is also where it is least predictable, hence making any prediction scheme unreliable. Rybach & Buntebarth (1982) have recently proposed an empirical relationship between seismic velocity and radiogenic heat production, but there are large departures from it. The Ivrea Zone study emphasizes the lack of relationship in a limited area, and the problem is much worse on the scale of a whole continent.

ON THE AVERAGE AMOUNT AND VERTICAL DISTRIBUTION OF RADIOACTIVITY

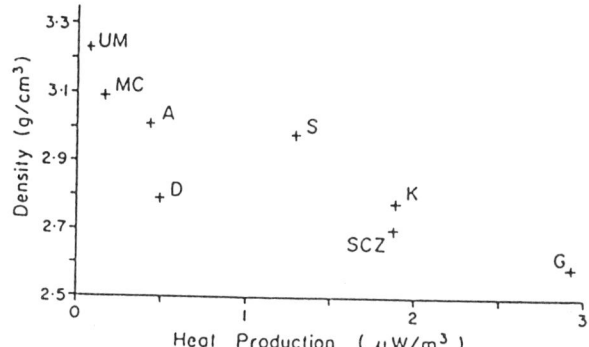

Fig.1. The relationship between mean heat production and mean density for different lithologies of the Ivrea and Strona Ceneri zones (from Galson, 1983). Note the large spread of values.

The aim of this paper is to summarize our present knowledge about crustal radioactivity. The first part focusses on estimates of the mean concentration of radiogenic elements and of the mean value of crustal heat generation. The second part deals with the vertical distribution of heat production. In the conclusion, we give ranges for both the average value of heat production and the mantle heat flow.

II. THE AVERAGE RADIOACTIVITY OF THE CONTINENTAL CRUST.

II.1. Straight averages.

The most obvious approach is to use all available data on samples retrieved from the continental crust (good reviews have been made by Rogers & Adams, 1969 a,b) and to compute an average of the values weighted according to the abundances of the various rock types. The most recent effort is due to Haack (1983) who also used constraints from heat flow data. His results are shown in Table 1. A similar and earlier kind of estimate due to Shaw is also given.

II.2. The andesite model.

It is now generally agreed that continental crust is extracted from the mantle in subduction zones. The exact process is complex, but is clearly associated with andesite genesis which produces typical continental compositions. Andesites are thus good candidates for the average continental rock prior to internal differentiation. Using radioactive data from andesite samples from all over the world, Taylor (1982) has obtained a crustal model (Table 1).

II.3. Crust/mantle models.

Pursuing the idea that continental crust is extracted from the mantle, it is possible to study the crust and mantle as two reservoirs in a closed system: the formation of enriched continental crust is made at the expense of the mantle which becomes depleted. Using the starting mantle composition which is known from meteorite studies, together with values for the depleted mantle, one can model the evolution of both the mantle and the crust reservoirs. There is a difficulty because the mantle is not homogeneous. The source of this heterogeneity is still debated and introduces an error in the procedure. Further constraints are brought by rare gas (Argon) data which allow bounds on the potassium concentration (Allègre et al., 1985a). The average crustal composition consistent with mantle values is given in Table 1 (from Allègre et al., 1985b). Note that it is similar to the previous estimates.

II.4. Measurements on vertical sections through the crust.

In some areas, it is possible to observe sections through the crust down to the Moho discontinuity and thus to estimate directly the total amount of heat produced over the whole crustal thickness.

Galson (1983) has recently completed such a study in the Ivrea Zone. He found that the heat production of felsic rocks is generally higher than that of mafic rocks from the same metamorphic grade (i.e. from the same stratigraphic position in the crust). The mean heat production of granulite facies rocks is 0.4 $\mu W/m^3$. At lower metamorphic grade, the predominance of felsic rocks leads to a mean heat production of 1.4 $\mu W/m^3$. He infers a total amount of heat generation of 42 mW/m^2 over the crustal thickness of 28 km. This corresponds to a mean heat production of 1.5 $\mu W/m^3$.

Nicolaysen et al.(1981) made a similar study in the Vredefort (South Africa) and reported an average heat production of 0.94 $\mu W/m^3$.

II.5. Estimates from heat flow data.

The above estimates make use of direct measurements on individual rock samples and rely on models of the chemical composition of the crust. It is possible to derive independent bounds from heat flow data.

Continental heat flow values vary by large amounts in young areas of the world in relation to the thermal perturbations which follow orogenic events (Sclater et al., 1980). In regions older than 800 My, the range of variations is smaller and can be accounted for solely by known differences in the radiogenic contents of surface rocks. Those regions are close to steady-state. The observed heat flow is thus the sum of the mantle heat flow Q_m and the heat produced by radioactive decay Q_g. The mean radioactivity of crustal rocks is thus given by:

$$A = \frac{Q_g}{h} = \frac{Q - Q_m}{h} \quad (1)$$

where h is the mean crustal thickness which we take to be 35 km. Q is the mean surface heat flow, for which Sclater et al.(1980) give a value of 46±15 mW/m^2 in provinces older than 800 My. This value is derived from measurements in boreholes and may be affected by the effects of recent glaciation. The true mean heat flow must be somewhat higher (Sclater et al., 1981). We take a reference value of 50 mW/m^2. Estimates for Q_m vary according to various authors. The upper bound is given by heat flow measurements on formations with no radioactivity such as anorthosite massifs. The value is quite constant at 26 mW/m^2. This yields a minimum value of 0.7 μW/m^3 for the mean heat production rate (through equation 1). The lower bound for Q_m is obviously zero. In this extreme case, the mean heat production rate is maximum at a value of 1.4 μW/m^3. This defines the range of values for A from the standpoint of heat flow measurements.

II.6. Summary.
The different estimates listed in Table 1 have been obtained using different methods. Yet the range is remarkably narrow, both for the individual concentrations of uranium, thorium and potassium and for the radiogenic heat production. The geochemical studies lead to a range of 0.7-1.0 μW/m^3, which is compatible with the constraints deduced from heat flow data. Furthermore, the lower bounds from both methods are identical. The results by Haack (1983) are derived from a more complicated analysis based partly on heat flow considerations and yield an anomalously high Th/U ratio. Excepting these, the three remaining geochemical estimates are in remarkable agreement. In comparison, the direct measurement made in the Ivrea Zone appears anomalously high. This is due to the heterogeneous character of the continental crust, which implies that a vertical section in a small area may not be representative of the crust on a large scale. This was recognized by Galson (1983) who found that the Ivrea Zone comprises an unusually high percentage of felsic rocks. We shall return to this representativity problem later.

III. THE VERTICAL DISTRIBUTION OF RADIOACTIVITY.

II.1. General considerations.
It is now well established that the lower crust is depleted in radioactive elements, both because it contains a large proportion of mafic rocks and because high grade rocks in general do not contain large amounts of radioactivity (Heier, 1979). This shows that the radiogenic heat production is not distributed uniformly in the crust. Before going into a detailed analysis, we stress that the

critical piece of information is the total amount of heat generation the crust and not its vertical distribution. We shall see however that the two pieces of information are not independent from each other. To illustrate the error introduced by the form of the vertical radioactivity distribution, we show in Figure 2 temperature profiles for two models with the same total amount of heat generation. Note that the temperature estimates differ by less than $80°C$, which corresponds to an error of 5%, or significantly less than errors introduced by other variables such as thermal conductivity.

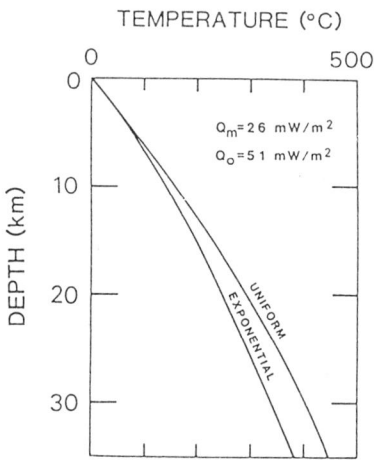

Fig.2. Temperature profiles through the crust in steady-state conditions for a thermal conductivity of 3.0 W/m.K. The total amount of heat generation is the same in both models ($25mW/m^2$). In the uniform model, radioactivity is constant at a value of 0.7 $\mu W/m^3$. In the exponential model, it decreases exponentially with an e-folding depth of 10 km.

III.2. The depth-scale of radioactive enrichment.

Little is known about the vertical distribution of radiogenic heat production. Direct evidence from boreholes is only available down to depths of about 4 km (Lachenbruch & Bunker, 1971). Field observations yield different answers (Nicolaysen et al., 1981; Galson, 1983). The difficulty is to obtain an average profile valid for a whole continental province. Present estimates rely entirely on an empirical relationship discovered by Birch et al.(1968) which relates heat flow to the radiogenic heat production at the surface (Fig.3). The relationship is linear and written as follows:

$$Q_0 = Q_r + D \cdot A_0 \qquad (2)$$

where Q_0 and A_0 are the surface values of heat flow and heat production. Q_r is the "reduced" heat flow and gives the heat flow for zero surface depth-scale of radioactive enrichment. The relationship has been observed in many regions of the world and in many different geologic environments (Jaupart et al., 1981). Values of depth-scale D vary, but are generally close to 10 km. This suggests that the radioactivity differences which are observed at the surface persist to great depths in the crust.

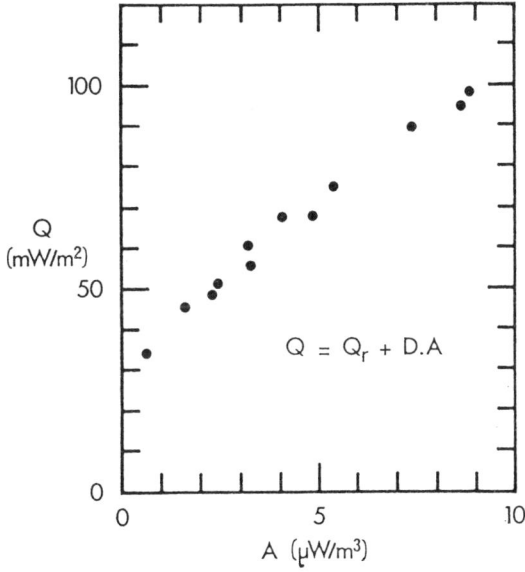

Fig.3. Heat flow versus radiogenic heat production in the Eastern USA province (which includes the State of New Hampshire) (from Birch et al., 1968). The data can be fit perfectly with a linear relationship yielding $Q_r=32\pm1$ mW/m^2 and D=7.8±0.4 km.

There are two basic models to interpret parameter D. In the first, radioactivity is distributed uniformly over a thickness D which is commonly assumed to be the average thickness of radioactive formations in the province (Roy et al., 1968). In the second model, some mechanism has acted to produce an exponential distribution

$A(z)=A_0 \cdot \exp(-z/D)$, where z is depth, over a thickness at least as large as 3D, or usually 30 km, which is close to the crustal thickness (Lachenbruch, 1970). This mechanism may be magmatic differentiation or the action of migrating metamorphic fluids. The two models differ essentially in the crustal depth at which the heat flow is equal to Q_r. In the uniform model, this depth is simply D, or about 10 km, and there remains about 25 km of crustal rocks of unknown radioactivity below the surface layer. In contrast, the exponential model is such that this depth must be at least 30 km. Thus, the reduced heat flow is simply equal to Q_m, the mantle heat flow. This explains the success of the exponential model, which has been widely used to calculate temperatures in the crust.

This paragraph shows the importance of using both heat flow and radioactivity data because it allows mutually consistent estimates of the mantle heat flow and the amount of crustal heat generation.

III.3. The interpretation of heat flow and radioactivity data.

To interpret the linear relation (2), several assumptions are needed. The first is to assume steady-state conditions, i.e. the continental lithosphere is in secular thermal equilibrium with the heat flow supplied at its base and the heat released by radioactive decay in crustal rocks. This is reasonable for most known heat flow provinces which are very old. The second assumption is that heat transfer is by conduction only, which is valid in cristalline terrains on a large scale. The third assumption is that heat transfer occurs along the vertical only, which corresponds to a zero-order approximation.

Under all these conditions, the heat equation is:

$$k \cdot \frac{d^2 T}{dz^2} + A = 0 \qquad (3)$$

where k is thermal conductivity (taken to be constant), T temperature, A heat production and z depth. Let d be the thickness over which heat production variations are related to those at the surface. In other words, over thickness d, the heat production can be expressed as follows:

$$A(z) = A_0 \cdot f(z) \qquad (4)$$

whatever the surface value A_0 is. Function $f(z)$ is simply the non-dimensional vertical distribution. d can be as large as the crustal thickness. The heat equation (3) can be integrated once over thickness d and yields:

$$Q(0) = Q(d) + \int_0^d A(z) \cdot dz \qquad (5)$$

where $Q(0)$ and $Q(d)$ are the values of heat flow at the surface and at depth d. Comparing (2) and (5), we get:

$$Q_r - Q(d) = \int_0^d A(z) \cdot dz - D \cdot A(0) \qquad (6)$$

By assumption, Q_r and $Q(d)$ are constant in a province, so that the left-hand side of equation (6) is constant. The right-hand side depends on the value of surface heat production which is variable. This shows that the only possibility is that both sides of equation (6) are equal to zero. Thus:

$$D \cdot A(0) = \int_0^d A(z) \cdot dz \qquad (7)$$

This can be written as:
$$D/d = \frac{1}{d} \cdot \int_0^d f(z) \cdot dz \qquad (8)$$

This shows that parameter D/d gives a measure of the form of the vertical distribution $f(z)$. D/d is equal to 1 if the distribution is uniform, and it is less than 1 if radioactivity decreases with depth (as in the exponential model).

We now study a test region where these simple ideas can be used. The State of New Hampshire has geological formations of many types and various ages, all extensively studied. Heat flow and radioactivity data exhibit a very good linear correlation which yields a D value of 7.8 km (Fig.3). There is good gravity and seismic coverage. The region is characterized by granitic formations which are markedly enriched in radioactive elements. These granites are tabular, rather thin and sometimes underlain by basic rocks (Sharp & Simmons, 1977). It is therefore possible to define their bottom without ambiguity. Thus, radioactive enrichment in the New Hampshire crust is characterized by local highs in thin units. The thickness estimates are small, always less than 5 km (Table 2). Hence thickness d (as defined above) must be less than 5 km, because the distribution of heat production is not continuous below plutons underlain by basic rocks. Because the measured D is 8 km, D/d is greater than 1. Similar results are obtained in other provinces throughout the world. This suggests that radioactivity increases with depth, in contradiction with most previously accepted models.

The most likely explanation is that surface samples are weathered and altered. Under these conditions, uranium can be lost in significant proportions (Rosholt et al., 1973). Thus the measured value of heat production underestimates the true value at depth. The vertical radioactivity distribution is best characterized as due to the combined effects of granite emplacement and late alteration near the surface. Hence, it can be described as increasing with depth, but in no relation to any deep seated process. There is abundant evidence for depletion. In some cases, up to 80% of the uranium has been leached from samples located close to the surface. Contrary to uranium, thorium is immobile and cannot be lost through alteration and weathering. Thus it retains its original distribution. Allowing for different distributions for uranium and thorium, it is possible to re-interpret heat flow and radioactivity data. The analysis is indeed compatible with U loss and yields thicknesses in reasonable agreement with other geophysical determinations (Jaupart, 1983).

As a first approximation, the amount of leachable uranium is proportional to its total amount in the rock (Labhart & Rybach, 1974). Alteration thus changes an initial linear relation into another one with a larger slope (Fig.4). But this requires in turn that there is such a thing as an initial linear relation in regions where pluton thicknesses appear quite variable (Table 2). What then is the cause of the linear relation? We have suggested in a recent paper the effects of horizontal heat conduction (Jaupart, 1983).

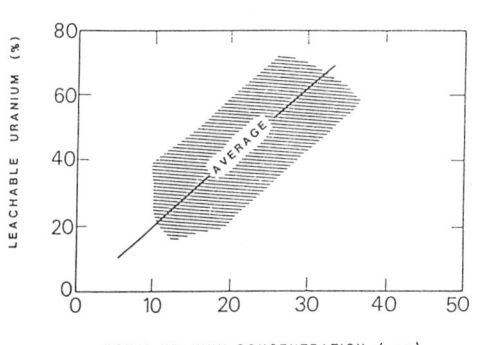 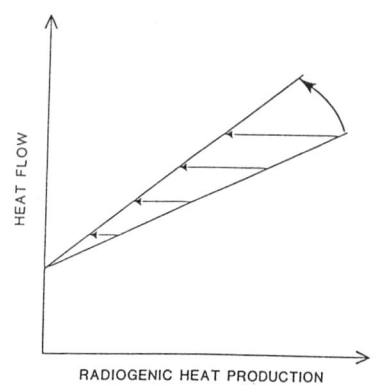

Fig.4. (a) Amount of leachable uranium versus total amount (from Labhart & Rybach, 1974). In general, the higher the U content, the higher the amount which can be lost. (b) Effect of U loss due to weathering and alteration on Q vs. A plots. The heat production measurement underestimates the true value at depth, hence leads to increase artificially depth-scale D.

III.4. Horizontal heat conduction in the upper crust.

The effects of horizontal conduction have been discussed at length by England et al.(1981) and later by Jaupart (1983). There are two simple ways to illustrate them. Consider first the case of an isolated pluton. Because it is of finite dimensions, it loses heat to its surroundings and the heat flow measured at its top will be less than that which would be expected if it was infinitely wide. Thus, the apparent thickness deduced from heat flow data is smaller than the true one (Fig.5). Further, for fixed radius, the apparent thickness reaches a finite limit (Fig.5). Hence, heat flow is poorly sensitive to thickness differences. Another way to look at the effects of horizontal conduction is to decompose the radioactivity and heat flow fields of a province in Fourier sine series and to look at each wavelength separately (Fig.6). Now the heat flow and radioactivity fields of New Hampshire are dominated by small wavelengths, of the order of the typical inter-unit distance which is

20 km. The smoothing effect is very large, exceeding 50%. The interpretation thus requires detailed modelling, like gravity studies for example, as pointed out by Simmons (1967). It must be kept in mind that uranium is sensitive to secondary processes which leave no trace on the gravity field.

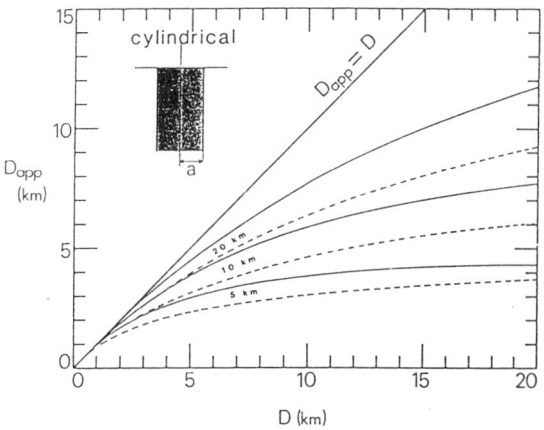

Fig.5. Plot of apparent thickness as a function of true thickness for a cylindrical pluton. Plain and dashed lines correspond to uniform and exponential models for the vertical distribution of radioactivity. Numbers along the curves are values of the radius.

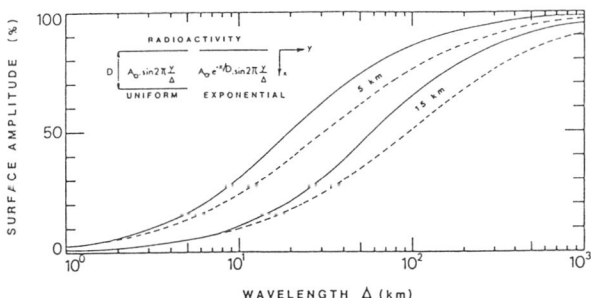

Fig.6. Amplitude of surface heat flow variations (expressed as a percentage of the value in the absence of horizontal conduction) as a function of wavelength. The plain and dashed curves correspond to the uniform and exponential models. Numbers along the curves are D values.

We conclude that horizontal conduction acts to obliterate thickness differences between neighbouring units. Below the thinnest units, redistribution of heat is effective and results in an almost uniform background heat flow. Deep plutons lose heat to their surroundings and appear thin. Thus, one ends up "seeing" a series of thin plutons standing on relatively high background. This creates a near-alignment in a Q_0 versus A_0 plot, whose slope is some average of the various thicknesses involved (Figure 7). The linear relation is an average which only holds in regions where pluton thicknesses are similar. In provinces where anomalously thin or thick formations can be found, there are departures from linearity (Jessop & Lewis, 1978).

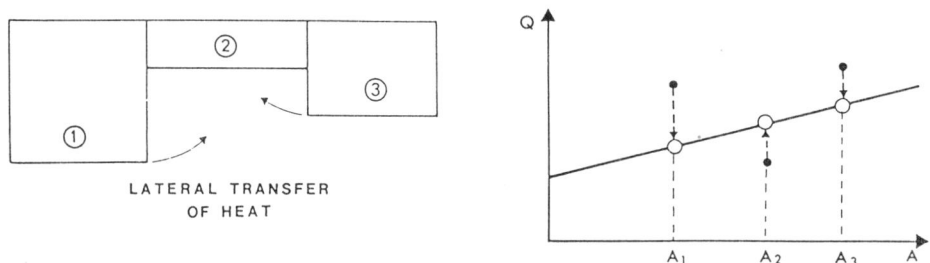

Fig.7. (a) Effect of deep horizontal heat transport on neighbouring formations. Unit (2) is thin and has its heat flow enhanced by contributions from the deeper units (1) and (3). (b) Schematic representation of the effects of horizontal conduction in a Q vs. A plot. The net result is to bring the points close to a single straight line.

III.3. Implications.

The very existence of the linear relation implies that heterogeneities located below the upper radioactive layer leave no detectable trace on the surface heat flow field. The uniformity of the background or reduced heat flow is well demonstrated in the northern part of the North-American continent. All the heat flow and radioactivity data from the the State of New York up to Newfoundland and the Maritime Provinces of Canada yield the same value of 32±3 mW/m^2 for Q_r (Fig.8). This uniformity holds for wavelengths ranging from 20 km, the typical inter-unit distance, to more than 1000 km, the total extent covered by available measurements. One must go to Virginia to find a slightly lower value of 27 mW/m^2 (Costain & Glover, 1979; Fig.8), which cannot even be considered significantly different on a statistical basis. It is clear that the lower crust is as heterogeneous as the upper crust (Fountain & Salisbury, 1981).

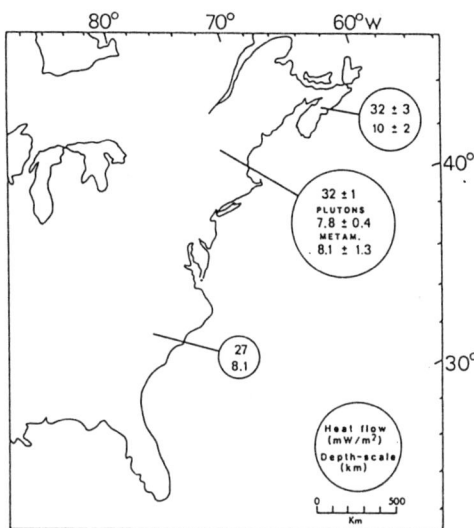

Fig.8. Map of part of the North-East American continent showing the values of Q_r (top number, in mW/m^2) and D (bottom number, in km). Note that the reduced heat flow is uniform over the whole area.

However, for typical dimensions of about 20 km, the averaging effect of heat conduction leads to smooth out the associated radioactivity differences. In fact, it can be shown that the surface heat flow integrates deep radioactivity contrasts over distances of more than 200 km (Jaupart, 1983). On this scale, the crust can be treated as homogeneous, which explains why the background heat flow is so uniform over the scale of a whole continent (Fig.8).

In fact, the very uniformity of the reduced heat flow across a province as complex as North-East America shows that it corresponds to an "average" crust. One must also note that the values of Q_r are strikingly constant throughout the world in provinces older than 300 My (the total range is 26-28 mW/m^2 once climatic corrections have been taken into account, Sclater et al., 1981). The average radioactive content of the crust is therefore the same in all these different provinces and must obviously be associated with a mean crustal structure valid for all of them.

We conclude that the interpretation of heat flow data must be made with a representative model of crustal heat production, i.e. a model which corresponds to an average over a whole continent. We have given in section 2 geochemical models of the crust which are consistent with heat flow data.

45

IV. CONCLUSION.

We have reviewed present knowledge about crustal radioactivity and arrived at a range of 0.7–1.0 $\mu W/m^3$ for the average heat production which is compatible with heat flow data. The range of values for the mantle heat flow comes out to be 15–26 mW/m^2. The lower bound of 0.7 $\mu W/m^3$ is probably not realistic as it requires that the reduced heat flow is equal to the mantle heat flow, hence implying that there is no radioactivity over a thickness of more than 20 km below the uppermost crust. Furthermore, it is due to Haack (1983) who used a peculiar method. At present, the best estimate is simply 1.0 $\mu W/m^3$ which corresponds to a mantle heat flow of 15 mW/m^2. There is unfortunately no way to determine the errors on these values. The important point to bear in mind is that it is not possible to vary them independently.

The vertical distribution of radioactivity is poorly constrained by present data. The linear heat flow relation of Birch is difficult to interpret and must be treated as a purely empirical way to relate in a rough way the heat flow and radioactivity fields of a province. It remains very useful as it is the only tool available to get rid of large heat flow variations due to shallow radioactivity contrasts. The reduced heat flow is the sum of the mantle heat flow and a large-scale average of heat production.

REFERENCES

ALLEGRE, CJ et al., 1985a, submitted to Earth Planet. Sci. Lett.
ALLEGRE, CJ et al., 1985b, submitted to Chemical Geology.
BIRCH, F, ROY, RF & DECKER, ER, 1968, in: Studies of Appalachian Geology, Interscience, New York, 437
BOTHNER, WA, 1974, Bull.Geol.Soc.Am. 85, 51
COSTAIN, JK & GLOVER, L, 1979, in: The Caledonides of the USA, p.215
ENGLAND, PC, OXBURGH, ER & RICHARDSON, SW, 1980, Geophys.J.RAS. 62
FOUNTAIN, DM & SALISBURY, MH, 1981, Earth Planet.Sci.Lett. 56, 263
GALSON, DA, 1983, Unpub. PhD Thesis, Cambridge Univ., 171 pp.
HAACK, U, 1983, Earth Planet.Sci.Lett. 62, 360.
HEIER, KS, 1979, Phil.Trans.R.Soc.A. 201, 413
JAUPART, C, SCLATER, JG & SIMMONS, G, 1980, Earth Planet.Sci.Lett. 52
JAUPART, C, 1983, Geophys.J.RAS. 75, 411
JOYNER, WB, 1963, Bull.Geol.Soc.Am. 74, 831
LABHART, TP & RYBACH, L, 1974, Geol.Rundsch. 63, 135
LACHENBRUCH, AH, 1970, J.Geophys.Res. 75, 3291
NICOLAYSEN, LL, HART, RJ & GALE, NH, 1981, J.Geophys.Res. 86, 10653
NIELSON, DL, et al., 1976, Mem.Geol.Soc.Am. 146, 301
ROSHOLT, JN, ZARTMAN, RE & NKOMO, IT, 1973, Bull.Geol.Soc.Am. 84, 989
ROGERS, JJW & ADAMS, JAS, 1969a&b, in: Handbook of Geochemistry, Springer Verlag, 90-1, 92-1
ROY, RF, BLACKWELL, DD & BIRCH, F, 1968, Earth Planet.Sci.Lett. 5, 1
RYBACH, L & BUNTEBARTH, G, 1982, Earth Planet.Sci.Lett. 57, 367
SCLATER, JG, JAUPART, C & GALSON, D, 1980, Rev.Geophys.Sp.Phys. 18
SCLATER, JG, PARSONS, B & JAUPART, C, 1981, J.Geophys.Res. 86, 11535
SHAW, DM, 1970, Precambrian Res. 10, 281
SHARP, JA & SIMMONS, G, 1978, AGU Abst.Prog. 10, 85
SIMMONS, G, 1967, Rev.Geophys. 5, 43
TAYLOR, SR, 1982, Phys.Earth.Planet.Int. 29, 233
WETTERAUER, R & BOTHNER, WA, 1977, EOS, Trans.AGU, 58, 542

TABLE 1
AVERAGE VALUES OF RADIOACTIVITY IN THE CONTINENTAL CRUST

	U (ppm)	Th (ppm)	K (%)	A ($\mu W/m^3$)	References
Straight Averages					
	0.66	4.43	2.19	0.69	Haack (1983)
	1.50	6.00	1.50	1.00	Shaw (1970) (*)
Andesite model					
	1.25	4.80	1.25	0.82	Taylor (1982)
Crust/mantle exchange					
	1.30	6.10	1.70	0.98	Allègre et al.(1985)
Exposed vertical sections					
Ivrea Zone	/	/	/	1.50	Galson (1984)
Vredefort	/	/	/	0.96	Nicolaysen et al. (1981)

(*) The K concentration has been estimated using a K/U ratio of 10^4.

TABLE 2
THICKNESSES OF THE MAJOR PLUTONS OF NEW HAMPSHIRE
(Determined from gravity studies)

Formation	Maximum thickness (km)	Reference
Exeter pluton	3	Bothner (1974)
Kinsman pluton	3	Nielsen et al.(1976)
Bethlehem pluton	3	"
Binary granite	3	"
White Mountains	5.5	Joyner (1963)
		Wetterauer & Bothner (1977)

L. ROYDEN[1]

A SIMPLE METHOD FOR ANALYZING SUBSIDENCE AND HEAT FLOW IN EXTENSIONAL BASINS

ABSTRACT

Equations describing subsidence and heat flow in extensional sedimentary basins must be consistent with subsidence and heat flow observations in the oceans. This constraint is sufficient to evaluate all constants in the expressions for basin subsidence and heat flow which result from a modified extension model (in which crustal thinning and lithospheric heating are to a certain degree independent). Subsidence and heat flow equations calibrated in this way are used to present a simple and fast graphical technique for determining the stretching (or thinning) factors for extensional basins, and for predicting surface heat flow within a basin directly from its subsidence history. Examples from the Pannonian Basin, the Vienna Basin, the Gulf of Suez and the Pattani Trough (Thailand) all show excellent agreement between predictions and observations. In addition, it is shown that heat flow through a sedimentary basin can be determined from the thermal subsidence of the basin alone, irrespective of stretching factors, pre-rift conditions, or initial subsidence.

I. INTRODUCTION

Lithospheric extension is now generally accepted as the mechanism of formation for a large class of sedimentary basins that develop within continental crust. Within the last few decades crustal extension has been proposed by many authors as a means of generating rift valleys and grabens (for example, Carey

---

(1) *Department of Earth, Atmospheric and Planetary Sciences, Massachusetts Institute of Technology, Cambridge, Massachusetts, USA.*

1958). McKenzie (1978) was the first to quantify the relationship between crustal thinning, basin subsidence, and surface heat flow. McKenzie's simple model has since been modified and expanded by other workers to include such features as sediment loading, thermal blanketing by sediments, lithospheric flexure, lateral heat conduction, and modified (or non-uniform) extensional processes. Even the simplest of these quantitative stretching models requires iterative solution on a hand calculator or small computer, and more involved models require correspondingly larger computer programs. In addition, many of the phyical parameters used in these analyses, such as lithospheric thickness, conductivity, density, etc., are poorly constrained and different authors have assumed different values and achieved somewhat different results.

The aim of this paper is to present a simple method for calibrating the subsidence and heat flow equations for rifted basins by comparison with well constrained subsidence- heat flow-age data from the oceans. This calibration is then applied to a one-dimensional modified (or non-uniform) extension model to provide a simple graphical method of analyzing subsidence and heat flow in stretched basins. Before proceeding, however, it is important to define what is meant by modified extension, why it is a useful concept in subsidence and heat flow analysis, and what minimum pieces of information are needed to constrain adequately the thermal history of a rifted basin.

II. MODIFIED EXTENSION

Modified extension (originally called non-uniform extension by Royden and Keen, 1980 and Sclater et al., 1980) differs from simple uniform extension proposed by McKenzie (1978) in that it allows crustal thinning to occur somewhat independently of heating within the lithosphere. It is frequently assumed that modified extension implies different amounts of stretching within the crust and the mantle lithosphere, possibly leading to space problems in areas adjacent to the extended region. This is not strictly correct. Modified extension assumes only that crustal thinning and lithospheric heating do not necessarily have a one-to-one relationship. There are at least two fundamentally different ways in which modified extension may occur.

(1) One type of modified extension does involve mechanical decoupling above horizontal, gently dipping detachment surfaces. Extension above the detachment surface may be larger or smaller than that below it. Such detachments have been observed in the field and on reflection seismic profiles, and can be shown to

exist at very shallow crustal levels as well as at mid- to deep crustal levels (Figure la). This type of mechanical decoupling may imply space problems in adjacent unstretched regions, but in most such cases it is not difficult to construct balanced cross-sections through areas where extension involves mechanical decoupling within the crust. Normal detachment faults can be shown to have up to 50 km of displacement in some areas, but cumulative displacement on some normal fault systems may be much larger.

(2) Another type of modified extension appears to be a poorly understood process acting at the lithosphere-asthenosphere boundary. Evidence for such a process comes entirely from quantitative studies of subsidence, uplift and heat flow within and adjacent to extensional basins. Such studies of the Pannonian Basin, the west coast of the Labrador Sea and the Gulf of Suez all suggest that crustal thinning and extension were accompanied by creation of a much larger thermal anomaly than can be explained by uniform extension of the crust and mantle lithosphere (Sclater et al., 1980; Royden et al., 1983a; Royden and Keen, 1980; Steckler, 1985). Indeed, in the Gulf of Suez and in the Pannonian Basin, the thermal anomaly is equivalent to stripping away most of the mantle lithosphere and replacing it by hot asthenosphere (see below). Keen (1984) and Buck (1984) have proposed that such heating may result from small-scale convection caused by strong horizontal thermal and density gradients at the margins of the extended regions (Figure lb). This mechanism appears plausible, but is difficult to prove. No space problem is implicit in this type of modified extension, although it does raise questions about the temperature and pressure dependence of viscosity in the mantle.

Under ideal conditions, either type of modified extension process can result in basin subsidence (or sedimentation patterns) and surface heat flow that are incompatible with uniform extension. However, in many basins, existing data sets do not allow one to distinguish between uniform and modified extension, either because of insufficient data or because the necessary information is not contained in the geologic record. Therefore, while modified extension may not be a rare occurrance, a strong case for its occurrance can only be made for a few basins as yet. In these few basins, analyses of subsidence and heat flow data can provide strong constraints on lithospheric deformation during rifting, but certainly do not yield unique determinations of rifting processes, particularly for the mantle lithosphere (see below).

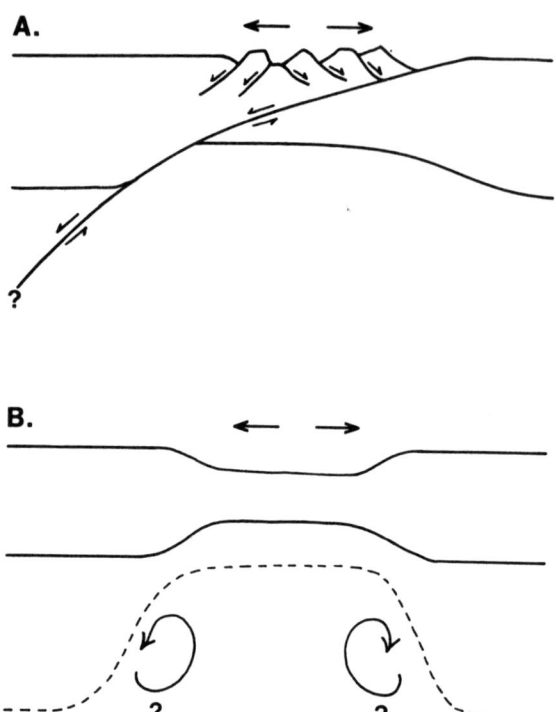

Figure 1. Two different ways in which modified (non-uniform) extension might occur. (A) Gently dipping detachment surface with extension occurring near the toe of the upper plate, but not in the lower plate immediately below. Upper plate extension must be balanced somewhere. Because material may also move perpendicular to the plane of the diagram, there are many ways to balance this geometry. For example, extension may be translated by strike-slip displacement into a neighboring thrust belt. Alternatively, upper plate extension may be balanced by extension at depth at some distance from the location of the upper plate extension. (B) Poorly understood processes acting near the lithosphere-asthenosphere boundary effectively removes lithospheric material and replaces it with hotter asthenosphere. Because this appears to occur rapidly, thermal conduction cannot be responsible. Buck (1984) and Keen (1984) have suggested that small-scale convection is driven by horizontal temperature gradients at the margins of the extended area.

Simple Method for Extensional Basin Analysis

Conceptually, there are two primary factors that control the subsidence history of a rifted basin: first, the amount of crustal thinning that occurs during extension; and second, the amount of heat added to the crust and upper mantle and (usually much less important) the way in which that heat is distributed with depth. After correcting for a number of secondary, though not necessarily insignificant, processes (such as horizontal conduction of heat, flexural properties of the lithosphere, loading and thermal blanketing by sedimentary rocks, etc.), these are the quantities that one tries to reconstruct from examination of subsidence and thermal data. It should be noted that magnitudes of crustal thinning and total lithospheric heating are the only constraints on lithospheric deformation that can be obtained from subsidence and heat flow data within a rifted basin. Realistic interpretations of lithospheric behavior must combine such analyses with structural data for the upper crust, geophysical measurements at deeper levels, regional tectonic data, and theoretical studies of mantle processes such as convection.

Extensional basin evolution can be divided into two main stages, which have characteristic sedimentological and structural features. The first stage corresponds to the rifting stage, as the crust and mantle lithosphere become attenuated and are replaced from below by passive upwelling of hot asthenosphere. At the same time, there is an initial change in elevation (usually subsidence, Figure 2). This occurs both as an isostatic response to net density changes resulting from crustal thinning (density increase) and from an increase of the average temperature within the upper mantle (density decrease). This will be called initial subsidence. The depth immediately after extension with be called the initial depth. Initial subsidence is a response to active extensional processes in the lithosphere, and the area of subsidence is usually well localized and fault bounded. Sediments deposited during this phase are often deposited very rapidly, usually contain synsedimentary faults and may have rotated bedding.

Following rifting, the second stage of subsidence is a passive process, corresponding to the cooling of the crust and upper mantle towards thermal equilibrium. Thermal contraction of lithospheric rocks results in long-term subsidence (of the order of 200 m.y.) that will be referred to here as thermal subsidence. Thermal subsidence is a function of the heat distribution in the lithosphere and is essentially independent of crustal thickness (Figure 2). It is usually of greater areal extent than that of the first phase and not usually confined o fault bounded troughs. Sediments are relatively flat-lying and undisturbed, often onlapping onto basement rocks adjacent to the initial rifts.

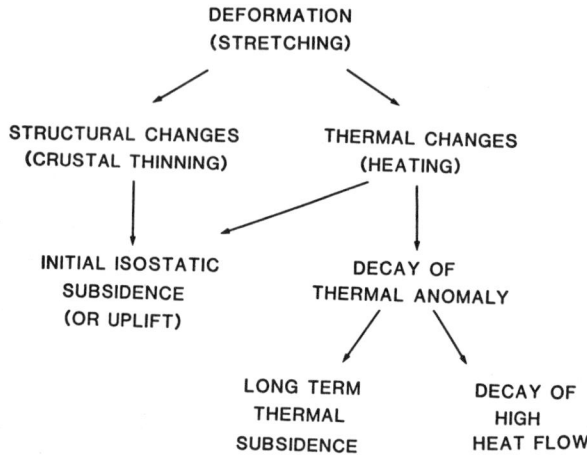

Figure 2. Simple flow chart diagramming the consequences of lithospheric stretching, which results in both thinned crust and a geotherm elevated above the background or equilibrium geotherm. Crustal thinning causes a density increase within a lithospheric column as lighter crustal rocks are replaced by denser mantle material. The raised geotherm causes a density decrease within the same column as colder, denser mantle lithosphere is replaced by hotter, less dense asthenosphere. If the net density of the lithospheric column decreases, initial uplift will occur; if the net density increases, initial subsidence will occur. The latter appears to be more common. As the elevated geotherm cools towards thermal equilibrium, the lithosphere cools, contracts, and subsides isostatically over about 200 to 300 m.y. Elevated surface heat flow present after stretching likewise decays. Note that surface heat flow and thermal subsidence both reflect how fast the lithosphere is cooling, and are thus linked.

During both phases, subsidence is amplified by isostatic compensation for sediment loading.

A long time after rifting (>200 m.y.), the lithosphere beneath the intended region approaches thermal equilibrium conditions, thermal subsidence ends, and the basin reaches its final depth. The final depth of the basin should depend only on the crustal thickness present beneath it because, by definition, the thermal anomaly associated with rifting has disappeared.

In order to calculate the amount of crustal attenuation and the amount of heating that occur during formation of a sedimentary basin, one must be able to (1) distinguish accurately between the

Simple Method for Extensional Basin Analysis

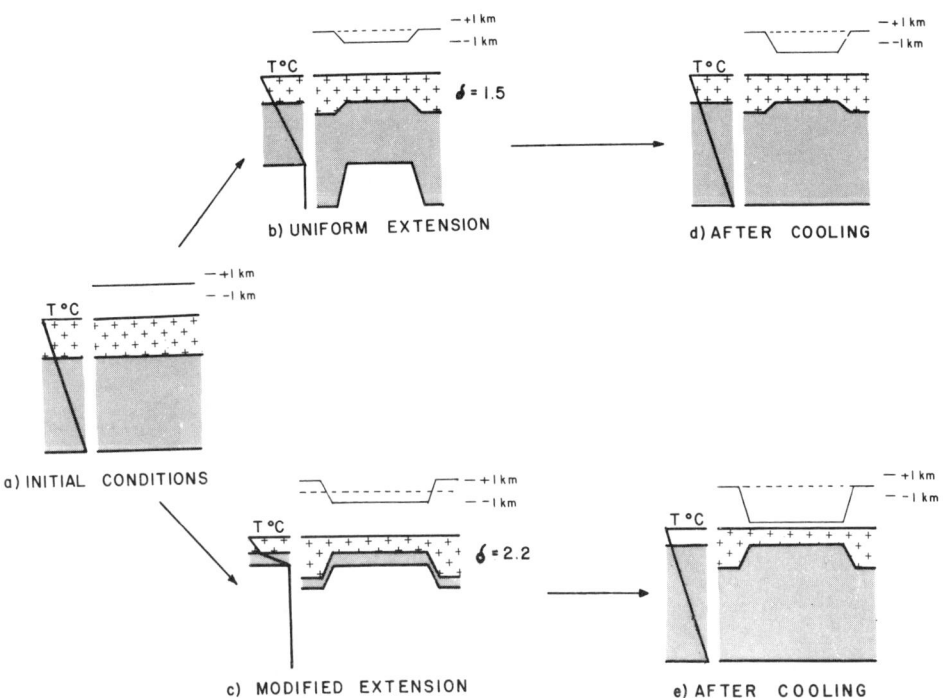

Figure 3. (a) Schematic diagram of continental crust (crosses) and subcrustal lithosphere (shaded) at equilibrium thermal conditions prior to extension. Initial crustal thickness is assumed to be 35 km and the base of the lithosphere is assumed to be maintained at a constant temperature throughout. Initial surface elevation (light line above diagram) is at sea level. (b) Uniform lithospheric extension by $\delta = \beta = 1.5$. Immediately after extension, thermal gradient is elevated by about 50% and initial basin subsidence is about 1 km (water loaded). (c) Modified lithospheric extension (crust is extended by $\delta = 2.2$ within the basin and unextended in uplifted blocks; subcrustal lithosphere is attenuated by about a factor of 10 everywhere). Immediately after extension, near surface temperatures are much higher than in Figure 3b even though basement elevation within the basin (water loaded) is identical. Outside the basin, basement is uplifted. (d) After cooling, temperature reflects thermal equilibrium conditions. Within the basin depth (water loaded) is roughly 2 km. (e) As for Figure 3d except that basement depth within the basin (water loaded) is roughly 3.5 km, so that even though initial subsidence was identical to that for uniform extension by $\delta = \beta = 1.5$, thermal subsidence is about 2.5 times greater. If no erosion has occurred, regions flanking the deep basin will return to sea level. Figure from Royden et al., 1983a.

initial and thermal phases of basin formation; (2) determine the magnitude of the initial subsidence, which includes knowing with some precision initial elevations prior to rifting; (3) determine rates of thermal subsidence or heat flow, preferably both; and (4) estimate the pre-extensional thermal structure (or thickness of the lithosphere). Figure 3 shows two possible ways in which a basin with an initial (water-loaded) depth of 1 km might have formed. However, the thermal subsidence and final depths are very different for the two cases. A geologist examining these basins shortly (10 to 20 m.y.) after extension, would not be able to distinguish between the two basins solely on the basis of total depth to basement and age. However, by measuring the amount of thermal subsidence and the heat flow in each basin, the difference would be readily apparent.

In the next section initial, thermal and final subsidence are simply related to extension parameters for modified extension, and those relationships are calibrated with oceanic data. Then in subsequent sections a graphical approach is presented for determining crustal thinning and lithospheric heating during extension.

III. FORMULATION AND CALIBRATION WITH OCEANIC STUDIES

In order to calculate the subsidence and heat flow histories of an extended region, crustal thinning and lithospheric heating need to be described by simple parameters. Crustal thinning can easily be described by a thinning factor δ (so that if the crustal thickness before stretching is h, the thickness after stretching is h/δ). This thinning factor does not necessarily imply anything about how thinning is distributed throughout the crust, but rather describes only the net crustal thinning.

TABLE I

Symbols and Values Used

h	crustal thickness before stretching	35 km
d	elevation before stretching	0 m
ℓ	lithospheric thickness	125 km
T	temperature (°C)	
T_m	temperature at the base of the lithosphere	
δ	crustal stretching factor	
β	subcrustal stretching factor	
x	depth (m)	
t	time since stretching (m.y.)	

Simple Method for Extensional Basin Analysis

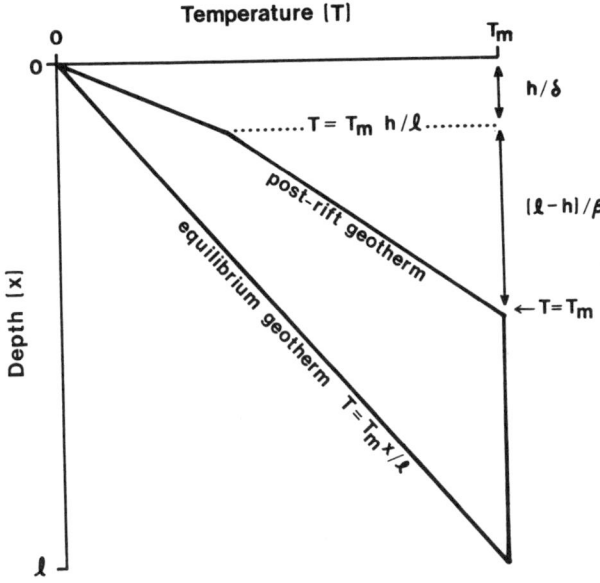

Figure 4. Schematic representation of the equilibrium geotherm, and the geotherm immediately after stretching. h is taken to be the crustal thickness before stretching, so that h/δ is the crustal thickness after stretching. Likewise, $(\ell -h)$ is the thickness of the mantle lithosphere before stretching, and $(\ell - h)/\beta$ is its thickness after. Symbols are listed in Table I.

The amount of heat distributed throughout the lithosphere is more difficult to describe by one parameter, because the resulting subsidence and heat flow will depend on exactly how temperature is distributed with depth. However, most geologically reasonable temperature structures can be approximated by two-legged geotherm such as that shown in Figure 4. Short wavelength differences from a two-legged geotherm will disappear quickly, and usually have little effect on subsidence and surface heat flow. Therefore, it is sufficient to describe heat distribution within the lithosphere with two parameters: δ to describe the assumed post-rift geotherm in the crust and β to describe the assumed post-rift geotherm in the mantle lithosphere (Figures 3 and 4). Note that δ also describes the net crustal thinning (see above), so that crustal heating and crustal thinning are assumed to be linked. β (mantle heating) is independent of δ. This simple parameterization clearly does not describe all possible geotherms, but it comes close to describing most geologically reasonable geotherms. (Note that $\beta = \delta$ corresponds to uniform extension.)

Analytical solution to the initial and thermal subsidence and the heat flow resulting from this parameterization are given exactly as a funciton of β and δ in Royden and Keen (1980). These expressions can be simplified and calibrated to agree with oceanic data as follows:

1) The final depth of a basin is a function of starting elevation and pre-rift elevation, and is linear in $1/\delta$ to within a few

57

percent. Then:

$$\text{final depth} = C_1 + C_2/\delta$$

C_1 can be evaluated by noting that in the oceans, where $\beta = \infty$, the final depth is about 7200 m. (This includes an 800 m correction for removing the oceanic crust isostatically. Uncertainty is estimated at a few hundred meter, so that 7200 ± 300 m should give a reasonable range of C_1, Parsons and Sclater (1977). If the pre-rift elevation was at a water loaded depth d below sea level, C_2 can be evaluated by noting that for $\delta = 1$ (no crustal stretching), the final depth must be the same as the pre-rift depth, d. Therefore:

$$\text{final depth} = 7200\text{m} + (d-7200\text{m})/\delta \qquad (1)$$

2) The total thermal subsidence (the thermal subsidence that occurs from immediately after rifting until an infinite time after rifting) must be proportional to the total amount of heat lost during cooling of the lithosphere. This is proportional to the difference in average temperature immediately after rifting and the average temperature a long time after rifting (the equilibrium geotherm). From analysis of Figure 4:

$$\text{total thermal subsidence} = C_3 \cdot \left[(1-1/\delta) + (1-h/\ell)^2(1/\delta-1/\beta)\right]$$

C_3 can be evaluated by noting that, in the oceans, where $\beta = \delta = \infty$, the total thermal subsidence is about 3900 m (Parsons and Sclater, 1977). Therefore:

$$\text{total thermal subsidence} = 3900\text{m}\left[(1-1/\delta) + (1-h/\ell)^2(1/\delta-1/\beta)\right] \qquad (2)$$

3) The initial depth immediately after rifting is equal to the final depth minus the total thermal subsidence, or:

$$\text{initial depth} = d/\delta + 3300\text{m}(1-1/\delta) - 3900\text{m}(1-h/\ell)^2(1/\delta-1/\beta) \qquad (3)$$

Simple Method for Extensional Basin Analysis

4) Time dependent depth, as given by Royden and Keen (1980) can be written as:

$$\text{depth} = C_4 - C_5 \sum_{\substack{n=1 \\ n,\text{odd}}}^{\infty} \left[((\delta-\beta)/n\pi)\sin n\pi H + (\beta/n\pi)\sin n\pi G \right] \cdot (1/n^2) \cdot \exp(-n^2 t/C_6)$$

where t is time in millions of years and:

$H = h/\ell\delta$

$G = (h/\ell)(1/\delta - 1/\beta) + 1/\beta$

C_4 can be evaluated by noting that at $t = \infty$, the expression after the summation sign is zero. Therefore C_4 is equal to the final depth given by equation (1). C_5 can be evaluated by noting that for the oceanic case, where $\beta = \delta = \infty$, the expression inside the square brackets is equal to 1. Therefore the total thermal subsidence between time $t = 0$ and time $t = \infty$ is given by:

$$\text{total thermal subsidence} = 3900\text{m} = C_5 \sum_{\substack{n=1 \\ n,\text{odd}}}^{\infty} 1/n^2 = C_5 \cdot \pi^2/8$$

giving $C_5 = 3200$m. C_6, the thermal time constant, is assumed to be the same as that in the oceans, or approximately 60 m.y. Thus:

$$\text{depth} = d/\delta + 7200\text{m}(1-1/\delta) - 3200\text{m} \sum_{\substack{n=1 \\ n,\text{odd}}}^{\infty} \left[((\delta-\beta)/n\pi)\sin n\pi H + (\beta/n\pi)\sin n\pi G \right] \cdot (1/n^2) \cdot \exp(-n^2 t/60) \quad (4)$$

5) Surface heat flow can likewise be written as:

$$\text{surface heat flow} = C_7 \left[1 + 2 \sum_{n=1}^{\infty} \left[((\delta-\beta)/n\pi)\sin n\pi H + (\beta/n\pi)\sin n\pi G \right] \cdot \exp(-n^2 t/C_6) \right] + (\text{radiogenic contributions from crustal rocks})$$

C_7 is the background heat flow and in the oceans is about 33 mW/m^2. The radiogenic contribution from crustal rocks is variable, but in this paper we use a value of 12 mW/m^2 as a reasonable value for continental crust. Thus:

Simple Method for Extensional Basin Analysis

$$\text{surface heat flow} = 33\text{mW/m}^2 \cdot \left[1 + 2\sum_{n=1}^{\infty}\left[((\delta-\beta)/n\pi)\sin n\pi H + (\beta/n\pi)\sin n\pi G\right] \cdot \exp(-n^2 t/60)\right] + 12\text{mW/m}^2 \quad (5)$$

Equations (1)-(5) thus give expressions for the water-loaded subsidence history and heat flow history of a sedimentary basin assuming that the basin was at thermal equilibrium prior to extension and knowing only the thinning factors β and δ, the pre-rift elevation d, and the pre-rift crustal thickness h. These expressions are convenient because, by calibrating all the parameters with the oceanic case ($\beta = \delta = \infty$), the subsidence and heat flow equations can be well calibrated even without knowing individual parameters such as crustal density, lithospheric thickness, etc. Equivalently, any combination of these individual physical parameters are acceptable and indistinguishable, provided that they give results consistent with observations in the oceans and thus yield equations (1)-(5).

IV. GRAPHICAL APPROACH

Equations (1)-(5) above can be easily presented in graphical form to allow simple and fast determinations of extension in a sedimentary basin. All of the plots presented in this paper assume $h/\ell = 0.3$ ($h \approx 35$ km, $\ell \approx 125$ km). Errors in $(1-h/\ell)^2$ of about 20% are introduced when h is taken to be 25 or 45 km. This should not contribute significant errors to the results of Eqs. (2)-(5), and the error is zero when $\beta = \delta$. As in the previous section, all subsidence must be water loaded, so that corrections for sediment loading and compaction are already assumed (for discussion see Sclater and Christie, 1980).

Figure 5a shows the initial depth that is present immediately after instantaneous stretching of continental lithosphere with pre-rift elevation at sea level. Note that either initial uplift or initial subsidence can occur. Figure 5a also shows that even when pre-rift elevations and pre-rift thermal structure are specified, there are a variety of different values of β and δ that give the same initial depth after stretching. Thus β and δ cannot be uniquely determined from the initial depth or subsidence alone (for example, Figure 3).

Figure 5b shows thermal subsidence, as a function of β and δ, that has occurred by 10 m.y. after the completion of active rifting. These plots show the amount of thermal subsidence, and

Simple Method for Extensional Basin Analysis

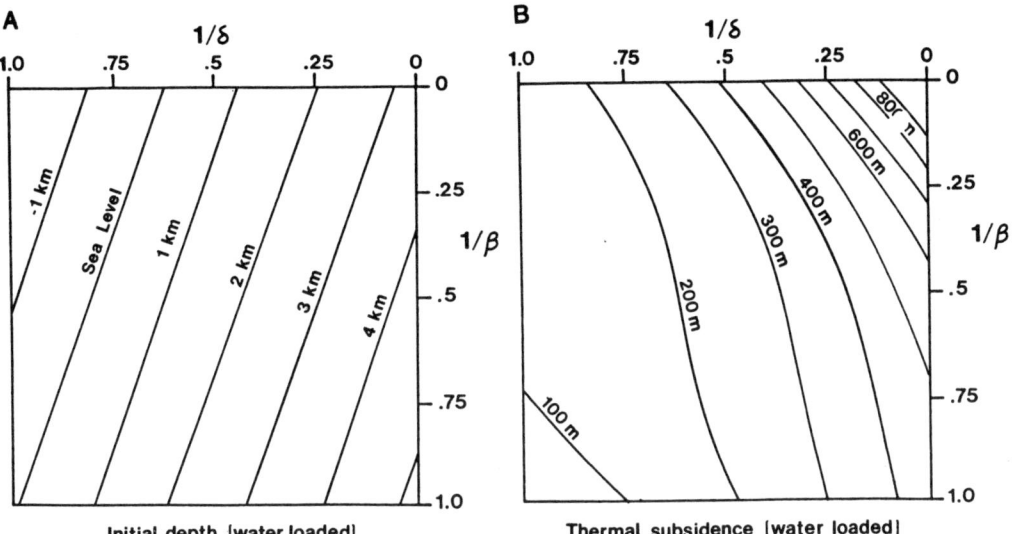

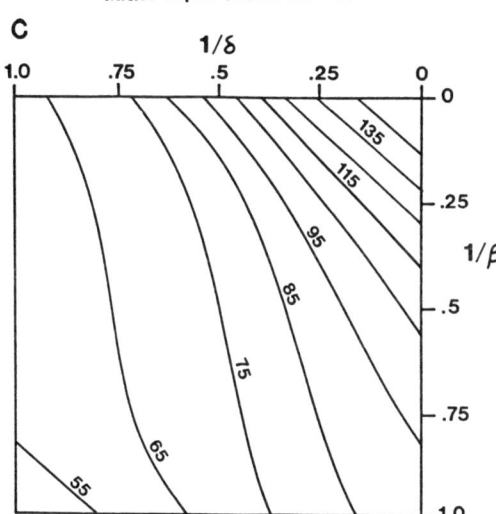

Figure 5. (a) Initial depth after stretching as a function of thinning parameters δ (crustal thinning) and β (subcrustal thinning). Elevation prior to stretching is assumed to be at sea level, and the pre-stretching geotherm is assumed to correspond to thermal equilibrium (see Figure 4). All elevations are water loaded, so that unloaded (or air loaded) uplift must be corrected for this effect. Sediment loaded subsidence within a basin must likewise be corrected for the effects of sediment loading. Extension is assumed to occur instantaneously. (b) Thermal subsidence, as a function of β and δ, that occurs between t = 0 (immediately after stretching) and t = 10 m.y. (10 m.y. after stretching). The pre-stretching geotherm is assumed to correspond to thermal equilibrium. All subsidence is assumed to be water loaded. (c) Surface heat flow, as a function of β and δ, at t = 10 m.y. The pre-stretching geotherm is assumed to correspond to thermal equilibrium Radiogenic heat production in the crust is assumed to contribute 12 mW/m^2 to the surface heat flow, so that in areas where the radiogenic contribution is greater or smaller than 12 mW/m^2, the values in this figure need to be adjusted accordingly. This plot also neglects possible effects of lateral heat conduction, thermal blanketing, and convection.

61

TABLE II

Initial and Thermal Subsidence (water loaded), Heat Flow and Age of Several Extensional Sedimentary Basins

Basin	Well	Age	Initial Subsidence (m)	Thermal Subsidence (m)	Heat Flow (mW/m^2)
Pannonian Basin[a]	1	10 Ma	1800 ± 300	440 ± 100	95 ± 10
	2	10 Ma	1300 ± 200	390 ± 80	95 ± 10
	3	10 Ma	920 ± 160	450 ± 70	95 ± 10
	4	10 Ma	670 ± 110	380 ± 40	95 ± 10
	5	10 Ma	500 ± 100	320 ± 20	95 ± 10
	6	10 Ma	270 ± 70	360 ± 20	95 ± 10
Vienna Basin[b]		15 Ma	0 to 1500	0 to 150	50 ± 7
Pattani Trough (Gulf of Thailand)[c]		12 Ma	2300 to 2800	300 to 400	90 ± 10
Gulf of Suez[d]		10-20 Ma	1000	?	60-84
		10-20 Ma	-1500	?	60-84

[a]Thermal subsidence based on extrapolation of thermal subsidence from 0-5 Ma and 0-8 Ma. Subsidence from Royden and Dövényi, in prep.; heat flow from Dövényi and Horváth, in prep., and Dövényi et al. (1983).

[b]Subsidence from Royden, in press; heat flow from Cermak, 1979.

[c]Data from Hellinger and Sclater (1983). Thermal subsidence of 300 to 400 m estimated from 1300 m of sediment loaded subsidence (already corrected for compaction) shown in Figure 13 of Hellinger and Sclater.

[d]Subsidence data from Steckler, in press. Unloaded uplift of 1000 m converted to 1500 m of water loaded uplift. Heat flow from Morgan et al. (1980).

Simple Method for Extensional Basin Analysis

not depth to basement. Depth to basement at any time would be given by the sum of the initial depth (Figure 5a) and the thermal subsidence at that time (Figure 5b). Figure 5a shows that even if thermal subsidence is reasonably well known, β and δ cannot be uniquely determined from the thermal subsidence alone.

Figure 5c shows surface heat flow, as a function of β and δ, at 10 m.y. after rifting. The background heat flow and radiogenic contributions were assumed to total 45 mW/m^2, so that in areas where these have higher or lower background heat flow or radiogenic heat production, this value may need to be adjusted accordingly. Note that β and δ cannot be determined uniquely from heat flow data, even if heat flow could be determined precisely.

Figures 5a-c are all plotted at the same scale, with identical β and δ axes. Thus by superimposing two or more of these plots, unique values of β and δ can be determined graphically from the subsidence history of a basin. The heat flow history can also be determined from the subsidence history by superimposing heat flow plots (Figure 5c) onto the subsidence plots (Figures 5a and b).

V. EXAMPLES

The Pannonian Basin, in East-Central Europe, is a young extensional sedimentary basin that formed between about 15 and 10 Ma, and has been undergoing thermal subsidence from 10 Ma until the present (see Royden et al., 1983a). In this basin, the initial and thermal subsidence can be distinguished, and subsidence for 6 well groups is summaried in Table II. (Wells within each group had similar subsidence histories. Uncertainties shown are mainly uncertainties in decompaction and unloading, not scatter in the data.) Well group 1 has an initial water loaded subsidence of about 1800 ± 300 m, depending on the decompaction parameters used, and a thermal water-loaded subsidence for 10 m.y. of 440 ± 100 m. Assuming that prior to extension the basement was near sea level, Figure 6a shows the initial subsidence (1800 m) plotted as a function of β and δ (taken from Figure 5a). Also plotted is the thermal subsidence (440 m) at t = 10 m.y. (taken from Figure 5b). The intersection of the initial subsidence curve and the thermal subsidence curve yield unique values of crustal thinning (δ = 2.8) and lithospheric heating (β = 7). When the uncertainties in the subsidence data are included, a field of permissible values of β and δ is defined, (δ = 2.4-3.6, β = 3-∞).

These values of β and δ can then be superimposed on the plot of heat flow at 10 m.y. after stretching to calculate predicted

heat flow values for this well group (Figure 6b). Well group 1
corresponds to theoretical heat flow values between 85 and
115 mW/m^2. The β and δ values for well groups 2-5 are plotted in
the same manner, and are also shown in Figure 6b. These well
groups correspond to theoretical heat flow values of 85 to
105 mW/m^2. The measured heat flow in the Pannonian Basin is about
95 ± 10 mW/m^2 (shaded region in Figure 6b) and shows little
systematic variation with depth to basement. Thus the theoretical
heat flow predicted from the subsidence history and the observed
heat flow are nearly identical, indicating that the thermal
subsidence within the basin and the surface heat flow are
consistent. In addition, Figure 6b shows that although the total
water loaded depth to basement varies from 2200 m to 500 m, the
predicted heat flow is nearly the same for all well groups. Thus
little variation in heat flow should be expected for shallow and
deep parts of the basin.

One can also observe that the uncertainties in determining β
and δ are much larger for well group 1 that for well group 6,
mainly because uncertainties in decompaction parameters have a
much larger effect for deeper wells. All of the well groups plot
in the vicinity of β = ∞, implying that the mantle lithosphere has
been almost entirely replaced by hot asthenosphere. This is
particularly clear for the shallower well groups, 3-6. The
allowable thinning parameters for well group 1 are almost
compatible with uniform extension, but it seems unlikely that well
group 1 is significantly different from groups 1-5.

Thinning parameters are also plotted for the Vienna Basin,
located in Austria and Czechloslovakia. The Vienna Basin belongs
to the same basin system as the Pannonina Basin, but is slightly
older. In the northeastern part of the basin extension ended at
about 15 Ma, although active faulting continues today in some
other parts of the basin. On the basis of tectonic and geometrical arguments, Royden et al. (1983b) and Royden (1985, in press)
have proposed that extension within the Vienna Basin is thin-
skinned, similar to the sketch shown in Figure 1a. (Extension is
apparently transformed by strike-slip faults to the adjacent
Carpathian thrust belt, where thrusting was active at the same
time as basin extensin occurred. Thus no space problem need be
created.)

Initial subsidence within the northeastern Vienna Basin
varies between about 100 and 1500 m. Thermal subsidence in the
same area is between 0 and 150 m, depending upon the corrections
added for regional uplift. These bounds are plotted in the lower
left corner of Figure 6b. This shows that both the amount of
crustal thinning (δ) and the amount of apparent subcrustal

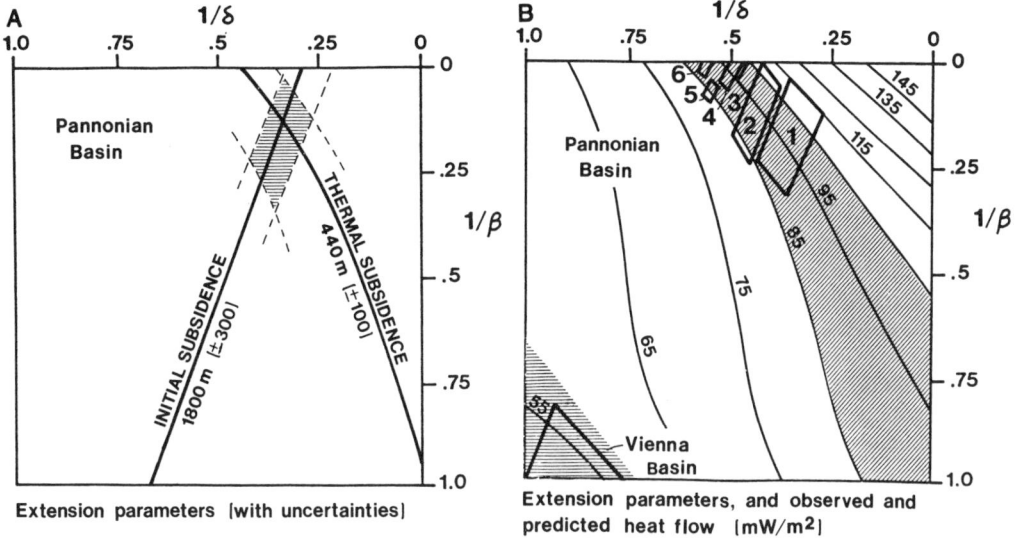

Figure 6. (a) Subsidence for well 1 in the Pannonian Basin as a function of δ (crustal thinning) and β (apparent sub-crustal thinning). Dark lines show initial subsidence of 1800 m (from Figure 5a) and thermal subsidence at t = 10 m.y. of 440 m (from Figure 5b). Intersection of the dark lines gives thinning parameters δ = 2.8, β = 7. Dashed lines show uncertainty of ±300 m for initial subsidence and ±100 m for thermal subsidence. Shaded region thus indicates all values of β and δ that are consistent with the initial and thermal subsidence data, including uncertainties. (b) Pannonian Basin: Dark boxes 1-6 show values of β and δ for wells 1-6 in the Pannonian Basin, calculated by the method shown in Figure 6a (Table II). Also shown are the theoretical heat flow values taken from Figure 5c. Thus the subsidence histories of the Pannonian Basin wells are consistent with theoretical heat flow of about 85-105 mW/m². Shaded area between 85-105 mW/m² indicates the measured heat flow in the Pannonian Basin. Therefore the theoretically predicted heat flow and the observed heat flow are in good agreement. Vienna Basin: Dark box shows values of β and δ that are consistent with the subsidence history of the Vienna Basin (Table II). Method is the same as for the Pannonian Basin. Shaded area shows observed heat flow of 50 ± 7 mW/m², in good agreement with the predicted theoretical heat flow.

thinning (β) are small, less than about 1.25. The theoretical heat flow consistent with the allowable values of β and δ is between 45 and 60 mW/m^2. Measured heat flow in the northern part of the Vienna gives 50 ± 7 mW/m^2 (Cermak, 1979), and so is in good agreement with the theoretical heat flow predicted from Figure 6b.

(Values of β, δ, and theoretical heat flow for the Vienna Basin cited above were calculated assuming t = 10 m.y. and assuming that thermal subsidence was 0-100 m for the first 10 m.y. of thermal subsidence.)

Thus the factor of two difference in heat flow and geothermal gradients between the Pannonian Basin and the Vienna Basin can be predicted from the different subsidence histories of the two basins, even though both basins are roughly the same age and depth. This also explains the great differences in depth to the hydrocarbon generation window, which is within basin sediments at about 2.5 to 3.0 km depth in the Pannonian Basin (Berczi et al., 1981), but within sub-basin rocks below about 5 or 6 km depth in the Vienna Basin (Kröll and Wessely, 1973).

Figure 7 shows a similar plot for two other extended regions in which stretching ended at about 10 Ma. The Pattani Trough in the Gulf of Thailand formed by crustal extension between about 40 and 12 Ma, and has been subsiding thermally since 12 Ma (Hellinger and Sclater, 1983). Hellinger and Sclater summarize the subsidence history and heat flow in the Pattani trough, and quote an initial (water loaded) subsidence of 2.8 km. Because the stretching phase lasted for 30 m.y. some thermal subsidence probably occurred during stretching, so that 2300 m to 2800 m would probably be realistic bounds on the initial subsidence due to crustal thinning. Hellinger and Sclater neglected to cite the (water loaded) thermal subsidence, but from unloading their decompacted subsidence curve in their Figure 13, I estimate about 300-400 m of thermal subsidence from 12 Ma until the present. These bounds are plotted in Figure 7 in the same way that the Pannonian Basin data are plotted in Figure 6b.

The uncertainties in the thinning parameters for the Pattani Trough data are larger than for the Pannonian Basin, and restrict δ to between about 2.5 and 4, and β to between about 1.7 and 3. The field of allowable β's and δ's indicates a theoretical heat flow between 80 and 95 mW/m^2. Measured heat flow in the Pattani Trough as cited by Hellinger and Sclater (1983) is 90 ± 10 mW/m^2, and so is in good agreement with theoretical values determined from the subsidence history.

Simple Method for Extensional Basin Analysis

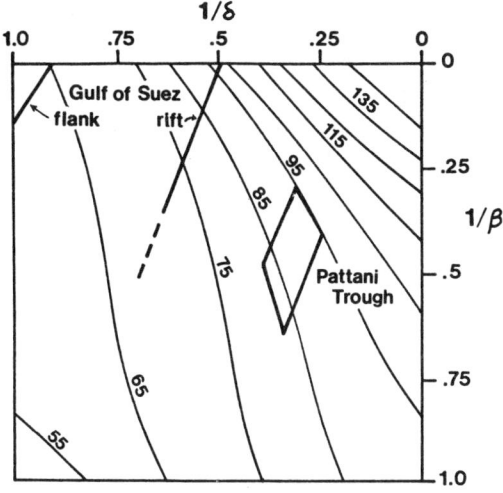

Extension parameters with predicted heat flow [mW/m²]

Observed Heat Flow	
Gulf of Suez	60-84 mW/m²
Pattani Trough	80-100 mW/m²

Figure 7. Pattani Trough (Gulf of Thailand): Values of β and δ calculated by the method shown in Figure 6a (box). Initial depth is between 2300 and 2800 m, and thermal subsidence (at t = 10 m.y.) is between 300 and 400 m (Table II). This corresponds to theoretical heat flow (from Figure 5c) between 80 and 95 mW/m². Measured heat flow in the Pattani Trough is 90 ± 10 mW/m² (shaded region), in good agreement with the theoretical predictions. Gulf of Suez: Initial (water loaded) depths of 1 km for the rift zone and -1.5 km (or 1 km unloaded uplift) for the rift flank are shown by dark lines. This gives $\delta < 1.1$ and $\beta > 6$ for the rift flank. If one assumes that $\beta > 4$ for the rift, δ must be between about 1.5 and 2. These values of β and δ correspond to a predicted theoretical heat flow of about 60 to 95 mW/m². Measured heat flow in the Gulf of Suez varies from 60 to 84 mW/m².

Hellinger and Sclater (1983) state that crustal thinning appears to be greater than sub-crustal thinning ($\delta > \beta$), and that extension within the Pattani Trough is thus non-uniform. Figure 7 indicates that this may indeed be the case, but it cannot be proven with the data presented in Table II, as the field of allowable β and δ values extends across the line $\beta = \delta$.

The Gulf of Suez is a post-Eocene rift, and most of the extension within the Gulf of Suez is probably Early to Middle Miocene (Garfunkle and Bartov as cited in Steckler, 1985), or approximately 10 to 20 Ma. Prior to extension, elevation of the ... region was near sea level, but during extension, the flanks of the rift were uplifted and are currently about 1 km above sea level. Approximately contemporaneously, the rift zone subsided to a (water loaded) depth of 1 km. At present I have no data discriminating initial and thermal subsidence within the rift zone.

Initial unloaded uplift of one kilometer for the rift flank corresponds roughly to 1500 m of water loaded uplift. This is plotted in Figure 7, and indicates that at the margins of the rift, crustal thinning is not large ($\delta < 1.1$) while the amount of heat added to the lithosphere is extremely high ($\beta > 6$). Even if a part of the flank uplift were due to lateral heat conduction of heat from rift zone (probably contributing less than one or two hundred meters), the amount of crustal stretching beneath the rfit flank must be less than about 20%. The amount of heating or apparent sub-crustal thinning is much greater than the crustal thinning.

The initial (water loaded) subsidence of 1 km within the deepest part of the rift is also plotted in Figure 7. If one assumes that β must be nearly as large within the rift as on the rift flank, this subsidence corresponds to a value of δ between about 1.5 and 2, in good agreement with estimates of crustal thinning made by Steckler (1985) ($\delta = 1.6$) for the deepest parts of the rift zone. Even without information on thermal subsidence within the rift zone, one can use Figure 7 to estimate theoretical heat flow values for the Gulf of Suez. Permissable values of β and δ for the rift flanks suggest a heat flow of about 60 mW/m^2 while for the deep rift zone they suggest an upper limit on the heat flow of about 95 mW/m^2. (This assumes rifting at 10 Ma. If the main rifting even is assumed to have occurred at 20 Ma, these values changes to 65 mW/m^2 and 90 mW/m^2.) Measured heat flow in the Gulf of Suez varies from 60 to 84 mW/m^2 (Morgan et al., 1980), in good agreement with the theoretical values. Thus, in some cases, estimates of heat flow can be made without information about thermal subsidence, particularly in rifts where flanking uplift occurs.

VI. THERMAL SUBSIDENCE VERSUS HEAT FLOW

Superposition of thermal subsidence as a function of β and δ, Figure 5b, and heat flow as a function of β and δ (Figure 5c),

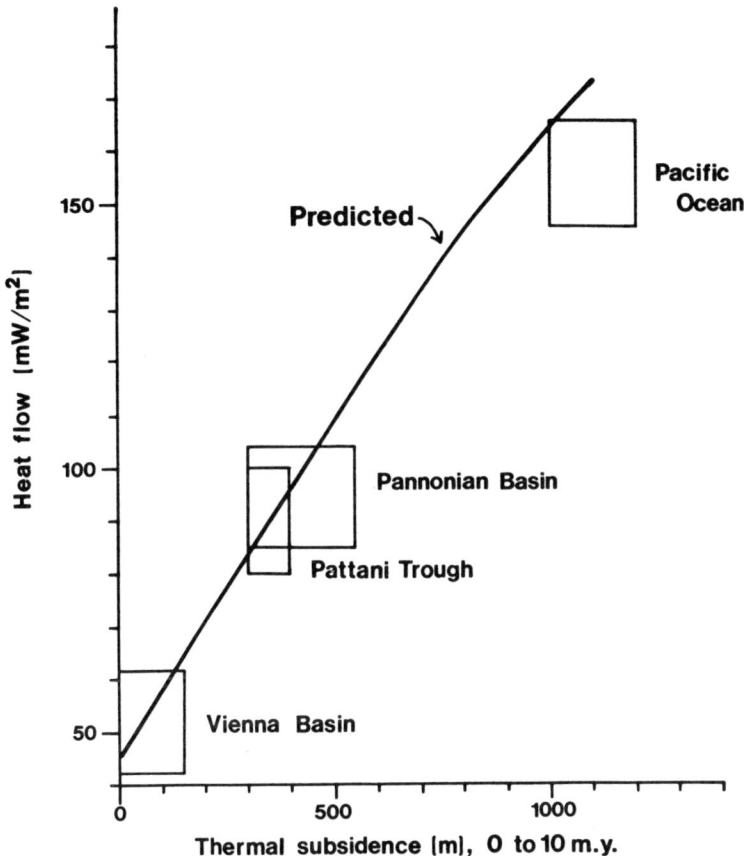

Figure 8. Relationship between thermal subsidence and surface heat flow at t = 10 m.y. Line shows the theoretical relationship taken from Figure 5d. Boxes show data from three basins discussed in this paper and from the Pacific Ocean (Sclater et al., 1976). Note that theoretical heat flow may have about ±10 mW/m² uncertainty associated with it, partly due to assumptions of radiogenic heat production and partly due to slight variations in predicted heat flow as a function of thermal subsidence.

shows that at t = 10 m.y., thermal subsidence and heat flow are closely linked. If one takes the relationship between thermal subsidence and heat flow for $\beta = \delta$ (uniform extension), then the same relationship will hold to about ±5 mW/m^2 irrespective of β and δ. In fact, superposition of Figures 5b and 5c shows that heat flow can be predicted from thermal subsidence alone, without knowing anything about β, δ, pre-rift elevation and thermal structure, or initial subsidence. This relationship is plotted in Figure 8.

Figure 8 also shows the observed thermal subsidence and the measured heat flow for the Pannonian and Vienna Basins and the Pattani Trough (Table II). In addition, subsidence and reliable heat flow data from the Pacific Ocean for ocean floor dated about 10 Ma is also plotted (from Sclater et al., 1976). The observations are in good agreement with the theoretical relationship between thermal subsidence and heat flow. At least ±10 mW/m^2 uncertainty should be assigned to the theoretical heat flow, because radiogenic contributions from crustal rocks may vary significantly. The mismatch between the Pacific heat flow and the theoretical heat flow curve is at least partly due to lower radiogenic contributions from oceanic crust (~5 mW/m^2?) than that assumed in Eq. (5) (12 mW/m^2).

VII. CONCLUSIONS AND DISCUSSION

The primary results of this paper can be summarized as follows:

1) Equations describing subsidence and heat flow in extensional basins must be consistent with observations of oceanic subsidence and heat flow. This constraint is sufficient to evaluate all constants in the expressions for basin subsidence and heat flow. Therefore it is not necessary to choose values for individual physical parameters, because all values that satisfy the oceanic case are equivalent to the expressions derived in this paper. Basin subsidence and heat flow are then functions of pre-rift elevation, pre-rift crustal thickness, crustal thinning, and apparent sub-crustal thinning (assuming that the crust is at thermal equilibrium prior to rifting).

2) These equations describing subsidence and heat flow can be presented graphically. By overlaying plots of subsidence and heat flow as functions of the thinning parameters, the acceptable thinning parameters for any basin are readily apparent. In this paper I used as examples data from basins where extension was completed at about 10 Ma, but identical techniques can be applied

to basins of any age. Plots of initial elevation after rifting can be constructed for different pre-rift elevations, as well as for pre-rift elevations at sea level. Because of the many uncertainties involved in basin reconstructions, this graphical approach is probably as accurate as one-dimensional computer models in most cases.

3) By including all uncertainties in initial and thermal subsidence (and heat flow), the size of the uncertainties in the thinning parameters are obvious. For example, the thinning parameters for well group 5 in the Pannonian Basin are extremely well constrained, while those for the Pattani Trough are only loosely bracketed (Figures 6 and 7). This approach also makes it easy to decide if the subsidence history of a particular basin is consistent or inconsistent with uniform lithospheric extension. For example, the Pattani Trough could have resulted from uniform extension but the Pannonian Basin clearly could not.

4) Superposition of Figures 5b and 5c shows that the heat flow through a ten million year old basin can be computed directly from the thermal subsidence of the basin, irrespective of thinning parameters, pre-rift conditions or initial subsidence. Although not shown in this paper, the same is true for basins of any age. Thus thermal reconstructions of extensional basins can be made directly from thermal subsidence histories alone (for example, Figure 8).

5) Heat flow through extensional basins older than about 50 m.y. is generally low (40-60 mW/m^2), and does not reflect significantly the earlier thermal history of the basin. Evaluation of thermal subsidence is therefore not particularly useful for predicting present day heat flow through these older basins. Thermal subsidence data can, however, be used to reconstruct the thermal histories of these older basins for earlier times when the surface heat flow had not yet decayed to its background value. By plotting the limiting values of thinning parameters for each basin or drill hole, one can easily estimate the permissible range of heat flow at any time. Such a technique could be used to predict the maturity of potential source rocks in extensional basins, and will also give the uncertainty range on those predictions.

Acknowledgements

This research was made possible by a Kerr-McGee Career Development Professorship at MIT and by donations from Texaco Inc. U.S.A. and Shell Research and Development Co. I would also like to thank the Institut Francais du Pétrole for help in defraying travel expenses.

References

BÉRCZI, I., J. KÓKAI, V. DANK and A. SOMFAI, 1981, Some new petroleum geological results obtained by hydrocarbon exploration wells drilled in the Hungarian part of the Pannonian Basin: Earth Evol. Sci., v. 1, no. 3, 301-306.

BUCK, W.R., 1984, Small-scale convection and the evolution of the lithosphere: Ph.D. Thesis, Massachusetts Institute of Technology, 256 pp.

CAREY, S.W., 1958, The tectonic approach to continental drift: Symp. Continental Drift, Hobart, p. 177-355.

CERMAK, V., 1979, Review of heat flow measurements in Czechoslovakia: in Cermak, V. and L. Rybach, eds., Terrestrial Heat Flow in Europe: Springer Verlag, New York, p. 152-160.

DÖVÉNYI, P., F. HORVÁTH, P. LIEBE, J. GÁLFI and I. ERKI, 1983, Geothermal conditions of Hungary: Geophys. Transactions, v. 29, no. 1, p. 3-114.

GARFUNKLE, Z. and Y. BARTOV, 1977, The tectonics of the Suez rift: Geol. Surv. Israel Bull., v. 71, p. 1-44.

HELLINGER S.J. and J.G. SCLATER, 1983, Some comments on two-layer extension models for the evolution of sedimentary basins: J. Geophys. Res., v. 88, no. B10, p. 8251-8269.

DÖVÉNYI, P. and F. HORVATH, in preparation, A review of temperature, thermal conductivity and heat flow data for the Pannonian Basin: AAPG Memoir.

KEEN, C.E., 1985, The dynamics of rifting: Deformation of the lithosphere by active and passive driving forces: Geophys. J. R. astr. Soc., v. 80, p. 95-120.

KRÖLL, A. and G. WESSELY, 1973, Neue Ergebnisse beim Tiefenaufschluss im Wiener Becken: Erdol, Erdgas Zeitschrift, v. 89, p. 400-413.

MCKENZIE, D., 1978, Some remarks on the development of sedimentary basins: Earth Planet. Sci. Lett., v. 40, p. 25-32.

MORGAN, P., C.A. SWANBERG, F.K. BOULOS, S.F. HENNIR, A.A. EL SAYED, and N.Z. BASTA, 1980, Geothermal studies in North Africa: in Proceedings of the International meetings held on the occasion of the Fifth Conference on African Geology (Issawi, Bahay, ed.), Egypt. Geol. Surv. Ann., vol. 10, p. 971-987.

PARSONS, B. and J.G. SCLATER, 1977, An analysis of the variation of ocean floor bathymetry and heat flow with age: J. Geophys. Res., v. 82, p. 802-825.

ROYDEN, L.H., in press, The Vienna Basin: A thin-skinned pull-apart basin: SEPM Spec. Publ.

ROYDEN, L. and C.E. KEEN, 1980, Rifting process and thermal evolution of the continental margin of eastern Canada determined from subsidence curves: Earth Planet. Sci. Lett., v. 51, p. 343-361.

ROYDEN, L., F. HORVÁTH, A. NAGYMAROSY and L. STEGENA, 1983a, Evolution of the Pannonian Basin system, 2. Subsidence and thermal history: Tectonics, v. 2, p. 91-137.

ROYDEN, L. and P. DÖVÉNYI, in prep., A critical analysis of extensional styles in the Pannonian Basin as determined from subsidence and heat flow data: AAPG Memoir.

SCLATER, J.G., J. CROWE, and R.N. ANDERSON, 1976, On the reliability of oceanic heat flow averages: J. Geophys. Res., v. 1, p. 2   300 .

SCLATER, J.G. and P.A.F. CHRISTIE, 1980, Continental stretching: an explanation for the post-mod-Cretaceous subsidence of the central North Sea Basin: J. Geophys. Res., v. 85, no. B7, p. 3711-3739.

SCLATER, J.G., L. ROYDEN, F. HORVÁTH, B.C. BURCHFIEL, S. SEMKEN, and L. STEGENA, 1980, The formation of the intra-Carpathian basins as determined from subsidence data: Earth Planet. Sci. Lett., v. 51, p. 139-162.

STECKLER, M.S., 1985, Uplift and extension at the Gulf of Suez - indications of induced mantle convection: submitted to Nature.

F. LUCAZEAU[1]

THE POST RIFT EVOLUTION
OF THE MASSIF CENTRAL (FRANCE)

INTRODUCTION

Continental rifts are generally associated with crustal extension, alkali volcanism, topographic doming, long wavelenght negative Bouguer anomalies and high heat flows ; it suggests that there is a genetic relation between rifting and deep seated processes in the upper mantle. Two mechanisms are proposed to explain their formations:
- active rifting, which involves deep convection in the mantle and consecutive subsequent thinning of the lithosphere by advection of hot asthenospheric material (Withjack,1978; Bott,1981; Spohn and Schubert, 1982; Neugebauer, 1983).
- Passive rifting, which involves differencial stresses in the lithosphere associated with plate motion or continental collisions (McKenzie, 1978; Molnar and Tapponnier, 1975; Vilotte et al., 1983; Oxburgh and Turcotte, 1974).
But the whole geological evolution of continental rifts is still not fully explained and the mechanisms of continental rifting are probably more complex. Taking the example of the Massif Central (located in the central part of France), for which various geological and geophysical data are available, we want to discuss in this paper the possible mechanisms of continental rifting in the West European platform.

I- OBSERVATIONS

1-Geological setting
- - - - - - - - - -
Several Hercynian provinces (Massif Central, Rhine graben, Bohemian massif) have been affected by rifting processes during Mid

---

[1] *Centre Géologique et Géophysique, Université des Sciences et Techniques du Languedoc, Montpellier, France.*

THE POST RIFT EVOLUTION OF THE MASSIF CENTRAL

Tertiary while compression occured at the same time in the Alps. The location of the grabens and the associated volcanisms in the context of alpine arc (Fig. 1) suggest that there is a possible genetic relation between the formation of this rift system and the continental collision (Tapponnier, 1977; Fleitout, 1984). The characteristics of the Massif Central are very similar to those observed in other parts of this rift system (Neugebauer, 1985).

Crustal extension in Massif Central

Rifting of the crust is associated with the formation of narrow N-S grabens filled by 500-1000 m of Stampian sediment (35-30 Ma) and with a moderate crustal thinning as proved by numerous refraction studies (Perrier and Ruegg, 1973). Crustal thinning doesn't exceed 20 percent as the minimum thickness of the crust is about 24 Km beneath the Limagne graben (Fig. 2c) whereas the normal thickness for the Hercynian crust is 30 km. The thinned crust corresponds to an anomalous upper mantle with P waves velocities as low as 7.2-7.4 Km/s at the bottom of the crust and increasing to 8.4 Km/s at 40-50 Km depth (Perrier and Ruegg, 1974), which seems to be a general characteristic of rifts (Fuchs et al, 1981).

Lithospheric structure (after S waves studies)

Studies of Rayleigh waves dispersion and attenuation (Souriau, 1981; Souriau et al, 1980) indicate that the lithosphere is thinned beneath the Massif Central to 50 Km whereas normal thickness is higher than 100 Km (Fig. 2d). Low quality factors suggest the possibility of partial melting at shallow depth at present time (Souriau et al., 1980). Note that different results are found by Panza et al., 1980: for these authors, lithospheric thickness is normal (i.e 90 Km) in the Massif Central.

Lithospheric structure (after peridotite xenoliths)

Peridotite xenoliths are remarkably abundant in the Cenozoic alkali basalts of the Massif Central, mostly in the volcanism younger than 5 Ma and consequently they can reflect the integrated sequence of evolution before their extraction (Nicolas et al., 1985). The studies of the texture of the xenoliths (Coisy and Nicolas, 1978) have shown that there exists a geographical distribution of the different types (Fig. 2b), centered on the area of thin crust and interpreted in term of diapiric ascent: equigranular textures, caused by large strain, are more abundant in the central part, whereas protogranular texture, corresponding to weaker or null strain are mostly found in the outer domains. Porphyroclastic types are characteristic of high shear strain deformation ($\dot{\varepsilon} = 10^{-13} s^{-1}$), and are

THE POST RIFT EVOLUTION OF THE MASSIF CENTRAL

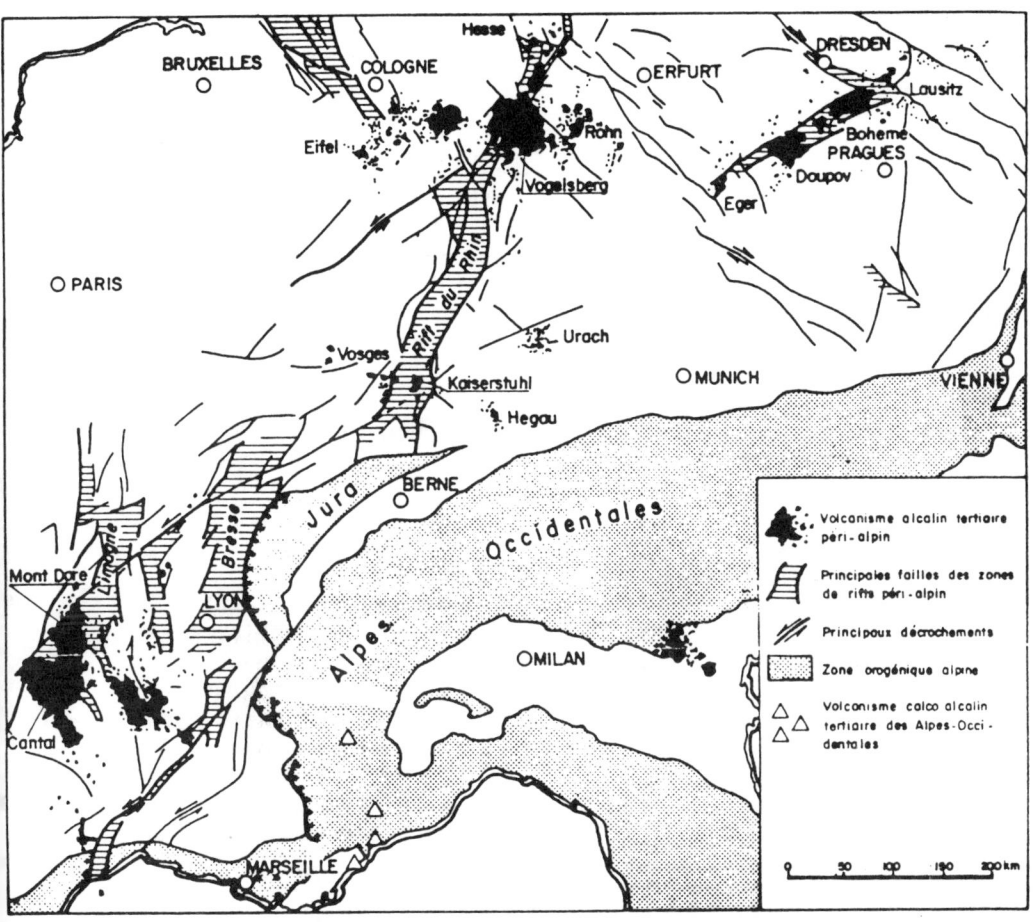

Figure 1: Location of the grabens (horizontal hatched zone) and associated alkali volcanism (black area) of the West European rift system (after Maury and Varet, 1980).

interpreted as representative of the skin of the diapir. Equilibration temperatures of the xenoliths, using various geothermometers, are between 950 and 1050 °C. The occurence of Opx-Cpx-Sp clusters in xenoliths is interpreted as produced by decompression melting (diapir uprise) but they indicate that peridotites have probably not been extracted during the main episode of magma release (Nicolas et al., 1985), but they have progressively cooled below the solidus (during about 1 Ma).

Crustal doming

Massif Central is associated with a doming of 1250 m amplitude and 200-300 Km wavelength (fig. 2e). According to geomorphological studies (Derruau, 1971), this topographic uplift is very recent (likely Villafranchian, i.e 4-2 Ma) and has been produced in a short period of time.

Gravity

Massif Central is characterized by a negative Bouguer anomaly of -50 to -60 mgals amplitude and a wavelength similar to the topography

Figure 2: Various geological and geophysical ▶ information about the Massif Central:
a- geological sketch. 1: Carboniferous sediment, 2: Cenozoic volcanism, 3: Paleozoic unmetamorphic rocks, 4: metamorphic basement, 5: Hercynian granites.
b- location of peridotite xenoliths. 1: protogranular type, 2: porphyroclastic type, 3: equigranular type (after Coisy,1977). The dotted area corresponds to the supposed diapiric area .
c- Isobaths of the Moho discontinuity in Km after Perrier and Ruegg (1973). The dotted area corresponds to the anomalous mantle.
d- Lithospheric thickness inferred from S waves studies (Souriau, 1976). The dotted area corresponds to a thin lithosphere (50-100 Km) and the hatched area to a normal lithosphere (more than 100 Km thick).
e- Average topography (equidistance of curves= 200m).
f- Bouguer anomaly (equidistance of curves = 10 mgals).
g- Diapir effect on the gravity field (equidistance of the curves =10 mgals). The 3 modelized profiles are represented.
h- Heat flow map of the Massif Central (equidistance of curves = 10 mW/m^2). The dots represent the measured values.

THE POST RIFT EVOLUTION OF THE MASSIF CENTRAL

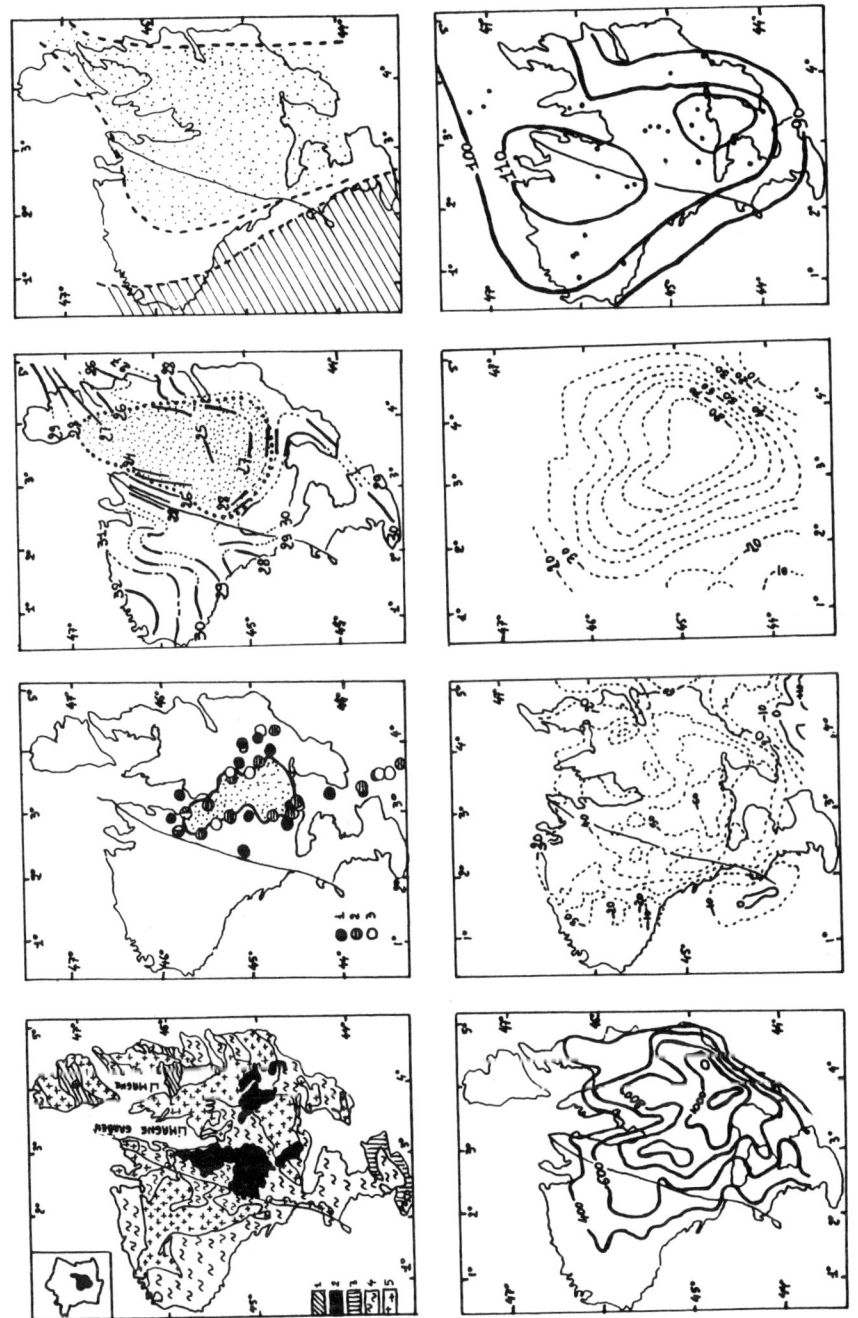

Figure 2.

(200-300 Km) (Fig. 2f). Perrier and Ruegg (1973) have shown that this anomaly could be explained by a mantellic low density body with a contrast of about 0.03 g/cm at 50 Km depth (which might correspond to the base of lithosphere as deduced from S waves studies). Using admittance function technics and removing crustal thinning effect, Lucazeau and Bayer (1982) have calculated the regional anomaly due to this mantle body. The amplitude becomes -80 to -90 mgals with a roughly circular-shaped anomaly (Fig. 2g).

Volcanism

Alkali volcanism in the Massif Central has occured in two distinct stages (Maury and Varet, 1980). The first stage, from Paleocene to Oligocene, consists of small occurences of undersaturated lavas (melilite or analcime nephelinites) erupted along the faulted Hercynian basement. According to geochemical models (Kay and Gast, 1973), this type of volcanism is initiated at great depths (deeper than the base of the lithosphere) with very low partial melting rates (less than 1 percent). During the graben formation, the volcanic activity stops (except in Languedoc) and restarts during early Miocene (20 Ma) with two paroxysms at the Miocene - Pliocene boundary (7 Ma) and during Villafranchian (4-2 Ma). There is a time dependant evolution of the chemistry of lavas from undersaturated conditions (nephelinites, basanites) to less alkaline lavas (olivine basalts and differenciated lavas); these latter are characteristic of higher partial melting rates (5-15 percent) at shallower depths in the lithosphere (base of the crust i.e 30 Km). For more recent volcanism (less than 5 Ma), there is a geographical distribution of lavas; the more evolved lavas are observed in the central part, whereas the more undersaturated lavas are mostly found in the outer parts. This geographical and time dependant distribution of the lavas is interpreted by a diapir ascent in the upper mantle (Maury and Varet, 1980).

Heat flow density measurements

Massif Central is characterized by very high heat flow values (fig. 2h) whereas it is normally ranging between 60 and 80 mW/m^2 in stable hercynian provinces. Heat production of the surface rocks is ranging between 2 μW/m^3 for metamorphic basement and orthogneisses, to more than 5 μW/m^3 for leucogranites; but there is no correlative increase of the surface heat flow. Therefore, it is not possible in this case to determine the deep component of heat flow by plotting heat flow versus heat production as normally done in stable provinces. Using heat production measurements of deep granulitic levels (xenoliths of these rocks have been carried up during recent volcanism episodes) and crustal seismic profiles (Lucazeau and Vasseur, 1981), it is possible to estimate the heat flow component arising from the mantle. It clearly shows an anomaleous value of 35 mW/m^2 beneath the graben and volcanic area.

II- MODEL OF EVOLUTION

The different observations summarized in the previous section clearly show that tectonic activity in the Massif Central has mostly occured in the period following the rifting stage (formation of the grabens): only few occurences of volcanism, with very small volume of lavas, are observed before. The small extension and thinning of the crust show that rifting was not a very intensive stage. On the other hand, the volcanic activity from Miocene to present time represents

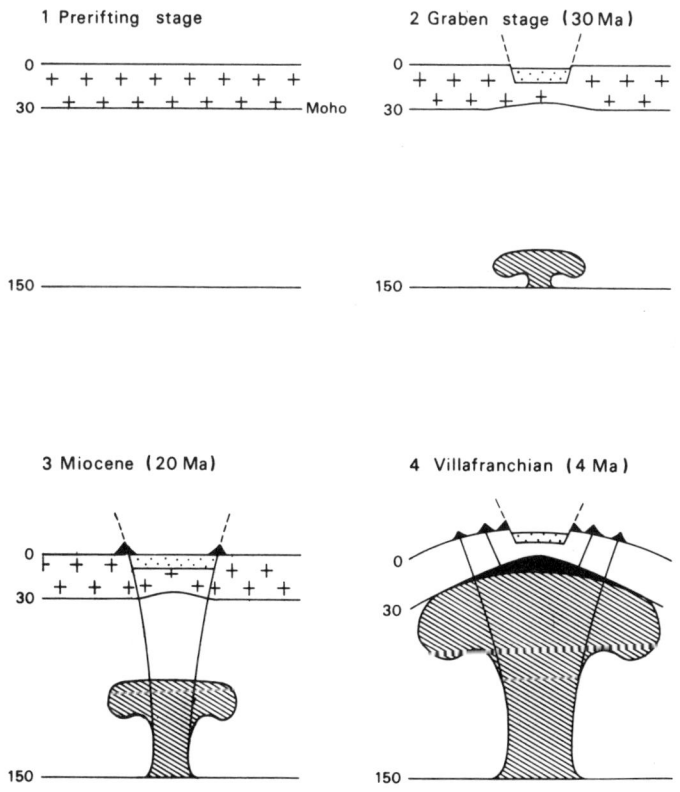

Figure 3: Possible evolution of the Massif Central following the rifting stage. The limits of the hatched area (diapir) have no physical reality.

THE POST RIFT EVOLUTION OF THE MASSIF CENTRAL

about 5000 Km^3 of lavas (i.e the same as Baikal rift), the peridotites present the same characteristics as those of Basin and Range province (Mercier, 1980), surface waves studies prove that partial melting still occur at shallow depths, conductive heat flow anomalies are very important and the Bouguer anomaly reflects the occurence of a very low density body in the mantle. This evolution is typical of a diapiric ascent of the mantle: we have tried, using a thermal kinematic model to explain the evolution of such a diapir and its effects on the thermal field, the gravity field, the topography and the compatibilty with partial melting rates inferred from

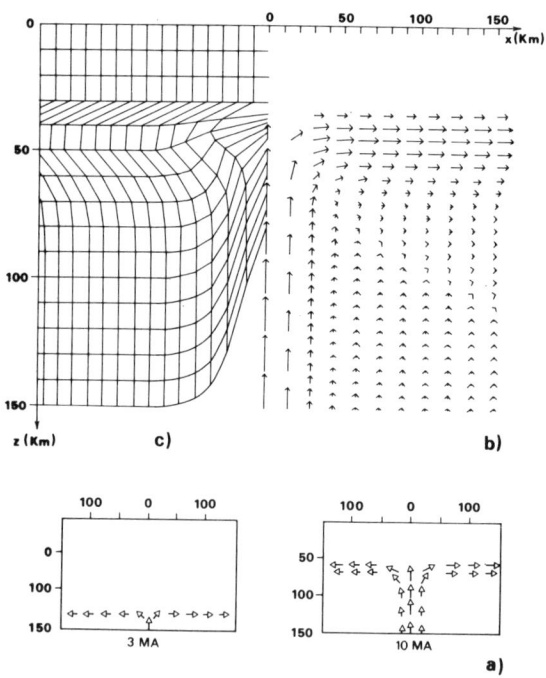

Figure 4: Velocity field used to simulate diapiric processes.
a) after 3 Ma and 10 Ma for a maximum velocity $V_m = 1$ cm/year,
b) after 12 Ma when the diapir reaches the base of the crust,
c) deformation corresponding to 12 Ma.

geochemistry of lavas. We propose the following scenario (Fig. 3):
1- before Oligocene, the lithosphere is assumed to be stable, at thermal equilibrium and with a constant crustal thickness.
2- During Oligocene, the lithosphere is stretched by a factor 1.2: a mechanical instability may be created at the base of the lithosphere which can induce a diapiric process.
3- from Miocene (20 Ma) to Pliocene, the mantle diapir progressively rises and widespread volcanism develops.
4- when the diapir reaches the base of the crust, the crust, more rigid than the mantle, offers a resistance to further ascent inducing lateral extension of the diapir. The pressure exerted at the base of the crust cause the observed doming.

The diapir ascent is represented by a time dependent, divergence free velocity field chosen arbitrarily (Fig. 4) ; the important parameters of this velocity field are adjusted to match the observations on heat flow, gravity, topography and partial melting.

Temperature field

The 2D time dependant equation is solved by a numerical method of finite elements taking into account the chosen velocity field in the convecting term. Initial conditions are given by a typical model for hercynian lithosphere, with a 3 layers crustal heat production distribution, a 30 km initial thickness of the crust and a 80 mWm^{-2} flow. Radioactive transferts in the mantle and latent heat effects due to partial melting are also taken into account.

Partial melting

Experimental studies on peridotitic material (Ito and Kennedy, 1967) show that dry solidus and liquidus temperatures are linearly related to pressure conditions (i.e. depth). In fact, H2O acts to significantly decrease the solidus temperature (Kushiro et al., 1968; Ringwood, 1975), but this effect is limited by CO2 (Eggler, 1978). Two extreme models of solidus were tested (the actual situation being probably intermediate) but the results were not significantly affected. Then, the melting rate model proposed by Ringwood (1975) was used.

Gravity anomaly

The gravity anomaly results from a lateral variation of density due to thermal expansion and partial melting. The pressure dependence of the liquid phase density into the melt is taken into account according to the model of Stolper et al. (1981).

THE POST RIFT EVOLUTION OF THE MASSIF CENTRAL

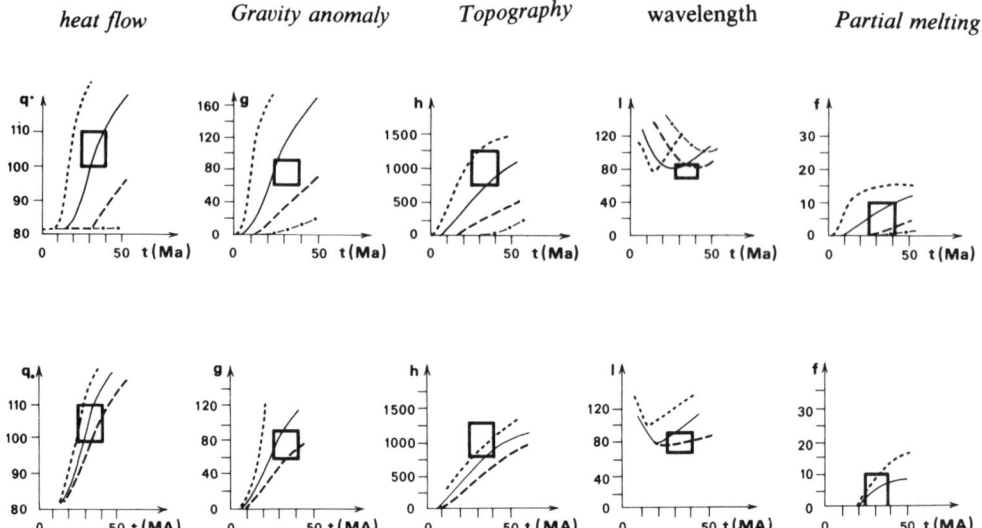

Fig. 5: Influence of parameters (a- vertical velocity V and b- width of the diapir) on the differents observations. Boxes correspond to the range of possibilities for the present time.
a- Velocities correspond to 0.25 cm/year (dashed-dotted line), 0.50 cm/year (dashed line), 1 cm/year (heavy line) and 2 cm/year (dotted line).
b- widths correspond to 20 km (dashed line), 40 km (heavy line) and 80 km (dotted line).

Topography
- - - - - -

Topography results from the load exerted by the low density body at the base of a thin elastic layer. A simple model of the elastic plate overlying a perfect fluid (Whithjack, 1979) is used to calculate the vertical displacement of the ground surface. The reference level is supposed to be 150 m as at present in Brittany.

Results and conclusions
- - - - - - - - - - - -

The most important parameters of the model are the velocity V_m which characterizes both the rising velocity of the diapir and the maximum vertical velocity of the magma in the asthenospheric body, the width of the diapir and the temperature and originated depth of the magma. We have made different computations in order to see the sensitivity of the present day observations (heat flow, Bouguer anomaly, topography, wavelength of the Bouguer anomaly, and partial melting) to these parameters. We have represented on Figure 5 the effects of the vertical velocity and the effect of the width of the diapir versus time for the different observations : the range of possibilities for present time are represented by the boxes. The effects of the vertical velocity is very important and only a velocity of 1 cm/year can explain the different observations. On the other hand, the width of the diapir doesn't seriously affect the different observations, except the wavelength of the Bouguer anomaly.

The best fit is obtained for the following values of parameters: vertical ascent of molten material V = 1 cm/year; width of diapir is 40 km; depth at which the diapir originates = 150 km; and temperature T = 1365 °C.

The resulting model is shown in Figure 6. Temperature field is represented for 25 Ma. The intersection between the geotherm and the solidus (here 0.1 % H2O solidus of Ringwood, 1975) gives the thermal boundary layer (TBL). Initially, the calculated depth is around 100 km, which is the expected value for a stable Hercynian domain (Souriau, 1976). After 25 Ma, the TBL depth becomes 50 km at the rift axis as previously suggested by Perrier and Ruegg (1973) and Souriau (1976 and 1981). Partial melting rates are acceptable values for the observed type of volcanism. The resulting heat flow anomaly and the topography are better explained for a longer time duration (30-40 Ma) than the gravity anomaly (25-30 Ma). However, the latter is more reliable and therefore the 25-30 Ma duration seems more satisfactory.

One can notice the high velocity of the diapir ascent (1 cm/year) but similar estimates were obtained by other authors (Aki, 1982 ; Neugebauer, 1983). However, it is the maximum value of the velocity and the average value is on the order of 1 mm/year magnitude.The 150 km depth for initiating the diapirism seems to be realistic according to recent ideas on alkaline volcanism (Carmichael et al., 1974 ; Ringwood, 1975). The occurence of garnet lherzolite xenoliths in the Massif Central volcanism (32 kbar, 1400 °C) (Berger, 1977 and 1979) also supports this estimated value.

In conclusion, the different anomalies which characterize the rift structure of the French Massif Central can be explained by a kinematic thermal model of diapir. But the main problem is to under stand how such diapir can form and evolve : it is therefore important to develop thermomechanical models which will precise how the formation of a diapir happens in such a context and develops in a period of time of 20-30 Ma, and in which conditions it can stop. In the context of basin development, it is important to know when diapiric processes have occured after the rifting stage since it could represent a substancial excess of heat in the early stages and create better conditions for the oil maturation.

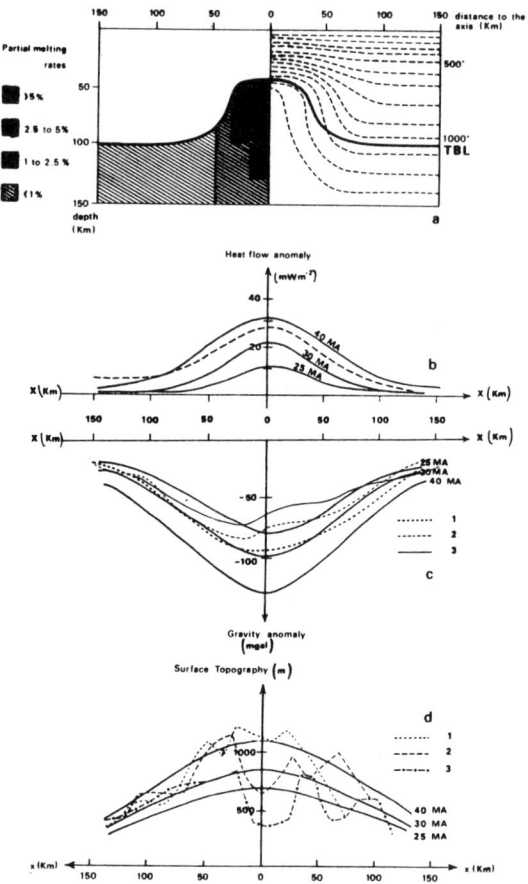

Figure 6: Results of the kinematic model of diapir.
a- Calculated isotherms for a time elapsed since the begining (25 Ma) ; on the right part, resulting partial melting accodring to the Ringwood's model (1975). The heavy line correspond to the thermal boundary layer (TBL) which represents the intersection between the solidus and the geotherm.
b- Calculated heat flow anomalies (heavy lines) for 25, 30 and 40 Ma. The expected anomaly (Lucazeau and Vasseur, 1982) is represented by the dashed line.
c- Gravity anomalies (heavy lines) computed for 25, 30 and 40 Ma. Profiles 1, 2, 3 correspond to observed profiles from Figure 2g.
d- Topography computed for 25, 30 and 40 Ma (heavy lines). Profiles 1, 2, 3 correspond to the previous observed profiles of the gravity anomaly.

REFERENCES

Aki, K., 1982. Three dimensional seismic inhomogeneities in the lithosphere and asthenosphere. Rev. Geophys. Space Phys., 20: 161-170.

Berger E., 1977. Sur la présence d'une lherzolite à grenats en enclave dans le basalte alcalin de la Vestide du Pal (Ardèche) : conditions d'équilibre, implications pétrogénétiques et géotectoniques. Compt. Rend. Acad. Sci., sér. D, 284 : 709-712.

Berger E., 1979. The role on partial melting of mantle diapirism, CO_2 and H_2O from the study of lherzolite nodules of intracontinental alkali basalts : examples of the French Massif Central. In : L.H. Ahrens 'ed.), Origin and Distribution of the Elements. Pergamon, Oxford, pp. 619-629.

Bott, M.H.P., 1981. Crustal doming and the mechanism of continental rifting. Tectonophysics, 73 : 1-8.

Coisy P., 1975. Structure et chimisme des péridotites en enclaves dans les basaltes du Massif Central. Modèles géodynamiques du manteau supérieur. Thèse Univ. Nantes, 118 pp.

Coisy P. and Nicolas A., 1978. Structure et géodynamique du manteau supérieur sous le Massif Central (France) d'après l'étude des enclaves de basaltes. Bull. Minéral., 101 : 424-436.

Derruau M., 1971. Sur la morphologie du Massif Central. In : Géologie, Géomorphologie et Structure profonde du Massif Central Français. Symp. J. Jung, Clermont Ferrand - Plein Air Service pp. 33-44.

Eggler D.H., 1978. The effect of CO_2 upon partial melting of peridotite in the system $Na_2O-CaO-Al_2o_3-MgO-SiO_2-Co_2$ to 35 kb with an analysis of melting in the peridotite H_2O-CO_2 system. Am. J. Sci., 278 : 305-343.

Fleitout L., 1984. Modélisation des contraintes tectoniques et des instabilités thermomécaniques dans la lithosphère. Univ. Paris Sud, Thèse, 433 pp.

Frey F.A., Green D.H. and Roy S.D., 1978. Integrated models of basalts petrogenesis : a study of quartz tholeiites to olivine melitites from southeastern Australia, utilizing geochemical evolution of the earth's crust and mantle. J. Petrol., 19 : 463-513.

Fuchs K., Bonjer K.P. and Prodehl C., 1981. The continental rift system of the Rhinegraben. Structure, physical properties and dynamical processes. Tectonophysics 73 : 79-90.

Ito, K. and Kennedy, G.C., 1967. Melting phase relations in a natural peridotite to 40 kilobars. Am. J. Sci., 265 : 519-538.

Kay, R.W. and Gast P.W., 1973. The rare earth content and origin of alkali-rich basalts. J. Geol., 81 : 653-682.

Kushiro, I., Syono, Y. and Akimoto, S., 1968. Melting of a peridotite nodule at high pressures and high water pressures. J. Geophys. Res., 73 : 6023-6029.

Lucazeau F. and Bayer R., 1982. Evolution thermique et géodynamique du Massif Central français depuis l'Oligocène. Ann. Géophys., 38 : 405-429.

Lucazeau F. and Vasseur G., 1981. Production de chaleur et régime thermique de la croûte du Massif Central. Ann. Géophys., 37 : 493-513.

Lucazeau F., Vasseur G., Kast Y. and Jolivet J., 1981. Données du flux de chaleur dans le Massif Central français. Ann. Géophys., 37 : 481-491.

Maury, R.C. and Varet J., 1980. Le volcanisme tertiaire et quaternaire de la France. In : A. Autran and J. Dercourt (Ed.) Evolution structurale de la France. Congr. Géol. Int., 26e, Coll. C7, Mém B.R.G.M. Fr., 107 : 137-159.

McKenzie D., 1978. Some remarks on the development of sedimentary basins. Earth Planet. Sci. Lett., 40, 25-32.

Mercier J.C.K., 1980. Single pyroxene thermobarometry. Tectonophysics, 70 : 1-29.

Molnar P. and Tapponnier P., 1975. Cenozoic tectonics of Asia : effects of a continental collision. Science, 189 : 419-426.

Neugebauer H.J., 1985. This volume.

Neugebauer, H.J., 1983. Mechanical aspects of continental rifting. Tectonophysics, 94 : 91-108.

Nicolas A., Lucazeau F., Bayer R., 1985. Peridotite xenoliths in Massif Central basalts : textural and geophysical evidence for asthenospheric diapirism. J. Wiley, in press.

Panza G.F., Mueller S. and Calcagnile G., 1980. The gross fractures of the lithosphere-asthenosphere system in Europe from seismic surface waves and body waves. Pure and Appl. Geophys., 118 : 1209-1213.

Perrier, G. and Ruegg, J.C., 1973. Structure profonde du Massif Central français. Ann. Geophys., 29 : 435-502.

Ringwood, A.E., 1975. Composition and Petrology of the Earth's Mantle. McGraw Hill, New York, 618 pp.

Souriau A., 1976. Structure profonde sous la France. Bull. Soc. Géol. Fr., XXIII : 65-82.

Souriau A., Correig, A.M. and Souriau M., 1980. Attenuation of Rayleigh wave across the volcanic area of the Massif Central, France. Phys. Earth Planet. Inter., 23 : 62-71.

Spohn T. and Schubert G., 1982. Convective thinning of the lithosphere. A mechanism for the initiation of continental rifting. J. Geophys. Res., 87 : 4669-4681.

Stolper, E., Walker, D., Hager, B.H. and Hays, J.F., 1981. Melt segregation of partially molten sources regions : the importance of melt density and source region size. J. Geophys. Res., 86 : 6261-6271.

Tapponnier P., 1977. Evolution tectonique du système alpin en Méditerranée : poinçonnement et écrasement rigide-plastique. Bull. Soc. Géol. Fr., 19 : 437-460.

Vasseur G., 1982. Synthèse des résultats du flux géothermique en France. Ann. Géophys., 38 : 189-201.

Vasseur G. and Lucazeau, F., 1982. Some aspects of heat flow in France. In : V. Cermak and R. Haenel (ed.), Geothermics and Geothermal Energy, Schweizerbart, Stuttgart, pp. 79-89.

Vilotte J.P., Daignières M., Madariaga R., 1982. Numerical modeling of intraplate deformation : simple mechanical models of continental collision. J. Geophys. Res., 87, B.13, 10709-10728.

Withjack, M., 1979. A convective heat trransfer model for lithospheric thinning and crustal uplift. J. Geophys. Res., 84 : 3008-3092.

H. J. NEUGEBAUER[1]

ON THE THERMAL EVOLUTION OF INTERIOR PLATFORM BASINS DYNAMIC THERMAL PROCESSES IN THE LITHOSPHERE

Sedimentary basins on the continental crust form in association with lasting vertical movements. The development of interior basins is in many cases accompanied by volcanic activity as well as early fault-controlled tectonics. It is thus obvious to include continental rift basins in our discussion. Furtheron reference is made to detailed observations of a well investigated fault-controlled basin, the Rhine Graben. Beside a summary of major characteristics of interior basins and their history some physical aspects of regional dynamic thermal models will be discussed.

OBSERVATIONS AND CONSTRAINTS

The thermal regime of a basin is depending on the quantity and distribution of heat producing elements in the crust and the sediments, the thermal conductivity structure, the chance for groundwater flow in the basin and finally the mantle heat flow into the base of the crust.

The generation of petroleum in a basin is indicative of a temperature range to which the source rocks have been exposed. Because the rate of oil generation depends exponentially on the temperature and only linear on the time. The chance that time and temperature compensate for each other is thus limited. For further discussion a temperature of 100° C will be taken as a representative value for the principal range of oil generation in a basin structure, Hunt (1979), Tissot & Welte (1984). Its regional implications for the thermal regime of the crust can be well accentuated with reference to calculated steady state thermal regimes.

[1] *Institut für Geophysik, Technische Universität Clausthal, Clausthal-Zellerfeld, Federal Republic of Germany.*

Steady State Thermal Structure of the Lithosphere

On the basis of a three layer lithosphere with specific thermal conductivity and radioactive heat generation for each layer steady state geotherms have been calculated by Willett et al. (1984). Geotherms were determined by corresponding surface heat flow values. In figure 1 the depth level of 100° C is related with the surface heat flow (solid line) as well as the corresponding equilibrium temperature of the lower crust at 30 km depth (dashed line) on the base of the above steady state models.

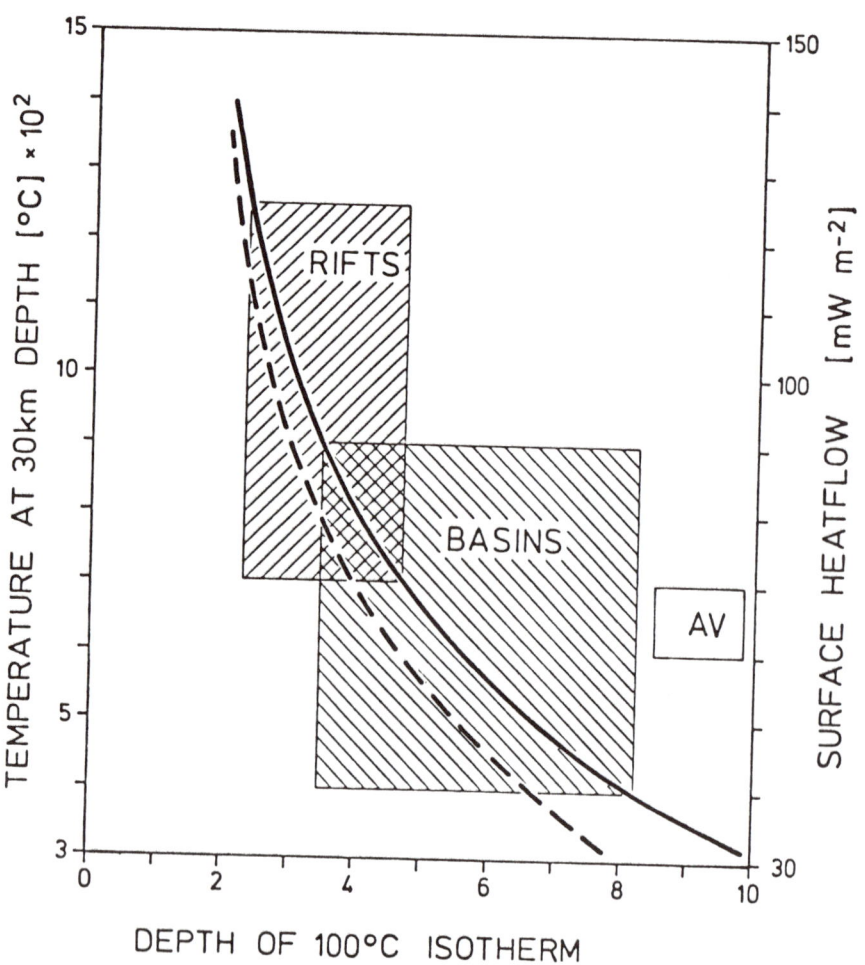

Figure 1. Compilation of temperatures and heat flow of steady state thermal models of the lithosphere. Measured present day heat flow data: Continental average (AV) after Chapman & Furlong (1977) in Morgan (1984); Basins after Willett et al. (1985) and Cenozoic Rifts after Morgan (1983).

This correspondence enables us to discuss present day heat flow at continental sites in the view of equilibrium thermal regimes of the crust and mantle. Thereafter the present day continental average heat flow between 60-70 mW m^{-2} - Late Paleozoic and Cenozoic-Mesozoic sites respectively - corresponds to a 100° C temperature level at about 5 km depth and 550° C at the lower crust at 30 km depth. Surface subsidence and sediment accumulation under continous thermal equilibrium would thus require 10^3-10^2 Ma providing rates between 5 to 50 m/Ma respectively before the appropriate temperature of petroleum generation in 5 km depth is reached.

The existence of basins of lower age and shallower depth compared with the steady state thermal conditions and the existence of hydrocarbons within this troughs leads to the general conclusions:

i   Subsidence with time is not a sufficient parameter to identify thermal history of basins.

ii There is obviously additional heat supply necessary during basin evolution.

The latter statement is well supported by the total range of present day heat flow of basins as well as that of Cenozoic continental rifts between 70-110 mW m^{-2}, Morgan (1983), figure 1. The corresponding steady state temperature exhibit 100° C at about 2 km depth and more than 1000° C at the lower crust. Those calculations will serve as an upper limit, because the conductive model overestimates temperatures for depth levels where convective heat transport is possible.

With respect to the rifted and nonrifted platform basins the question must be raised, whether they share the same mechanism of heat supply or not.

Subsidence Patterns

Basin subsidence has been documented at many sites up to several hundred million years - Michigan Basin 400 Ma, North Sea, Paris Basin 200 Ma -. Subsidence occurs in alternating periods of law - 5-10 m/Ma - and rapid - 25-100 m/Ma - rates, interrupted by periods of inversion and redistribution of sediments. Unconformities of basin subsidence are independent of the total age of a basin structure. This is well documented for the Cenozoic rift trough of the southern upper Rhine Graben, figure 3 (top).

The tectonic mode of interior basin subsidence varies from fault-controlled rifts to non-faulted basins with intermediate types with an early tectonic phase. It is an open question whether fault-tectonics during interior basin evolution is an option depending on the dynamical conditions of the rocks. A less well recognized feature of basin subsidence is the systematic circular to oval geographical

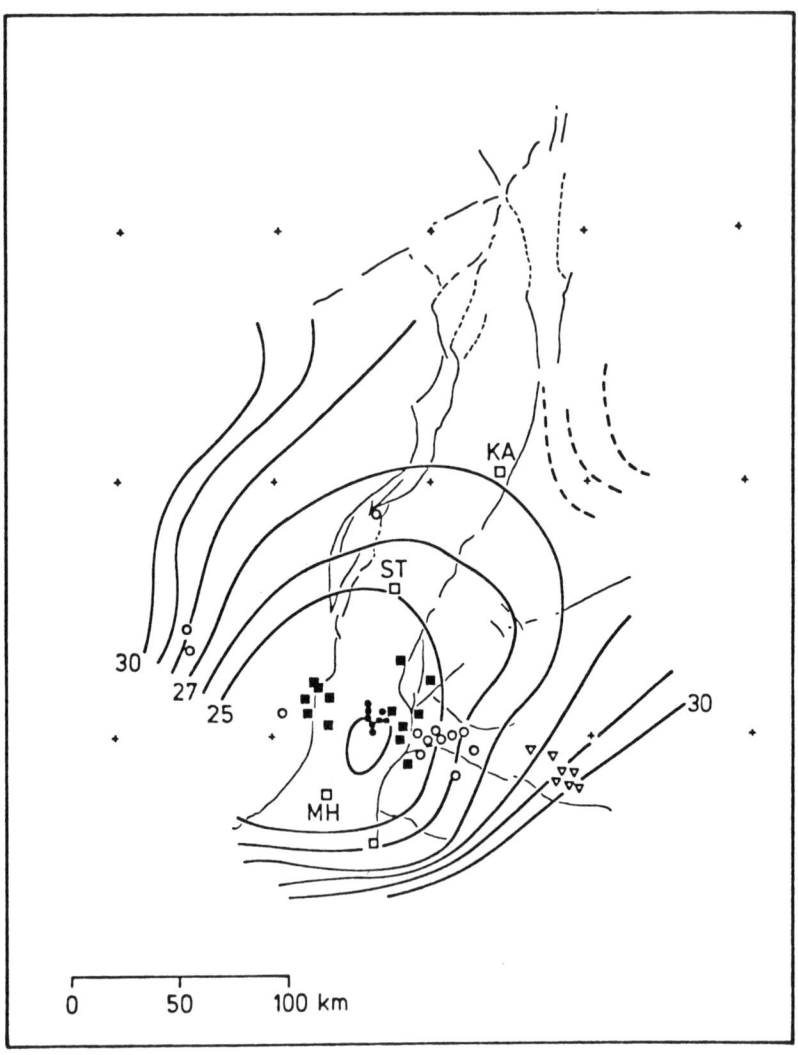

Figure 2. Upper Rhine Graben main boundary faults. Depth to crust-mantle boundary contours in km after Edel et al. (1975). Volcanic sites of known age and barometric depth of origin after Neugebauer & Walter (1983) - circles, squares and triangles - KA- Karlsruhe, ST-Strasbourg, MH-Muhlhose.

shape of platform basins. The aspect ratio of extreme basin diameters varies mainly between 1:1 to 1:2. This holds even for basins with an early tectonic linear structure. Moreover rift structures like the linear Upper Rhine Graben appear to be composed of such individual tectonic elements. This can be demonstrated by means of a part of the Graben in figure 2, where the transitional Moho forms an oval structure associated with the linear faulted trough at the surface. The northern extension of the graben gives another example of the same situation.

Crustal Thinning

The existence of a respective "counterpart" of interior basins and rifts at the crust-mantle boundary is a common observation. A former interpretation of the transitional lower crust in terms of an upwelling of the mantle and a upward shift of the crust-mantle boundary has been improved by means of refined seismic experiments, synthetic seismogramm analysis, laboratory measurements as well as investigations of exposed sites of presumably lower crust. Thereafter those regions of the lower crust are most likely in accord with a layered lithologic structure where mafic and ultramafic commulates are interleaved with lower crustal granulates, Deichmann & Ansorge (1983), Hale & Thompson (1982), Karson et al. (1984), Finlayson & Matur (1984) and O'Reilly & Griffin (1985).

The lateral extend of this so-called "thinned lower crust" is usually congruent with the geographical shape of the basin or exceeds young rift structures like in the example of figure 2. With respect to the 30 km depth level of the Moho in figure 2 the volume of the contoured transitional crust-mantle boundary compares with the volume of the sediments accumulated in the graben structure above by a ratio of 10:1. Effective crustal stretching by fault-control has been determined to be less than 3 kilometers assuming a detachment level at 20 km depth, Richter (1985).

The discussed nature, size and shape of the anomalous lower crust underlaying basins and rift structures suggest an origin different from the common approach by horizontal stretching.

Igneous Activity

Igneous activity is well known from young rift structures but less well documented for old basins. Although the analysis of borehole samples revealed frequent volcanic events during basin evolutions in Central Europe since Mesozoic times, Ziegler (1982). Because of the lack of adequate data sequences for various basins we might focus onto a well documented period of igneous activity in space and time for the Rhine Rift system, figure 2. The situation is representative for the entire rift zone and we will draw some conclusions which might apply equally well to other basin structures.

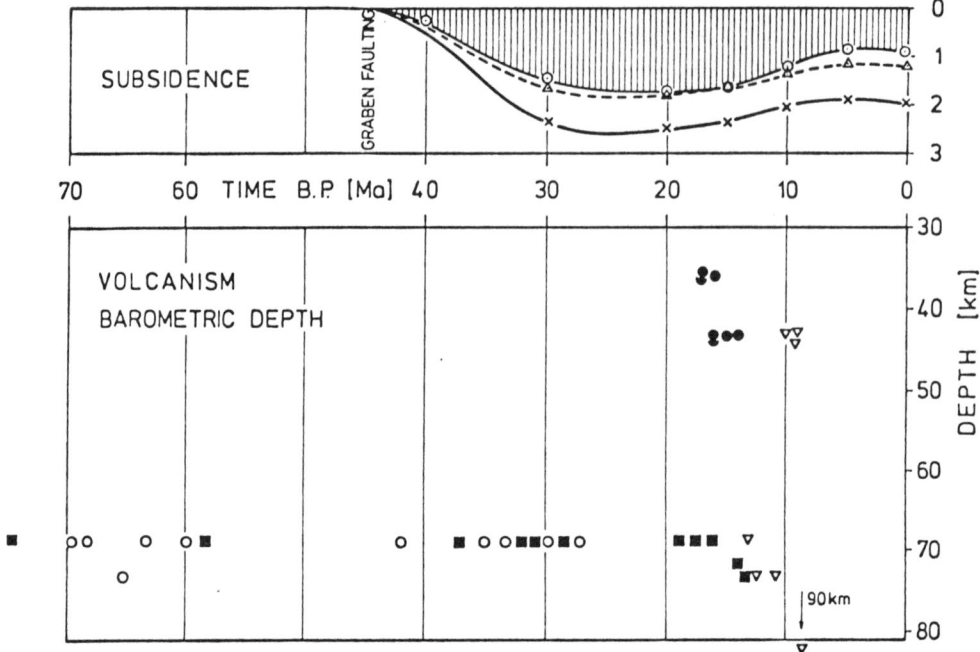

Figure 3. Subsidence of the base of the Rhine Graben as a function of time (top) after Roll (1979); sites: Kaiserstuhl-circles, Muhlhouse-triangles, Strasbourg-crosses. Barometric depth of source region of volcanic rocks as a function of time after Neugebauer & Walter (1983). Geographical position of samples indicated in figure 2.

i There is a clear correspondence between volcanic foci positions and the lateral range of the transitional lower crust.

ii Volcanic activity precedes and accompanies majour fault tectonics of the upper crustal layers, figure 3.

iii Barometric determination of the depth of origin of the dated volcanic samples reveals a lasting igneous activity from deep sources especially beneath the crest of the lower crustal transition zone. The largest volume errupted ever since in this region is associated with a shallow depth magma source at central position with respect to the transitional lower crust. This event occurs long time after the onset of volcanic activity within the graben, figures 2 and 3.

iv The main volcanic event - Kaiserstuhl - is accompanied by an unconformity of graben wedge subsidence, figure 3, top.

v The uprise of magma implies both the transport of matter and heat convectively into a 'crustal' position. Magmas usually have temperatures between 900-1100° C.

Consequently a long period of volcanic activity from a deep
source must be identified with a period of heating of the overlaying lithosphere.

One might speculate on the controlling parameters determining the
ratio between extrusive and intrusive igneous events in such an area.

DYNAMIC THERMAL PROCESSES

The thermal evolution of interior basins of rifted and/or nonrifted nature can not be describe quantitatively neglecting the dynamics of basin development. Mathematical-physical models have a
descriptive as well as a predictive component. The quantitative representation of basic principles provide large families of well defined
solutions due to the variation of the involved parameters. Those
solutions allow to find out the controlling parameters of the princi-
pals investigated. On the other side observations must be taken as
the sum of a complex history. Here one has to find the typical parameters which are usually not equally well expressed at all of the
individual structures found. This leads frequently to controversial
opinions about the characteristic nature of an observed phenomenon.

The aim of a quantitative model is to express the typical features
of an observed phenomenon in terms of controlling parameters of an
adequate physical concept. A successful model will thus link a minimum of theoretical representation with a maximum of prediction.

The thermal evolution of interior basins is dominantly influenced
by

- mechanisms of heat supply in excess to a 'normal' crust
- mechanisms of heat transport
- dynamics of structural changes of the crust and upper mantle
- structural and temporal distribution of physical parameters.

Quantitative approximations of the thermal evolution of basins
will be discussed on the basis of the above aspects and finally in
the view of the characteristic observations presented in the first
part of the paper.

Lithospheric Stretching

A one-dimensional kinematic approach suggested by McKenzie (1978)
was designed to describe the uprise of isotherms into the lithosphere
in response to lateral stretching of the lithosphere under isostatic
conditions. Subsequent basin subsidence is assumed to be a function
of thermal contraction with time.

The approximation of thermal basin data by this model led to the
necessity to introduce a layered lithosphere with an increasing amount
of stretching with depth, Royden & Keen (1980). This adjustment of

isotherm positions to surface heat flow does, however, no longer depend on a physical meaning of stretching. Because stretching coefficients up to 100 across the lithosphere will rather point towards an alternative mechanism. Furtheron the stretching concept is hard to adopt for basins with respect to

- the circular shape of interior basins
- cases of missing fault-control
- shape and nature of the transitional lower crust
- the limitation of volcanic activity in space and time
- the development of basins exceeding time constants for cooling.

Stretching of the shallow parts of the lithosphere seam to be rather the consequence of a deep seated process than a cause for the evolution of platform basins.

Thermal Mantle Plume

The basic idea of this concept is the upward movement of the lithosphere-asthenosphere boundary which is defined as a phase-boundary representing the onset of partial melting. The upwelling from an equilibrium position is driven by an increase of heat flux into the base of the lithosphere which is supplied by a so-called mantle plum. The difference between the heat flow of the plume and the conductive heat loss provides thermal energy to convert lithosphere into plume material of the temperature $T_p(z)$. This approach of heat supply by a moving phase-boundary is known as Stefan-problem.

The time dependent thermal disturbance of a starting geotherm T_o is shown in figure 4. The surface temperature of 10° C and T_o are based on steady state conductive heat transfer trough a layered half space and correspond to a surface heat flow of 60 mW m^{-2}. The assumed plume strength δ is five times the initial equilibrium heat flux of 31.4 mW m^{-2} at the base of the lithospheric layer, Neugebauer et al. (1983).

The rate of uprise of the phase-boundary as well as its final equilibrium position are controlled by the plume strength δ. According to the given model parameters $\delta = 5$ yields a "thinning" of the lithosphere from 100 to 20-30 km after about 30 Ma, figure 4. For $\delta < 5$ the phase-boundary reaches only half of the layer thickness within 60 to more than 100 million years. A plume strength greater than five - that means greater 150 mW m^{-2} - causes unrealistic small final thicknesses of the lithosphere.

Corresponding excess surface heat flow is a function of the apparent temperature gradient with depth, figure 5. A relative time delay of this quantity with respect to the thinning process is obvious.

The most reasonable strength ($\delta = 5$) with respect to lithospheric thinning causes an excess heat flow of 100 mW m^{-2} at the surface. This quantity, however, exceeds observations drastically even those for Cenocoic rifts, figure 1.

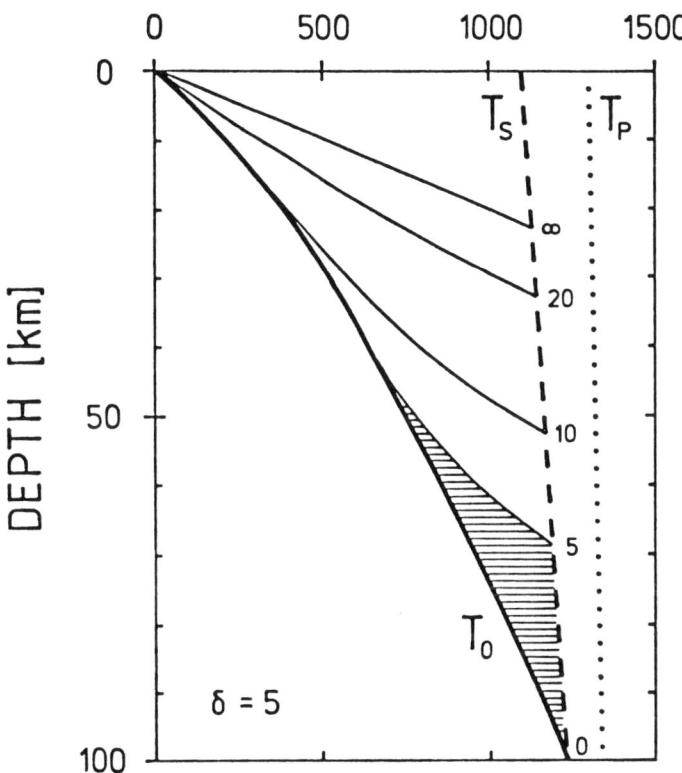

Figure 4. Disturbance of initial temperature T_0 by a step like increase of mantle heat flow. Amount of increase $\delta = 5$. T_p - plume temperature, T_s - mantle solidus.

The thermal potential of this concept appears to be to high compared with the limiting parameters of interior basins. However, the main problem araises from the circumstances that there is no modelimanent interruption or change of strength with time. Every additional limitations for a better adjustment are arbitrary assumptions. Compared with the long history of many basins the suggested time windows of activity initiated by a moving plate over a fixed plume will not be adequate. Finally the existence of partially molten material below a moving phase-boundary will represent extremely unstable situation with respect to inverted densities for the melt.

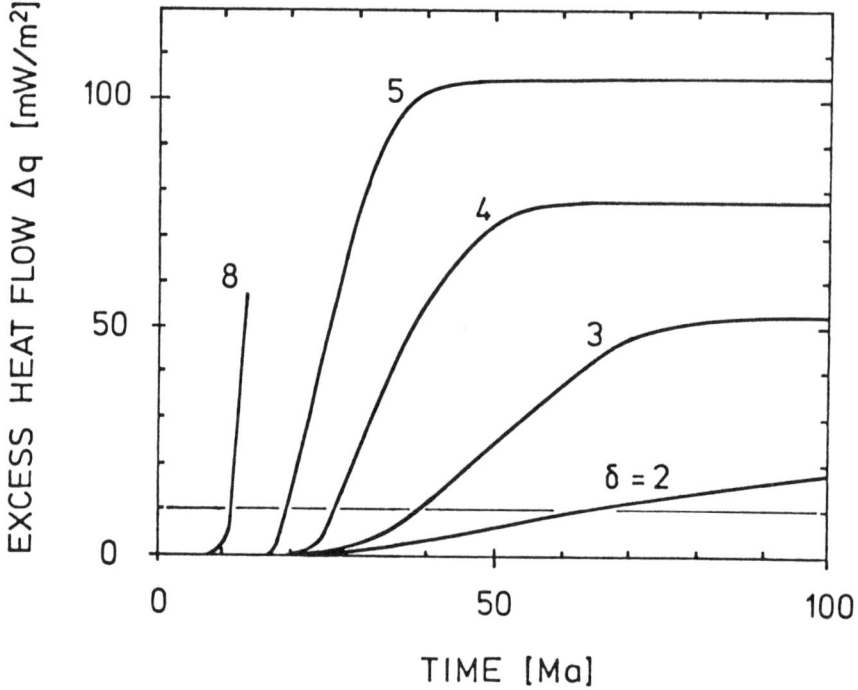

Figure 5. Calculated excess heat flow in response to various plumes strength (curve parameter).

Diapirs and Intrusions

A third categorie of physical concepts which are appropriate to describe the evolution of platform basins is the diapiric intrusion of material under upper mantle physical conditions. It combines the transport of a limited amount of heat and matter over a limited depth range within a limited period of time.

The physical principle requires a local inversion of density and in addition appropriate mechanical properties of rocks which allow deformation and transport of the material in order to attain an equilibrium mass distribution in the gravity field.

Density inversions within the earth's crust and upper mantle are associated for instance with thermal anomalies, partial melting of rocks or metasomatic processes. The latter are equivalent to a thermal event. Diapiric intrusions on the earth are observed for a wide variety of physical and mechanical conditions. Salt domes represent a well known member of minor thermal influence involved into its evolution. The ascent of basaltic magmas or the emplacement of granitic batholiths on the other side are examples where the temperature is of great importance. Although between both examples viscosities of magmas and of the environment differ many orders of magnitude from each other.

Theoretical investigations on the development of Rayleigh-Taylor instabilities and quantitative, numerical simulations of diapiric intrusions yield a number of controlling parameters for this physical concept and with respect to upper mantle conditions, Neugebauer (1983), Neugebauer et al. (1983), Reuther et al. (1985).

i The rate of diapiric uprise is controlled by the viscosity of the overlaying layers (crust and lower lithosphere for instance) and the amount of the inverse density contrast.

ii For specific rates minimum inverse densities for a given viscosity of the overburden have been derived. For a purely dynamic situation and a rate of 10 km Ma^{-1} for instance the minimum density contrast varies between $0.0001 \leqslant \Delta\rho \leqslant 0.1$ g cm^{-3} and the corresponding viscosity range of the lithospheric layer is $10^{19} \leqslant \mu_L \leqslant 10^{22}$ poise.

iii The depth level of stagnation of diapiric uprise is controlled by the change of viscosity across the lithospheric layer. More than three orders of magnitude of local contrast are required.

iv Convective transport of heat requires high rates of uprise in competition of cooling by conduction. The minimum rate is in the order of 10^2 km Ma^{-1}.

Figure 6 demonstrates the disturbance of a homogenous temperature field by a diapiric intrusion of less dense material. The heating of the zones adjacent to the head of the diapir is governed by the visosity contrast between the two media. For a low local ratio the crustal region becomes heated convectively as well, a high ratio of viscosities reduces the heat supply only to conductive transport outside the diapir.

Thereafter the ability of a diapiric intrusion to stagnate its rapid uprise at a particular depth range provides a rather restrictive control of the otherwise very efficient convective heat transport towards the surface. This control is predominantly dynamical although the viscosity has been assumed to be exponentially temperature dependent. The change of the controlling mechanism of heating from convection to conduction within the zone ahead of the diapir is responsible for a moderate heat supply and a time delay. Within this frame the efficiency of the heat supply by diapirs is affected by the volume of the intrusion and thus its thermal capacity in addition.

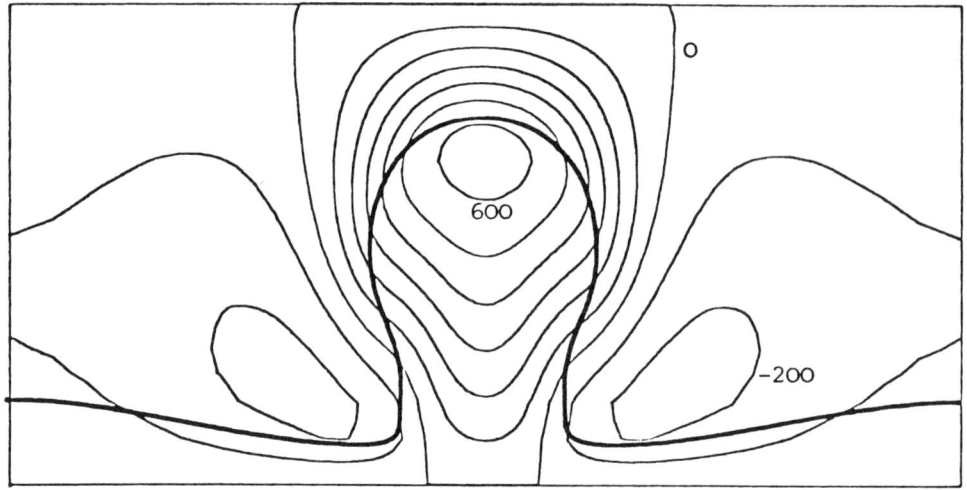

Figure 6. Difference between initial temperature field with linear increase with depth and the disturbed field at a specific time in response to the diapir: Diapir boundary - heavy line, density contrast - 0.1 g cm^{-3}, variation of viscosity top to bottom 10^1, temperature difference - thin lines, contour intervals 100° C (after Reuther et al. 1985).

Discussion

The final discussion will be concerned with a summary of the typical observations of the first part and a modified diapir-concept.

(1) The thermal evolution of many interior basins indicates excess heat supply with upper and lower limits by the existence of hydrocarbons.

(2) Basin subsidence is restricted in depth, however, a lasting process over possibly several hundred million years.

(3) Subsidence with time is commonly interrupted by unconformities. Correlation with voluminous volcanic eruptions has been observed.

(4) Fault-controlled crustal extension ranges from low to zero magnitude.

(5) Interior basins exhibit predominantly circular to oval geographic shape even for the cases with early linear tectonic phases.

(6) Interior basins are typically underlayn by 'transitional' lower crust of presumable lamelar structure, mixed lithology, intermediate seismic velocity and density.

(7) Volcanic activity is geographically limited to the area of transitional lower crust.

(8) Volcanic activity preceding basins subsidence has been observed.

(9) Igneous activity by thermal and petrological reasons indicates the transport of matter <u>and</u> heat convectively through the lithosphere.

(10) Traces of heterogenous lithosphere associated with regions of volcanic activity have been revealed by means of seismic wave travel time residuals.

In the view of the above characteristic features related with the thermal and dynamical evolution of platform basins the concept of diapiric intrusion of matter into the lithosphere appears to be most appropriate. Although we can demonstrate that stagnation of diapiric uprise is likely for lithospheric conditions we might be forced to overcome the view of one diapiric event which is associated with the formation of a basin. The strongest arguments against this view come from the extreme thermal efficiency of such a diapir even if its uprise is stopped below or at the lower crust. The seggregation of large volumes of magma or partial molten material with inverse density will not be an instantanous 'event' but rather a more or less continous process over long periods of time. Thus the view of small volumes of material seggregated or molten with time would provide a base of multiple intrusive events. Surface expression of this mechanism in terms of volcanic activity could be interpreted as an indication. The moderate heating of basin structures over long time with sequential epochs of thermal expansion due to heating and cooling and corresponding basin movement could easily be related. Activity in terms magma uprise once started will attract subsequent intrusions because of the preheated region in the lower lithosphere. This gives a simple explanation for the concentric shape of basins as a consequence of a bell shaped thermal dome in the lower lithosphere, the formation of a congruent space for mafic and ultramafic intrusives is a consequence. Late voluminous volcanic events will easily 'use' such a prepared lower lithosphere to move the source into a shallow subcrustal position. Fault-controlled crustal extensions appears than as a possible consequence of a moderately warmed lower lithosphere.

The numerical model of a circular body with inverted density in an environment of increasing temperature and decreasing viscosity with depth demonstrates the principle of small intrusions to many respects, figure 7.

The shape of the limited volume changes during its uprise from a circle to a laterally extended lense due to the increase in absolute viscosity and viscosity contrast as well. The isotherms become distorted upwards forming a thermal wake behind the blob.

This view of a piecewise heating of the lithosphere by the uprise of small volumes of material with slight density inversion provides a mechanism to comprise structural, dynamical and thermal constraints even for long periods of time withouth the use of unrestricted processes of catastrophic nature.

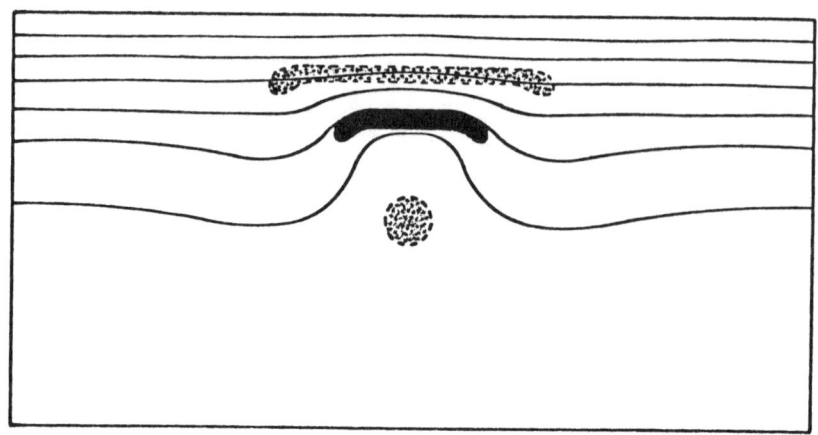

Figure 7: Diapiric uprise of a circular body. Temperature from top to bottom 1200° C, viscosity decrease with depth three orders of magnitude. Disturbed isotherms correspond to the stage in time indicated by the black lense.

ACKNOWLEDGMENT

We wish to thank D.S. Chapman for contributing modern thermal standards to this investigations. Thanks to Barbara Stietzel for typing the manuscript. This work was supported by Deutsche Forschungsgemeinschaft.

REFERENCES

DEICHMANN, N. & ANSORGE, J. (1983): Evidence for Lamination in the lower continental crust beneath the Black Forest (Southwestern Germany). J. Geophys. 52, 109-118

EDEL, J.B., FUCHS, K., GELBKE, C. & PRODEHL, C. (1975): Deep structure of the Southern Rhinegraben area from seismic refraction investigations. J. Geophys. 41, 333-356

FINLAYSON, D.M. & MATHUR, S.P. (1984): Seismic refraction and reflection features of the lithosphere in northern and eastern Australia, and continental growth. Annales Geophysicae 2, 711-722

HALE, L.D. & THOMPSON, G.A. (1982): The seismic reflection character of the continental mohorovicic discontinuity. J. Geophys. Res. 87, 4625-4635

HUNT, J.M. (1979): Petroleum Geochemistry and Geology. W.H. Freeman, San Francisco, pp. 617

KARSON, J.A. & COLLINS, J.A. (1984): Geologic and seismic velocity structure of the crust/mantle transition in the Bay of Islands ophialite complex. J. Geophys. Res. 89, 6126-6138

McKENZIE, D. (1978): Some remarks on the development of sedimentary basins. Earth Plan. Sci. Lett. 40, 25-32

MORGAN, P. (1983): Constraints on rift thermal processes from heat flow and uplift. Tectonophysics 94, 277-298

MORGAN, P. (1984): The thermal structure and thermal evolution of the continental lithosphere. Physics and Chemistry of the Earth

NEUGEBAUER, H.J. (1983): Mechanical aspects of continental rifting. Tectonophysics 94, 91-108

NEUGEBAUER, H.J., WOIDT, W.-D. & WALLNER, H. (1983): Uplift, Volcanism and Tectonics: Evidence for mantle diapirs of the Rhenish Massif, in (K. Fuchs et al. ed.) Plateau Uplift, Springer, Berlin, 381-403

NEUGEBAUER, H.J. & WALTER, R. (1983): Volcanism, Lithospheric thinning and tectonic implications. Terra cognita 3, 117

O'REILLY, S.Y. & GRIFFIN, W.L. (1985): A xenolith-derived geotherm for southeastern Australia and its geophysical implications. Tectonophysics 111, 41-63

REUTHER, C., NEUGEBAUER, H.J. & CHRISTENSEN, U. (1985): Diapiric intrusions and the transport of matter and heat. 23rd Gen. Ass. IASPEI, Tokyo

RICHTER, A. (1985): Quantitative investigations on the tectonic development of sedimentary basins. Thesis, Technische Universität Clausthal

ROYDEN, L. & KEEN, C.E. (1980): Rifting process and thermal evolution of the continental margin of eastern Canada determined from subsidence curves. Earth and Plan. Sci. Lett. 51, 345-361

TISSOT, B.P. & WELTE, D.H. (1984): Petroleum Formation and Occurrence. 2nd ed. Springer, Berlin, pp. 699

WILLETT, S.D., CHAPMAN, D.S. & NEUGEBAUER, H.J. (1984): Mechanical response of the continental lithosphere to surface loading: effect of thermal regimes. Annales Geophysicae 2, 679-688

WILLETT, S.D., CHAPMAN, D.S. & NEUGEBAUER, H.J. (1985): A thermo-mechanical model of continental lithosphere. Nature 314, 520-523

ZIEGLER, P.A. (1982): Geological atlas of western and central Europe. Elsevier, Amsterdam, pp. 130

I. MORETTI[1], C. FROIDEVAUX[2]

PHYSICAL MODELS OF EXTENSIONAL TECTONICS

ABSTRACT

Geological and geophysical evidences (in particular thermal and gravity anomalies) suggest the presence of hot mantle material below the zones of continental rifting. Using a thermo-mechanical numerical model we quantify the lithospheric thinning caused by such a deep thermal anomaly. The mantle is assumed to be an incompressible non-Newtonian fluid with temperature -and pressure - dependant viscosity. The index power law is 3 in the mantle and 7 in the crust. This yields a viscosity minimum at the base of the lithosphere and therefore favours rapid convective thinning. The corresponding mass displacements disturb the lithosphere. The stresses at the bottom of the crust are used to quantify the vertical movements and the tectonic regime within the continental crust. The heat flow anomalie is also computed.

The present work indicates that these deep geodynamic processes can induce well localized extensive tectonic stresses and the thinning of the lithosphere on a short time scale. This analysis also suggests that the subsidence in the rift valley is mainly due to crustal thinning while the uplift of the shoulders of the rift is a consequence of the advection of hot material from the asthenospheric channel. One of these phenomena should not be considered as caused by the other even if an elastic response is assumed to the lithosphere. A regional extensional stress enhances the convective phenomena and favours the crustal thinning but is not necessary. A decrease in this far-field force drasticaly increases the uplift of the shoulders and reduce the width of the rift valley. This case may be representative of the African rift.

---
(1) *Institut Français du Pétrole, Rueil-Malmaison, France.*
(2) *Laboratoire de Géophysique, Université Paris Sud, Orsay, France.*

INTRODUCTION

Extensional tectonics is localized in well defined provinces. In continents the latter exhibit characteristic structural features cartooned in Figure 1 (Ramberg and Morgan, 1984). The rift valley is 50 to 150 km wide - 60 km for Suez (Garfunkel and Bartov, 1977) and 110 km for Baikal (Puzyrev et al., 1978). It can be filled by up to 10 km of sediments 5 for Suez and 7 for Afars (Tiercelin and Favre, 1978). On its flanks we can observe broad shoulders with some 300 km of lateral extension. The elevation of which reaches 2000 meters above the valley floor -1800 m, for Kenya (Baker et al., 1972) and 1100 m for Rio Grande (Chapin, 1979). The transition from the central valley to the shoulders is quite abrupt and shows large normal faults with kilometric throws (for Suez, see Chenet et Letouzey, 1983). Locally the surface heat flow can be very high because of the presence of magmatic intrusions (Morgan, 1982). However its average value on the shoulders does not exceed the average continental value by more than 50%. The deep seismic velocity structure points to a thinner crust under the central valley, typically by 20-50 %, -35/50 km for Baikal (Puzyrev et al., 1978) 30/40 km for East Africa (Long, 1976) for the 20/40 km Rhinegraben (Mueller, 1978). It also suggests the presence of hot mantle material down to 300 km and over a halfwidth broader than

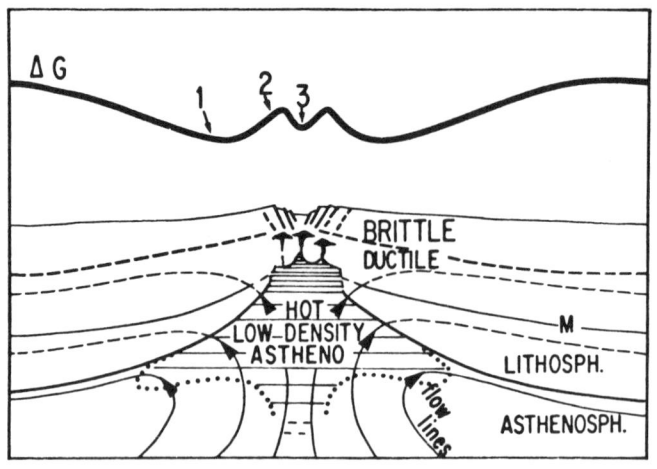

Wavelengths:
Anomaly:

1	1000 km
2	200 km
3	50 km

Fig. 1 - Schematic cross section of a continental rift with qualitative Bouguer gravity anomaly profile (after Ramberg and Morgan 1984)

150 km. This low density body generates a broad negative gravity Bouguer anomaly as shown in Figure 1 (Ramberg and Morgan, 1984). Two more features are found in this gravity signal; a positive contribution related to crustal thinning and another negative anomaly, much narrower than the first one, and due to the sedimentary filling of the central valley.

The geological record indicates that the rifting process can take place fairly rapidly, say within 10 Ma. Such a rapid lithospheric thinning has ruled out the idea of a simple conductive perturbation of the thermal structure caused by an asthenospheric temperature anomaly. Indeed such a perturbation can only penetrate through the lithosphere with a time constant of the order of 100 Ma.
Two mechanisms have been put forward to achieve thinning with the right time scale: on the one hand mechanical stretching of both the crustal and mantle lithosphere can be postulated (Mc Kenzie, 1978). However both the Baikal and the East African rifts offer a more complex pattern: the crust is thinned by 25%, whereas the mantle lithosphere is completly replaced by hot asthenospheric mantle. In the same way, the uplifts bordering the Gulf of Suez indicate lithospheric heating greatly in excess of that produced by uniform extension of the lithosphere (Steckler, 1985). On the other hand, a possible upward advection of hot material has been investigated (Neugebauer, 1982; Mareschal, 1982; Turcotte and Emerman, 1983; Spohn and Schubert, 1983).

Presently the best answer is that the thermal structure of the lithosphere can be destabilized on a short time scale (Fleitout and al., 1984; Yven and Fleitout, 1985). The latter model is based on a temperature- and pressure-dependent rheology. This yields a viscosity minimum at the base of the lithosphere and therefore favours rapid convective thinning. The present paper adopts a similar approach.

The quantification of vertical displacements, central subsidence vs shoulders uplift, leads to a sensitive issue. What mechanisms can account for the short range transition from the central graben to the broad uplifted shoulders? Models based on the elastic response of the lithosphere (Chamot-Rooke, 1904) do not provide a satisfactory explanation. The elastic loading approach has been quite successful for oceanic plates. Indeed the bending seaward of the trenches induces a bulge of moderate amplitude, about 300 m, compared with several kilometers for the depth of the trench (Carey et Dubois, 1981). The distance between the deep trench and the elastic bulge amounts to some 300 km (Watts, 1978). Neither of these numbers fits the situation for the continental rifts. Indeed the elevation of the shoulders is much more important relative to the depth of the rift valley, and the maximum topography is located as close as some 50 km away from it. These characteristics preclude the use of elastic models, i.e. of regional compensation schemes, within the tectonically active part of the rift. The compensation being local, this also requires separate mechanical causes for the foundering of the graben and for the uplift of its shoulders. The purpose of the present paper is to analyse the internal dynamics of the deep rift structure with its hot thermal root, and to make quantitative assessments of the induced observables such as vertical displacements, tectonic stress and thermal evolution.

The tectonic evolution results from different competing effects. A deep thermal anomaly generates lithospheric thinning. At the surface this yields a broad thermal doming and extensive deviatoric stresses. On the contrary crustal thinning due to this local stress field and to possible far field extensional forces induces subsidence. In this paper, both phenomena are modelized. The models are based on various possible rheological laws for the lithosphere. The computed response is strongly dependent upon this choice.

MODEL DESCRIPTION

Fig. 2 depicts the general physical aspects of our model which consists of a vertical 600 km depth and 625 km broad cross section. We have adopted a quasi-steady state model consisting of a serie of secondary convective rolls extending between the base of the continental lithosphere and 600 km depth (Fleitout and Yuen, 1984). A thermal perturbation in the asthenosphere enhances this convective flow, the resulting mass displacement and stresses thin both lithosphere and crust.

To study the evolution of this perturbation we solve the conservation equations for mass, momemtum and energy in a infinite Prandtl number fluid:

$$\nabla \underline{u} = 0 \qquad (1)$$
$$\nabla \underline{\tau} \cdot \nabla p + \rho g \alpha (T - T_o) = 0 \qquad (2)$$
$$\frac{\delta T}{\delta t} = K \nabla^2 T - u \cdot \nabla T \qquad (3)$$

$\underline{u}$ is the velocity field, t the time, p the perturbation pressure, $\underline{\tau}$ the deviatoric stress tensor, K the thermal diffusity, T the temperature in absolute degrees, g the gravitational acceleration (9.8 m/s^2).

The density is given by
$$\rho = \rho_o (1 - \alpha (T - T_o))$$

in the mantle ρ_o = 3300 kg/m³, in the crust ρ_o = 2800 kg/m³

α is the coefficient of volumetric thermal expansion (3.10^{-5} K^{-1}) and T_o is a reference temperature (1300°C in the mantle and 200°C in the crust).

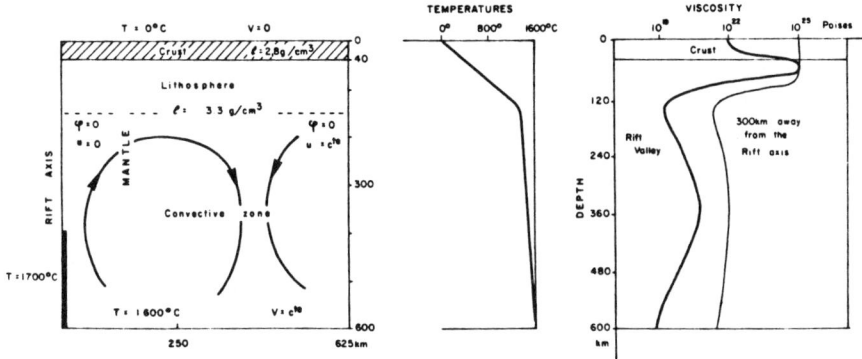

Fig. 2 - Schematic diagram of the model and boundary conditions. ρ is the density at the temperature reference (see text). T is the temperature, φ the lateral heat flow, u the horizontal velocity and v the vertical velocity. The initial temperature gradient is 10°C/km until 130 km depth and 0.64°/km in the convective zone.

Viscosity law:

Fleitout and Yuen (1984) show the importance of the choice of the viscosity law for the study of secondary convection. The purpose of this paper is to examine the coupling between lithospheric and crustal thinning. The mechanical behavior of the mantle will be considered as known and, according to Fleitout's and Yuen's (1984) conclusions, we assume a temperature and pressure dependent rheology

$$\eta(T, z) = A \exp\left(\frac{Q}{RT} - \frac{Q}{RT_r}\right) \exp\left(\frac{z}{B} + \frac{z^2}{C}\right)$$

for the Newtonian case. R is the gas constant, Q the activation energy and T_r a reference temperature. The parameter values of A, B, C, Q are the same as those used by Fleitout and Yuen (1984) and Fleitout, Yuen and Froidevaux (1984).

For non-Newtonian cases the viscosity also depends on the second invariant of the deviatoric stress tensor

$$\tau_2 = \left(\frac{1}{2}(\tau xx^2 + \tau zz^2) + \tau xz^2\right)^{1/2}$$

111

and

$$\eta(T, z, \tau_2) = \frac{\eta(T, z)}{D \tau_2^{n-1}}$$

n = 1 corresponds to the Newtonian case.

Experimental deformations of olivine (Carter, 1976) show that n = 3, but values 1 and 5 will be tested for the mantle.

When n increases, the deformation is concentrated in the weakest zone. We can therefore model the brittle behavior of the crust by a non-Newtonian viscosity with a high exponent n (7 in our study).

The numerical code used to solve the thermo-mechanical equations was originaly written by Anderson and Bridwell (1980) and has been adapted by us for a Cray x MP computer. Typically, we use a mesh of 806 elements with 31 points along the vertical and 26 in the horizontal direction. This box contains two different materials, i.e. the crust and the mantle. The crust is initially 40 km thick.

Boundary conditions:

At t = 0, the thermal profile is a continental profil. The gradient in the lithosphere is 10° C/km: in the asthenosphere it is only 0.64° C/km.

We impose a small horizontal perturbation

$$T(x, z) = T(z) (1 + \gamma \exp \frac{x^2}{A^2}) \qquad A = 100 \text{ km}$$

On the right-hand side of the box we assume that

$$\frac{\delta T}{\delta x} = 0 \quad \text{and} \quad \tau xz = 0$$

and the imposed horizontal stretching velocity varies between 0 and 10 mm/y.

The left side of the box is assumed to be the symmetry axis of the rift structure, therefore between 0 and 400 km the appropriate boundary condition for temperature (i.e. zero horizontal heat-flow) reads:

$$\frac{\delta T}{\delta x} = 0 \qquad \text{and} \qquad \tau xz = 0$$

At greater depth (between 400 and 600 km) a thermal perturbation is simulated by fixing T = 1700°C on the corresponding vertical segment of the box wall.

On the top of the box the vertical velocity is zero and the temperature is 0°C. On the bottom, the vertical velocity is an imposed function of the extension velocity. Indeed, the surface of the box should remain constant to satisfy mass conservation (1). The thermal condition is a base temperature of 1600°C.

Results

A first case is presented with the following specific values: a lateral velocity v = 6 mm/yr, a non Newtonian crust with n = 7, a non Newtonian mantle with n = 3.

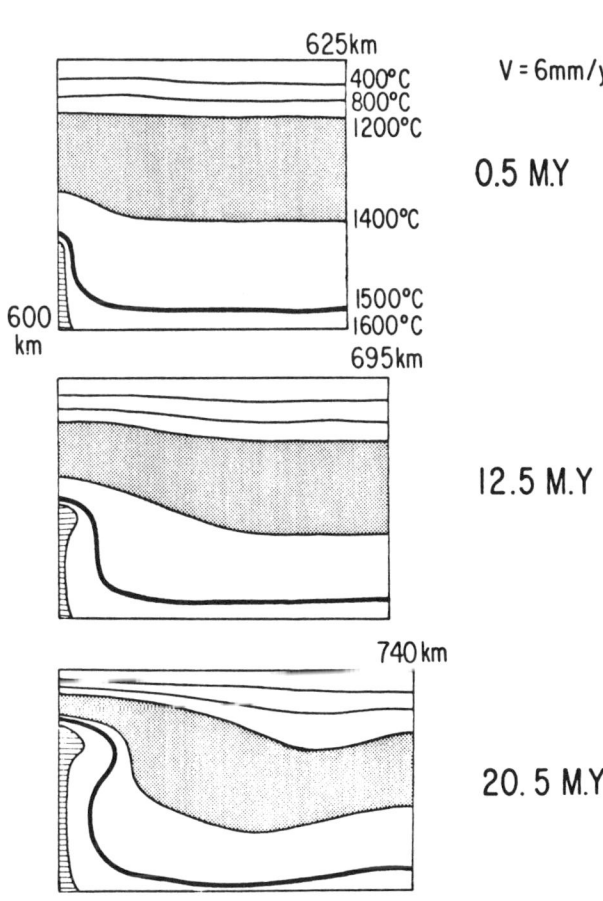

Fig. 3 - Thermal evolution for a opening velocity of 6 mm/yr.

Fig. 3 shows the thermal evolution: the convection strongly perturbs the configuration of isothermes. The warm material migrates from the central part (here on the left) to the cold side near the lithosphere - asthenosphere boundary. On the otherhand some cold return flow occurs at the bottom of the box. The shape of the lower isotherms indicates that advective processes dominate the evolution at depth.

Fig. 4a shows the vertical movements in function of time. There are only tectonic effects. Here both sedimentation in the rift valley or erosion on the shoulders are ignored.

The initial doming is due to the initial thermal disturbance ($\gamma = 0.02$). After 5 My subsidence due to the

113

crustal thinning balances this uplift, and the topography disappears. The shoulders and the rift valley become visible after about ten million years. Then these phenomena accelerate, and the central graben becomes narrower (100 km in this case). At 21 My, the shoulders are 500 meters high and 500 km wide. The corresponding heat flow variations are shown in Figure 4b and the rate of horizontal deformation in Figure 5. This figure also shows the rate of deformation, obviously constant, that can be predicted using the homogeneous thinning postulated by McKenzie (1978) for the study of sedimentary basins. In our study the concentration of deformation stems from the non-Newtonian rheology of the crust and of the mantle. Viscosity decreases rapidly vertically above the deep thermal anomaly because both the temperature and the deviational stress increase. Crustal thinning is considerable and this zone subsides in spite of the uprising of hot (i.e. light) asthenospheric material.

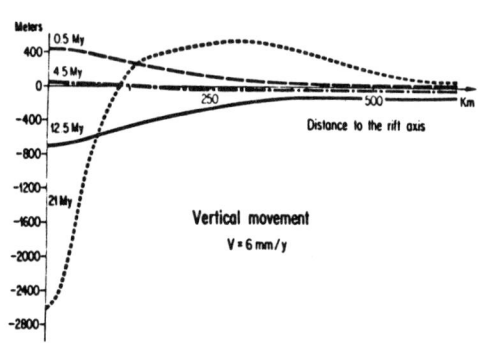

Fig. 4a - Vertical movements as a function of time. The crust and the mantle have a non-Newtonian rheology and the power law index are 7 and 3 respectively. The opening velocity is 6 mm/yr.

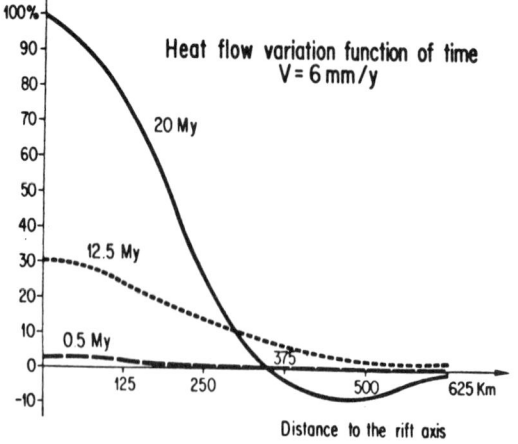

4b - Heat flow variations with time associated with the same parameters. The reference value is the initial heat flow on the right side of the box.

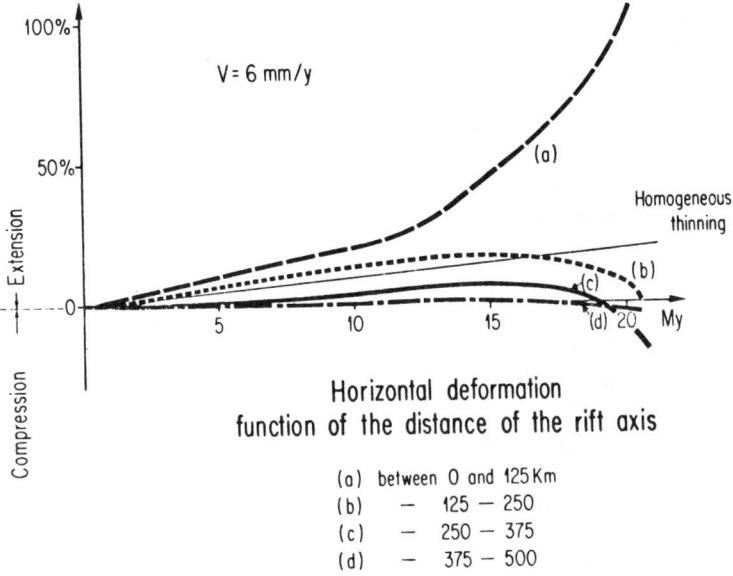

Fig. 5 - Horizontal deformation with the same hypothesis as in Figure 3. The curve "homogeneous thinning" is the ratio between the present and initial width of the box, it corresponds to the case of an average deformation.

This hot material migrates from the center of the rift to the cold side near the lithosphere-asthenosphere boundary. Indeed, this boundary corresponds to a low viscosity zone. It generates a thermal erosion of the base of the lithosphere. This material's advection induced the shoulder's uplift. In the present case the amplitude of this uplift is limited because we assume a great stretching velocity (6 mm/yr). If this velocity is only 3 mm/yr, the amount of the shoulder's uplift after 20 My would be 900 meters (and the subsidence of the rift valley would be 1700 meters).

We have calculated the tectonic subsidence, i.e. the movements due to deep causes. Assuming local isostasy, we can evaluate the resulting overall subsidence, including sedimentation in the rift valley. A tectonic graben, 2600 m deep, filled with sediments with a specific gravity of 2, induces a 13 km deep valley (or 7 km of sediment and 3 km of water). During 21 My, this corresponds to a velocity of sedimentation velocity of 0.6 mm/yr. This value is compatible with geological data (0.9 mm/yr in Kenya - Tiercelin and Favre, 1978; 0.3 mm/yr along the Gulf of Suez for late Miocene evaporites and 1 mm/yr for the Middle Miocene muds (Scott and Govean, 1984). The final depth (13 km or 7 km and 3 km of water) is great because the assumed opening velocity is great. With this hypothesis, the final state with a crustal thinning of 27 km (β = 3) and a horizontal extension rate of 105% corresponds almost to a continental margin (Gulf of Lions β_{max} = 5, present thickness of sediments 8 km, 3 km water, distance between the oceanic crust and the "normal" continental crust about 250 km, horizontal extension rate = 170 % in this central zone (Le Douaran et al., 1984). Results with a smaller opening velocity or without regional extension have also been examined.

Influence of rheological parameters

In the mantle

Previous publications (Fleitout and Yuen, 1984; Yuen and Fleitout, 1985) suggest that data (gravity, topography, heat-flow) are more compatible with a non-Newtonian rheology in the upper mantle. Figure 6 depicts the crustal thinning in the central graben for different mantle rheologies. The opening velocity is 3 mm/yr, and the crust is non-Newtonian with exponent 5. The convective destabilization of the lithosphere is faster for a non-Newtonian fluid, and the corresponding rate of crustal thinning is greater and more compatible with the timing of rifting (10-20 My).

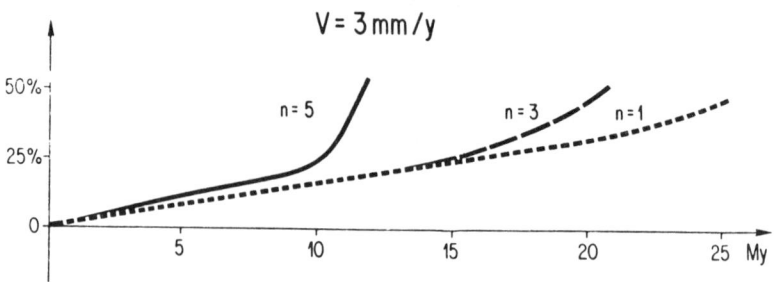

Fig. 6 - Crustal thinning in the center of the rift as a function of the mantle rheology. The crust is non-Newtonian with power law exponent 5. The opening velocity is 3 mm/yr. Thermal hypothesis are the same as in Figure 3. The initial thickness of the crust is 40 km. With the hypothesis of homogeneous behavior, the average thinning after 20 My would be about 10%.

In the crust

Our model is very sensitive to crustal rheology. Indeed, the crust is subjected to different stresses: an extensional stress due to regional tectonics and to asthenospheric flow at the bottom of the lithosphere, and a vertical stress due to the uprising of the mantle. Crustal thinning, and vertical movements, i.e. heat flow variations, etc., depend on the crustal behavior in this stress field. If the crust is strong, crustal thinning is weak. On the contrary, if the crust is brittle, thinning is larger and localized. Non-Newtonian rheology, with a high exponent, simulates this second behavior which better fits the data. Indeed, geological records prove that the deformation zone is limited in space. Figure 7 shows the topography after 21 My for different behaviors: a Newtonian (i.e. resistant) crust and two non-Newtonian crusts. The mantle is non-Newtonian with n = 3, and the opening velocity is 6 mm/y. The difference between curves (a) and (c) is very important. With a Newtonian crust the rift valley is always wide, about 450 km; the shoulder's uplift and the graben's subsidence are small.

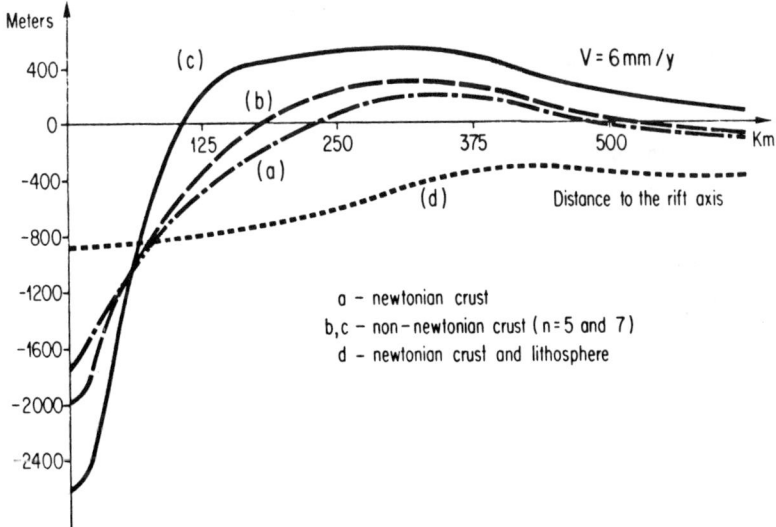

Fig. 7 - Topography after 21 My function of the crustal rheology. The mantle is non-Newtonian with a power law index of 3, except for the d-curve. Thermal hypothesis and regional stress are the same as in Figure 3.

PHYSICAL MODELS OF EXTENSIONAL TECTONICS

For comparison this figure also shows the vertical movements with Newtonian rheology for both crust and mantle, with all other hypothesis being similar (case d). It seems to be hopeless to try to simulate geological deformation in active tectonic zones with Newtonian rheologies.

Influence of the opening velocity

In the previous graphics, we have always assumed a regional extension and imposed a horizontal velocity at the right side of our box. In fact, this hypothesis is not necessary. If there is no regional extension, the propagation of the deep thermal anomaly is sufficient to thin the crust in the central zone (Fig. 8). A thickening of the crust then occurs in the border regions (under the shoulders) because of crustal mass preservation in the box. We think that the preservation of crustal mass might actually not be as effective as is indicated by the seismic data for the Bay of Biscay (Chénet et al, 1983) or for the Gulf of Lions (Le Douaran et al., 1984). Anyway, even if this mass preservation does exist, crust thickening may occur over a large distance. In this case the shoulder's width exceeds that of our box. Nevertheless we can conclude that, with realistic rheclogies continental rifting can be modeled assuming only a deep thermal anomaly.

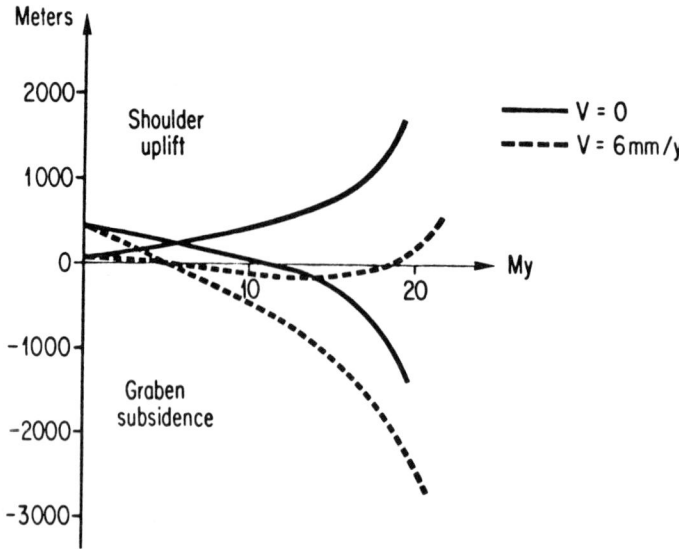

Fig. 8 - Vertical movements for two different opening velocities. All parameters are the same as in Figure 3. The graben subsidence corresponds to the displacement at the rift axis. The "shoulder uplift" is the vertical displacement 300 km away from this axis. It is not the maximum of the uplift.

This case also represents the rift evolution, when the regional stress ceases to be extensional, such as, for exemple, along the Gulf of Suez after the opening up of the Gulf of Aqaba (Bartov et al., 1980; Mart and Hall, 1984). Comparing results in the different cases - opening velocity of 6, 3 or 0 mm/y - we can predict that a decrease in the extensional velocity causes a reduction of the subsidence of the central graben and a large increase in the shoulder uplift. During the first 10 million years the rate of subsidence was 88 m/My for v = 6 mm/yr and only 44 m/My for v = 0. Likewise the breadth of the shoulders increased. For example, 375 km from the rift axis, vertical movements are about zero if the opening velocity was 6 mm/yr, and the uprising velocity was 50 m/My (after 10 My) without any regional extension. Geological data in the Suez Rift indicate two phases there. During the early Miocene the graben subsidence was great, and it decreased after that but the border surrounding it continued until the present time, and the topography in the Sinaï exceeds 2000 meters. In summary, if the regional stress is largely extensional the central graben is deep: if there is no regional extension the shoulder uplift is great (> 1500 m after 18 My) and affects a broad region.

Influence of the deep thermal anomaly

If there is no thermal anomaly, the crust and the mantle are homogeneous, and then the thinning is also homogeneous. We suppose a little initial thermal perturbation of 1%:

$$T(x,z) = T(z) \left(1 + 0.01 \exp \frac{-x^2}{A^2}\right)$$

with A = 100 km. The opening velocity is again 6 mm/yr and power the law indices are again 3 in the mantle, and 7 in the crust. During 30 My the deformation is homogeneous. Later boudinage phenomena occur. These are possible in a non-Newtonian fluid when the variations of viscosity are sufficent (Ricard and Froidevaux, 1985). The deformation after 48 My are depicted Figure 9.

This period of 30 My required for the initiation of boundinage depends on the initial heterogeneity of the crust or of the lithosphere. With our rheological law, a temperature perturbation of 1% induces a variation of the viscosity inferior to 10. If the initial thermal perturbation is 3%, boudinage appears after 25 My.

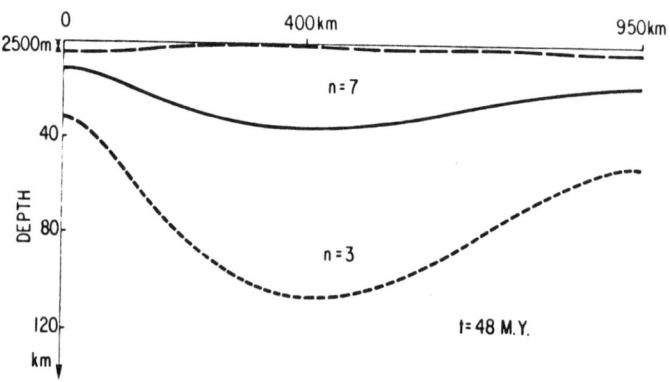

Fig. 9 - Deformation after 48 My without any initial deep thermal anomaly. The crust and the mantle are non-Newtonian, power-law index 7 and 3 respectively. The opening velocity is 6 mm/yr. The initial thermal perturbation of 1% on the right side of the box. The curve a is initially at 120 km depth and the curve b corresponds to the Moho, initially at 40 km depth. The maximum vertical offset of the surface topography at this time is about 2500 m.

CONCLUSION

The propagation of an asthenospheric thermal anomaly can cause continental rifting. It causes the formation of two zones: a limited region with crustal thinning and subsidence bounded by large uplifted regions, the shoulders.

The lithosphere is thinned out by the mass displacement of mantellic material. This phenomenon is associated with secondary convection. This thinning out is faster (about 10 My) than purely conductive thinning (about 100 My). A extensive regional stress enhances these convective phenomena and favors crustal thinning, but is not necessary.

This paper shows the importance of the rheological laws. With Newtonian rheologies in both crust and lithosphere, a deep thermal anomaly, initialy between 400 and 600 km, induces just a few meters of thermal doming at the surface. Lithospheric thinning was slow and crustal thinning almost non-existent because the crust is very strong with this hypothesis. Non-Newtonian rheologies can be used to quantify the focusing of the deformation and to study the coupling between both crustal and lithospheric thinnings and the regional tectonic stress.

ACKNOWLEDGMENTS

We have benefited from stimulating discussions with L. Fleitout and P.Y. Chénet; P. Weil and B. Ranger critically read the manuscript.

BIBLIOGRAPHY

Anderson C. and Bridwell R., 1980 - A finit element method for studying the transient non-linear thermal creep of geological structures. Intern. Journ. of numerical and analytical methods in geomechanics, 4, 255-276.

Baker, B., Mohr, P. and Williams, L., 1972 - Geology of the Eastern rift system of Africa, Geol. Soc. Amer., sp. pap., 136, 67 p.

Bartov, Y., Steinitz, G., Eyal, M. and Eyal, Y., 1980 - Sinistral movement along the Gulf of Aqaba - its age and relation to the opening of the Red Sea. Nature, 285, 220-222.

Carey E. and Dubois J., 1981 - Behaviour of the ocean lithosphere at subduction zones; plastic yield strengh from a finit-element method. Tectonophysics, 74, 99-110.

Carter, N., 1976 - Steady state flow of rocks. Rev. of Geophy. and Sp. Phys., 14, 301-360.

Chamot-Rooke, 1984 - Modèle de plaque mince élastique avec paramètre flexural variable, DEA de géotectonique, Université P. & M. Curie.

Chapin C., 1979 - Evolution of the Rio Grande rift, Riecker (Ed.) in Rio Grande Rift: tectonics and magmatism.

Chénet, P.Y., Montadert, L., Gaviaud, H., and Roberts, D., 1982 - in Studies in continental margin geology, AAPG Mem., n°34, 703-715.

Chénet P.Y. and Letouzey J., 1983 - Tectonique de la zone comprise entre Abu Durha et Gebel mezzazat (Sinaï, Egypte) dans le contexte de l'évolution du rift de Suez, Bull Centres Rech. Explor. Prod. Elf-Aquitaine, 7, 1, 201-215.

Fleitout L. and Yuen D., 1984 - Steady-state, secondary convection beneath lithospheric plates with temperature and pressure dependent viscosity, J.G.R., 89, 9227-9244

Fleitout L. Yuen D. and Froidevaux C., 1984 - Thermomechanical models of lithospheric deformation: application to the tectonics of western Europe. Annales geophysical.

Garfunkel, Z. and Bartov, Y., 1977 - The tectonics of the Suez rift, Geol. Survey of Israel, 71, 1-41.

Mc Kenzie D., 1978 - Some remarks on the development of sedimentary basins, Earth Planet Sciences Letters, 40, 25-32.

Le Douaran, S., Burrus, J. and Avedik, F., 1984 - Deep structure of the North-Western mediterranean basin: results of a two-ship seismic survey. Marine geology, 55, 325-345.

Long R., 1976 - The deep structure of the East African rift and its relation to Afar, in Afar between continental and oceanic rifting, Schweizerbart.

Mareschal J.C., 1983 - Mechanism of uplift preceding rifting, Tectonophysics, 94, 51-66.

Mart,Y. and Hall, J., 1984 - Structural trends in the northern Red Sea, JGR, 89, 352-364.

Morgan, P., 1982 - Heat flow in rift zones, Palmasson G. (Ed), Continental and oceanic rifts, geodynamics series, **8**.

Mueller S., 1978 - Evolution of the earth's crust, Ramberg and Newmann (Ed.), Tectonics and geophysics of continental rifts.

Neugebauer H., 1983 - Mechanical aspects of continental rifting, Tectonophysics, 94, 91-108.

Puzyrev N.N., Mandelbaum M.M., Krylov S.V., Mishenkin B.P., Petrik G.V. and Krupskaya G.V., 1978 - Deep structure of the Baikal and other continental rift zones from seismic data, Tectonophysics, 45, 15-22.

Ramberg, I. and Morgan, P., 1984 - Physical characteristics and evolutionary trends of continental rifts. In press.

Ricard, Y. and Froidevaux, C., 1985 - Lithospheric boudinage and stretching instability. In press.

Scott, R. and Govean, F., 1983 - Early depositional history of a rift basin: Miocene in the western Sinaï. VII EGPC Seminar. Cairo, 1984.

Spohn T. and Schubert G., 1983 - Convective thinning of the lithosphere: a mechanism for rifting and mid plate volcanism on earth, Venus and mars. Tectonophysics, 94, 67-90.

Steckler, M., 1985 - Uplift and extension at the Gulf of Suez - indications of induced mantle convection submitted to Nature.

Tiercelin J.J. and Favre H., 1978 - Rates of sedimentation and vertical subsidence in meorifts and paleorifts, Tectonics and geophysics of continental rifts.

Turcotte D. and Emerman S., 1983 - Mechanisms of active and passive rifting, Tectonophysics, 94, 39-50.

Watts, A., 1978 - An analysis of isostasy in the world(s ocean 1. Hawaiian-Emperor seamount chain, J.G.R., 83, 5989.

Yuen, D. and Fleitout, L., 1985 - Thinning of the lithosphere by small-scale convective destabilization. Nature, 313, 125-128.

CHAPTER 2
HEAT TRANSFER IN SEDIMENTS

PART A

POROSITY AND COMPACTION

K. MAGARA[1]

POROSITY-DEPTH RELATIONSHIP DURING COMPACTION IN HYDROSTATIC AND NON-HYDROSTATIC CASES

ABSTRACT

Porosity-depth relationships of shales have been established in many different sedimentary basins of the world. A few important observations can be made on the relationships; 1. At shallower depths, porosity reduction is relatively rapid, whereas it slows at depth, due possibly to the combined effect of increasing grain-to-grain contact area of shales, decreasing shale permeability, and increasing water viscosity with compaction and depth. 2. Curves are sometimes shifted to lower porosity values at a given depth, due mainly to the effects of uplift and erosion in the geological past. 3. In relatively deep sections, shale porosity may be abnormally high (undercompacted shale), if pore-fluid pressure is abnormal.

The concept of physical balance among the total stress, effective stress, and fluid pressure in the subsurface can explain the association of undercompacted shales and abnormal fluid pressure quite well. However, the amount of fluid pressure estimated from this concept is insufficient to fully explain the measured abnormal fluid pressure in the Gulf of Mexico and Mackenzie Delta basins. Some other causes, such as aquathermal pressuring and generation of hydrocarbons from organic matter, may have further increased the fluid pressure there. The effect of smectite-illite conversion on the pressure increase is considered to be insignificant.

In an undercompacted zone which was caused by the com-

---
(1) *Faculty of Earth Sciences, King Abdulaziz University, Jeddah, Saudi Arabia.*

bination of relatively rapid sedimentation (or burial) rate, low average permeability (or thick shale in most cases), and deep total burial of relatively young sedimentary rocks, geothermal gradient is usually greater than that in a normally compacted interval, because of relatively low thermal conductivity of the undercompacted shales. Such a high geothermal gradient in a closed and undercompacted zone will further increase the fluid pressure to an extremely high level by both the aquathermal and hydrocarbon generation effects.

INTRODUCTION

Since publication of a classical paper on estimation of abnormal fluid pressure from wire-line log data by Hottman and Johnson(1965), this method has been almost universally accepted in the petroleum industry for monitoring abnormal pressure which could risk safe drilling operations This method is primarily based on an intimate relationship among the total stress (overburden pressure), effective stress (grain-to-grain contact pressure), and fluid pressure in the subsurface which was previously defined and studied by Terzaghi and Peck(1948) and Hubbert and Rubey (1959). There are also many other versions of similar techniques developed by other geologists and engineers working in the industry.

Although Hottman and Johnson(1965) begin their paper with theoretical discussions of the subsurface stress and pressure relationship, the paper in fact simply attempted to develop empirical relationships among some well-log-responses and measured abnormal fluid pressure for practi- applications.

In 1975, Magara reexamined the significance of the stress and pressure relationship on the basis of the actually measured subsurface data by Hottman and Johnson. He then concluded that the subsurface stress conditions, which were interpreted from the well-log-responses, are insufficient to fully explain the measured abnormally high fluid pressure in the Gulf Coast region.

Some other causes must thus have played a role in further increasing pressure there. Possibilities are the aquathermal pressuring effect formally proposed by Barker (1972) and hydrocarbon generation effect proposed by many geologists and geochemists including Hedberg(1980).

POROSITY-DEPTH RELATIONSHIP IN HYDRO AND NON-

SHALE POROSITY-DEPTH RELATIONSHIP

Porosity-depth curves of shales have been established in many different parts of the world. Figure 1 shows such curves from ten basins or regions.

The curves of some basin are significantly shifted to left (or lower porosities at a given depth) from others in Figure 1, suggesting possibilities of uplift and surface erosion in the geological past. If a sedimentary sequence was uplifted and its uppermost part was removed by erosion, more compacted sections move upward to shallower depths. Therefore, porosity at a given depth after erosion will become less than that of the non-erosion case.

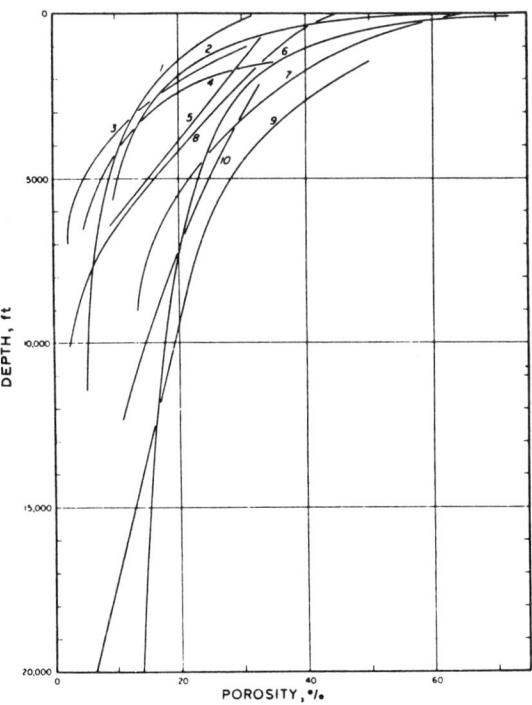

Fig. 1. Relationship between porosity and depth for shales(from Rieke and Chilingarian, 1974): 1=Proshlyakov(1960), 2=Meade(1966), 3=Athy(1930), 4=Hosoi(1963), 5=Hedberg(1936), 6=Dickinson(1953), 7=Magara(1968), 8=Weller(1959), 9=Ham(1966), 10=Foster and Whalen(1966).

POROSITY-DEPTH RELATIONSHIP IN HYDRO AND NON-

Magara(1976) proposed a method of estimating thickness of erosion, using sonic log data (Fig. 2). The shallow "normal compaction trend" is extrapolated to the original sonic transit time value of the sediment at the time of deposition (200 μs/ft in this case), or to the original depositional surface. The difference between the estimated original surface and the present surface is the thickness of the removed section by erosion.

Another interesting feature of Figure 1 is that the rate of porosity decrease is not constant but changes with depth; it is relatively rapid at shallower depths and decreases at greater depths of burial.* Three possible reasons for changing rate of porosity reduction may be considered.

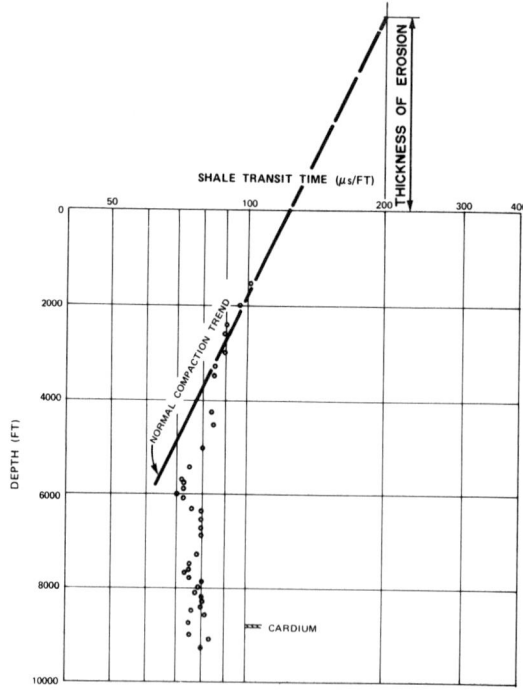

Fig. 2.

Shale transit time -depth plot of a well in Western Canada. Method of estimating erosion thickness is shown.

---

*Some of higher porosity values at depth may be due to the effect of undercompaction and abnormal fluid pressure which will be discussed later.

POROSITY-DEPTH RELATIONSHIP IN HYDRO AND NON-

1. Increasing grain-to-grain contact area with depth

At shallower burial depths, relatively weak grain-to-grain contacts can be expected. The area of contacts between the grains per given cross-sectional area or given volume of shale is relatively small at this stage. If a certain amount of overburden pressure is added at this stage, the shale will compact at a relatively high rate because the pressure applied per given area is relatively large.

At a deeper burial stage, on the contrary, for a given increase of the overburden pressure, the pressure increase applied to a given contact area will be relatively small because of a larger grain-to-grain contact area at that time. The rate of porosity reduction would accordingly be smaller at this later stage.

2. Decreasing permeability with depth

About the shale permeability, a few have been reported. Figure 3 shows a cross plot of the laboratory-measured porosity and permeability values of shales from the U.S.A., Canada, and Japan, and the esimated porosity-permeability relationship (CRETACEOUS SHALE) compiled by Magara(1969). According to this figure, permeability decreases as porosity decreases. Because porosity normally decreases with depth, permeability also will decrease with increasing depth of burial.

The rate of fluid expulsion is mainly controlled by permeability, so that it would be faster at shallower depths than at deeper depths, resulting in faster compaction or porosity reduction at shallower depths than at deeper.

3. Increasing viscosity of water with depth

Low(1976) studied viscosity of water in montmorillonite as shown by the dashed line in Figure 4. The water density values from Martin(1962) are also included in this figure. Because Low's viscosity and Martin's density values are plotted against the g H_2O/g clay and the distance in nm from the clay surface, both of which can be converted into porosity, their relationships can be expressed in terms of changing porosity; both values are relatively high when porosity is low (or g H_2O/g clay and distance from the clay surface are small).

Youn(1974) studied lattice, bound, and free waters in

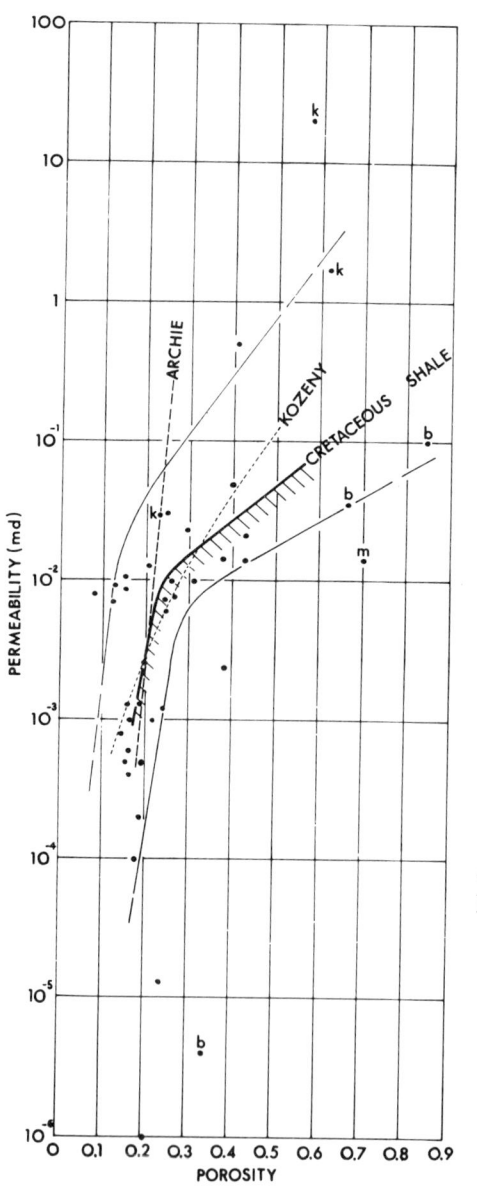

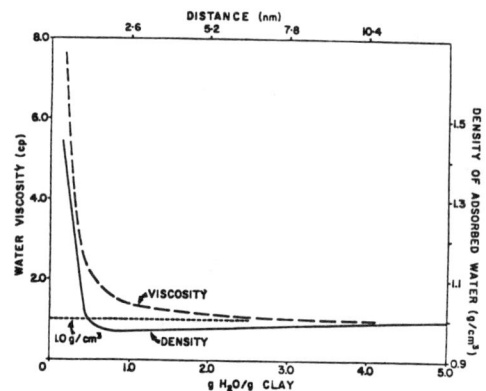

Fig. 4. Density(Martin,1962) and viscosity(Low,1976) of adsorbed or structured water in montmorillonite.

Fig. 3. Comparison of porosity-permeability relationships. Solid circles show data from Canada, U.S.A., and Japan. k=kaolinite, m=montmorillonite, b=bentonite (Magara, 1969).

the Tertiary shales of the Mackenzie Delta basin, Canada. A thick diagonal line in Figure 5 shows his empirical relationship between the shale porosities from side-wall neutron porosity(S.N.P.) and formation density(F.D.C.) logs. Bulk density scale is also shown along with the porosity scale at the bottom of the diagram.

The equal-porosity relationship is shown by a thin diagonal line (EQUAL POROSITY LINE) in Figure 5. When Youn's empirical relation is compared with the equal-porosity line, one can realize that the porosity determined from the neutron log is 8-20 % higher than that from the density log.

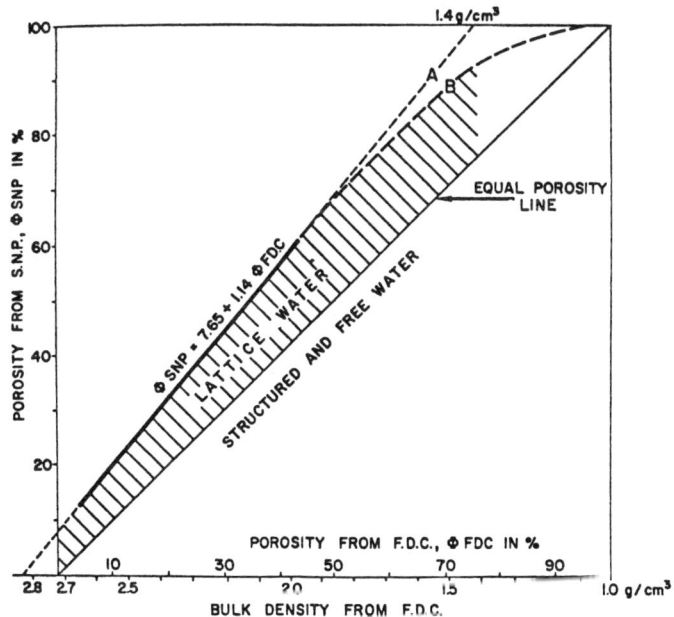

Fig. 5. Porosity cross plot of Beaufort shales, Canada, from sidewall neutron and formation-density logs. Data from Youn(1974).

This difference may be due to the fact that the neutron log reads all the hydrogen atoms as porosity, while in the density log the porosity is calculated on the basis of the matrix or grain density (in this case, 2.72 g/cm^3). Shale matrix or grains contain some lattice water (mainly OH water) which is part of the clay minerals.

It may then be summarized that porosity from the neutron log includes the total of the lattice, structured (or bound), and free waters, but porosity derived from the density log includes the bound and free pore waters only.

In the very early stages of compaction (F.D.C. porosity >50 %), the amount of the lattice water shown by the shaded area is unchanged; compaction is primarily caused by expulsion of relatively free pore water, and therefore is relatively fast.

In the later stages of compaction (F.D.C. porosity < 50 %), the amount of the lattice water decreases with compaction. This suggests that at these later stages chemical and mineralogical changes of shales take place in addition to the mechanical compaction and fluid expulsion, and even some lattice water may be converted into other forms of water (bound or free).

Low's (1976) viscosity and Martin's (1962) density relationships were transferred to the porosity diagram and are shown in Figure 6. The viscosity and density lines in this figure intercept the bottom axis (porosity from F.D.C.) at different porosity values, suggesting that as the shales compact, they must expel more viscous and dense water.

Fluid flow would be slower at greater compaction stages, based on Darcy equation,

$$q = -\frac{k}{\mu}\left(\frac{dp}{dz}\right) \quad \dots\dots\dots\dots\dots\dots\dots\dots\dots\dots(1)$$

where q is volume of fluid moving per unit area per unit time, k permeability, μ viscosity, and (dp/dz) fluid pressure gradient. Equation (1) suggests that the value q decreases as μ increases, other factors being constant. Therefore, the rate of shale compaction would slow down at these later compaction stages.

In summary, the rate of shale porosity reduction slows down due probably to the combined effect of increasing grain-to-grain contact area, decreasing permeability, and

POROSITY-DEPTH RELATIONSHIP IN HYDRO AND NON-

increasing water viscosity with compaction and depth.

SANDSTONE POROSITY-DEPTH RELATIONSHIP

Porosity-depth relationships of sandstones can be approximated by straight lines on normal(arithmetic) graph paper (Proshlyakov, 1960, Maxwell, 1964, and Galloway, 1974). In other words, the rate of porosity reduction is relatively constant from shallower to deeper burial depths. Therefore, the previous explanations made for shales seem not to be applicable for sandstones.

As both Maxwell and Galloway suggested, the effect of temperature on sandstone porosity seems to be quite significant, suggesting that some chemical and diagenetic effects rather than the physical effect primarily controls the porosity of sandstones.

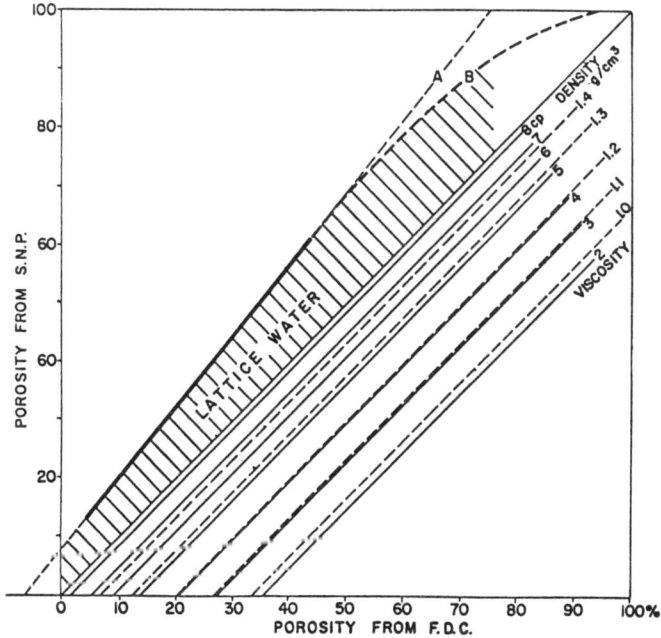

Fig. 6. Porosity cross plot of Beaufort shales, Canada, combines with possible variations of density and viscosity of adsorbed or structured water derived from Martin(1962) and Low(1976).

POROSITY-DEPTH RELATIONSHIP IN HYDRO AND NON-

NORMAL COMPACTION AND UNDERCOMPACTION OF SHALE

Based on Athy's(1930) porosity-depth relationship of the Paleozoic shales in Oklahoma, Hubbert and Rubey(1959) defined following mathematical function,

$$\phi = \phi_0 \cdot e^{-cZ} \quad \ldots\ldots\ldots\ldots\ldots\ldots\ldots\ldots\ldots\ldots(2)$$

where ϕ is shale porosity at depth Z, ϕ_0 original shale porosity at surface (Z=0), e base of the Napierian logarithms, and c constant of dimension (length)$^{-1}$.

This function can be shown as a curved line on normal (arithmetic) graph paper (e.g. curve 3 in Fig.1), and as a straight line called "NORMAL COMPACTION TREND" on semilog paper (porosity - logarithmic scale, and depth - arithmetic scale, see upper part of Fig. 7). The slope of this straight line is related to the value c in equation(2)

As explained earlier, in deeper intervals of relatively young sedimentary basins (usually Tertiary or Cretaceous), the shale porosity values are greater than those equivalent to the normal compaction trend (see lower part of Fig.7). Shales in this deeper interval are called "UNDERCOMPACTED SHALES", which are usually associated with abnormally high fluid pressures.

The stress-pressure relationship in the subsurface can be shown by Terzaghi equation as follows,

$$S = \sigma + p \quad \ldots\ldots\ldots\ldots\ldots\ldots\ldots\ldots\ldots(3)$$

where S is total stress(or overburden pressure), σ effective stress(or grain-to-grain contact pressure), and p pore-fluid pressure. The value σ of shale is depended on its porosity ϕ ; σ increases as ϕ decreases and vice versa (Hubbert and Rubey, 1959).

Therefore, under given overburden pressure, S(or given depth), subnormal σ (abnormal ϕ or undercompacted shale) must be associated with abnormally high fluid pressure, p.

The association of the undercompacted shale and abnormal fluid pressure can also be explained by a burial-compaction model shown in Figure 8. In this model, compaction history of a shale at depth Z is considered. Compaction of this shale in the earlier stages (surface to depth Z_e) was normal, resulting in normal pore-fluid expulsion. Fluid pressure during these stages was hydrostatic.

At depth Z_e, fluid expulsion was arrested completely, so that during subsequent burial to Z there was no further compaction (σ was unchanged). Z_e may be called "ISOLATION DEPTH" at which pore-fluid was isolated from the surrounding rocks.

The increase in pore-pressure during burial from Z_e to Z is equivalent to the increase in overburden pressure. This is because the σ value did not increase when the total overburden pressure increased from Z_e to Z, so that the increased overburden pressure must be carried entirely by the pore-fluid.

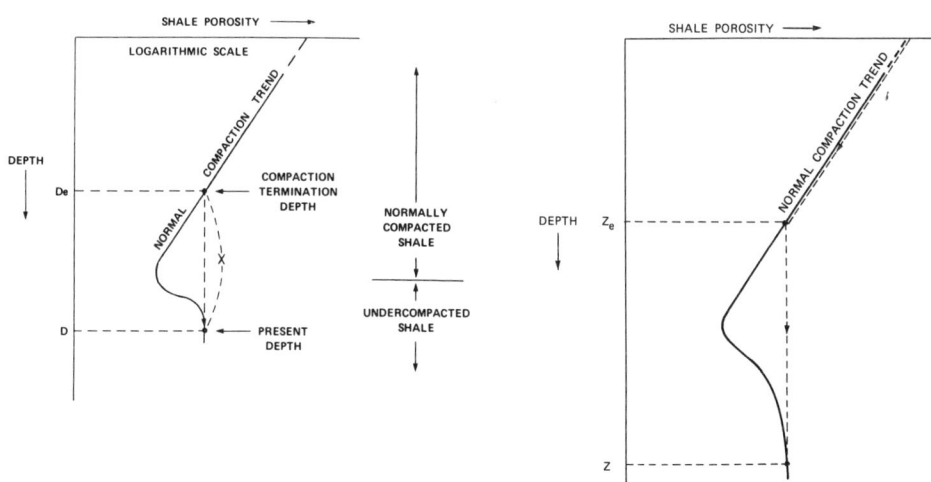

Fig. 7. Schematic shale porosity-depth plot, showing normally compacted and undercompacted shales.

Fig. 8. Schematic shale porosity-depth plot, showing history of shale porosity change. Refer to text for explanation.

The abnormal fluid pressure, p, at depth Z in this model can be given as follows (see also Fig.8),

$$p = \rho_w \cdot g \cdot Z_e + \bar{\rho}_{bw} \cdot g \cdot (Z - Z_e) \quad \ldots\ldots\ldots\ldots (4)$$

where ρ_w is density of formation water, $\bar{\rho}_{bw}$ average bulk density of sedimentary rocks, and g gravity acceleration. This equation means that the abnormal fluid pressure at Z is given as the total of the hydrostatic pressure between the surface and Z_e, and of the overburden pressure between Z_e and Z.

Above equation (4) can also be derived from Terzaghi equation (3). A common factor applicable to the both equations is that temperature is unchanged (or constant). However, in the true subsurface, temperature usually increases with depth based on the geothermal gradient of the area.

In fact, Magara(1975) applied the above concept to the actual subsurface data from the Gulf Coast area, and concluded that fluid pressure calculated from Terzaghi's concept is insufficient to explain the measured fluid pressure there. Therefore, some other causes must be involved in increasing fluid pressure to these high levels.

AQUATHERMAL EFFECT

Figure 9 shows a plot of the depth increase since isolation $(Z - Z_e)$ and the corresponding measured fluid-pressure-increase $(p - p_e)$ from Hottman and Johnson's(1965) Gulf Coast data. If the fluid pressure increase is solely due to the increase of overburden pressure since isolation, as predicted from Terzaghi equation, one would anticipate it to be shown by line A (1 psi/ft - approximate overburden pressure gradient, Fig.9). However, the actually measured fluid-pressure-increases (dots in Fig.9) are plotted above line A, suggesting that some other causes must have contributed to the pressure increases.

There may be two possible causes considered; aquathermal effect proposed by Barker(1972), and hydrocarbon generation effect proposed by many geologists and geochemists including Hedberg(1980). Both effects are closely related with the temperature increase in the subsurface. Magara(1978) further proved that the aquathermal effect is important in the Mackenzie Delta, Canada, as well.

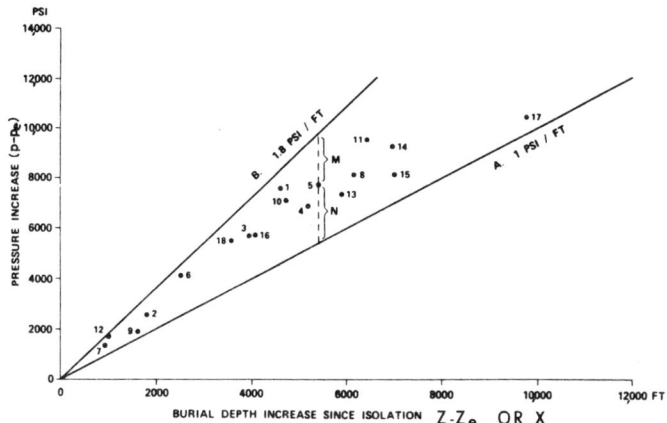

Fig. 9. Relationship between increases of burial depth($Z-Z_e$) and of pressure($p-p_e$) since isolation of pore-fluids. Numbers refer to the wells in Hottman and Johnson's table 1(1965).

In summary, the development of an isolated system, which is indicated by undercompaction, is a necessary condition for abnormal pressuring. However, the degree of the fluid pressure increase is depended upon other factors - aquathermal and hydrocarbon generation effects.

EFFECT OF EROSION

Suppose that the area was uplifted and the uppermost part of the sedimentary sequence was removed by erosion. Deeper part of the sequence had been undercompacted and abnormally pressured.

During and after the erosion, temperature of the remaining sequence would have dropped, reducing the aquathermal effect significantly. Therefore, at the present time, undercompacted shales may be associated with normal or even subnormal fluid pressure, if the erosion was significant. Note that, in Figure 10, porosity of the undercompacted shales of curve B (with erosion) is less than that of the normally compacted shales of curve A (without erosion). The effective stress of the former shales is greater than that of the latter shales, so that pore-pressure of the former can be less than that of the latter.

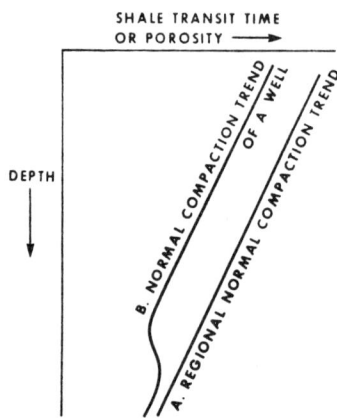

Fig.10. Schematic shale transit-time vs. depth plots. Regional normal compaction trend(A) and plot after erosion(B).

POSSIBLE EFFECT OF SMECTITE-ILLITE CONVERSION

According to Martin's(1962) data(Fig.4), the density of adsorbed water in montmorillonite(or smectite) can be as high as about 1.4 g/cm^3. If this water is released to become free water of 1.0 g/cm^3 density, there would be a significant volume expansion. If such an expansion takes place under isolated or closed conditions, fluid pressure will increase to an abnormally high level, as suggested by Powers(1967).

However, Martin's data also suggest that other water away from the dense water in the clay has density less than 1 g/cm^3 (see Fig.4). When the former water becomes free, the latter water would also become free water, causing possible shrinkage or contraction. Both phenomena occur within the clay, so that one effect may reduce or even cancel another effect, resulting in almost no net effect. If such is a case, there would be virtually no pore-fluid pressure change before and after the conversion.

Another idea regarding the clay mineral conversion is related to so-called collapse of the clay minerals. If this happens, a larger proportion of the overburden pressure may have to be carried by pore-fluid, causing high pressure. However, if this is true, the shale density should increase after the conversion and collapse. Note that abnormal fluid pressure is usually associated with undercompacted shales whose density is subnormal. Therefore, the subsurface data do not support the idea of clay-mineral-collapse.

POROSITY-DEPTH RELATIONSHIP IN HYDRO AND NON-

In any situation, there is no quantitative relationship between the clay mineral conversion and the generated fluid pressure available at present, so that there would be no practical application of this concept in predicting or estimating fluid pressure. Another problem with this concept is that, at the present time, it is virtually impossible to know the amount of the original smectite before the conversion. This further suggests that it is not possible to estimate the amount of conversion, which seems to be essential in this hypothesis.

HEAT FLOW AND GEOTHERMAL GRADIENT IN NORMALLY COMPACTED AND UNDERCOMPACTED INTERVALS

As described above, increasing subsurface temperature with isolation could cause abnormal fluid pressure. It is also known that, in many undercompacted and abnormally pressured zones, the geothermal gradients are also abnormally high (e.g. Gulf Coast, Schmidt, 1973). Such an association can be explained by the concept of conduction heat flow shown by Lewis and Rose(1970) as follows,

$$Q = \left(\frac{k_h}{L}\right) \cdot A \cdot \Delta T \quad \ldots\ldots\ldots\ldots\ldots\ldots\ldots (5)$$

where Q is heat flow, k_h coefficient of thermal conductivity of the substance, L length(thickness) of the substance, A cross-sectional area of the substance, and ΔT temperature difference between both ends of the substance.

If the heat flow through the substance of a unit cross-sectional area is considered, equation (5) can be simplified as follows,

$$q = k_h \cdot \frac{\Delta T}{L} \quad \ldots\ldots\ldots\ldots\ldots\ldots\ldots\ldots (6)$$

where q is heat flow per unit cross sectional area, and $\Delta T/L$ gradient of temperature change over distance (geothermal gradient).

Lewis and Rose(1970) showed the thermal conductivities of four substances as follows,

Water : 0.363 Btu/(h.ft.°F)

Slate : 1.138 Btu/(h.ft.°F)

Granite : 1.210 Btu/(h.ft.°F)

Sandstone : 1.330 Btu/(h.ft.°F)

There are relatively small differences among the thermal conductivities of slate, granite, and sandstone, while the difference between water and other(solid) substances is much greater. Water has only about one-third the thermal conductivity of the rock substances, suggesting an important insulating capacity of water in sedimentary rocks.

Lewis and Rose(1970) mentioned that the coefficient of thermal conductivity of a water-saturated, porous sedimentary rock can be given as,

$$k_h = k_{hma} \cdot \left(\frac{k_{hw}}{k_{hma}}\right)^\phi \quad \ldots \ldots \ldots \ldots \ldots \ldots (7)$$

where k_{hma} is thermal conductivity of the rock grains or matrix, k_{hw} thermal conductivity of water, and ϕ porosity of the sedimentary rock.

Thermal conductivities of two different shales, which consist of rock matrix similar to slate and water, with 10 and 20 % porosity each can be given as,

for 10 % porosity : k_h = 1.015 Btu/(h.ft.°F)

for 20 % porosity : k_h = 0.906 Btu/(h.ft.°F)

If there is a sequence which is composed of two shaly formations of 10 % porosity(normally compacted) and 20 % porosity(undercompacted), respectively, and if heat is flowing vertically upward, we obtain

$$q_{10\%} = q_{20\%} \quad \ldots \ldots \ldots \ldots \ldots \ldots \ldots \ldots \ldots (8)$$

or

$$\left(k_h \cdot \frac{\Delta T}{L}\right)_{10\%} = \left(k_h \cdot \frac{\Delta T}{L}\right)_{20\%}$$

(see equation 6).

Therefore, if the geothermal gradient through the 10 % shale is 1.0 F/100 ft, the gradient through the 20 % shale can be given as,

$$\left(\frac{\Delta T}{L}\right)_{20\%} = \frac{1.015}{0.906} \times 1\,°F/100\ ft = 1.12\,°F/100\,ft$$

The above sample calculation suggests that the geothermal gradient of an undercompacted zone is higher than that of a more normally compacted zone, provided that lithology is constant.

A slight difference in lithology between these two zones may have relatively little effect on geothermal gradient, because of the comparatively small differences in thermal conductivity among the different rock materials already mentioned. However, if there is no significant difference in the level of compaction, lithological variation can be the principal cause of a change in geothermal gradient.

In summary, a high geothermal gradient in a closed and undercompacted zone will further increase the fluid pressure to a very high level by the aquathermal effect as well as the hydrocarbon generation effect.

REFERENCES

Athy, L. F., 1930, Density, porosity and compaction of sedimentary rocks, AAPG, v.14, p.1-24.

Barker, C., 1972, Aquathermal pressuring: role of temperature in development of abnormal-pressure zones, AAPG, v.56, p.2068-2071.

Dickinson, G., 1953, Geological aspects of abnormal reservoir pressures in Gulf Coast, Louisiana, AAPG, v.37, p.410-432.

Foster, J. B., and Whalen, H., 1966, Estimation of formation pressures from electrical surveys - offshore Louisiana, Petrol. Technology Jour., v.18, p.165-171.

Galloway, W. E., 1974, Deposition and diagenetic alteration of sandstone in northeast Pacific arc-related basins; Implications for graywacke genesis, GSA, v.85, p.379-390.

Ham, H. H., 1966, New charts help estimate formation pressures, Oil and Gas Jour., v.64, p.58-63.

Hedberg, H. D., 1936, Gravitational compaction of clays and shales, Am. Jour. Sci., v.31, p.241-287.

Hedberg, H. D., 1980, Methane generation and petroleum migration, In Problems of Petroleum Migration, AAPG Studies in Geology No.10 (Eds. W. H. Roberts, III, and R. J. Cordell), p.179-206.

Hosoi, H., 1963, First migration of petroleum in Akita and Yamagata Prefectures, Japanese Assoc. Mineralogists, Petrologists and Econ. Geologists Jour., v.49, p.43-55, p.101-114.

Hottman, C. E., and Johnson, R. K., 1965, Estimation of formation pressure from log-derived shale properties, Petrol. Technology Jour., v.17, p.717-722.

Hubbert, M. K., and Rubey, W. W., 1959, Role of fluid pressure in mechanics of overthrust faulting, GSA, v.70, p.115-206.

Lewis, C. R., and Rose, R. C., 1970, A theory relating high temperatures and overpressures, Petrol. Technology Jour., v.22, p.11-16.

Low, P. F., 1976, Viscosity of interlayer water in montmorilonite, Soil. Sci. Soc. America Proc., v.40, p.500-505.

Magara, K., 1968, Compaction and migration of fluids in Miocene mudstone, Nagaoka Plain, Japan, AAPG, v.52, p.2466-2501.

Magara, K., 1969, Porosity-permeability relationship of shale, Can. Well Logging Soc. Jour., v.2, p.47-93.

Magara, K., 1975, Importance of aquathermal pressuring effect in Gulf Coast, AAPG, v.59, p.2037-2045.

Magara, K., 1976, Thickness of removed sediments, paleopore pressure, and paleotemperature, southwestern part of Western Canada Basin, AAPG, v.60, p.554-565.

Magara, K., 1978, Compaction and Fluid Migration - Practical Petroleum Geology, Elsevier, Amsterdam, 319p.

Martin, R. T., 1962, Adsorbed water on clay: a review, Clays Clay Miner., 9(Proc. 9th Natl. Conf. Clays Clay Minerals, 1960), Pergamon, New York, N. Y., p.28-270.

Maxwell, J. C., 1964, Influence of depth, temperature, and geologic age on porosity of quartzose sandstone, AAPG, v.48, p.697-709.

Meade, R. H., 1966, Factors influencing the early stages of compaction of clays and sands - review, Sed. Petrology Jour., v.36, p.1085-1101.

Powers, M. C., 1967, Fluid-release mechanisms in compacting marine mudrocks and their importance in oil exploration, AAPG, v.51, p.1240-1254.

Proshlyakov, B. K., 1960, Reservoir properties of rocks as a function of their depth and lithology, Neol. Neft. Gaza, v.12, p.24-29.

Rieke, H. H., III, and Chilingarian, G. V., 1974, Compaction of argillaceous sediments, Elsevier, Amsterdam, 424p.

Schmidt, G. W., 1973, Interstitial water composition and geochemistry of deep Gulf Coast shales and sandstones, AAPG, v.57, p.321-337.

Terzaghi, K., and Peck, R. B., 1948, Soil Mechanics in Engineering Practice, Wiley, New York, N. Y., 566p.

Weller, J. M., 1959, Compaction of sediments, AAPG, v.43, p.273-310.

Youn, S. H., 1974, Comparison of porosity and density values of shales from cores and well logs, M. A. Thesis, Univ. Tulsa, Oklahoma.

N.C. DUTTA[1]

SHALE COMPACTION, BURIAL DIAGENESIS, AND GEOPRESSURES: A DYNAMIC MODEL, SOLUTION AND SOME RESULTS*

ABSTRACT

We present here a model of geopressures based on the compaction of low permeability sediments and show by examples how the density and pore-pressures of sedimentary rocks can be evaluated based upon some key geologic parameters such as depositional history, burial rates, and geothermal gradients. This model differs from earlier works in several aspects: (1) it is applicable to sediments with "variable compaction coefficient;" (2) it does not assume that the viscosity (η), and the effective stress (σ) are independent of temperature (T); (3) it allows for arbitrary initial conditions, boundary conditions, and loading processes in convenient ways; and (4) moving boundaries are dealt with using a Lagrangian technique. Numerical solution of this model is carried out by a finite-difference algorithm using the implicit difference technique. The nonlinearity of the partial differential equation is handled by a fast and accurate iteration scheme. We present several numerical examples of time histories of the distribution with depth of bulk density, fluid pressure gradient (FPG), and effective stress of compacting sand-shale sequences. These examples are chosen to simulate approximately the depositional history and environment of the Gulf Coast basin to show how the geopressure profiles vary with (1) sediment depositional sequence and thickness, (2) sedimentation rate, (3) geothermal gradient, and (4) permeability-porosity relation for shales.

---

* Based on a paper presented by the authors during the 53rd Annual International Meeting of SEG, September 1983, in Las Vegas.

(1) *ARCO Resources Technology, Plano, Texas, USA.*

The diagenetic transformation of smectite to illite (S/I) contributes to geopressuring due to release of bound water in the pore system. In this paper, we briefly discuss how time and temperature affects smectite diagenesis in the mixed-layer clay, assuming that the composition of smectite remains unchanged throughout the burial history. We propose a kinetic model for the S/I reaction based on a first order rate theory, determine the activation energy of the reaction ($\approx 19.3^{\pm}.7$ kcal/mole) by using X-ray derived data from the published literature on smectite diagenesis from several Gulf Coast wells, and test the model predictions. We find that the model predicts the correct shape and extent of the clay reaction as observed in several wells from the Texas-Louisiana Gulf Coast area. The model includes the temperature-transients due to rapid burial of sediments which have variable thermal properties. The model has a predictive capability; it shows that for younger sediments, the midpoint of the S/I diagenesis is at depths much deeper than those for older sediments. Guided by the kinetic model of smectite diagenesis, we have re-evaluated the time evolution of geopressure. We have found that S/I transformation contributes significantly to the fluid pressure buildup in the geopressured environment, but that by itself alone, cannot account for the observed magnitude of pore pressure in overpressured sediments.

A significant conclusion that has emerged from our study is that higher geothermal gradients lead to significantly higher pore pressures, greater undercompaction of sediments and very low effective stress (≤ 150 psi) for a range of geologically significant burial rates. The effective stress in the shales is further lowered by smectite diagenesis in the depth range of hydrocarbon exploration in geopressured environment of the Texas-Louisiana Gulf Coast area.

The model of shale compaction presented here is applicable in relatively simple depositional environments (e.g., absence of faulting and multidimensional flow). Nonetheless, it provides an interpretive tool for velocity and density analysis in geopressured clastic sediments. From the present study, we have been able to identify several key areas where further experimental and modeling effort should be concentrated. These are: (1) shale compaction data in high temperature, low effective stress region; (2) permeability-porosity data for shales; (3) a comprehensive treatment of chemical diagenesis of shales and adjacent sands, and (4) a deterministic model of basin building which includes temperature evolution along with fluid pressure buildup in a coupled manner.

SHALE COMPACTION, BURIAL DIAGENESIS, AND GEOPRESSURES

I. INTRODUCTION

Detection and quantitative evaluation of prospects in areas of abnormally high pore fluid pressures (geopressures) are critical to exploration, drilling and production operations for hydrocarbon resources. These higher than normal pore pressures are encountered worldwide in formations ranging in age from the Cenozoic era (Pleistocene age) to as old as the Paleozoic era (Cambrian age). Several, and often a multitude, of superimposed factors may be the cause of high pore fluid pressures in porous rocks in excess of hydrostatic pressure. The phenomena are related to geologic, physical, geochemical, and mechanical processes. Seismic interpretation in geopressured areas (such as in the clastic basin of the Texas-Louisiana Gulf Coast) requires an understanding of how density and velocity of geopressured rocks vary with such key geological parameters as sedimentation rates, geothermal gradient, sediment deposition sequence and thickness, etc. With this objective, we developed a model of the compaction of fine-grained, low-permeability sediments (shales) and its relation to the evolution of geopressures. This model is an extension of unpublished work of the late R. L. Chuoke of Shell Development Co., and incorporates effects of temperature on shale compaction.

In the published literature, there are a number of mechanisms which produce excess fluid pressures. Some important ones are: (1) continuous loading and compaction of fine-grained, low permeability sediments (Dickinson, 1953; Bredehoeft and Hanshaw, 1968); (2) clay mineral transformation, such as smectite to illite conversion (Powers, 1967; Hower et al., 1976); (3) local tectonic compression accompanying structural growth (Hubbert and Rubey, 1959); (4) fluid volume expansion or contraction caused by temperature changes (Levorson, 1954; Barker, 1972); and (5) hydrocarbon generation (Meissner, 1978; Timko and Fertl, 1971). It is our opinion that no single mechanism alone can explain the observed geopressuring in basins throughout the world and that in most cases some of these mechanisms act simultaneously. Nonetheless, two mechanisms of geopressure have received most attention in the literature. These are rapid burial of thick units of normally pressured shales and burial diagenesis of clay, especially the conversion of smectite to illite at elevated temperatures and pressures and subsequent release of bound water. In this paper, we suggest that the dominant cause of geopressuring in the Gulf Coast environment is shale compaction, followed by the burial diagenesis (smectite dehydration) of shales. The first part of this paper deals with a physical and quantitative model of shale compaction in a basin which is undergoing a nonsteady-state subsidence due to fluid flow and active sedimentation. We show by examples how the density and pore pressures of sedimentary rocks can be evaluated based upon some key geologic parameters such as depositional history, burial rates, and geothermal gradients. This model differs from all earlier works in several aspects: (1) it is applicable to sediments with "variable compaction coefficients;" (2) it does not assume that the viscosity (η), and the effective stress (σ) are independent of temperature (T); (3) it allows for arbitrary initial conditions, boundary conditions, and loading processes in convenient ways; and (4) moving boundaries are dealt with using a Lagrangian technique. Numerical solutions of this model are carried out by a finite-difference algorithm using the implicit difference technique. The quantitative results depend critically on the shale compaction and permeability data. We present several numerical examples of time histories of the distribution with depth of bulk density, fluid pressure gradient (FPG), and effective stress of compacting sand-shale sequences. These examples are chosen to simulate approximately the depositional history and environment of the Gulf coast basin to show how the geopressure profiles vary with (1) sediment depositional sequence and thickness, (2) sedimentation rate, (3) geothermal gradient, and (4) permeability-porosity relation for shales.

In the second part of this paper we discuss the well known diagenetic transformation of smectite (swelling clay) to illite (sedimentary mica) through an intermediate mixed-layer smectite-illite (S/I) clay and study its effect on geopressuring in the Gulf Coast Tertiary Province. We model the S/I transformation in Gulf Coast shales by using the chemical kinetics of first order reactions, determine the activation energy of the reaction ($\approx 19.3 \pm .7$) kcal/mole) by using X-ray derived data on shale cuttings from the published literature, and test the model predictions. We find that the model predicts the correct shape and the extent of the clay reaction as observed in several wells from the Texas-Louisiana Gulf Coast area. The model has a predictive capability; it shows that for younger sediments (Plio-Pleistocene age), the midpoint of the S/I diagenesis is at depths much deeper than those for older sediments (Eocene and older). Further, the reaction rate depends strongly on the temperature, the main phase of the reaction taking place in a temperature window of $\approx 70-100°C$. The main source of uncertainty in our model prediction is the lack of precise knowledge of the true formation temperature. The kinetic model of shale burial diagenesis and the attendant dehydration mechanism is incorporated in our model of geopressure due to shale compaction. We find that smectite dehydration contributes significantly to the buildup of geopressures, but this mechanism alone cannot explain the observed anomalously high fluid pressures in geopressured environments of the Gulf Coast province.

SHALE COMPACTION, BURIAL DIAGENESIS, AND GEOPRESSURES

II. SHALE COMPACTION MODELS AND QUANTITATIVE RESULTS

A. Physical Models

In Figure 1a, we show the physical model for the shale-on-shale compaction. The model considers sediments continuously accumulating in a fluid environment on an impermeable base which subsides over geologic time. The depositional environment is assumed to be a level surface and its areal extent is always much greater than its thickness. In the beginning of the basin creation (time, t=0), we assume that there is a thin layer of water-saturated shales (mud) on the top of an impermeable base. In the final analysis, the thickness of this "initial layer" is insignificant. Beginning at time zero, shale layers are deposited on the top of this initial layer at a given rate, always assuming that the layers remain horizontal. The physical characteristics of the freshly deposited shale (mud) are the same throughout the area but may vary during the depositional history. As the unit becomes thick, the pore volume reduces and simultaneously water is expelled from the deeply buried shale. Since the horizontal extent of the layers is much larger than the vertical extent, we may assume that the expelled water goes upward to the surface. Furthermore, the depositional environment is such that the shale does not extrude laterally and is not subjected to lateral compression. Hence the shale matrix material is also confined to vertical motion. If the pore water in the shale can escape freely, the rate of compaction is rapid and a hydrostatic pressure condition prevails throughout the growing stratigraphic column. However, if the escape of the pore water is hindered by the low permeability of the sediment, the rate of compaction becomes slow. The shale becomes undercompacted and overpressured.

The mechanism just described is dynamic and inherently nonlinear since the time evolution of the fundamental field variable, porosity, depends on the permeability and effective stress which in turn depend on the porosity. From the porosity distribution at a given time, one can derive other physical quantities as a function of depth in the shale column, i.e., fluid pressure, bulk density, fluid flux, etc.

The porosity is taken to depend only on the effective stress and the temperature, although it is straightforward to include other effects such as age, lithology, and time history, provided that one has reliable data pertaining to these quantities. For water we include the dependence of its viscosity on temperature but neglect the small variation of its density with temperature, pressure and salinity and changes of viscosity with salinity. We also

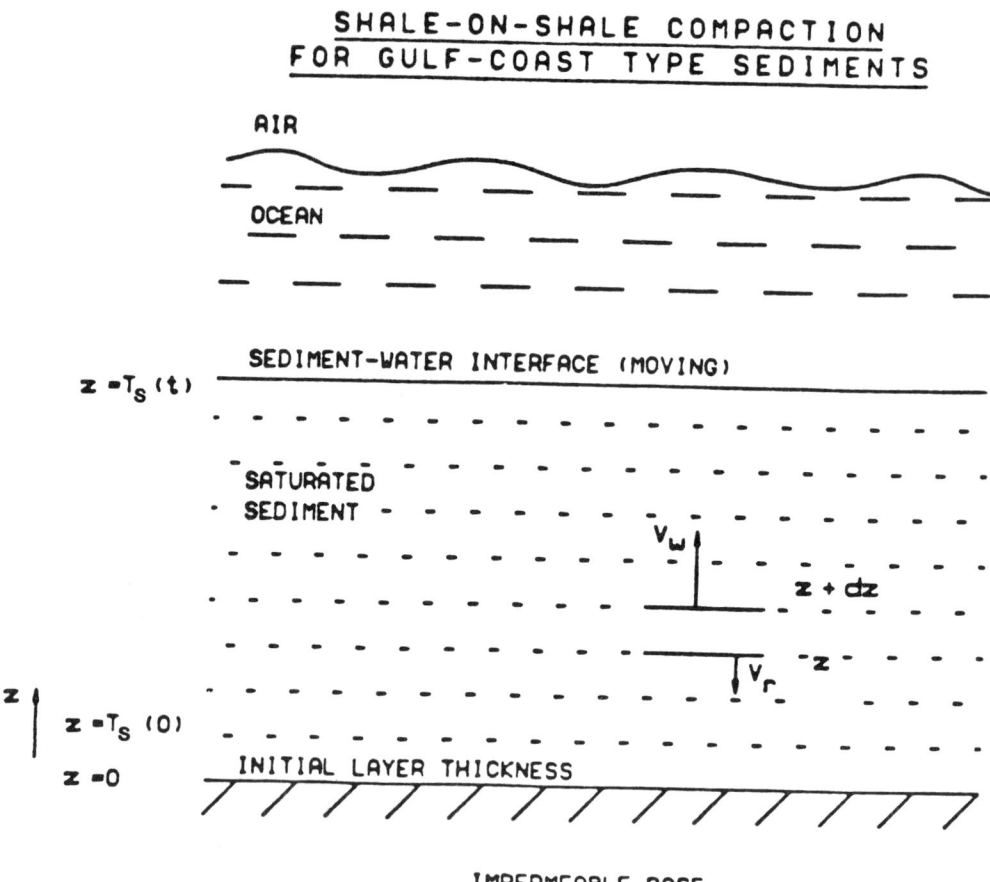

FIG. 1A. The physical model of the geopressure buildup due to gravitational compaction of shale-on-shale. Water-saturated sediment of constant porosity is continuously being deposited at a uniform rate at the sediment-water interfaces, $T_s(t)$, which is an upper moving interface. The z-coordinate is fixed at the top of the impermeable base. Velocity of the rock is v_r whereas v_w is the velocity of the pore water with respect to the fixed coordinate system.

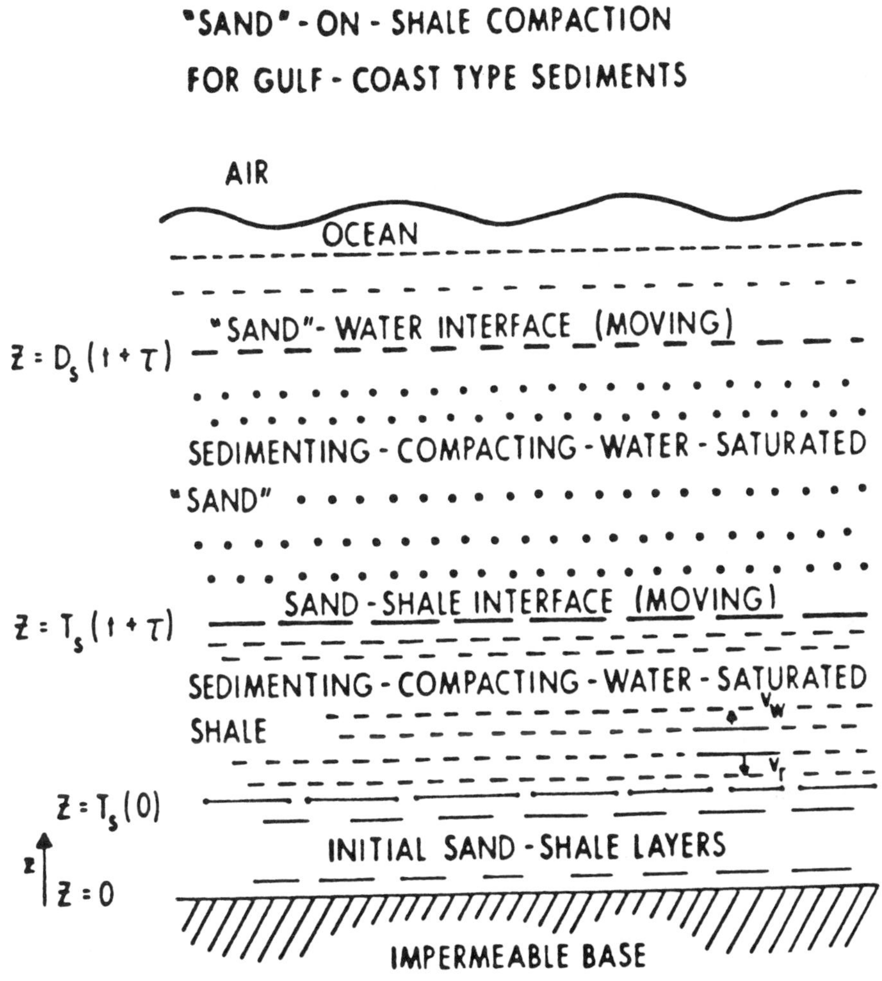

"SAND"-ON-SHALE COMPACTION
FOR GULF-COAST TYPE SEDIMENTS

```
                        AIR
            ~~~~~~~~~~~~~~~~~~~~~~~~~
                      OCEAN
            - - - - - - - - - - - - - -
            - - - - - - - - - - - - - -
Z = D_s(t+τ) ─── "SAND"- WATER INTERFACE (MOVING) ─ ─ ─
            . . . . . . . . . . . . . . . . . . . . .
            SEDIMENTING - COMPACTING - WATER - SATURATED
            "SAND" . . . . . . . . . . . . . . . . . .
            . . . . . . . . . . . . . . . . . . . . .
Z = T_s(t+τ) ──── SAND - SHALE INTERFACE (MOVING) ────
            ─ ─ ─ ─ ─ ─ ─ ─ ─ ─ ─ ─ ─ ─ ─ ─ ─ ─ ─ ─ ─
            SEDIMENTING - COMPACTING - WATER - SATURATED
            SHALE          ─ ─ ─ ─ ─ ─ · v_w ─ ─
            ─ ─ ─ ─ ─ ─ ─ ─ ─ ─ ─ ─ ─ ↓ v_r ─ ─
Z = T_s(0)  ━━━━━━━━━━━━━━━━━━━━━━━━━━━━━━━━━━━━━━━
                   INITIAL SAND - SHALE LAYERS
  ↑z
Z = 0       ////////////////////////////////////////
                      IMPERMEABLE BASE
```

FIG. 1b. The physical model of the geopressure buildup due to gravitational compaction of sand-on-shale sequence. Water-saturated shale of constant porosity, φ_s, is continuously being deposited at a uniform rate Γ on an impermeable base. The z-coordinate is fixed at the top of the impermeable base. At the end of time, t, water-saturated "sand" of constant porosity, φ_s, is deposited at a uniform rate, γ, on top of the shale. The moving coordinates of the sand-shale interface and the sand-water interface at the end of an additional time τ are denoted by $T_s(t + \tau)$ and $D_s(t + \tau)$, respectively. The velocity of the rock is v_r whereas v_w is the velocity of the pore water with respect to fixed coordinate system.

ignore the thermal expansion and finite compressibility of water and the rock grains, since they are small compared to the changes in the bulk volume of the shale due to the decrease in porosity and expulsion of water with increasing stress on the rock matrix.

Having located the top of the growing shale column after shale deposition for a time interval, t, we switch to the second model: the "sand-on-shale" deposition. This model is shown in Figure 1b. In this model, we start depositing normally pressured sand-shale layers (to be referred to as "sands") on top of the freshly sedimented overpressured shale. By "sand," we mean any saturated sediment which is infinitely more permeable than the "shale" underneath. Consequently, as the water expelled from the shales moves through the sand (which is assumed to be connected everywhere) we do not allow any backpressure to build up in the sands near the sand-shale interface and the sands always remain hydropressured. However, sands do get normally compacted due to gravity and therefore one has to locate moving sand-water interface at any given time. This is achieved in this model by employing the same Lagrangian formulation of the problem as for the shale-on-shale compaction problem (for example, see Ozisik, 1968). Further, for the "sand" we use the same compaction model as for shale and thus at the sand-shale interface porosity is continuous. The fluid pressure at that interface is also continuous. However, the fluid pressure in the "sand" is given by the hydrostatic pressure, whereas the fluid pressure in the shale column below the interface will vary in accordance with the particular details of initial configuration of fluid pressure in the shale at the time of commencement of sand deposition and the particular loading process for the sand.

In a coordinate system fixed at the base, the governing compaction equation relating porosity (ϕ), height (z), and time (t) is,

$$\frac{\partial}{\partial z}[q(1-\phi)] = \frac{\partial \phi}{\partial t} , \qquad (1)$$

where q is Darcy's flow velocity defined as

$$q = \frac{\kappa}{\eta}[g(\rho_r - \rho_w) + \frac{d\sigma}{dz}] \qquad (2)$$

Here κ is the permeability, η is the viscosity, ρ_r and ρ_w are grain and water densities, assumed constant, and σ is the "effective stress," defined as the difference between the overburden stress and the fluid stress. This equation is dynamic and nonlinear, and involves moving boundaries. We developed a finite-difference algorithm (using implicit differences) in which nonlinearity is handled by an iterative scheme and moving boundaries are dealt with by a Lagrangian technique. Numerical solutions are obtained for the two models just described: (1) shale-on-shale compaction, (b) sand-on-shale compaction. In both cases, we can prescribe arbitrary initial and boundary conditions, loading processes, and a set of constitutive relations. The latter are relations among (1) σ, temperature (T) and void ratio (ε); (2) viscosity (η) and T; and (3) permeability (κ) and ε.

Our model differs from all published models (Bredehoeft and Hanshaw, 1968; Sharp and Kortenhof, 1982; Bishop, 1979; Smith et al., 1979) in an important respect; it is applicable to sediments with <u>variable</u> and <u>temperature dependent</u> compaction and flow properties. From our studies, we find that temperature plays an important role on fluid pressure profiles in geopressured formations. Temperature dependence of the void ratio is such as to produce excess water from compacting shales and its flow must be accommodated by available permeability of the sediment. Consequently, we do <u>not</u> employ the common assumption of constant compaction coefficient in our solution of the geopressure model.

B. Results

We present several examples chosen to simulate approximately the depositional history and environment of the Gulf Coast basin to show how the magnitude and profiles of bulk density (ρ_b in g/cm^3), fluid pressure gradient (FPG in psi/ft), and effective stress (σ in psi) vary with (1) sediment depositional sequence and thickness, (2) sedimentation rate, (3) geothermal gradient, and (4) permeability of shales. A typical example is shown in Figure 2 for the sand-on-shale compaction model. Here, we start with approximately 5000 ft of normally pressured shale on an impermeable base and deposit water-saturated shale of ≈57 percent porosity on its top at the rate of 1 ft of solid (zero porosity) shale/1000 years. The geothermal gradient is taken to be 1.3° F/100 ft. At the end of the shale deposition period (10 million years), when the shale has already been geopressured, we start depositing "sand" at the same rate. The resulting time histories of ρ_b, FPG, and σ are shown, respectively, in Figures 2a, 2b, and 2c, as functions of depth after 5, 10, and 15 million years of sand deposition. From these figures, we observe a transition zone in the geopressured shale below the hydropressured sand. This transition zone is characterized by (1) a reduction in shale bulk density with depth, (2) a rapid buildup of FPG with depth, (3) a sharp drop in σ with depth, and (4) development of a permeability barrier. At the onset of sand deposition, the parameter changes within the transition zone are gradual. However, as the shale is buried deeper, these changes become sharp. The width of this zone and the shape of parameter variation within it are dependent on factors such as the

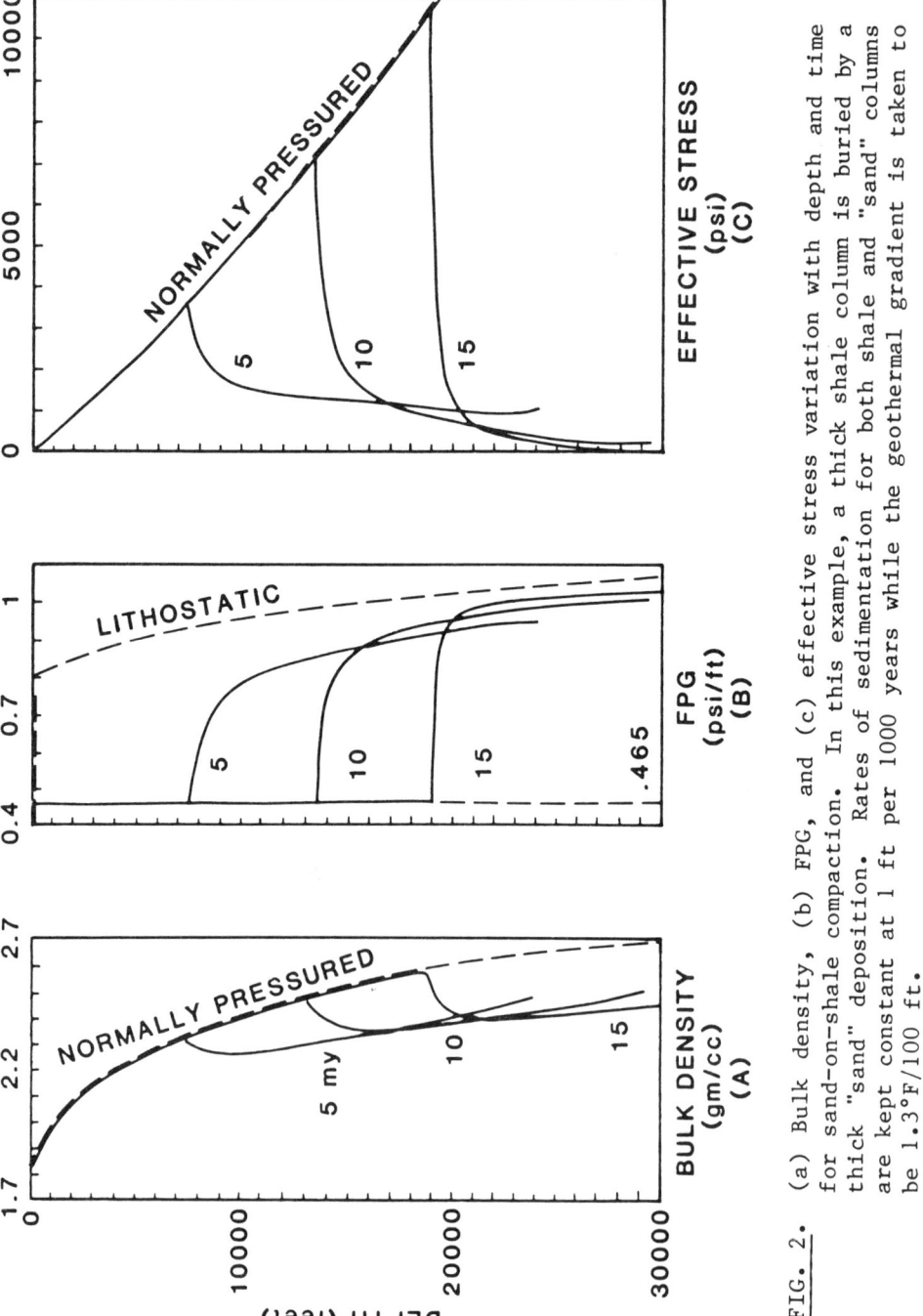

FIG. 2. (a) Bulk density, (b) FPG, and (c) effective stress variation with depth and time for sand-on-shale compaction. In this example, a thick shale column is buried by a thick "sand" deposition. Rates of sedimentation for both shale and "sand" columns are kept constant at 1 ft per 1000 years while the geothermal gradient is taken to be 1.3°F/100 ft.

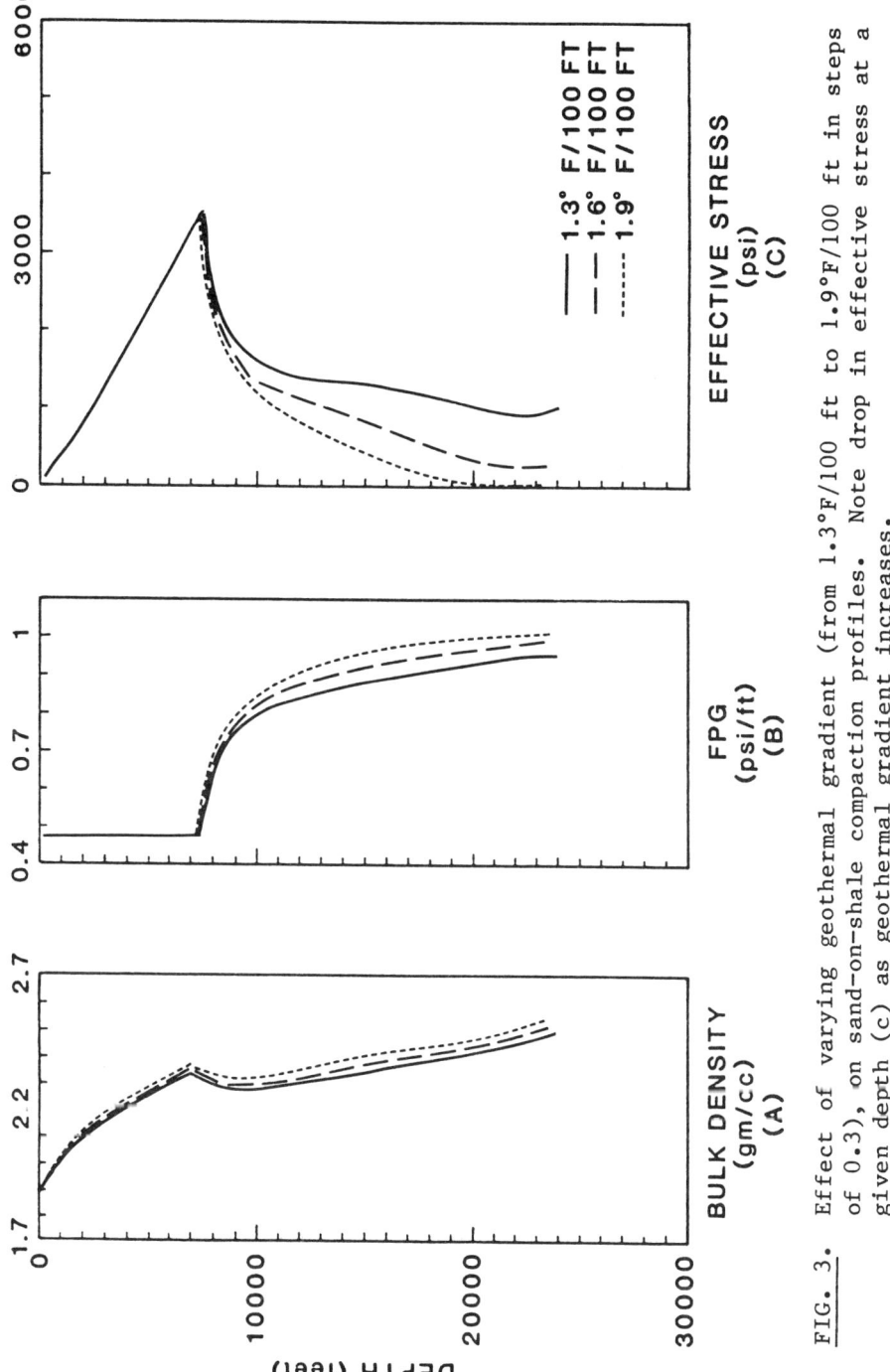

FIG. 3. Effect of varying geothermal gradient (from 1.3°F/100 ft to 1.9°F/100 ft in steps of 0.3), on sand-on-shale compaction profiles. Note drop in effective stress at a given depth (c) as geothermal gradient increases.

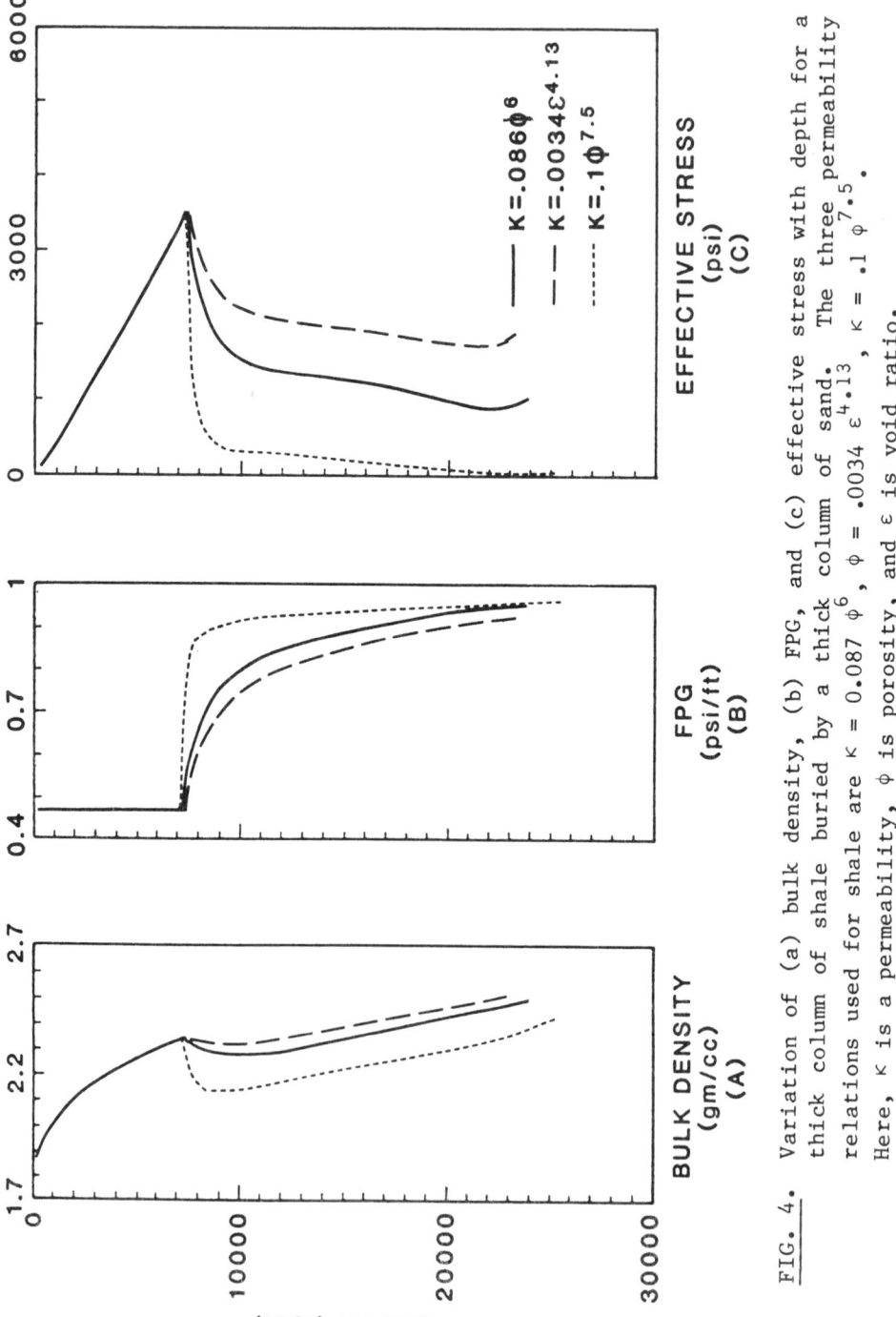

FIG. 4. Variation of (a) bulk density, (b) FPG, and (c) effective stress with depth for a thick column of shale buried by a thick column of sand. The three permeability relations used for shale are $\kappa = 0.087 \phi^6$, $\phi = .0034 \varepsilon^{4.13}$, $\kappa = .1 \phi^{7.5}$. Here, κ is a permeability, ϕ is porosity, and ε is void ratio.

permeability, geothermal gradient and the past pressure history of the buried shale. In the Gulf Coast region, the observed width of transition zones between hydropressured and geopressured sequences typically ranges between 1000 - 3000 ft. All of the features shown in these figures are known to occur in sedimentary basins.

It is possible to obtain more than one transition zone in a stratigraphic column. In that case one has several bands of sands and shales, each with its characteristic high or low permeability. The present model can be applied to include such complex depositional sequences by some simple modifications of the boundary conditions at sand-shale interfaces.

The geothermal gradient plays an important role in the pore pressure evolution in shales because the shale void ratio decreases as temperature increases at a constant effective stress. In Figure 3 we show the effect of varying geothermal gradient (from 1.3° F/100 ft to 1.9° F/100 ft in steps of 0.3) on sand-on-shale compaction profiles, keeping other parameters the same as in Figure 2. We find that the magnitude of geopressuring at a given depth increases as temperature rises at that depth, thus causing a large drop in σ. At a depth of 20 000 ft and a geothermal gradient of 1.9 °F/100 ft, the effective stress is below 100 psi and the fluid pressure gradient is close to 1 psi/ft. Rocks subjected to such low effective stresses can be speculated to sustain open fractures (at these depths) due to any lateral stress variation and hence can provide a pathway for hydrocarbon migration.

Our results are very sensitive to the permeability model of shales. Lacking reliable data on κ versus ϕ of shales, we constructed three analytical models (see Figure 4) and examined the resulting geopressure profiles. The results are shown in Figure 4 for the same remaining parameters as those employed in Figure 2. The wide variability of geopressuring in these three permeability models clearly demonstrates the importance of having reliable shale permeability data, if we are to make a quantitative use of any compaction model. Such data are urgently needed.

III. SHALE DIAGENESIS AND GEOPRESSURE

A. Smectite to Illite Reaction

Among the mineral reactions that take place during burial of sediments in the Gulf Coast Tertiary Province, the diagenetic transformation of smectite (swelling clay) to illite (sedimentary mica) through an intermediate mixed-layer smectite-illite (S/I) clay is probably the most important reaction. Our knowledge of the nature of this clay mineral reaction has mainly come from the pioneering work of Hower and his colleagues (Hower et al., 1976). These workers documented, by examining changes in several Gulf Coast wells, that the main compositional changes in S/I burial diagenetic sequence are: (1) increase of illite layers in the mixed-layer clay, (2) gain of interlayer potassium, (3) increasing substituting of aluminum for silicon in the tetrahedral layer, (4) reduction of octahedral iron from ferric to ferrous, and (5) release of Mg^{++}, Fe^{++}, Ca^{++}, Si^{++++}, and Na^+, and water. Hower's (1981) description of the overall S/I reaction is as follows:

Smectite + K-feldspar+ Mica -> Smectite/Illite + chlorite +
quartz + water.
(Swelling) (Mixed-layer S/I)

It should be noted that the exact nature of the S/I reaction is not well understood and that there are other possible reactions, all of which lead to transformation of smectite to illite (Boles and Franks, 1979; Foster, 1981). In the present work, no attempt is made to establish the exact nature of the reaction. The main focus of the work has been to model the overall chemical reaction and estimate the effect of the released bound water on the fluid pressure gradients in shales under subsurface conditions. We model the S/I reaction as proposed by Hower (1981) using the first order kinetic theory which includes the effect of both time and temperature, determine the activation energy of the reaction by using X-ray derived data on smectite diagenesis from several Gulf Coast wells, test the model predictions and incorporate it in the shale compaction model derived earlier to find the role it plays in affecting the magnitude of geopressure. We stress that a more thorough understanding of this important geological phenomenon than has been presented here is crucial, not only because this reaction contributes to geopressure but it also affects mechanical properties of shales and sands adjacent to shales. The latter is achieved by sandstone cementation via mass transfer in sand-shale sequences (Boles and Franks, 1979; Freed, 1982) in geopressured reservoirs.

Some of the factors that may control the S/I reaction extent are: (1) temperature, (2) time, (3) pressure, (4) pore water chemistry, and (5) composition of starting smectite. The first two factors are dominant one. The remaining factors are purely speculative at this time because we have no data available to either support or discard them. We must, however, point out the importance of the last factor, namely the effect of variable starting material of shale on the diagenetic reaction. Both Foster (1981) and Bruce (1983) have raised this point in their work, although no direct evidence was presented in their publications. Nonetheless, the possibility remains that these authors are correct in their speculations, in which case one must sort out the influence of inherited detrital mineralogy from diagenetically formed minerals.

B. Kinetic Theory of S/I Reaction and Comparison with the X-ray Data From Gulf Coast Wells

We assume, following Hower et al. (1976), Hower (1981), and Boles and Franks (1979), that whatever the detailed nature of the S/I reaction, the reaction proceeds because of the availability of potassium (either from decomposition of K-feldspar or discrete illite). We further assume that the reaction is first order, it is controlled by an activation energy, E, which is constant, and a frequency factor, A, consistent with E, and the rate of water release is controlled by the rate of S/I reaction. As a result of these assumptions, one can show that the fraction of smectite, $N(t)$, as a given time, t, is given by,

$$N(t) = N(o) \exp \left[-\int_0^t A \exp \left(-E/RT(t) \right) dt \right], \tag{3}$$

where $N(o)$ is the initial fraction of smectite and $T(t)$ indicates a specific temperature (T) - time (t) burial history and R is the universal gas constant.

Paleo-temperature reconstruction of a given stratigraphic column is an essential step in calculating the extent of the diagenetic reaction using Equation (3). The reconstruction of the subsurface temperature of sedimentary basins is the subject of this symposium. In our work, we have estimated the paleo-subsurface temperatures by numerically solving the nonlinear, one-dimensional heat diffusion equation with appropriate initial and boundary conditions. The model uses lithology information of a given stratigraphic column from the available suite of wireline logs and burial history records obtained from paleontologic data. The program includes a complete treatment of temperature transients due to the deposition and erosion of sediments at the surface. It also accounts for the variable depth dependent geothermal gradients due to lithologies with varying thermal properties and compaction states. Thus, the thermal history reconstruction model used here does not assume that the overall present day temperature gradient is also applicable to past periods.

We have determined the two parameters, A and E of Equation (3) by using the observed S/I diagenetic depth profiles from three calibration wells: DOE Well No. 1 in Brazoria County, Texas (Freed, 1981); Chevron No. 1 Farwell in West Baton Rouge,

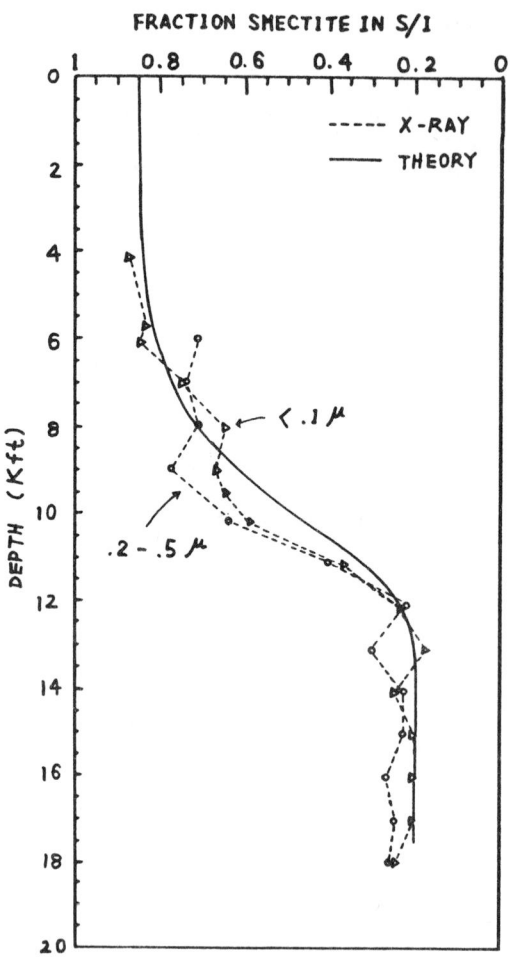

FIG. 5. X-ray derived smectite proportions in mixed-layer S/I clay in CWRU No. 6 Well, Harris County, Texas, as reported by Hower et al. (1976) for two clay-size fractions. The solid line is the predicted S/I diagenetic-depth profile from the present theory.

Louisiana (Hower, 1981); and Pan American No. A5, Manchester Field. LA (Schmidt, 1973). The procedure used was a nonlinear least-square parameter optimization technique using the calculated time-temperature histories of these wells as described above. Our current estimate for these parameters are,

$$A = .4 \times 10^5 \text{ (yr)}^{-1},$$
$$E = 19.3 \pm .7 \text{ kcal/mole}.$$

In Figures 5 through 7, we show the comparison between the predicted S/I diagenetic-depth profile using the parameters given above and the observed profiles from X-ray data for a number of Gulf Coast wells which are reported in the literature. These are CWRU No. 6 (Hower et al., 1976); Gulf Oil Corporation No. 2 Texas State Lease 53034 (Freed, 1982); and Well-C, Offshore Louisiana (Perry, 1969). The agreement between the kinetic model and the data is good. The model predicts the onset of the reaction and its overall extent fairly consistently. The major uncertainty in the model prediction is the lack of knowledge of true equilibrium formation temperatures. We found that the readings from the maximum logging temperatures cannot be used to predict the rate of the S/I reaction reliably. Another major source of uncertainty is the lack of precise knowledge of age-depth relations. The effect of compaction was not found to be too significant.

Since the present model is essentially kinetic, we expect to have a significant effect of reaction time (due to geologic burial rates) on the diagenetic-depth profiles. From the model we expect that in "older" rocks, which have had a much longer time at a given temperature, the diagenetic reaction would proceed further. By the same token, the opposite is true for "younger" rocks. Perry (1969) observed this effect from two wells in the Gulf Coast. This effect is also evident when we compare the profiles for Well-C (Plio-Miocene age, Figure 7) and CWRU No. 6 (Oligocene age, Figure 6). The predicted profiles are in overall agreement with the X-ray data. We believe that these example plus other model studies not discussed here show a clear evidence of the kinetic nature of the shale diagenesis.

C. Shale Diagenesis and Geopressure

We have incorporated the kinetic model of S/I reaction in the theory of shale compaction presented earlier, and evaluated the resulting effect on the geopressure profiles. Desorption isotherms by Van Olphen (1963) on smectite show that it retains two interlayers of water at effective stresses and temperatures well beyond those where illitization occurs in the subsurface. Perry and Hower (1972) showed that if smectite retains these two interlayers

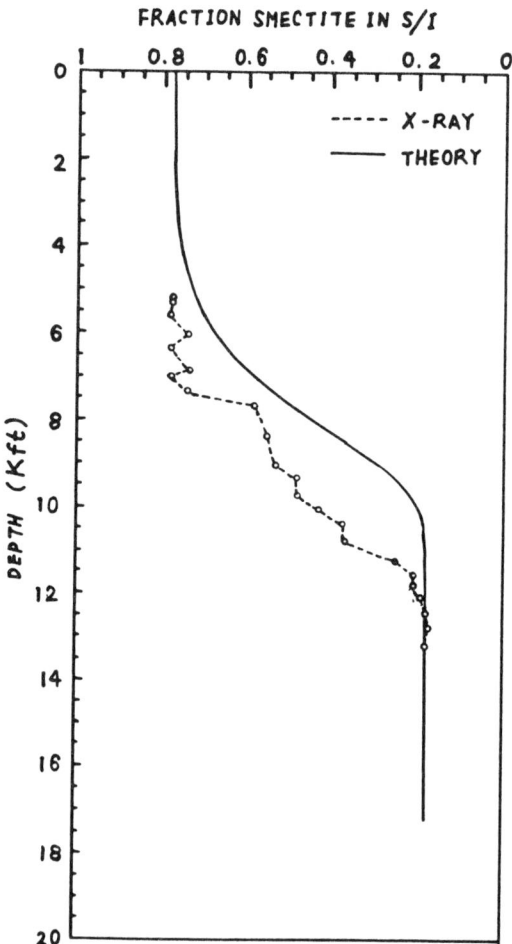

FIG. 6. Estimated smectite proportions within mixed-layer smectite-illite system for Gulf Oil Corporation, No. 2 Texas State Lease 53034, Brazoria County, Texas, as reported by Freed (1981). The solid line is the predicted S/I diagenetic-depth profile from the present theory.

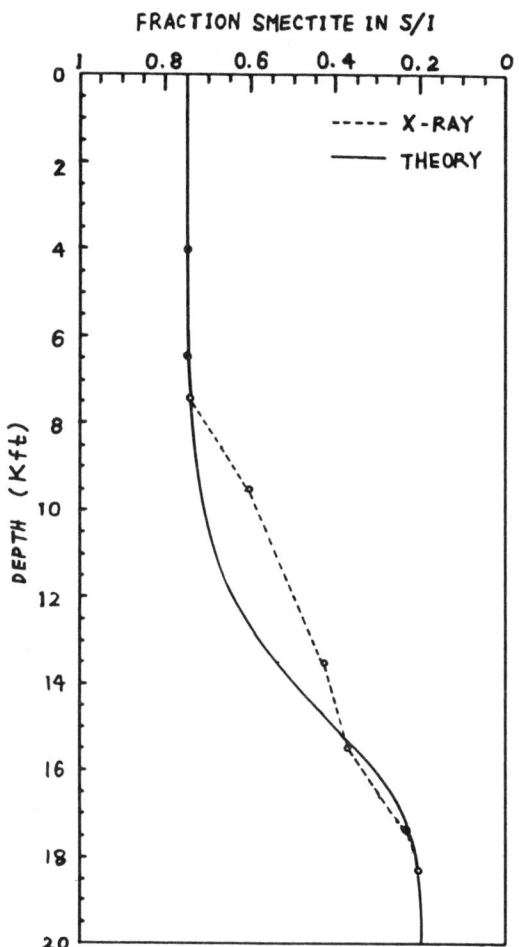

FIG. 7. X-ray derived smectite proportions in mixed-layer smectite-illite system for Well-G, Offshore Louisiana, as reported by Perry (1969). The solid line is the predicted S/I diagenetic-depth profile from the present theory.

of water at subsurface temperatures, the water makes up about 35% of the volume of the crystallite and further, the smectite dehydration in a diagenetic sequence releases about 15 volume percent pore water. If this diagenetic reaction occurs in a permeable shaley sand or a laminated sand-shale sequence with lateral leakage, the extra water may bleed away with no geopressuring. However, if the reaction occurs in a thick shale which will already be geopressured due to compaction disequilibrium, the extra water will increase the degree of geopressuring.

In the present model, the effect of the shale mineralogy change with burial diagenesis was incorporated by including a cation exchange capacity (CEC) term in the constitutive relation for shale void ratio, temperature and effective stress. This term is an explicit function of time (t)-temperature (T) history of sediment. There is some evidence from isothermal compaction studies on clays that indicate that at fixed temperature and stress, the void ratio is proportional to the cation exchange capacity and otherwise independent of mineralogy. Our constitutive relation was designed to reflect this dependence and was found to adequately represent the variation of shale void ratio with both amount of clay and type of clay. Typically, shallow shale samples have a CEC of 22 to 24 meq/100 gm. After illitization, it reduces to 7 to 14 meq/100 gm. We believe that the diagenesis also affects the permeability of shales, but have not yet incorporated its effect in the present model because of the lack of data in this regard.

The effect of shale diagenesis on computed geopressure profiles was studied systematically by varying sedimentation rates and geothermal gradients. In each case, we accounted for the temperature transients. We observed that the computed geopressure profiles depicted a diagenetic transition zone. This zone is characterized by a rise in the bulk density and FPG, and a significant drop in the effective stress. The location of the minimum in the effective stress is strongly dependent on the burial rate and the maximum geothermal gradient encountered during the burial. From the computed results we find that the shale diagenesis contributes significantly to overpressuring but that it cannot be the dominant mechanism. The overpressuring due to mechanical compaction disequilibrium is still the dominant cause. It alone can cause fluid pressure gradients of the magnitudes observed by drilling experience in the Gulf Coast, for a range of geologically significant burial rates and geothermal gradients appropriate for the Gulf Coast Tertiary Province.

SHALE COMPACTION, BURIAL DIAGENESIS, AND GEOPRESSURES

IV. DISCUSSIONS AND CONCLUSIONS

In this paper we presented a mathematical model for the gravitational compaction of shale as a mechanism of geopressure taking into account the temperature effects in the constitutive relations and the flow equations. The resulting nonlinear partial differential equations are solved by a finite difference algorithm using an implicit scheme. The moving boundary problems are dealt with exactly by a Lagrangian coordinate transformation technique. The model differs from all published models (Bredehoeft and Hanshaw, 1968; Sharp and Kortenhof, 1982; Bishop, 1979; Smith et al., 1979) in an important aspect: it is applicable to sediments with variable and temperature dependent compaction and flow properties. The temperature plays an important role on fluid pressure profiles in geopressured formations in that the increasing temperature further lowers the effective stress on geopressured rocks drastically. The numerical results presented in the paper can provide a guide in assessing the importance of various geologic parameters which control the distribution of excess pore pressure and effective stress on rocks. The continued sedimentation - compaction process of shale can lead to overpressuring at all depths. The persistence of high fluid pressure as predicted by the model indicates that the model can explain the existence of high pressures in older sedimentary basins as well as younger ones. However, in any actual situation, one would expect to encounter stresses that would be relieved by faulting. The development of faults still proses a challenge to the mathematical description.

We also proposed a kinetic model of smectite to illite transformation and estimated the rate of this reaction using geological parameters, such as burial rates and geothermal gradients. The model predicts the correct shape and extent of the reaction for Gulf Coast environment using the thermal history created by available paleontologic age, and present day equilibrium formation temperature data and the solution of a nonlinear heat diffusion equation to obtain paleo-subsurface temperatures. For Pleistocene through Late Miocene age rocks, the model predicts the clay diagenesis to take place at depths much greater than the older rocks, such as Oligocene-Upper Cretaceous age. In geopressured conditions, the diagenetic transformation of shales causes greater lithification and higher pore pressure. Although the dehydration of smectite certainly contributes to the magnitude of overpressuring at burial depths of about 8000 ft or more in the Gulf Coast area, it cannot be the sole source of geopressure buildup. The compaction disequilibrium of buried argillaceous sediments is still the dominant source of geopressuring phenomenon.

The model may be employed to ascertain the pressure history of sedimentary basins to estimate how thickness of strata is modified during burial history and to provide insight into hydrocarbon migration and fracture occurrence. From the relations obtained for the variation of effective stress with depth for geopressured shales, one can speculate that the existence of shear failure and fracture development in certain shale sequences is likely. The depth at which such fractures occur would depend upon the geothermal gradient and the sedimentation rate, among other factors. The likelihood of such fracture could have strong influence on conjectures about the primary and secondary migration of hydrocarbons.

As result of our analysis of the present model, we have been able to identify several key areas where further investigations and research are need. Some of these areas are as follows:

1. The results of the compaction model are found to be very sensitive to the details of the constitutive relations among porosity, effective stress, temperature, and mineralogy of shales and the relation among shale permeability, porosity, and mineral constituents. In particular, it has become clear from the parametric studies that the relations governing the porosity and permeability of shales at high temperature and under low effective stress have a very strong effect on the model description of shale under hard geopressures. That is one aspect where data have been sparse and this work would benefit from additional reliable experiments. Another aspect deals with diagenesis. Diagenesis of clay minerals seems to have significant effect on both porosity and permeability of shales and may indeed be important to the properties of other lithologies, e.g., the cementation of sandstones. Experimental studies of these factors on cores taken from geopressured formations should be useful. Finally, one should continue to test the validity of the transfer of results from laboratory shale studies to in situ conditions using wireline well log and formation pore pressure data where they become available.

2. The present one-dimensional model is a simplification of field conditions. For many real field problems it may be necessary to develop a two- or three-dimensional model that permits lateral fluid flow.

3. The model of basin building should be as deterministic as possible. It should, for example, include coupled effects of mass as well as energy (heat) transport due to convection and conduction. This becomes important in modeling basins in those epochs when the growth of the basin is very rapid.

4. The compaction of sediments is very sensitive to the distribution of in situ stresses. In this study, the global stress distribution is assumed to be uniform horizontally. Modification

should be considered in areas with complex geologic structures because the local stress concentration could be highly variable from place to place. Further, these forces could vary with time. Reconstruction of past tectonic forces presents a formidable problem but certain common features, such as growth faulting, may follow a recurrent and predictable pattern of development.

REFERENCES

BARKER, C., 1972, Aquathermal pressuring - role of temperatures in developments of abnormal pressure zones: Am. Assoc. Petrol. Geol. Bull., 56, 2068-2071.

BISHOP, R. S., 1979, Calculated compaction states of thick abnormally pressured shales: Am. Assoc. Petrol. Geol., 63, 918-933.

BREDEHOEFT, J. D., and HANSHAW, B. B., 1968, On the maintenance of anomalous fluid pressures: 1. Thick sedimentary sequences: Geol. Soc. Am. Bull., 79, 1097-1106.

BRUCE, C. H., 1982, Relation of illite/smectite diagenesis and development of structure in the northern Gulf of Mexico basin: presented at the AAPG Research Conference on Role of Clay Minerals in Hydrocarbon Exploration, October 10-13, 1982, Santa Fe.

BOLES, J. R., and FRANKS, S. G., 1979, Clay diagenesis in Wilcox sandstones of southwest Texas: Implications of smectite diagenesis on sandstone cementation: Jour. Sed. Petrol., 49, 55-70.

DICKINSON, G., 1953, Geophysical aspects of abnormal reservoir pressures in Gulf Coast, Louisiana: Am. Assoc. Petrol. Geol. Bull., 37, 410-432.

FOSTER, W.R., 1981, The smectite-illite transformation: its rote in generating and maintaining geopressure: paper presented at the 94th Annual Meeting of the Geol. Soc. Am., November 2-5, 1981, Cincinnati.

FREED, R. L., 1982, Clay diagenesis and abnormally high fluid pressure: presented at the 52nd Annual Meeting of SEG, October 17-21, 1982, Dallas.

HOWER, J. ESLINGER, E. V., HOWER, M. E, and PERRY, E. A., 1976, Mechanism of burial metamorphism of argillaceous sediment: I. Mineral and chemical evidence: Geo. Soc. Am. Bull., 87, 725-737.

HOWER, J., 1981, The influence of mineral diagenetic reaction on the pore water chemistry of shale: presented at the 94th Annual Meeting of the Geol. Soc. Am., November 2-5, 1981, Cincinnati.

HUBBERT, M. K., and RUBEY, W. W., 1959, Role of fluid pressure in mechanics of fluid-filled porous solids and application to overthrust faulting: Geol. Soc. Am. Bull., 70, 115-166.

LEVORSEN, A. I., 1954, Geology of petroleum: W. W. Freeman & Co.

MEISSNER, F. F., 1978, Petroleum geology of the Bakken formation, Williston basin, North Dakota and Montana. In: The economic geology of the Williston basin, Montana, North Dakota, South Dakota, Saskatchewan, Manitoba: Montana Geol. Soc., Billings, 207-227.

OZISIK, N. M. 1968, Boundary value problems of heat conduction, International Textbook Co., Pennsylvania.

PERRY, E. A., Jr., 1969, Burial diagenesis in Gulf Coast pelitic sediments, Ph.D. dissertation, Case Western Univ.

PERRY, E. A., Jr., and HOWER, J., 1972, Late stage dehydration in deeply buried pelitic sediments, Am. Assoc. Petrol. Geol. Bull., 56, 2013-2021.

POWERS, M. C., 1967, Fluid release mechanisms in compacting marine mudrocks and their importance in oil exploration: Am. Assoc. Petrol. Geol. Bull., 51, 1240-1254.

SHARP, J. M., and KORTENHOF, M. H., 1982, Numerical model of shale compaction, aquathermal pressuring, and hydraulic fracturing: presented at the AAPG Research Conference on Role of Clay Minerals in Hydrocarbon Exploration, October 10-13, 1982, Santa Fe.

TIMKO, K. J., and FERTL, W. H., 1971, Relationship between hydrocarbon accumulation and geopressures and its economic significance: Jour. Petr. Techn., 23, 923-933.

VAN OLPHEN, H., 1963, Compaction of clay sediments in the range of molecular particle distances: Clays and Clays Min., 11, 178-187.

B. DOLIGEZ, F. BESSIS, J. BURRUS[1],
P. UNGERER, P.Y. CHÉNET[1]

INTEGRATED NUMERICAL SIMULATION OF THE SEDIMENTATION HEAT TRANSFER, HYDROCARBON FORMATION AND FLUID MIGRATION IN A SEDIMENTARY BASIN: THE THEMIS MODEL

ABSTRACT

The geological phenomena intervening during the history of a sedimentary basin are generally interactive, and modelling the involved physical and chemical processes is an accurate way to quantify them. The 2D integrated autoadjusted THEMIS model has been built for this purpose, with a special focus on the hydrocarbon formation and migration processes. The main principles of the model are exposed here, and the influence of the various critical physical and chemical laws is discussed. An application of an hydrodynamic study in the North Sea Basin is presented.

INTRODUCTION

The geological phenomena occurring during the history of a sedimentary basin, and especially those leading to the accumulation of hydrocarbon pools, involve the combination of numerous physical and chemical processes. In this respect, time and temperature are of primarily importance in the hydrocarbon generation processes, while the pressure history will be the main factor governing the dynamics of hydrocarbons and water circulation. At last, the fluid dynamics, the sedimentation and the geodynamic setting may have an important role in the temperature distribution within the basin.

---

(1) *Institut Français du Pétrole, Rueil-Malmaison, France.*

The use of deterministic computer models is probably the most efficient way to study the interaction of these processes and, as a consequence, to better understand the geological phenomena (Chénet et al., 1983). In addition, if the validity of the physical and chemical laws used in the models is ensured by successful modeling in well documented areas, several predictions can be made with a good degree of confidence.

Several models dealing with a few basic physical and chemical principles have already been able to account for some geological features of the sedimentary basins. Ungerer et al. (1984) and Bessis (1986) have shown how the backstripping of the successive sedimentary layers in a sedimentary basin, with a quiet tectonic history, may help in quantifying the successive paleobathymetric profiles during the basin history. Subsidence reconstructions in rifted-type sedimentary basins have evidenced the lithospheric thermal contraction phenomena under the basin. Mc Kenzie (1978); Steckler and Watts (1980); and Keen and Beaumont (1982) have also shown how the heat-flow and the subsidence are interrelated in such type of basins. Royden et al. (1980), Mc Kenzie (1980) and Chénet (1984) among many others, have shown how to obtain the temperature history of a sedimentary column from the heat-flow studies in these basins and to deduce the position of the hydrocarbon window.

Tissot and Espitalié (1975) have successfully described the formation of hydrocarbons, as a result of six parallel kinetic chemical reactions where the organic matter is cracked into oil. Secondary craking of oil into gas was described by a single reaction. More recently, Ungerer et al. (this volume) have shown that this kinetic process could account both for the laboratory experiment of organic matter maturation of immature samples and for the present day maturity level in source-rocks. At last, Durand et al. (1983) could reproduce the HC migration phenomena on cross sections for several basins, taking into account the capillarity of hydrocarbons and the diphasic flow of HC and water linked with the pressure regime driven by the progressive compaction of the sediments with burial.

THE THEMIS MODEL

The purpose of the THEMIS model is to integrate within the horizontal and vertical dimension and with time the various physical and chemical laws that were previously evidenced. Several improvements were included in the description of the geological phenomena.

THE THEMIS MODEL

Structure of the model

As the amount of data and the kind of investigation may vary in each example, the THEMIS model has been built in several modules, that may be used separately or combined together (fig. 1). Each module corresponds to the simulation of one or several geological phenomena involved in the evolution and the description of the petroleum potential of the basin. They may be chain-like organized, as a given module generally requires the description of the phenomena considered in the preceding ones. One may of course neglect one or several phenomena depending of the kind of study to be made.

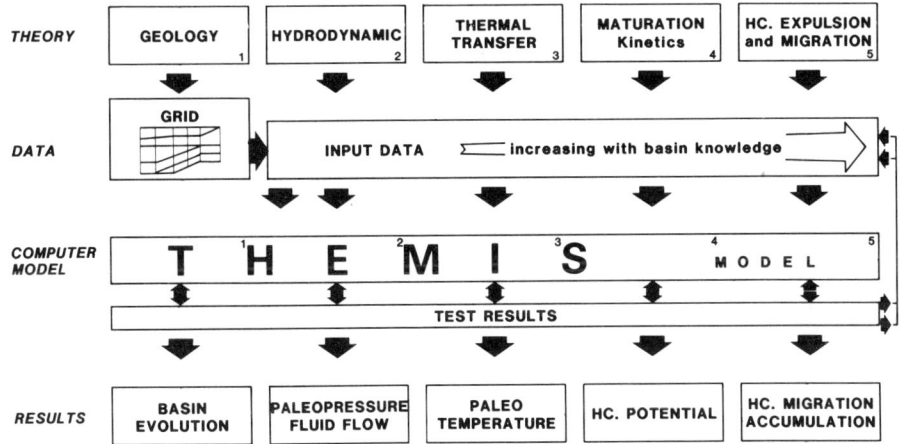

Fig. 1 - Structure of the THEMIS model. The modules and their corresponding results

The first module corresponds to a "backstripping study" of a given cross-section of the basin. It consists in removing the successive sedimentary layers at each recorded stratigraphic event on the cross section and decompacting the remaining layers. It gives the characteristics of the grid system that constitute the frame of the following modules. The subsidence history of the cross section may be reconstructed.

The second module reconstructs the fluid dynamic history within the cross section, including the pressure distribution and the fluid velocities.

THE THEMIS MODEL

The history of the thermal phenomena is also computed in a separate module. Within this one, the continental crust and the upper mantle, up to 100 km in depth, may be considered in the temperature calculations and rifting simulations. The history of fluid flow is reconstructed and its effect on the heat transfer within the sediments may be taken into account.

The hydrocarbon saturation in the various parts of the basin may also be obtained (module 4 of fig. 1) as a result of the kinetic chemical reactions involved in the organic matter maturation and driven by the temperature history.

At last, the HC expulsion, migration and accumulation is described, combining the flow of hydrocarbons and water. This module integrates the previous calculations of HC generation and may take into account the thermal transfer study.

1) Reconstitution of the sedimentation by backstripping

The mesh construction (fig. 2).

The method consists in removing the successive sedimentary layers recorded stratigraphically on a given cross-section. For each stratigraphic limit, the shape of the basin is computed, taking into account the state of compaction of the already deposited strata, the paleobathymetry, and the sea-level variations (Bessis, 1986). In order to compute the sedimentation rate it is assumed that the volume variation with burial of a given layer is only due to the expulsion of the water from the porous sediments and that the present day porosity depth relationship together with the volume of solid matrix remains constant with time. In the case of an erosion, when the burial decreases, the non eroded levels remain compacted. In this approach, the dissolution and recristallisation phenomena during the compaction and eventual under compaction processes are neglected.

This module computes the sedimentation rates along the cross-section and between the stratigraphic boundaries. These rates are approximate as the pressure effects on compaction are neglected. This appears to be satisfactory (Ungerer et al., 1984) in terms of thickness reconstruction.

THE THEMIS MODEL

A grid is automatically prepared for the geological section under investigation. (Fig. 2b, 2c) In the grid, each row corresponds to a lithostratigraphic interval. Its thickness may be zero in the present day situation if erosion or non deposition occurred in the geological history of the basin. Each element of the grid is attached to the solid matrix of a given layer. Its physical properties (density, permability, conductivity, etc.) and its geometry will vary as a function of the lithologic type of the solid matrix and the porosity. Elements can be triangle or quadriangle for "pinch-out", truncations and sedimentary wedges.

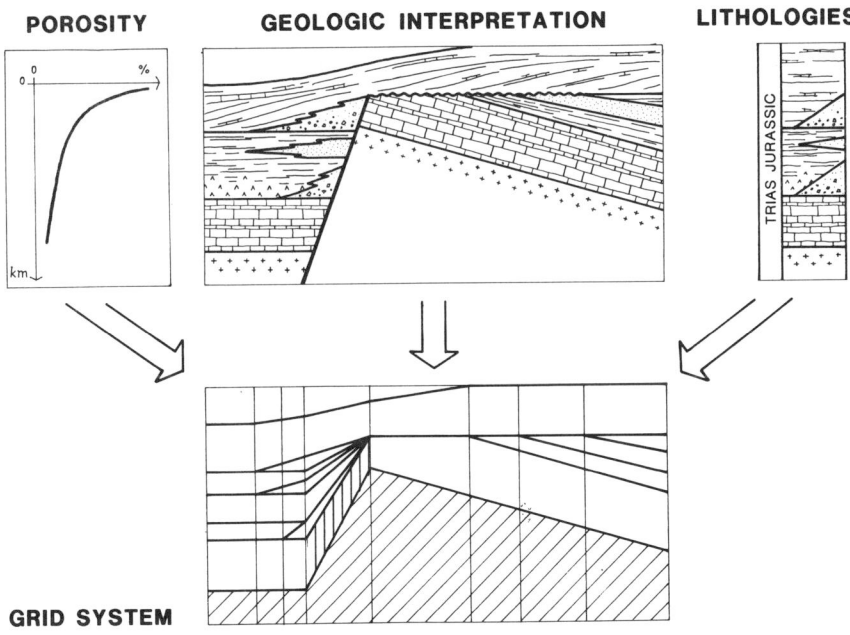

Fig. 2 - The mesh construction

The mesh reproduces the main geologic features of the cross section. The sandy bodies near the main normal fault are represented by triangular elements. The faulted area is represented by a specific elongated element (hatched). The stratigraphic truncations an the hanging wall of the tilted block are also considered.

The number of rows will vary between these stratigraphical boundaries, in order to represent the successive geological configurations. The displacement of a given sedimented element will be guided along the vertical axis. During a given period, the thickness of the upper row increases or decreases in case of sedimentation and erosion respectively. The choice of such a deformable and moving grid enables a convenient description of the successive geometries of the basin. The mesh may be detailed in crucial horizons, such as source rocks and reservoirs. At last, it facilitates the formulation of the transport equations (heat transfer, fluid flows) and the mass balance for solids solved in the various modules.

The equation governing the thickness H and depth Z variations with time will be

$$\int_{z(t)}^{z(t)+H(t)} [1 - \phi(z)] \, dz = \int_{z(0)}^{Z(0)+H(0)} [1 - \phi(z)] \, dz \quad (1)$$

(mass conservation)

The tectonic subsidence is also computed, assuming an isostatic response of the lithosphere to the sediments and water loading (see Chénet et al., 1983 and Bessis, 1986 for more details). It represents the subsidence of the substratum of the basin that would have occurred without the sediments and sea water loading effect. The tectonic subsidence pattern emphasizes the deep internal evolution of the basin. In particular, the knowledge of the subsidence rate in rift-type basins permits to calibrate the deep thermal regime in the lithosphere below these basins, as both phenomena are linked.

2) Pressure and fluid flow history in a sedimentary basin

The compaction due to the progressive burial of the sediments with time may be considered as the main driving force for the fluid flows in sedimentary basins. The sedimentary load (Sg) is supported both by the solid fraction of the rocks and by the fluids, which can be expressed by the law of effective stress (TERZAGHI, 1948).

$$S_g = \sigma + P \quad (2)$$

The porosity ϕ is given by an empirical law as a function of the effective stress (Fig. 3):

THE THEMIS MODEL

$$\Phi = \Phi(\sigma) \qquad (3)$$

This law has been calibrated in well known areas where pressure conditions are hydrostatic. The concept of irreversibility of the compaction, which states that once the effective stress decreases, in overpressured zones for instance, the porosity increases much less than when normal compaction occurs, was also considered.

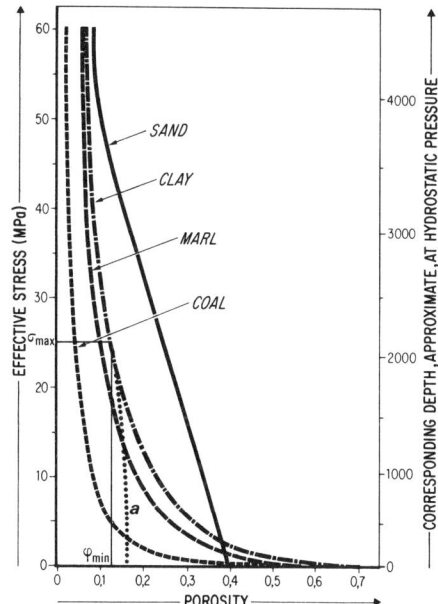

Fig. 3 - Effective stress-porosity relationships for different lithologic types

Empirical stress porosity relationship for several lithologic types. In case of the release of the effective stress, after reaching a maximum σ max, the porosity increases above φ min, following a specific relationship ("a" curve).

The fluid flow, considered here as monophasic (water), is described by the Darcy's law.

$$\vec{V} = \frac{-\bar{\bar{K}} \cdot \rho_w \cdot g}{\mu} \; \overrightarrow{\text{grad}} \; \left(\frac{P}{\rho_w g} - Z\right) \qquad (4)$$

The permeability tensor $\bar{\bar{K}}$ is anisotropic, in order to deal correctly with the case of alternating permeable and unpermeable layers, such as sand-shale sequences.

$$\bar{\bar{K}} = \begin{vmatrix} K & 0 \\ 0 & \theta K \end{vmatrix} \qquad (5)$$

The permeability is given by the Kozeny-Carman formula, relating the porosity and specific surface of the rock.

$$K = \frac{0.2 \cdot \phi^3}{(1-\phi)^2 S_o^2} \qquad (6)$$

In the case of overpressure conditions, the permeability is allowed to increase in order to simulate hydraulic fracturing. The fracturation threshold is given by the pressure-stress relation (empirical):

if $P/S_g > f$, then K is replaced by

$$K\left[1 + F(P/S_g)\right] \qquad (7)$$

where F is a function which increases infinitely when the fluid pressure reaches the geostatic limit.

3) Thermal phenomena

The thermal history of a sedimentary basin depends on the evolution of the crust and the mantle at depth, and on the subsequent heat transfer within the sediments.

Within the basin itself, the thermal transfer is mainly of conductive origin, as shown by the existence of a geothermal gradient. The convective heat transport, linked with the fluid motions, may however strongly disturb the geothermal gradient. Previous studies have shown that the convection of heat by vertical expulsion of fluids during the compaction is of secondary importance, if the sedimentation rate is less than 1000 m/yr (Perrin, 1983; Chénet, 1984; Hermanrud (this volume). However, 2D convective phenomena may not be negligible if a significant fluid flow is concentrated in high permeability zones. Oxburgh (1985) and Gosnold (this volume) for instance, have proposed various examples in which the regional water circulation has perturb the heat flow pattern. This phenomenon has been considered in the THEMIS model, coupling the fluid velocity analysis and the equation of heat transport when imposing proper hydraulic charges at the lateral boundaries of the model. The natural convection of the fluids linked with their buoyancy changes with temperature have not been considered. It may have a certain influence on the temperature field, especially in high permeability zones (Perrin, 1983).

The velocity of filtration for the fluids is obtained by the Darcy's law in a reference frame fixed with respect to the elements boundaries.

$$\vec{u} = \Phi (\vec{V}_w - \vec{V}_s) \qquad (8)$$

Then, the equation of heat transfer within the sediments will be:

$$\frac{\partial}{\partial t}\left[(\rho C_b)T\right] = -\text{div}\left[\Phi \vec{V}_w (\rho C_w)T + \vec{V}_s (1-\Phi)(\rho C_s)T\right] + Q + \text{div}(\lambda \overrightarrow{\text{grad}}\, T) \qquad (9)$$

total heat variation — convection forced by compaction or regional fluid flow — heat generation — conduction

The heat capacity of the bulk sediment is additive:

$$\rho C = \rho C_w \Phi + \left[\sum P_i (\rho C)_i\right](1 - \Phi) \quad \text{J/kg °C} \qquad (10)$$

depending on the proportion of the lithologies P_i

The thermal conductivity is able to vary with porosity, lithology and temperature.

$$\lambda = \lambda_b \left(\frac{\lambda_w}{\lambda_b}\right)^{\Phi} (1 + \alpha T)^{-1} \quad \text{W/m°C} \qquad (11)$$

A given history of heat-flow or temperature can be imposed at the base of the model. In fact the sedimentary cover, the crust and the mantle have to be integrated in a single system, because if the sedimentation rate is sufficently high for instance, the input of cold and poorly conductive sediments will tend to lower significantly the heat flow up to great depths (blanketing effect) and the analytical heat-flow should be reduced, as discussed by Bessis and Burrus (this volume). In addition, the heat flow input below the substratum of the sedimentary basin is governed by the radioactive heat generation within the continental crust (Jaupart, 1984) and the geodynamic setting of the basin.

THE THEMIS MODEL

In the rifted basins, the geodynamic evolution has been tentatively explained by the occurrence of a deep thermal anomaly that may affect the lithosphere as a whole (Mc Kenzie, 1978; Moretti and Froidevaux, this volume). This thermal anomaly, whose cause may be of various origins (stretching and/or thinning of the lithosphere), is responsible for the subsidence of the basin during and after the rifting phase.

The crust and mantle attenuation and heating during a rifting event may be incorporated in the THEMIS model. The part of the grid system representing the crust and the mantle is deformed following a thinning velocity field, deduced from the total amount of crust and mantle thinning ratios after rifting and the duration of rifting. After this, the crust/mantle system is supposed to recover progressively its original thermal state.

The equation of heat in the crust mantle medium will be:

$$(RC)\frac{\partial T}{\partial t} = \text{div}(\Lambda \vec{\text{grad}}\, T) - \text{div}(RCT\,\vec{V}) + Q \qquad (12)$$

total heat / heat / heat advection / radioactive
variation with time / conduction / (rifting) / generation

The thinning velocity field is given by:

$$\text{div}\,\vec{V} = 0 \text{ (mass conservation)}$$

$$\frac{\partial V_x}{\partial z} = 0 \text{ (a vertical axis remains vertical during extension)} \qquad (13)$$

The total or tectonic subsidence pattern is calculated, computing the thermal dilatation or contraction of the whole system, and its isostatic response to the water and sedimentary load.

4) The formation of hydrocarbons

The formation of hydrocarbons is described by mean of the kinetic model initially proposed by Tissot and Espitalié (1975). The amount of hydrocarbons generated is given as a function of time, temperature and kind of organic matter. It is described by a series of n parallel chemical reactions obeying a kinetic of order 1 and following the Arrhenius law.

$$\frac{d\xi_i}{dt} = -A_i \exp\left[-\frac{E_i}{RT}\right]\xi_i \qquad (14)$$

$$q = \sum_1^n (\dot\xi_{io} - \dot\xi_i)$$

The values of the initial petroleum potential may be deduced from Rock-Eval pyrolysis on immature samples of the various source-rock types encountered in the sedimentary basin. Furthermore, the distribution of the frequency factors and the activation energies may be obtained by a specific adjustment model (OPTIM) that may be performed after the Rock-Eval pyrolysis (Ungerer et al, this volume).

The gas formation is also considered, by a unique secondary cracking reaction, with an efficiency factor.

$$\frac{dq}{dt} = A_g e^{-E_g/RT} q \cdot \alpha_g \qquad (15)$$

5) The migration of hydrocarbons

The migration and accumulation of the hydrocarbons may also be considered as a consequence of the pressure build up in the sedimentary basin (Durand et al., 1984). Namely, the generation of petroleum, which results from the maturation of the sedimentary organic matter (Tissot and Welte, 1978), tends to increase the fluid pressure in the source-rocks. When the hydrocarbon saturation becomes high enough, the hydrocarbon phase becomes continuous in the source-rocks and the total fluid pressure due to the oil generation and the burial may overcome the capillary forces. The migration can then start. Once the hydrocarbons have reached more permeable beds, the buoyancy forces become the driving forces. At last, HC accumulation may occur in permeable beds once the buoyancy forces do not overcome the fluid retention forces in the overlaying seals.

THE THEMIS MODEL

The flow of water and HC is described with the formalism of diphasic flow in a porous media. However, the model cannot presently deal with three fluid phases (oil/gas/water) but oil/water and gas/water simulations may be performed. The first case, where gas migration is neglected, corresponds to the oil migration pattern before an important burial of the source-rocks, which would lead to gas formation. The second case corresponds to the gas migration, assuming that gas only is expelled from the source-rock, the oil remaining sealed.

The motion of fluids is obtained after the Darcy's law. This law has been extended to the polyphasic flow through the relative permeability concept (Marle, 1972) which is commonly used in reservoir engineering. This formalism allows to describe the various steps of the hydrocarbons migration with the same physical principles.

The modified Darcy's law is:

$$\vec{V}_w = \overline{\overline{K}} \frac{k_{rw} \cdot \rho_w \cdot g}{\mu_w} \text{ grad}\left(\frac{P}{\rho_w g} - Z\right) \quad (16)$$

for the water

$$\vec{V}_h = \overline{\overline{K}} \frac{k_{rh} \cdot \rho_h \cdot g}{\mu_h} \text{ grad}\left(\frac{P_h}{\rho_h g} - z\right) \quad (17)$$

for the hydrocarbons.

The dynamic viscosity is obtained after the Bingham formula, as a function of the temperature.

$$\mu_h = a\, e^{-b/T} \quad (18)$$

$$\mu_w = a \left/ \left[b\,(T-c) \sqrt{d + (T-c)^2} - e \right] \right.$$

a, b, c, d and e: empirical constants

THE THEMIS MODEL

The pressure in the HC and in the water phases are assumed to follow the capillarity equations

$$P_h - P = P_c \qquad P_c = 2\gamma/R \qquad (19)$$

During the primary migration of hydrocarbons away from the source-rocks, the saturation of hydrocarbons has become high enough, and their relative permeability, together with pressure, are sufficiently high to permit an effective expulsion. During the secondary migration stage, the same concept predicts that the migrating hydrocarbons will use prefered paths where the permeability is high and will tend to invade almost completly the pore volumes, when the saturation goes above a given value, because the water relative permeability is then very low. To account also for the possible saturation of HC in the migration avenues, their relative permeability is supposed to be not zero beyond a few per cent saturation (Fig. 4). On the other hand the buoyancy forces will tend to expell the hydrocarbons from the permeable layers which stay structurally beyond the traps. At last, the dismigration process is accounted for, by considering rather low permeability layers, such as silts, where buoyancy forces may overcome the capillary forces.

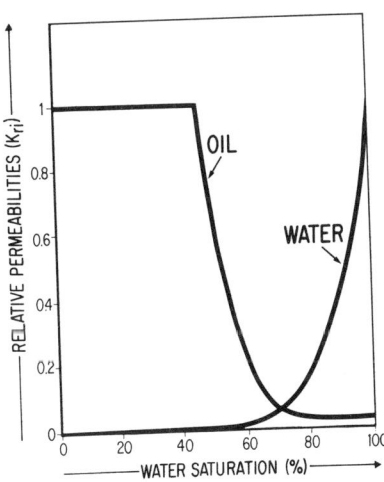

Fig. 4 - The relative permeabilities relationship

6) Algorithm and boundary conditions

At each time step, a given row is increasing in thickness when sedimentation occurs, or decreasing in the case of an erosion. A new row may also be deposited after a new recorded stratigraphic event. Solving the equations for mass balance of water, hydrocarbons and solid matrix for each element of the grid, the variations of porosity in all the existing elements of the successive rows may be known. Once the porosities variations are known, the volumes, pressures, positions as well as the water and hydrocarbons velocities through each element may be computed. In a second system, the temperature are obtained, solving the thermal equations.

This kind of numerical solution in Lagrangian coordinates is similar to the finite difference implicit method (Carnahan et al., 1969). The time step is automatically adjusted in order to ensure the stability of the results. At the end of a model run, an automatic procedure compares the computed geometry of the basin with the real geometry. If the divergence between both geometries is too high, an additional run is automatically performed, with a corrected estimate of the compaction trend, and an adjustement of the sedimentation rates.

The geometric conditions are imposed by the successive paleobathymetric profiles at the top of the mesh during the geological history of the basin. A simple backstripping study with the first module of the model may help in the determination of these profiles. The number of rows and their thickness, representing the underlying lithosphere, is choosen according to the thermal study to be done.

The hydraulic conditions on the side of the studied corss-section may consider either lateral or no lateral flow. At the bottom of the sea, hydrostatic conditions are supposed. For aerial deposits, the hydraulic head is equal to the height above sea-level. Lateral regional fluid flow, due to meteoric supply for instance, may also be imposed. This kind of conditions is often difficult to quantify at present time and obviously very difficult to quantify for the passed million years, as paleoclimates, as well as paleorelief are involved.

Thermal boundaries conditions are taken as follow :

. imposed temperature at the base of the sea (eventually corrected for variations through time),
. $T = 1330°C$ at the base of the lithosphere, or given heat flow, or

temperature history at the base of the sediments,
. zero lateral heat-flow, or imposed lateral heat-flow history.

7) A geological example: Viking graben (North Sea)

The geological example has been choosen in the North Sea where a good geological control on the peroleum geology is already available (Illing and Hobson, 1980). The seismic cross-section through the Viking graben, the central structure and the Horda platform has been modelized, since it offers some interesting characteristic features of the North Sea basin, enabling to focus on some advances of the THEMIS model (Fig. 5).

The geological evolution in this area may be summarized as follow (Badley et al., 1984): (1) after a rifting phase of Permian age, thick continental deposits cover the entire area during triassic times (Buntsandstein, Muschelkalk and Keuper). The magnitude of rift faulting and the thickness of the deposits is not well documented. (2) After the deposition of lower jurassic Statfjord sands and Dunlin clays formations, the major rifting phase initiated, with important normal faulting, giving birth to the present day block structure. The Brent sands and coals and upper jurassic Kimmeridge and Heather Clays deposited at this time, in a deepening basin. The end of rifting is marked by a major unconformity encompassing part of lower cretaceous times. At last, from the Cretaceous up to present time, generalized subsidence occurred, the Viking trough being progressively invaded by the sequences prograding westward from the continental shelf.

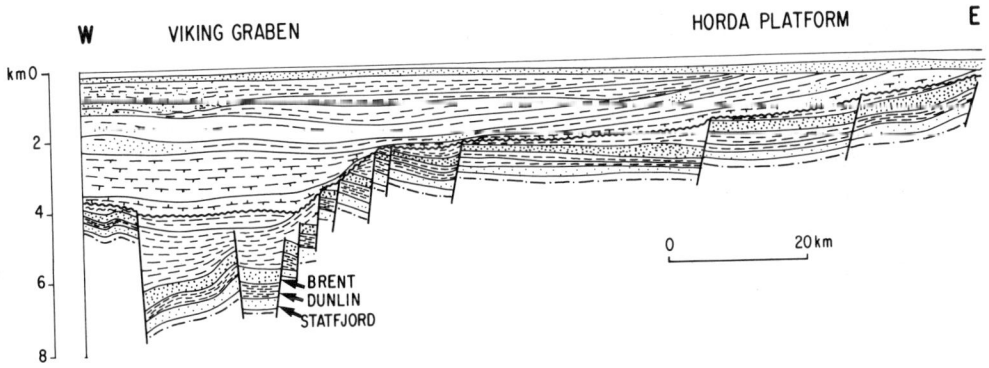

Fig. 5 - The structure

THE THEMIS MODEL

On the meshed geological cross-section (Fig. 6), the faulted areas bordering the Viking graben have been represented by steeply dipping thin layers, in order to take into account their specific permeability properties. The bottom layers correspond to the triassic series, for which we assume a 2 km thickness, below the Viking graben. In the central area, the lateral discontinuities due to the number of step faults (see Fig. 5) have been represented by triangular elements. The successive onlaps of the cretaceous series against the structure have been also represented by triangular elements, as well as the truncations of the cretaceous and tertiairy prograding series against the Quaternary units on the Horda platform.

Each element of the mesh has homogeneous physical and chemical properties but they will not be studied in detail here. The Statfjord and Brent sands have been considered as high permeability rows, while Brent coals are considered as the most efficient source-rocks. The Dunlin, Heather and Kimmeridge formations are impermeable clays, which behave in a minor way as source rocks. In the post jurassic series, the lower cretaceous marls have been distinguished, as well as the paleocene chalks and the sand/shale ratios in the Tertiairy.

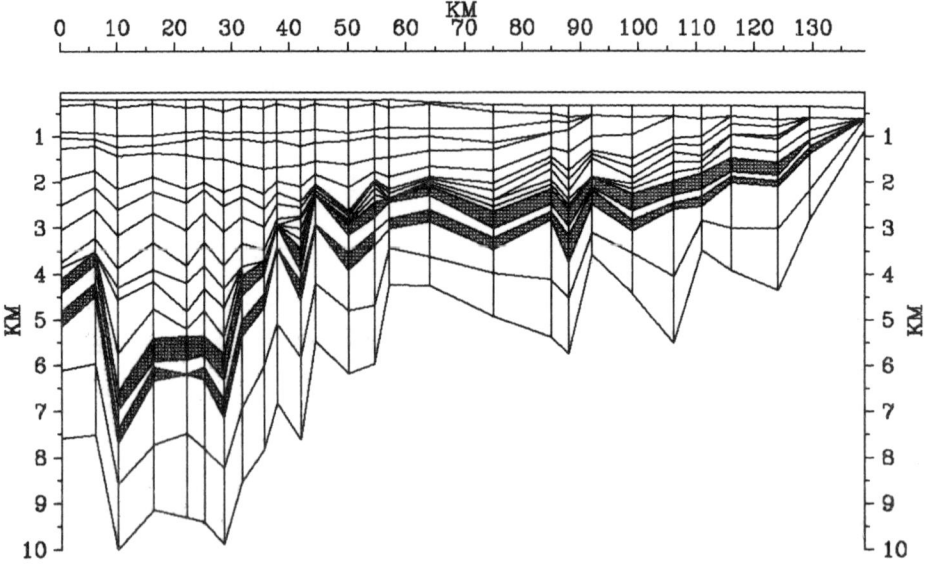

Fig 6 - The meshed cross section.

THE THEMIS MODEL

1. The backstripping study

The reconstruction of the successive palinspastic cross sections of the basin were made assuming that the geometry of the substratum remained almost undeformed after the period of active faulting during triassic and jurassic times. The shape of the cretaceous and tertiairy prograding sequences gave an estimate of the successive paleoslopes. For the Jurassic, we assumed a progressive deepening of the Viking trough, up to the beginning of the Cretaceous, were the reflectors of ·this formation onlap the rift unconformity, suggesting the existence of a depression. On the structure, the sedimentation gaps during the same period suggest subaerial erosions. On the Horda platform, reduced synrift subsidence occurred, with an overall tilting of a few degrees of the prerift substratum.

2. Pressure and fluid flow history

The progressive pressure build up may be reconstructed through time. Significant excess pressure conditions are predicted in the graben since the upper Cretaceous (88 My). They may be related to the absence of drainage areas on the flank of the graben and to the presence of a sufficently impervious cover since this period. On Fig. 7a, showing the excess pressure situation 65 My ago, the sandy layer of the Brent formation in the graben are still under hydrostatic conditions, as they are probably connected with the sea bottom at this time. In the present day situation, the same layer is now overpressured (Fig. 7b), but the fluid flow rate is important (Fig. 8), as it concentrates both the fluid expelled from the hanging wall and the foot wall. On the flank of the structure, toward the graben, a significant flow through the Cretaceous cover is predicted, as a probable consequence of hydraulic fracturing in this area. On the Horda platform, the main fluid flow rates are concentrated in the sandy layers (Brent and Statfjord formations), and generally directed toward the structural highs.

THE THEMIS MODEL

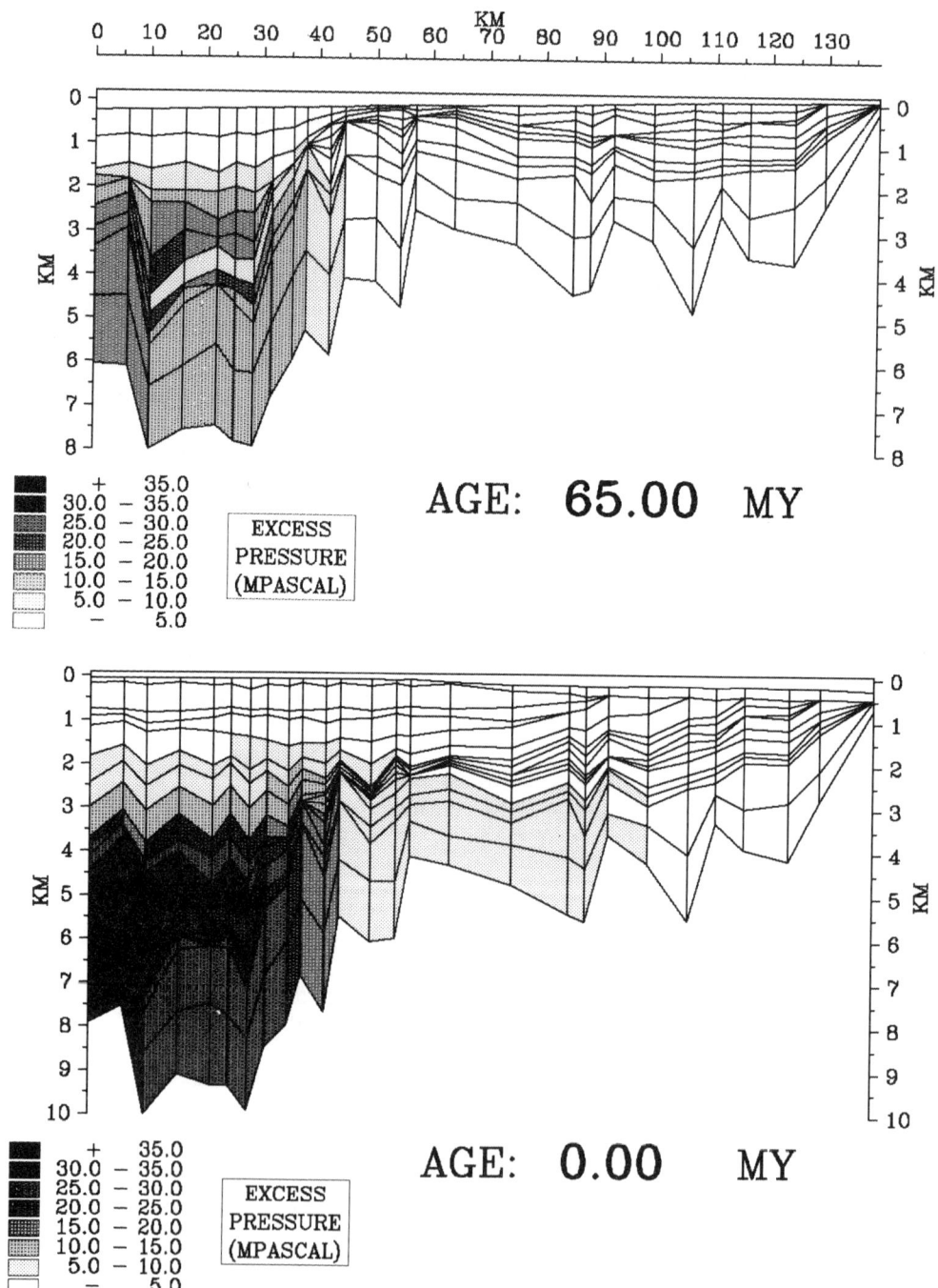

Fig. 7 – The excess pressure reconstruction 65 My before and at present time

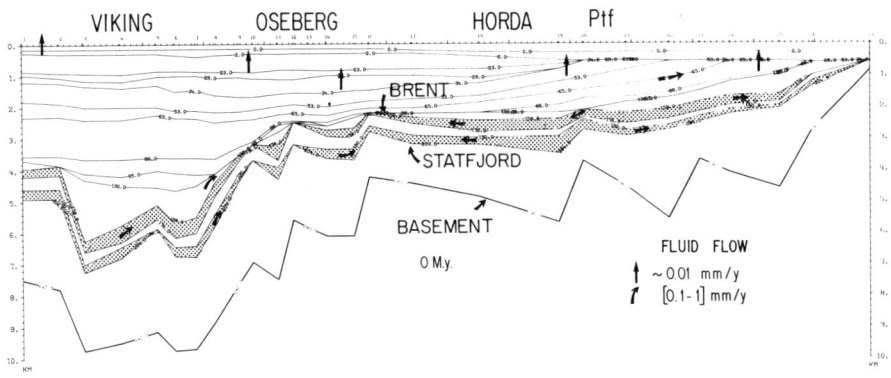

Fig. 8 - The fluid flow pattern at present time

CONCLUSION

The THEMIS model for basin analysis implies the combination of numerous physical and chemical laws, expressed by equations. Part of these equations rely upon well known physical phenomena, such as fluid flow in a porous media or thermal transfer by conduction and convection. In addition, several equations come from empirical (or semi-empirical) considerations. These equations may be considered as valid if they enable to reproduce satisfactorily the present day situation in the sedimentary basins. As most of them have been successfully used in previous models, we may be confident in their combination, as the description of the phenomena is appropriate. Furthermore, the application of the model to a specific case suggests various questions on the present day situation of the basin, such as, the occurrence of hydraulic fracturing and the amount of overpressuring in the North Sea example. These predictions may be readily tested against the available data and have important consequence on the understanding of the geological, physical and chemical history of the basin.

THE THEMIS MODEL

LIST OF THE VARIABLES

Fluids flow

ϕ	porosity
Z	depth
$\rho_{w,s,m}$	density of the water, sediments and lithospheric mantle
g	gravity acceleration
V	velocity of the fluid (m/s)
$K_{rw,rh}$	relative permeability of the water, HC
$\underline{\underline{K}}$	permeability tensor of the sediments
P, P_h	pressure of the fluid, HC (Pa)
P_c	capillarity pressure (Pa)
θ	anisotropy coefficient
So	specific area of the rock (m^2/m^3)
f	fracturation parameter ($0 < f < 1$)
$\mu_{w,h}$	dynamic viscosity of the water, HC
γ	capillarity
R	radius access to pore (m)
Sg	total or geostatic stress (Pa)
σ	effective stress (Pa)
Ts	thickness of the sediments

HC Formation

Sat	fluid saturation (%)
ξ_i	residual petroleum potential (mg/g C org)
ξ_{io}	initial petroleum potential (mg/g C org)
q	amount of HC formed (kg)
A_i	arrhenius constant)
E_i	Activation energy) for n primary reactions $o < i < n$ (cal/mole)
R	constant = 2 cal/mole ° k
A_g	arrhenius constant)
E_g	activation energy) for secondary reaction
α_g	reaction efficiency ($0 < \alpha_g < 1$)

Heat transfer

T	temperature
Q	heat source
RC	heat capacity of the lithosphere (J/m^3)
Λ	thermal conductivity of the lithosphere (W/m °C)
λ	conductivity of the bulk sediment (W/m °C)
(ρc)	heat capacity of the bulk sediment (J/m^3)
u	filtration velocity of water/solid (m/s)
$(\rho C)w$	heat capacity of water (J/m^3)

THE THEMIS MODEL

ACKNOWLEDGEMENTS

We thank G. Bessereau, and R. Vially from IFP for useful discussion on the geology of the North Sea.

BIBLIOGRAPHY

Badley, M.E., Egeberg, T., Nipen, O., 1984 - Development of rift basins illustrated by the structural evolution of the oseberg feature, Block 30/6, offshore Norway. J. Geol. Soc. London, vol. 141, 1984, p. 639-649.

Beaumont, C., Keen, C.E., Boutillier, R., 1982 - On the evolution of rifted continental margins, comparison of model and observations from the Nova Scottian margin. J.R. astr. soc. 70, p. 667-715.

Bessis, F., 1986 - Some remarks on subsidence study of sedimentary basins. Application to the gulf of Lions margin (Western Mediterranean). Marine Petroleum Geol. (in press).

Carnahan, B., Luther, H.A., O'Wilkes, 1969 - Applied numerical methods. Wiley, N.Y., 604 p.

Chénet, P.Y., 1984 - Thermal transfer in sedimentary basins. Paleotemperatures reconstruction and maturation studies in the Gulf of Lion margin in thermal phenomena in sedimentary basins. B. Durand (Ed.), Ed. Technip.

Chénet, P.Y., Bessis, F., Ungerer, P., Nogaret, E. and Perrin, J.F., 1983 - How geological mathematical models can reduce exploration risks. Proc. 11th World pet. Congress, London. SP7 (in french), p. 385-404.

Durand, B., Ungerer, P., Chiarelli, A., Oudin, J.L., 1984 Modélisation de la migration de l'huile. Application à deux exemples de bassins sédimentaires. Proc. of the 11th world Petr. congr., London, PD1 (1), p. 1-13.

Gosnold, W.D., 1986 - Heat-flow studies in sedimentary basins. This volume.

Hermanrud, C., 1986 - On the importance to petroleum generation of heat transfer due to compaction. Related fluid convection in a sedimentary basin. This volume.

Illing, L.V. and Hobson, G. (Eds), 1980 - Petroleum geology of the continental shelf of north-west Europe. Heyden and sons. 521 p.

Jaupart, C., Sclater, J.G., Simmons, G., 1982 - Heat-flow studies: constraints on the distribution of uranium, thorium and potassium, in the continental crust. Earth Planet Sci. Letter, 52, p. 328-344.

Mc. Kenzie, 1978 - Some remarks on the development of sedimentary basins. Earth Planet Sci. Letter, **40**, p. 25-32.

1981 - The variation of temperature with time and HC maturation in sedimentary basins formed by extension. Earth and Planetary Sci. Letters, **55**, p. 87-98.

Marle, 1972 - Les écoulements polyphasiques en milieu poreux. Cours de production, tome IV, Ed. Technip.

Moretti, I., Froidevaux, C., 1986 - Lithospheric thinning and extensional tectonics. This volume.

Perrin, J.F., 1983 - Modélisation du champ thermique dans les bassins sédimentaires. Application au bassin de la Mahakam-Indonésie. Thèse de Doct. Ing. Université Bordeaux I.

Royden, L., Sclatter, J.G., Von Herzen, R.P., 1980 - Continental margin subsidence and heat-flow: important parameters in formation of petroleum hydrocarbons. AAPG Bull., **64**, P. 173-187.

Sclater, J.G., Christie, P.A.F., 1980 - Continental stretching; an explanation of the post mid-cretaceous subsidence of the central North Sea Basin. J. of Geophysical Research, **85**, B.7, p. 3711-3739.

Smith, L., Chapman, D.S., 1983 - On the thermal effects of groundwater flow. Regional scale system. J. of Geophysical Research, **88**, B.1, p. 593-608.

Terzaghi, K., Peck, R.B., 1948 - Soil mechanics in engineering practice. Wiley N.Y., 2nd ed., 84 p.

Tissot, B., Espitalié, J., 1975 - L'évolution thermique de la matière organique des sédiments : applications d'une simulation mathématique. Rev. Inst. F. du Pétrole, **30**, p. 734-777.

Tissot, B., Welte, D.H., 1978 - Petroleum formation and occurrence. Springer verlag, Berlin.

Ungerer, P., Bessis, F., Chénet, P.Y., Durand, B., Nogaret, E., Chiarelli, A., Oudin, J.L., Perrin, J.F., 1984 - Geological and geochemical models in oil exploration: principles and practical examples. AAPG Memoir 35, Demaison G., Murris R.J. ed., p. 53-78.

Ungerer, P., Espitalié, J., Marquis, F., Durand, B., 1986 - Use of kinetic models of organic matter evolution for the reconstruction of paleotemperatures. Application to the case of the Gironville well (France). This volume.

Watts, A. and Steckler, M., 1981 - Subsidence and tectonics of Atlantic type continental margins, Oceanologica acta, N° SP, Colloque C3. Geologie des marges continentales, 26th, IGC, p. 143-154.

PART B

HEAT FLOW AND WATER FLOW

W. D. GOSNOLD[1], D. W. FISCHER[2]

HEAT FLOW STUDIES IN SEDIMENTARY BASINS

Abstract

In continental heat flow studies, sedimentary basins are usually avoided because of difficulties in obtaining thermal conductivity measurements and because temperature gradients may contain advective signals caused by moving groundwater. These problems are superimposed in the Denver, Kennedy and Williston Basins where complex geothermal gradients derive both from large contrasts among thermal conductivities of strata and from regional groundwater flow.

The occurrence and magnitude of advective heat flow within the Denver, Kennedy and Williston Basins is conceptually consistent with simple models that relate groundwater flow to the piezometric surface and to subsurface structures, i.e., folds and faults. An advective heat flow of $+25$ mW/m^2 has been determined for an area in the eastern margin of the Denver Basin, and quantities of $+35$ mW/m^2 and $+10$ mW/m^2 have been determined respectively for parts of the southeastern and northeastern parts of the Williston Basin. A detailed analysis of bottom hole temperatures obtained from drill holes in the area of the billings anticline in the Williston Basin indicates that information on subsurface structures and groundwater flow may be obtained from heat flow studies. Additional information that may be derived from these heat flow studies includes: the occurrence and nature of geothermal resources, oil source rock maturation and secondary migration of petroleum, formation and deposition of strata-bound ores.

Introduction

Sedimentary basins provide challenging problems for heat flow workers in both applied and theoretical studies. The challenges for

---

(1) *Mining and Mineral Resources Research Institute, University of North Dakota, Grand Forks, North Dakota, USA.*
(2) *North Dakota Geological Survey, Grand Forks, North Dakota, USA.*

applied studies arise in that the surface heat flow in basins contains signals from several sources and sinks in addition to the heat flow in the upper crust. These signals derive from basinal properties such as geological development, hydrogeology, lithology, and structure, and pose problems because they may not be discernable to the heat flow worker. Researchers conducting heat flow studies focused on tectonic implications often find difficulty in obtaining reliable heat flow data due to the nature of sedimentary basins. This is the case for the Great Plains Province of the United States. Previous interpretations of heat flow in the Great Plains (Blackwell, 1969; Combs and Simmons, 1973; Sass et al., 1976; Scattolini, 1978) have not resolved the wide spread occurrence of high heat flow with the stable tectonic history of the province. Recent studies (Gosnold et al., 1982; Gosnold, 1985) have attributed most of the high heat flow to advective sources that derive from regional aquifers. The heat flow disturbance in the Great Plains is essentially the same as that reported for the Alberta Basin (Majorowicz and Jessop, 1981; Lam and Jones, 1984), where recharge areas have low heat flow and discharge areas have high heat flow. Other heat sources in the Great Plains, such as radioactive heat production in basement rocks, have been documented in only a few areas (Blackwell and Steele, 1981; Gosnold, 1982) due to the lack of data.

From a theoretical standpoint, there has been considerable interest in analyzing the formation and thermal history of sedimentary basins (McKenzie, 1978; Turcotte and Ahern, 1977; Haxby, et al., 1976; Ahern and Mrkvicka, 1984; Nunn et al., 1984 Garner and Turcotte, 1984). These studies attribute basin formation to various thermo-mechanical models such as crustal extension or the emplacement of large dense masses within the upper mantle and lower crust. In the general case, subsidence of the sedimentary basin is treated as the response of the crust to a thermal event wherein the subsidence curve appears to match the theoretical cooling curve. Although the thermo-mechanical models offer plausible explanations for the formation and subsidence history of sedimentary basins, they rely on the virtually untestable hypothesis of a past thermal event. In keeping with the tenet of maintaining multiple working hypotheses, we also find plausible explanations for the formation of sedimentary basins due to crustal flexure from sediment loading (Quinlan and Beaumont, 1984) and to a combination of loading and crustal phase changes (Fowler and Nisbet, 1985).

A blend of applied and theoretical studies of the thermal histories of sedimentary basins has dealt with the temperatures of the sediments and coal or hydrocarbon generation in response to paleotemperatures (Lopatin, 1971; Waples, 1980; Klemme, 1975; Hitchon; 1984; Schmoker and Hester, 1984). These studies generally have not used heat flow data but have approximated a component of

HEAT FLOW STUDIES IN SEDIMENTARY BASINS

heat flow, i.e., geothermal gradients, either by assumption or by calculations based on bottom hole temperature data. In most instances the geothermal gradient has been assumed to be linear and nonvarying in time (Waples, 1980; Schmoker and Hester, 1984). These simplifying assumptions about the geothermal gradient can lead to serious errors because the actual gradient is not only nonlinear in a stable basin (Gosnold, 1984), but it also varies with time in an active basin due to compaction and diagenesis of the sediments (Sclater and Christie, 1980).

Clearly applied and theoretical studies of the thermal parameters of sedimentary basins are growing fields of interest, and there is much need for refinement of these analyses. In this perspective, we have attempted to extend applied heat flow studies in yet another way. While the applied heat flow studies of the Alberta Basin (Majorowic and Jessop, 1981) and the Great Plains (Gosnold, 1982) are beginning to account for some heat flow anomalies in stable intracratonic basins, it is intriguing to ask whether heat flow studies could provide useful information on the nature of sedimentary basins. If heat flow anomalies in basins can be attributed to certain properties of the basins, i.e., structure and groundwater flow, can we deduce those properties from the heat flow? The question is intriguing, and we feel that it is certainly worth investigation. We first approach it by examining the results of studies of some of the relations between heat flow and basin properties in several sedimentary basins in the Great Plains. Then we present a detailed analysis of an area in the Williston Basin in North Dakota. To begin we describe the general geology and heat flow of the Great Plains.

Geologic Setting of the Great Plains

The Great Plains physiographic province extends north from central Texas to the Northwest Territories of Canada including an area of about $5 \times 10\,6$ km^2 between the Rocky Mountains and the Interior Lowlands (Figure 1). Sedimentary rocks within the province represent depositional sequences for all of the Phanerozoic. The tectonically stable nature of the crust underlying the area of the Great Plains province during the Phanerozoic is indicated by the areal continuity of these platform sediments. For example, virtually all of the Cretaceous formations are continuous throughout the northern and central Great Plains. The surface of the Precambrian basement is largely horizontal, but it is deformed by major structures such as the Black Hills and the Denver, Powder River, and Williston Basins, as well as numerous smaller structures (King, 1969).

HEAT FLOW STUDIES IN SEDIMENTARY BASINS

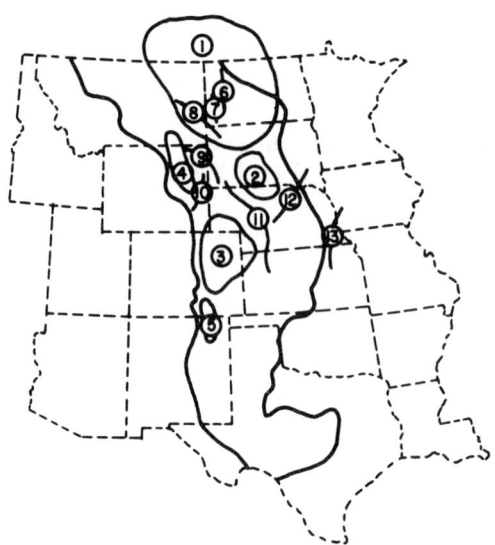

Figure 1 - Location and structural features of the Great Plains. Province boundary shown by dark line. Strucutres: 1-Williston Basin, 2-Kennedy Basin, 3-Denver Basin, 4-Powder River Basin, 5-Raton Bain, 6-Nesson Anticline, 7-Billings Anticline, 8-Cedar Creek Anticline, 9-Black Hills Uplift, 10-Hartville Uplift, 11-Chadron-Cambridge Arch, 12-Siouxana Arch, 13-Nemaha Ridge.

In a tectonic framework, the Great Plains form a broad transition zone between the Rocky Mountain Laramide orogenic belt and the Central Lowlands which have been stable since the Precambrian. Two distinguishing topographic features, i.e., uplifted basement blocks in the northern Plains and a regional eastward slope with elevation changing from about 1600 metres in the west to about 300 metres in the east, imply recent yet mild tectonism in the province. Our interpretation of the tectonic setting based on the tectonic map of North America (King, 1969) is as follows. Deformation of the Precambrian basement underlying the Great Plains is abundantly evident close to the Rocky Mountains but is known only on the Nemaha Ridge in the eastern half of the Plains (see Figure 1). The tectonic style of basement deformation is small scale block faulting similar to block faulting in the Rocky Mountains. Block faulting is especially prominent in eastern Wyoming, western South Dakota, and eastern Montana where uplifted blocks and adjacent basins characterize the boundary between the Plains and the Rockies. In Montana, domal uplifts expose Precambrian basement rocks in the Big Belt, Little Belt, Little Rockies, and Big Snowy Mountains. Volcanic and intrusive activity occurred during the Tertiary in the northern Great Plains, and in or near the Black Hills as recently as the Pliocene. This mild tectonic episode appears to have extended eastward in the province involving structures such as the Chadron and Miles City arches, the Nesson and Cedar Creek anticlines, and a number of other basement faults in Montana and North Dakota. The easternmost deformation of the basement in the Great Plains province occurs along the Nemaha ridge in Kansas and Nebraska. The age of most recent movement on the Nemaha ridge is thought to be Cretaceous (Steeples, 1982). Otherwise the province has been tectonically stable since Precambrian time.

Previous Heat Flow Studies

Published heat flow data for the Great Plains province are few and interpretations of these data vary. In the first interpretations, Blackwell (1969) and Roy et al., (1972) extended the zone of high heat flow associated with the southern Rocky Mountains northward into the Black Hills (Figure 2). This interpretation was based on three heat flow data in South Dakota and on the concept of continental heat flow provinces (Roy et al., 1972). Roy et al., (1972) suggested that heat flow in the remainder of the Great Plains should be similar to that observed in the eastern U.S., i.e., about 50 mW m^{-1}.

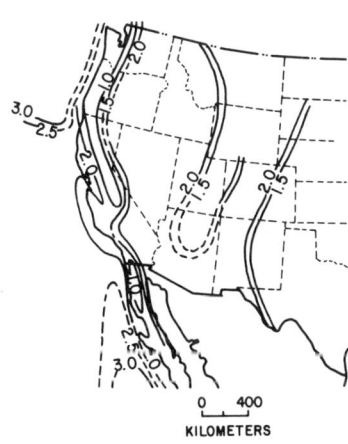

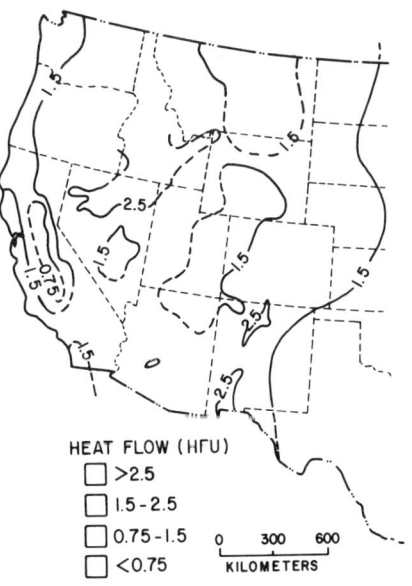

Figure 2 - Heat flow contour map of western United States adapted from Roy et. al., 1972. (1 HFU = 41.8 mW m^{-2})

Figure 3 - Heat flow contour map of western United States adapted from Sass et. al., 1976.

In subsequent studies high heat flow was observed at sites in North and South Dakota (Combs and Simmons, 1973) and in Kansas and Colorado (Sass et al., 1976). On the basis of these additional data, and from the general coincidence between continental heat flow provinces and physiographic provinces, Sass et al. (1976) and Lachenbruch and Sass, (1978) inferred that heat flow greater than 1.5 HFU (63 mW m^{-2}) characterized much of the central and northern Great Plains (Figure 3).

The interpretations of heat flow in the Great Plains by Roy et al., (1972) and Sass et al., (1976) differ in several ways other than the obvious conclusions. Blackwell (1978) pointed out that the former map was an attempt to show the general conditions existing in the upper mantle below an average continental crust, whereas the latter map was an attempt to show surface heat flow conditions. Blackwell (1978) further suggested that neither map can satisfy all needs for interpretation of heat flow in a region. One map is too general for delineation of local anomalies, and the other map does not distinguish between regional and local anomalies. Blackwell (1978) offered a revised heat flow map which is interpretative in nature. This map (Figure 4) includes additional data from the northern Great Plains (Scattolini, 1977) and also shows much of the Great Plains as having high heat flow.

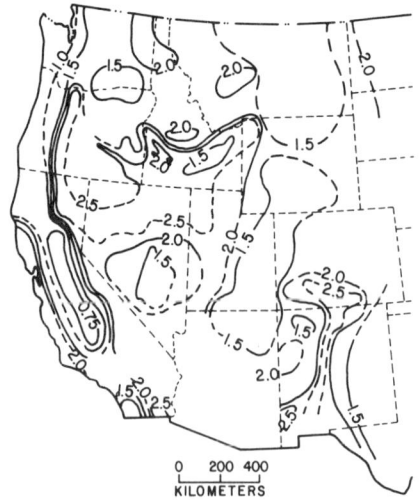 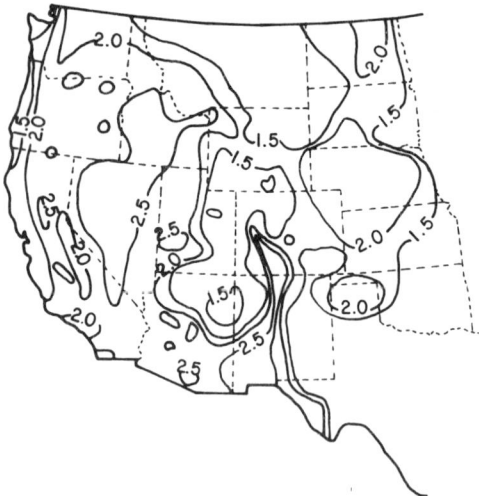

Figure 4 - Heat flow contour map of western United States adapted from Blackwell (1978).

Figure 5 - Heat flow contour map of western United States based on silica temperatures. Adapted from Swanberg and Morgan (1978).

HEAT FLOW STUDIES IN SEDIMENTARY BASINS

Swanberg and Morgan (1981) used an unconventional approach to heat flow investigation on a regional basis by applying an empirical linear relation between the results of silica geothermometry and regional heat flow (Swanberg and Morgan, 1979). A significant advantage of Swanberg and Morgan's (1981) approach to regional heat flow analysis is that the number of existing silica data is several orders of magnitude greater than the number of conventional heat flow data. However, exactly what is represented by the silica-heat-flow analysis is uncertain. The data may be significantly influenced by groundwater circulation, thus they may contain considerable "noise." Blackwell and Steele (1981) suggest that the silica heat flow data may be more appropriately related to geothermal gradients than to heat flow. Nevertheless the heat flow map based on silica geothermometry (Swanberg and Morgan, 1981) must be seriously considered in any analysis of heat flow in the Great plains because the map correctly predicts the occurrence of high heat flow within large regions from North Dakota to Texas (Figure 5).

In theory the Great Plains should exhibit low heat flow similar to that of the eastern U.S. except where Cenozoic tectonism occurred in the northern Great Plains. The wide spread occurrences of high heat flow probably indicate local sources which could be anomalously radioactive basement rocks or heat transported by moving groundwater. Recent studies (Gosnold, 1982, 1985) suggest that both types of sources may occur within the Great Plains, but that heat advection in regional aquifers may dominate the thermal regime of large areas.

Sources and Sinks

At least three localities within the high heat flow areas in the Great Plains show evidence for advective heat flow disturbances. These localities are the eastern flank of the Denver Basin, the Kennedy Basin, and parts of the Williston Basin. Discovery of the advective heat flow disturbances came about through heat flow studies conducted for geothermal exploration (Gosnold and Eversoll, 1982; Gosnold, 1984; Harris et al., 1981). The magnitudes of the disturbances are estimated to be on the order of 15 mW m^{-2} in the Denver Basin, 20 mW m^{-2} in the Kennedy Basin, and 12 mW m^{-2} in the Williston Basin. Details of the these advective systems were discussed by Gosnold (1985) and are summarized here.

The advective heat flow disturbance in the Denver Basin occurs along the eastern flank of the Basin just west of the Chadron Arch. Geothermal gradients in the area range from 41 K km^{-1} in the west to 60 K km^{-1} in the east. The trend in temperature gradient patterns corresponds to the predicted effect of heat advection in eastward groundwater flow in a regional aquifer. In this case several members of the Dakota Group, the basal Lakota and the "D" and "J" sands, contribute to a complex aquifer system. Groundwater flow in this system is toward the east at a rate of about 1 meter per year and is driven by an eastward sloping potentiometric gradient (Helgeson, et al., 1982). The nature of the heat flow disturbance is similar to that reported for the Alberta Basin (Majorowicz and Jessop, 1981; Lam and Jones, 1984) where the discharge areas show high geothermal gradients and the recharge areas have low geothermal gradients. In the Denver Basin, the high heat flow zone ends west of the Chadron Arch where the sedimentary section is very nearly flat lying. No disturbance over the southern part of the Chadron Arch has been detected in what is still a sparse data set.

Simple models matching heat flow and groundwater flow in the Dakota Group in the Denver Basin were proposed by Gosnold (1985). These models were based on groundwater flow within the Dakota aquifer only and predicted that flow rates on the order of 1.2 metres per year could cause the observed heat flow anomaly. A complicating factor not considered in these models is that the aquifer is not confined, and it receives recharge through regional fracture systems (Helgeson et al., 1982; Belitz and Bredehoeft, 1983). The occurrence of recharge through a regional fracture system severely limits the applicability of simple models for matching heat flow and groundwater flow in the aquifer. The effects on the temperature regime by downward moving water in fractures could be modeled in conjunction with models of flow within aquifers but no such attempts have been made. Many problems arise in the realm of aquifer recharge, not the least of which is that, in the central and northern Great Plains, most of the recharge to aquifers is periodic and occurs in the springtime.

Advective heat flow disturbances in the Kennedy Basin occur in the recharge areas of the Madison and Dakota aquifers on the eastern flanks of the Black Hills, in relation to structures in the basin, and in the discharge areas of both aquifers. Background heat flow in the area of the Blackhills is considered to be greater than 1.5 HFU (63 mW m^{-2}), and two heat flow measurements of 0.6 HFU (25 mW m^{-2}) were attributed to the effects of downward moving groundwater (Roy et al., 1972). Several manifestations of advective heat flow appear on a geothermal gradient map of South Dakota by Schoon and McGregor (1974). A generalized cross section derived from that map (Figure 6) shows that low geothermal gradients occur in the recharge areas of these two regional aquifers and extend along the western descending limb of the Kennedy Basin. A small geothermal gradient high occurs over a structure that may be an uplifted basement block just east of the Blackhills. There are two other prominent geothermal gradient highs, one over the region where the Madison aquifer discharges into the overlying Dakota aquifer, and the other

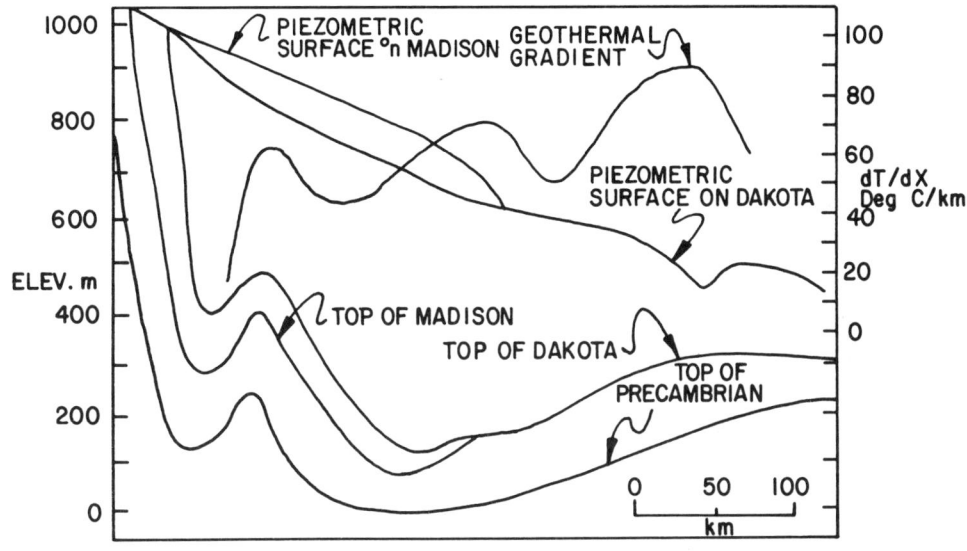

Figure 6 - Generalized cross section of Kennedy Basin.

in the discharge area of the Dakota aquifer along the Missouri River.

In the Northern Great Plains, at least six major regional aquifers contribute to the surface heat flow signal. Major studies of the five lower aquifers, Madison (Mississippian), Duperow (Devonian), Interlake (Silurian), Red River (Ordovician), and Deadwood (Cambrian-Ordovician) by Downey (1984), Konikow (1976) and MacCary (1984) provide an overview of the relation between basin hydrology and subsurface temperatures. In general these aquifers are recharged at high elevations in Montana, Wyoming, and South Dakota and discharge hundreds of kilometres to the northeast in eastern North Dakota and southern Manitoba (Figures 7 & 8). High density brines appear to divert fresh water flow around the central part of the Williston Basin causing flow to the north, south and upward into overlying formations (Downey, 1984). Fracture systems penetrate the entire geologic section (Downey, 1984) and may influence surface heat flow by their effects on vertical and horizontal fluid flow. High porosity zones in the Madison and underlying aquifers (Downey, 1984) correspond with geothermal gradient highs shown by Konikow (1976) and Harris et al. (1981).

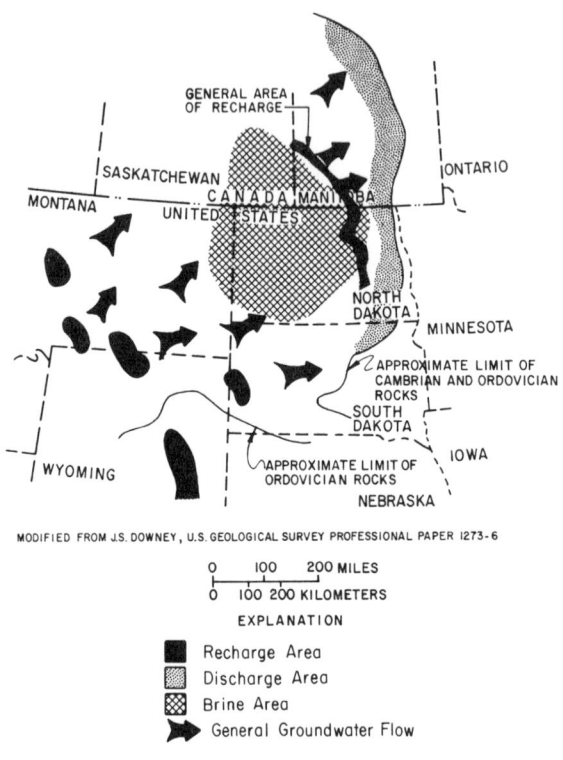

Figure 7 - Flow directions, recharge areas and discharge areas in the Cambro-Ordovician Aquifer. Adapted from Downey (1984).

answers to these questions we have conducted a detailed study of the Billings Anticline, a structure in the Williston Basin.

These correlations between basin hydrogeology and temperature patterns in regional aquifers suggest that thermal structure could be predicted on the basis of hydrogeologic data. Could the inverse case also be true? Could heat flow studies provide insight into hydrogeologic and structural properties of basins? In pursuit of

HEAT FLOW STUDIES IN SEDIMENTARY BASINS

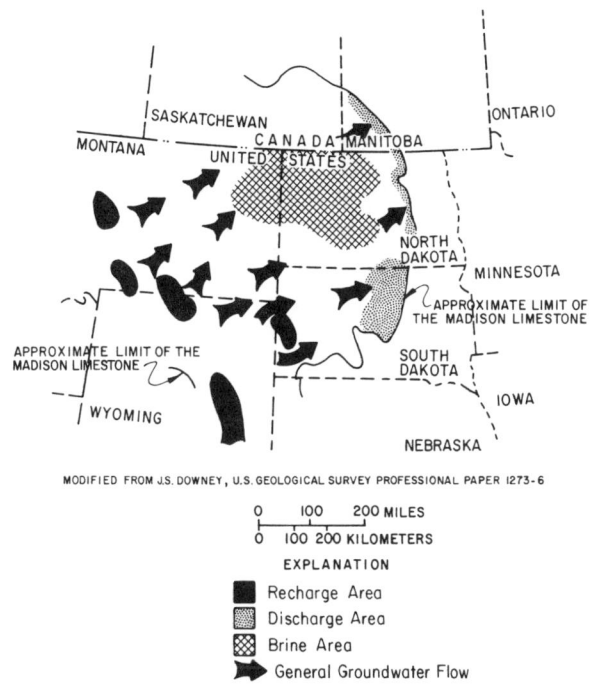

Figure 8 - Flow directions, recharge areas and discharge areas in the Madison Aquifer. Adapted from Downey (1984).

Analysis of the Billings Anticline

Well logs from more than 1000 drill holes in area of the Billings Anticline in the Williston Basin (see Figure 1) provided data for a detailed study of the relation between regional groundwater flow, structure and heat flow. A data base including up to nineteen formation tops from the Pierre Shale (Cretaceous) to the Precambrian, and temperatures from 416 drill stem tests was compiled for analysis. The stratigraphic data were used to produce structure contour maps for the "Inyan Kara (Cretaceous), Charles Salt (Mississippian), Bakken (Devonian), and Red River (Ordovician)," base of the Charles Salt, Bakken, and Red River Formations. These formations are unambiguous picks on the well logs, thus they provide good structural control for mapping. The structure of the Billings Anticline is shown as a three dimensional surface on the Bakken Formation in Figure 9. The view in this figure is toward the southwest, and the figure is tilted 30 degrees to the northeast.

209

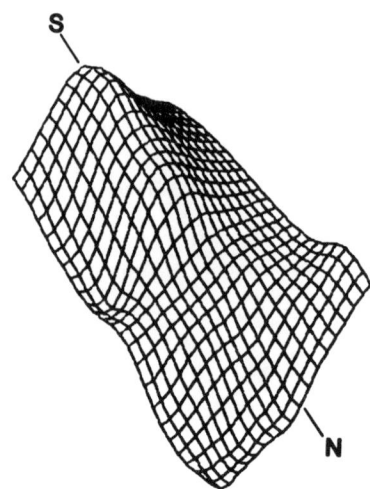

Figure 9 - Three-dimensional surface representation Structure of the Bakken (Devonian) Formation. The view is toward the southwest and the surface is tilted 30 degrees to the northeast. Area = 20KM X 20KM

Regional groundwater flow is eastward across the Billings Anticline in the Inyan Kara, Madison, Duperow, and Red River Aquifers (Downey, 1984). According to Downey (1984), most of the groundwater flow is in the upper aquifers. Groundwater flow in the lowermost aquifers, i.e., Red River and Deadwood, may be blocked by dense brine accumulations.

There are two possible effects of the Billings Anticline structure on subsurface temperatures. If sufficient thermal conductivity contrasts exist, refraction of heat due to the structures could occur. Alternately or in addition, regional groundwater flow could cause high heat flow on structural highs and low heat flow on structural lows. Both cases were tested by constructing two-dimensional finite-difference heat flow models based on stratigraphic data. The models indicate that no significant thermal refraction occurs due to the structure. The reason for this is that the thermal conductivities of the formations affected by the structure, i.e., the Paleozoic carbonates, do not form the necessary contrasting relationships. Analysis of the flowing groundwater model (Figure 10) indicates that a significant thermal anomaly should be produced.

Proper tests for these models would involve considerable expense in obtaining adequate heat flow data. However some data are readily available from oil exploration wells as bottom hole temperatures. Problems in the accuracy of these data are well known

(Chapman et al., 1985; Drury, 1984; Gosnold, 1982). In many cases, the corrections applied to BHT data to obtain true formation temperatures may render the data even less accurate (Drury, 1984). Use of these data as a general data set to show regional trends carries less risk of inaccuracy than attempting to determine actual formation temperatures (Gosnold, 1982; Chapman et al., 1985), and this is the approach we have taken.

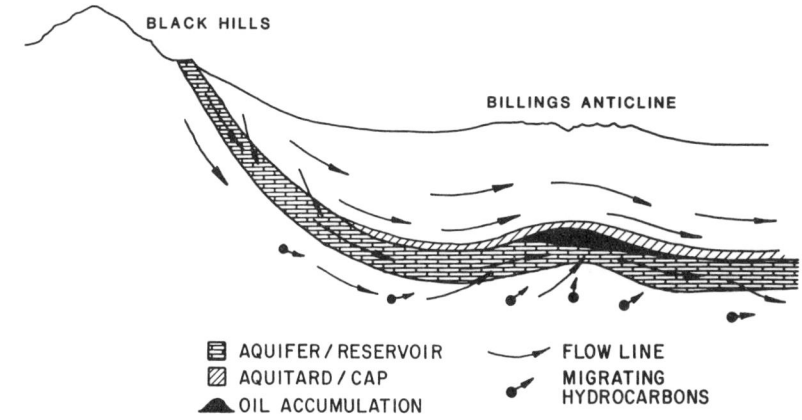

Figure 10 - Conceptual illustration of regional groundwater flow across the Billings Anticline in the Williston Basin.

We chose to use temperatures logged during first-run, conventional bottom hole drill stem tests, and not to attempt any corrections to the temperatures. By selecting these data we eliminated temperatures logged closer to equilibrium conditions, but we avoided mixing equilibrium temperatures with disturbed temperatures.

The BHT data were converted to temperature gradients for analysis of regional trends. Because the geologic section in the study area is essentially lithologically uniform, a two-point temperature gradient calculation gives the same regional trends for gradients as would be obtained calculating the temperature increase across each lithology individually. We evaluated the temperature gradients at two levels. The first data set (Figure 11) is for temperatures recorded at depths less than 3 km and the second (Figure 12) is for temperatures recorded at depths greater than 3 km. This essentially separates shallow data which should be influenced by all underlying aquifers and structures from data that are influenced only by the Cambro-Ordovician aquifer. Comparison of Figures 9, 11, and 12 shows that structural trends are closely matched by the temperature gradient trends for the shallower aquifers but not for the deeper aquifers.

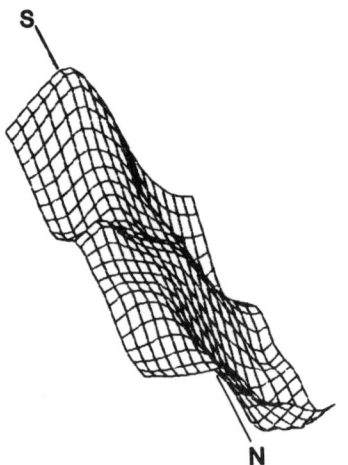

Figure 11 - Three-dimensional surface representation of temperature gradients calculated from BHT data for the Madison (Mississippian) Formation.

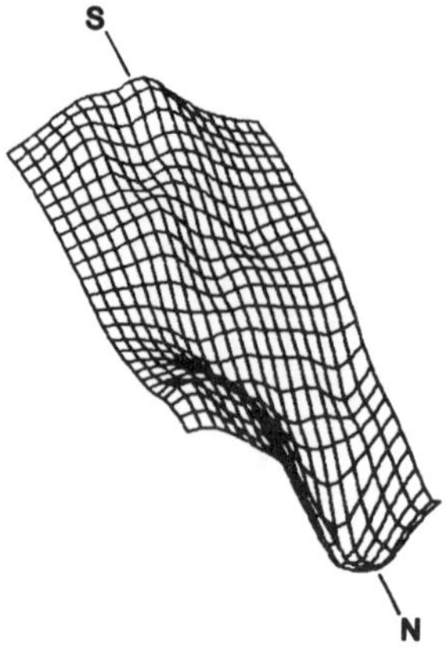

Figure 12 - Three-dimensional surface representation of temperature gradients calculated from BHT data for the Cambro-Ordovician Aquifer.

These results are interpreted to indicate that groundwater flow in regional aquifers directly affects subsurface temperatures in the area of the Billings Anticline. The correlation between trends in the temperature gradient data and the structure of the Billings Anticline suggest that shallow heat flow data might be used to study deep structures where such regional groundwater flow occurs. The Great Plains province might be particularly suited to this type of study because of the thermal insulating effect of the thick Cretaceous shales that cover the entire province.

Implications for Economic Mineral Deposits

Regional groundwater flow in aquifers that also form traps for hydrocarbon accumulations certainly raises interest in the possibility of secondary migration and deposition in hydrodynamic traps. Toth (1984) strongly supports the concept of petroleum deposits in hydrodynamic traps created by local and regional groundwater flow. Hitchon (1984) suggests that a genetic link exists between hydrocarbon occurrence and the fluid regime of the Alberta Basin. In the Great Plains, we find widespread occurrence of thermal anomalies caused by groundwater flow in aquifers which also contain hydrocarbons. Hydrocarbon occurrences on the eastern flank of the Denver Basin coincide with the high heat flow zone. In fact the eastern limits of hydrocarbon accumulations and high heat flow are a close match. Hydrocarbon occurrences in the Billings Anticline area also match the geothermal gradient highs, and wells on the eastern flank of the structure generally are more productive than those on the crest of the structure. Because of the close link between these hydrodynamic conditions and surface heat flow, we suggest that applied heat flow studies could be useful in exploration for hydrocarbons.

Conclusions

A major portion of the high heat flow signal in the Great Plains province of the United States is caused by advective heat sources in regional aquifers. Local and regional recharge and discharge to aquifer systems affect surface heat flow in predictable ways. Groundwater flow in regional aquifers generates heat flow signals that may be used to determine subsurface structures. In light of these findings it seems that applied heat flow studies may be useful in exploration for hydrocarbons.

Ahern, J.L., and Mrkvicka, S.R., 1984, A mechanical and thermal model for the evolution of the Williston Basin, Tectonics, 3, p. 79-102.

Beaumont, C., 1978, The evolution of sedimentary basins on a visco-elastic lithosphere: Theory and examples, Geophys. J. Roy. Astr. Soc., 55, p. 471-498.

Belitz, K.E., and Bredehoeft, J.D. 1983, Hydrodynamics Denver basin, an explanation of subnormal fluid pressures, (abstract) A.A.P.G. Bull., 67: p. 422.

Blackwell, D.D. 1969, Heat flow in the northwestern United States Journal of Geophysical Research, v. 74, p. 992-1007.

Blackwell, D.D., and Steele, J.L., 1981, Heat flow and geothermal potential of Kansas, Final report for Kansas State Agency Contract 949, pp. 69.

Blackwell, D.D., 1978, Heat flow and energy loss in the eastern United States, Geol. Soc. America Memoir 152, p. 175-208.

Combs, J., and Simmons, G., 1973, Terrestrial heat flow determinations in the North Central United States, Journal of Geophysical Research, v. 78, p. 441-461.

Downey, J.S., 1984b, Geohydrology of the Madison and associated aquifers in parts of Montana, Nebraska, North Dakota, South Dakota, and Wyoming, U.S. Geol. Surv. Prof. Paper 1273-G, pp. 47.

Drury, M.J., 1984, On a possible source of error in extracting equilibrium formation temperatures from borehole BHT data, Geothermics, v. 13, p. 175-180.

Fowler, C.M.R., and Nisbet, E.G., 1985, The subsidence of the Williston Basin, Can. J. Earth Sci., v. 22, p. 408-415.

Garner, D.L., and Turcotte, D.L., 1984, The thermal and mechanical evolution of the Anadarko Basin, Tectonophysics, 107, p. 1-24.

Gosnold, W.D., Jr., and Eversoll, D.A., 1982, Geothermal resources of Nebraska, 1:500,000 scale map, National Geophysical and Solar-Terrestrial Data Center, National Oceanic and Atmospheric Administration, Boulder, CO.

Gosnold, W.D., Jr., 1985, Heat flow and groundwater flow in the Great Plains of the United States, Journal of Geodynamics, (in press).

Gosnold, W.D., Jr., 1984, Heat flow as an indicator of regional groundwater migration in the Great Plains, (abstract) A.A.P.G. Bull.

Gosnold, W.D., Jr., 1982, A heat flow anomaly in the Great Plains, EOS, v. 63, p. 1091.

Harris, K.l., Howell, F.L., Winczewski, L.M., Wartman, B.L., Umphrey, B.L., and Anderson, S.B., 1981. An evaluation of hydrothermal resources of North Dakota. Phase II Final Technical Report, E.E.S. Bull. No. 81-05-EES-02, Grand Forks N.D., pp. 291.

Haxby, W.F., Turcotte, D.L., and Bird, J.M., 1976, Thermal and mechanical evolution of the Michigan Basin, Tectonophysics, v. 36, p. 57-75.

Helgeson, J.D., Jorgenson, D.G., Leonard, R.B., and Signor, D.C., 1982, Regional study of the Dakota aquifer (Darton's Dakota revisited). Groundwater, 20: p. 410-414.

Hitchon, B. 1984, Geothermal gradients, hydrodynamics, and hydrocarbon occurrences, Alberta, Canada, A.A.P.G. Bull, v. 68, p. 713-743.

Klemme, J.D., 1975, Geothermal gradients, heat flow, and hydrocarbon recovery, IN A.G. Fischer and S. Judson, eds., Petroleum and Global Tectonics, Princeton Univ. Press, p. 251-304.

King, P. B., 1969, Tectonic map of North America, U.S. Geol. Survey.

Konikow, L.F., 1976, Preliminary digital model of ground-water flow in the Madison Group, Powder River Basin and adjacent areas, Wyoming, Montana, South Dakota, North Dakota, and Nebraska, U.S. Geol. Surv. Water-Resources Investigations 63-75, pp. 44.

Lachenbruch, A.H., and Sass, J.H., 1977, Heat flow in the United States and the thermal regime of the crust, IN The Earth's Crust, edited by J.G. Heacock, American Geophysical Union Monograph 20, p. 626-675.

Lam, H.L., and Jones, F.W., 1984, Geothermal gradients of Alberta in western Canada, Geothermics, v. 13, p. 181-192.

Lopatin, N.V., 1971, Temperature and geologic time as factors in coalification, Akad. Nauk SSSR Izv. Ser. Geol., no. 3, p. 95-106.

MacCary, L.M., 1984, Apparent water resistivity, porosity, and water temperature of the Madison Limestone and underlying rocks of Montana, Nebraska, North Dakota, South Dakota, and Wyoming, U.S. Geol. Surv. Prof. Paper 1273-D, pp. 14.

Majorowicz, J.A., and Jessop, A.M., 1981, Present heat flow and a preliminary paleogeothermal history of the central Prairies Basin, Canada, Geothermics, 10: p. 81-93.

McKenzie, D., 1978, Some remarks on the development of sedimentary basins, Earth Planet. Sci. Lett., 40, p. 25-32.

Nunn, J.A., Sleep, N.H., and Moore, E.W., 1984, Thermal subsidence and generation of hydrocarbons in the Michigan Basin, Amer. Assoc. Petrol. Geol. Bull., 68, p. 296-315.

Quinlan, G.M., and Beaumont, C. 1984, Appalachian thrusting, lithospheric flexure and the Paleozoic stratigraphy of the Eastern interior of North America, Tectonics, 3, Can. J. Earth. Sci., 21, p. 973-996.

Roy, R.F., Blackwell, D.D., and Decker, E.R., 1972, Continental heat flow, IN Nature of The Solid Earth, edited by E.C. Robertson, McGraw Hill, NY.

Sass, J.H., Diment, W.H., Lachenbruch, A.H., Marshall, B.V., Monore, R.J., Moses, T.H., Jr., and Urban, T.C., 1976, A new heat-flow contour map of the United States, U.S. Geol. Survey Open-File Report, 76-756, 24 p.

Sass, J.H., Blackwell, D.D., Chapman, D.S., Costain, J.K., Decker, E.R., Lawver, L.A., and Swanberg, C.A., 1981, Heat flow from the crust of the United States, IN Touloukian, Y.S., Judd, W.R., and Roy, R.F., ed., Physical Properties of Rocks and Minerals, v. II-2. McGraw-Hill Book Company, p. 503-522.

Scattolini, R., 1978, Heat flow and heat production studies in North Dakota. Ph. D. Thesis, University of North Dakota, Grand Forks.

Schmoker, J.W., and Hester, T.C., 1984, Organic Carbon in Bakken Formation, United States Portion of Williston Basin, A.A.P.G. Bull., v. 67, No. 12, p. 2165-2174.

Schoon, R.A., and McGregor, D.J., 1974. Geothermal potentials in South Dakota, South Dakota Geological Survey Report of Investigations, 110, 76 p.

Sclater, J. G., and Christie, P.A.F., 1980, Continental stretching: An explanation of the Post-Mid-Cretaceous subsidence of the central North Sea Basin, J. Geophys. Res., v. 85, p. 3711-3739.

Steeples, D., 1982, Structure of the Salina-Forest City inner basin boundary from seismic studies, University of Missouri-Rolla, Journal No. 3, p. 55-80.

Swanberg, C.A., and Morgan, P., 1979, The linear relation between temperatures based on the silica content of groundwater and regional heat flow, Pure and Applied Geophysics, v. 117, p. 227-241.

Swanberg, C.A. and Morgan, P., 1981, Heat-flow map of the United States based on silica geothermometry, IN Touloukian, Y.S., Judd, W.R., and Roy, R.F., ed., Physical Properties of Rocks and Minerals, v. II-2. McGraw-Hill Book Company, pp. 540-548.

Toth, J., 1980, Cross-formational gravity-flow of groundwater: A mechanism of the transport and accumulation of petroleum (The generalized hydraulic theory of petroleum migration), IN Roberts, W.H., and Cordell, R.J., eds., Problems of Petroleum Migration, A.A.P.G. Studies in Geology No. 10, p. 121-168.

Turcotte, D.L., and Ahern, J.L., 1977, On the thermal subsidence history of sedimentary basins, J. Geophys. Res., 82, p. 3762-3766.

Waples, D.W., 1980, Time and temperature in petroleum formation: Application of Lopatin's method to petroleum exploration, A.A.P.G. Bull., v. 64, p. 916-926.

Chapman, D.S., T.H. Keho, M.S. Bauer, and M.D. Picard, 1984, Heat flow in the Uinta Basin determined from bottom hole temperature (BHT) data, Geophysics, v. 49, p. 453-466.

* Reference added to proofs.

M.N. LUHESHI, D. JACKSON[1]

CONDUCTIVE AND CONVECTIVE HEAT TRANSFER IN SEDIMENTARY BASINS

1.0 INTRODUCTION

An accurate value of the present day basal heat flow is of crucial importance in the study of hydrocarbon maturation. Conventional estimates of heat flow are based on measurements of temperatures in boreholes. These temperatures are then used to evaluate basal heat flow assuming that only vertical steady state conduction is operative. Many theoretical studies of the thermal effects of lateral conductive heterogeneities and fluid flow have been published (see Smith and Chapman, 1983 for a review). These studies show that ignoring these effects can have disastrous consequences on evaluating a representative basal heat flow at any given site.

2.0 HEAT FLOW VARIATIONS IN ALBERTA

This paper describes an attempt to model the heat flow variations in part of the Alberta Basin (for a review see Lam and Jones, 1984 and Hitchon, 1984). On the basis of mainly uncorrected BHT (Bottom Hole Temperature) data various temperature gradient maps of Alberta have been produced which show several anomalies.

---

(1) *Geophysical Division, BP Exploration Company, Ltd, London, UK.*

CONDUCTIVE AND CONVECTIVE HEAT TRANSFER

The variation in heat flow is too large to be accounted for by changes in the basal input. An apparent correlation between variations in hydraulic head with the heat flow anomalies has lead to the belief that the heat flow in the sediment fill is largely controlled by water flow under the influence of various topographic features.

The intention of this study was to see to what extent it is possible to model these heat flow variations using a numerical model produced by Smith and Chapman (1983).

3.0 THE SMITH AND CHAPMAN MODEL

This model consists of a self-consistent numerical solution of the coupled heat and fluid flow equations and includes the following features,

(i) 2-D geometry,

(ii) thermal conductivity is variable and may be anisotropic,

(iii) permeability is variable and anisotropic,

(iv) the viscosity of the fluid may vary with temperature,

(v) the fluid density is a function of temperature (i.e. a buoyancy force is allowed for).

The boundary conditions of the model are,

(i) the surface temperature is specified (but can vary with position),

(ii) the basal heat flow is specified (but can vary with position),

(iii) zero horizontal fluid and heat flux at the side boundaries.

4.0 GEOLOGICAL DATA

The section considered here is marked on the map in Fig. 1. The topography follows the section given by Hitchon (1984, see Fig. 18).

CONDUCTIVE AND CONVECTIVE HEAT TRANSFER

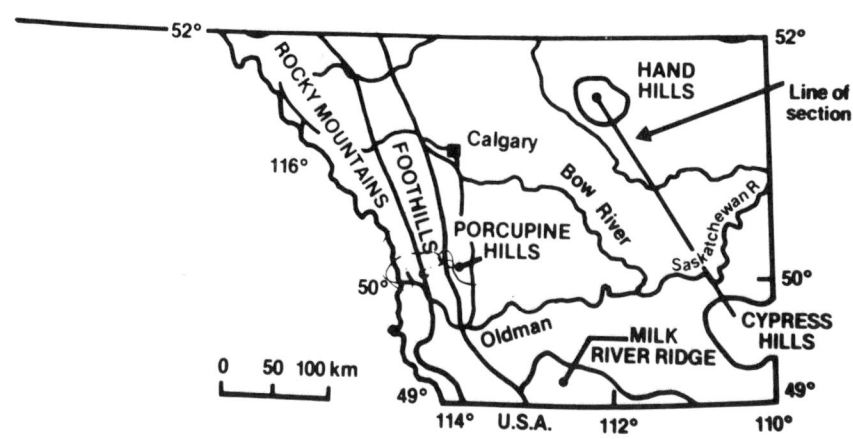

Fig. 1 : Location map showing the model section in Southern Alberta

The groundwater flow regime is assumed to be dominated by the highlands of the Cypress and the Hand Hills. Any hydraulic disturbance due to the Rockies foothills is taken to be be small along the line of section considered here. Schwartz et. al. (1981) have argued that the tectonic events which formed the Rockies may have disrupted the East/West hydraulic continuity, thus leading to a dominantly North/South water flow (roughly parallel to our line of section).

The permeabilities (Fig. 2) were extracted from maps given by Schwartz et. al. (1981). The Schwartz et. al. maps do not cover the whole of the study area and hence it has been necessary to extrapolate values on the depth model. The permeability is highest in the Upper Cretaceous.

The thermal conductivities (Fig. 3) were estimated from data in Majorowicz et. al. (1984) and Majorowicz and Jessop (1981). Above the Paleozoic unconformity the conductivity is generally low apart from the areas which are taken to be predominantly sandy. Below the paleozoic the section contains anhydrites, hence the higher conductivity.

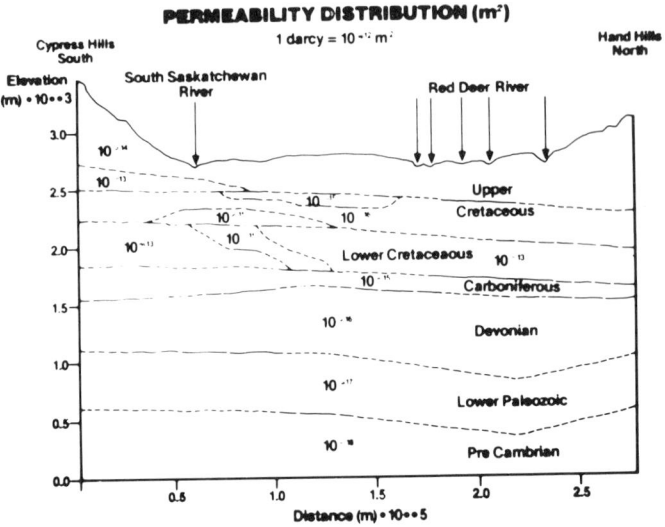

Fig 2 : Permeability distribution

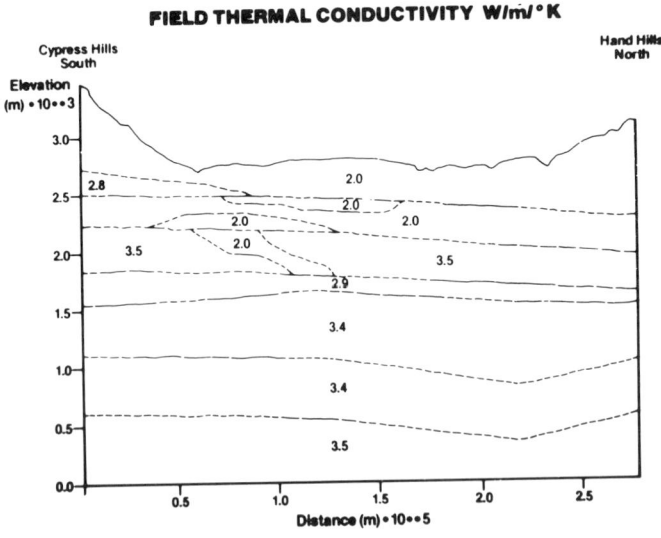

Fig 3 : Thermal conductivity

CONDUCTIVE AND CONVECTIVE HEAT TRANSFER

5.0 MODELLING

The analysis was performed in three parts,

- assuming zero water flow and uniform thermal conductivity,
- zero water flow and variable thermal conductivity,
- non-zero water flow and variable thermal conductivity

The third case is the one which should be comparable with the field data. The first two cases are included purely for the sake of isolating the different physical mechanisms affecting the heat flow.

In all cases the basal heat flow has been assumed to be 60 mW/m^2 and the surface temperature as 6°C. The water viscosity and density were evaluated using equations (9) and (10) in Smith and Chapman (1983).

5.1 Uniform thermal conductivity

This case was considered to isolate the effects of topography and to allow for comparison with the more reasonable heterogeneous models. The whole section was assumed to have a uniform thermal conductivity of 2.5 W/m/K and the permeability was taken as zero.

The only active heat transfer mechanism is conduction. The temperature contours are shown in Fig. 4 and are seen to closely follow the outline of the surface as expected. There is a very small 2-D effect due to variations in topography.

5.2 Non-uniform thermal conductivity

For this example the thermal conductivity in the section is as shown in Fig. 3. The isotherms are shown in Fig. 5 and clearly display a greater variability than Fig. 4. There is a small variation in the heat flow across the section (about 5% from the background basal value of 60 mW/m^2), which is due to refraction.

5.3 Variable thermal conductivity, non-zero water flow

Fig. 6 shows the full solution to the steady state problem. The section has the permeability structure shown in Fig. 2. The isotherms are severely depressed at the southern end of the section below the Cypress Hills and are raised beneath the discharge at the South Saskatchewan River valley.

223

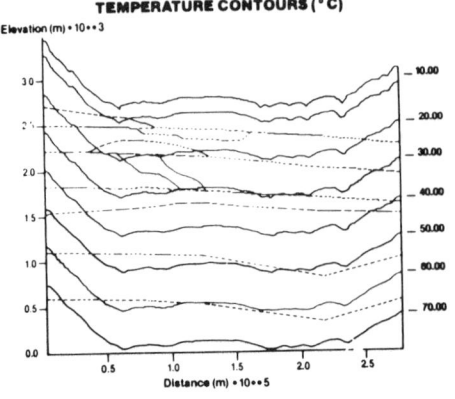

Fig. 4 : Zero water flow, uniform thermal conductivity

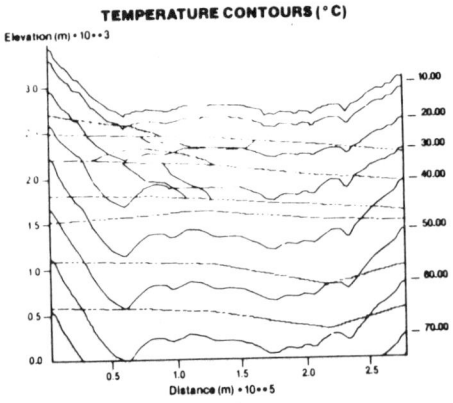

Fig. 5 : Zero water flow, variable thermal conductivity

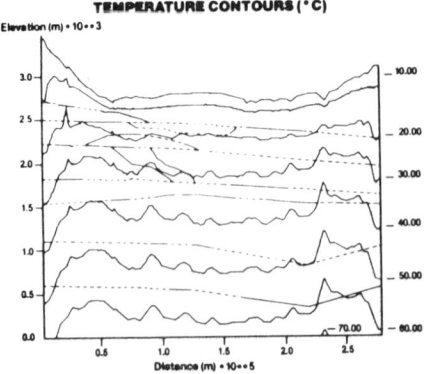

Fig. 6 : Full model allowing for water flow

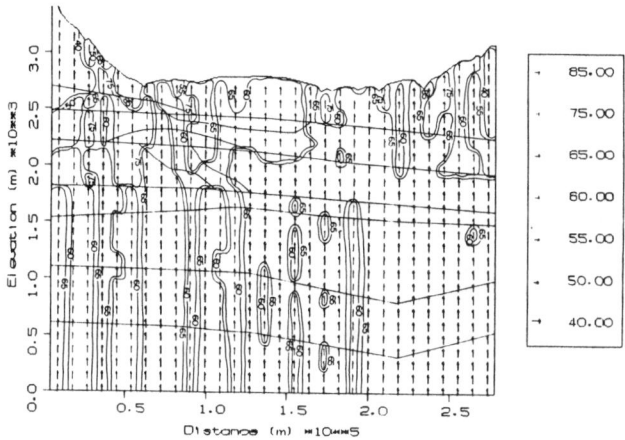

Fig 7 : Vertical heat flux including the effects of water flow

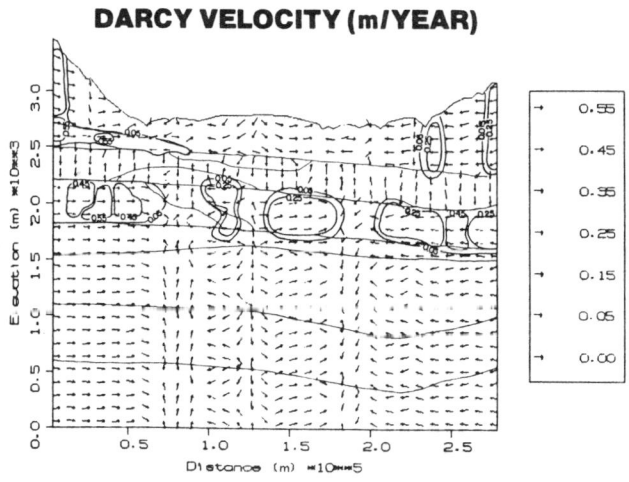

Fig 8 : The Darcy velocity for the model of Fig. 3.

CONDUCTIVE AND CONVECTIVE HEAT TRANSFER

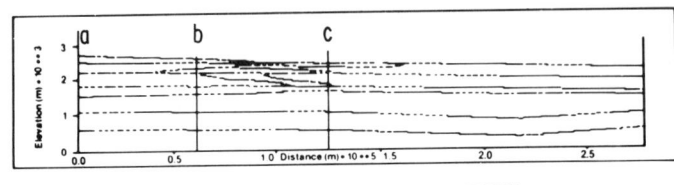

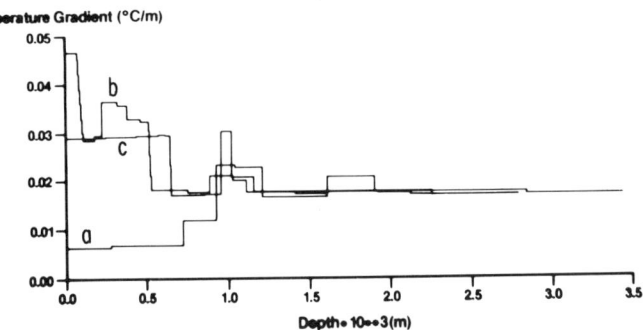

Fig 9 : Typical temperature gradient profiles for Fig. 6.

CONDUCTIVE AND CONVECTIVE HEAT TRANSFER

A similar effect is seen at the northern end of the section with the exception of the 10 degree contour which has not been lowered by as much as might have been expected. The reason for this is that the water flow in this area is low because of the small permeability, implying that the relatively cool water does not penetrate far into the section. The water flow rate at the very north of the section increases with depth down to the Carboniferous leading to the localised lowering of the temperature seen at elevations below 2.5 km (see Fig. 6).

Fig. 7 shows the vertical heat flux across the section. The variability of heat flow is very much greater than in the case of zero water flow. The heat flux is seen to vary from less than 40 mW/m^2 (at the recharge areas below the Cypress and Hand Hills) to 80 mW/m^2 (at the discharge points near the river valleys).

The correlation of heat flux variation with fluid flow is clearly shown by comparing Fig. 7 with a display of the Darcy velocity shown in Fig. 8. The water flow is seen to be concentrated mainly above the Carboniferous, with velocities reaching a maximum of 0.55 m/year.

The influence of the water flow is clearly seen in the temperature gradient/depth profiles shown in Figure 9. The increase of gradient with depth below the Cypress Hills is typical of the effect of water flow. Similarly the decrease of gradient beneath the South Saskatchewan River is indicative of upward flowing water.

5.4 Comparison with heat flow data

Majorowicz et. al. 1984 have produced maps of average heat flow for Southern Alberta which show variations for intervals above and below the Paleozoic surface. Measured heat flow values based on these maps along the line of section shown in Figure 1 are reproduced in Figures 10 and 11 (for intervals above and below the Paleozoic respectively). The basal heat flow input to the model is shown as a horizontal line at .06 W/m^2. The model predictions are for an elevation of 2500m (above the Paleozoic) and 1500m (below the Paleozoic).

Clearly there is a large discrepancy between model and measured data. In particular for the interval above the Paleozoic the measured values are typically half the model predictions (see Figure 10).

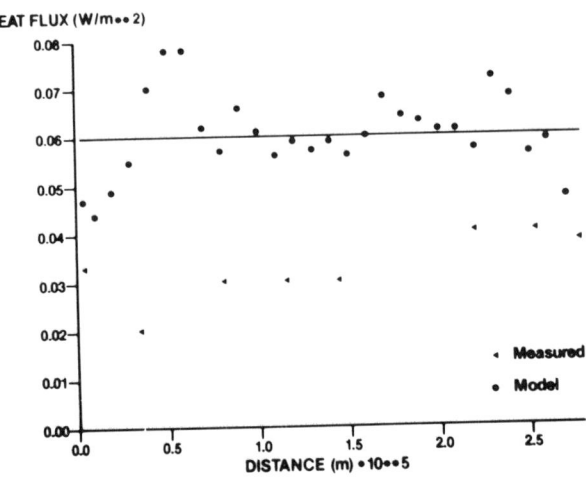

Fig. 10 : Comparison of measured and predicted heat flow

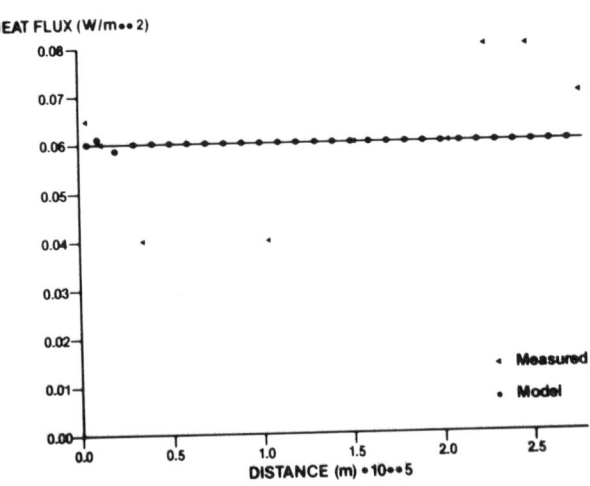

Fig. 11 : Comparison of measured and predicted heat flow

CONDUCTIVE AND CONVECTIVE HEAT TRANSFER

Figure 11 shows that below the Paleozoic the model predicts very little variation in heat flow, whereas the Majorowicz et. al. data show changes of about +/- .02 W/m 2.

The Measured values shown in Figures 10 and 11 are based on temperature gradients evaluated using uncorrected BHT values. Experience has shown that such raw BHT data can differ by typically 10-30% from the real equilibrium formation temperature (at 100 degrees Centigrade) and consequently we can expect the heat flow maps to have errors of roughly this magnitude (Luheshi, 1983).

Corrections for drilling disturbance may not improve matters. In the authors' experience the commonly used BHT buildup models generally produce estimates of equilibrium temperature that can be as much as 15-20% in error depending on circumstances.

The raw BHT data appear to indicate a temperature 'break' at the Paleozoic surface, with the gradient increasing with depth in areas of high topography (in contradiction to the decrease expected from the thermal conductivity data, see Majorowicz et. al. 1984). Given that this effect is real then this would imply water flow in the deeper part of the section.

It is possible to make the water flow penetrate deeper into the section by assuming higher permeabilities in the Carboniferous and the Devonian. Increasing the permeability in the Carboniferous and the Devonian to 100 millidarcies (probably too high for the lithologies unless fracturing is invoked) has a a strong effect on the heat flow above the Paleozoic giving a better match with the data in the recharge zones (see Figure 12). The discharge zone (between about 50 to 200 km.) still shows a significantly higher heat flow. The heat flow below the Paleozoic (Figure 13) is not greatly changed from the results in Figure 11.

Increasing the permeability in the Paleozoic to 100 millidarcies induces flow velocities in these beds of up to 35 cm/year and increases the maximum flow rate in the section from 55 cm/year to 75 cm/year.

5.5 Discussion

The model results show that the temperature distribution in the section is almost entirely dominated by convection effects above the Paleozoic. The contribution of refraction is relatively small. The disturbance of the temperature contours in the Pre-Cambrian is due to the convection in the shallower beds. The heat flow solution in the Pre-Cambrian is almost entirely conductive.

CONDUCTIVE AND CONVECTIVE HEAT TRANSFER

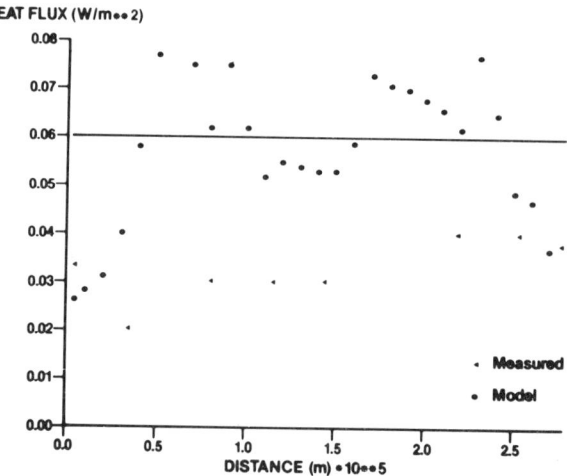

Fig. 12 : Comparison of measured and predicted heat flow; Devonian and Carboniferous beds with 100 md permeability

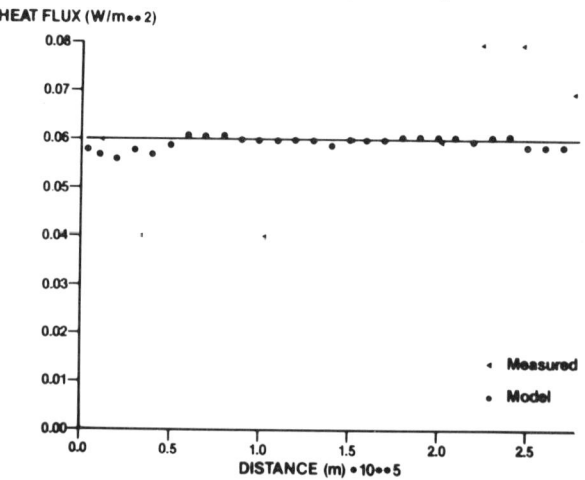

Fig. 13 : Comparison of measured and predicted heat flow; Devonian and Carboniferous beds with 100 md permeability

CONDUCTIVE AND CONVECTIVE HEAT TRANSFER

The model broadly agrees with expectations in that the section is cooled in the recharge areas and warmed in the discharge regions. The results also show variability which reflects relatively small scale local variations in topography. This indicates a need for an accurate map of the water table elevation which forms one of the boundary conditions of the model.

Comparison of the model heat flow with published data shows a large disagreement above the Paleozoic. The BHT derived heat flows indicate a generally very low heat flow above the unconformity across the whole section. The bulk lowering of the heat flow across the unconformity indicates a strong regional down dipping water flow at this level. Such a flow is not included in this model because of the side boundary conditions.

The most likely source of this deeper water in the Carboniferous and Devonian beds is the Rockies Foothills either to the West or to the South of the section considered here. There is a basement high to the west of the Cypress Hills (the Sweetgrass Arch) which probably precludes water flow with the necessary geometry from the Foothills in the West. This and the results given by Schwartz et. al (1981) indicate a mainly northerly flow in the Devonian. The implication of this is that the model section should be extended to the south to include the recharge for the Devonian aquifer. The modelling indicates a flow rate of about 30 cm/year is needed in the deeper aquifer to obtain the reduction in heat flow indicated by the data. This implies the need for a high hydraulic head gradient and/or high permeabilities.

A sensitivity study, not reported here, shows that the fluid flow is very sensitive to the variation of permeability in the section. This is probably the single most notable uncertainty in the present data set. The fluid flow patterns will be strongly affected by more detailed structure and variations in lithology. Since the heat flow is dominated by convection then changes in the geometry and lithology of the section are bound to have a dramatic effect on the temperature field.

The model study has also shown a great sensitivity to the thermal conductivity structure in the section indicating a corresponding need for more detailed maps of this parameter.

An interesting point that is revealed by the fluid flow solution is the existence of a stagnant zone that appears to coincide with the gas fields in the area (see Figure 8 at distances of 100 to 225 km and elevations of 2250 to 2500m). This correlation between low flow zones and the hydrocarbon reservoirs has been noted by other authors (e.g. Hitchon 1984), but here we have been able to produce a quantitative model of the effect. The low flow velocities (typically less than 5mm/year) and their

CONDUCTIVE AND CONVECTIVE HEAT TRANSFER

direction may indicate a hydrodynamic contribution to the trapping mechanism for these fields. Without more detailed maps and knowledge of the local geology one can only speculate on this point.

6.0 CONCLUSIONS

The thermal regime along the line of section shown in Figure 1 is dominated by water flow above the Paleozoic surface. The heat flow pattern predicted by the model is broadly in agreement with expectation (high in discharge zones and low in recharge zones) but detailed comparison with published maps is speculative to a great degree.

The temperatures predicted by the model are very sensitive to the basal heat flow assumptions. This sensitivity extends beyond global changes in temperature gradient in that changes in heat flow modify the water flow (through variations in viscosity and density) thus affecting the shape and extent of anomalies. Given accurate temperature/depth records at various locations along the section it should be possible to assess from these the structure of the basal heat flow. Due to uncertainties in other parameters, though, such an estimate will always have a degree of uncertainty, i.e. the inversion from measured temperatures to heat flow values will not be unique.

The water flow predicted by the model supports the contention by Hitchon (1984) that there is a connection between the hydrocarbon traps in the area and the hydrological state of the section.

This study has confirmed the need for the interpretation of individual temperature data sets in their own local geological setting. Convective effects can clearly be extremely important under certain conditions.

It would appear that using the conventional 1-dimensional steady state model to evaluate basal heat flow from borehole data can be very misleading. This reinforces the need for using accurate numerical models such as that of Smith and Chapman (1983) and Woodbury and Smith (1985) to study the borehole data more carefully.

The comparison with field data has so far been disappointing. The main reason for this is the lack of a strong regional water flow in the Devonian as modelled here. The obvious next step is to include the recharge for the Devonian beds by looking at a section extended further south and also at an East West section to include the Rockies Foothills to the west.

This work has shown that it is possible to model the heat and fluid flow in geological sections of arbitrary heterogeneity. It has also shown that such modelling is essential for the interpretation of detailed temperature data. In order to resolve relatively small scale variations ideally requires accurate temperature/deth surveys conducted in shut-in boreholes.

ACKNOWLEDGEMENTS

The authors wish to express their thanks to L. Smith and D.S. Chapman for kindly providing a copy of the computer code on which much of this work is based. We also thank Dr. Ian Hutchison for helpful discussions and comments on this paper.

In addition the authors wish to thank the Chairman and Board of Directors of The British Petroleum Company plc for permission to publish this paper.

REFERENCES

HITCHON, B., 1984; " Geothermal Gradients, Hydrodynamics, and Hydrocarbon Occurrences, Alberta, Canada". AAPG, 66, 713-743, 1984.

LAM, H.L. AND JONES, F.W., 1984; "Geothermal Gradients of Alberta in Western Canada". Geothermics, 13, 181-192, 1984.

LUHESHI, M.N, 1983: "Estimation of Formation temperature from borehole measurements". Geophys. J. R. Ast. Soc. (1983), 74, 747-776.

MAJOROWICZ, J.A. AND JESSOP, A.M., 1981; "Regional Heat Flow Patterns in the Western Canadian Sedimentary Basin". Tectonophysics, 74, 209-238, 1981.

MAJOROWICZ, J.A., JONES, F.W., LAM, H.L AND JESSOP, A.M., 1984; "The variability of Heat Flow Both Regional and With Depth in Southern Alberta, Canada: Effect of Groundwater Flow?". Tectonophysics, 106, 1-29, 1984.

SCHWARTZ, F.W., MUEHLENBACHS, K. AND CHORLEY, D.W., 1981; "Flow-system Controls of the Chemical Evolution of Groundwater". Journal of Hydrology, 54, 225-243, 1981.

SMITH, L. AND CHAPMAN, D.S., 1983; "On the Thermal Effects of Groundwater Flow, 1. Regional Scale Systems". Journal of Geophysical Research, 88, 593-608, 1983.

WOODBURY, A.D. AND SMITH L., 1985; "On the Thermal Effects of Three-Dimensional Groundwater Flow". Journal of Geophysical Research, 90, 759-767, 1985.

P. GOBLET, E. LEDOUX, G. de MARSILY[1]

POSSIBILITIES OF ABNORMAL FLOW IN SEDIMENTARY BASINS: SOME EXAMPLES

INTRODUCTION

The overall patterns of water flowing from the continents to the oceans in sedimentary basins show a general drainage with a vertical movement of the flow, mainly downward in the upstream zones and upward around the coasts and in the sea.

This general pattern may, however, be significantly disrupted and there are, at present, several hydrogeological examples of such disruptions. Special geological conditions, climatic changes or variations in the sealevel can cause modifications in the underground flow which upset the general pattern during time spans that are not insignificant on the geological time scale.

This paper endeavours to describe this type of phenomenon using two examples based on regional hydrogeologic data.

The first example, taken from the Aquitaine Basin in south-western France, describes fairly localized geological anomalies that have a regional influence affecting the hydrogeology of the entire basin.

The second one comes from West Africa and shows how the rising of the sealevel, which occurred during the Quaternary, is responsible for freakish flows both in the superficial and in the deep aquifers several hundreds of kilometers inland.

Our aim in presenting these two examples, chosen among many others, is to demonstrate what precautions one should take when using a model to reconstruct or extrapolate underground water flow over long periods of time. The general patterns, realistic as to their main

(1) *Centre d'Informatique Géologique, Ecole des Mines de Paris, Fontainebleau, France.*

POSSIBILITIES OF ABNORMAL FLOW IN SEDIMENTARY BASINS

features, may, when based on fixed hypotheses, be flawed and lead to unrealistic representations in the modelled structure and thus produce erroneous simulations.

This problem is especially acute when one is dealing with predictions on a model over very long periods of time. This is the case, for instance, of the assessment of the risks involved in disposing of high-level radioactive waste in deep geological formations where the geologist is expected to reason on the scale of geological time. This is also true in the opposite case when one attempts to reconstruct the circulation of fluids during geological time in order to better understand the formation of various mineral deposits. The study of the appearance of these mineral deposits shows that in most cases they are caused by geological anomalies which have given rise to abnormal concentrations. Thus, any attempt at modelization that does not take account of these special conditions would be doomed to failure. The rule also applies to the case of petroleum accumulations formed by the maturation of organic matter at depth and concentration due to the migration of the oil along favourable pathways.

I. ABNORMAL CIRCULATION IN THE AQUITAINE BASIN

The Aquitaine Basin consists of a sedimentary formation exceeding three thousand meters in thickness and extending over nearly 100 000 km2 in south-western France. The deep geological structure is, by now, well-known thanks to data from oil prospecting. These data were used by hydrogeologists in 1973-74 to attempt an assessment of the total water resources in the basin by building a mathematical simulation model integrating the tertiary and secondary aquifers as far as the Upper Jurassic (Besbes et al., 1974).

The fitting of the model was the result of a cooperation between the University of Bordeaux, the French Geological Survey, the Agency for the Adour-Garonne Basin and the Paris School of Mines using the observed piezometric data and information concerning the history of the exploitation of the different superposed aquifers. Two sectors were identified where the flow anomalies attributed to geological causes have a regional impact.

A. GENERAL HYDROGEOLOGICAL PATTERN

On the regional scale the aquifer system of the Aquitaine Basin may be represented as a stack of 8 superposed aquifers occupying the permeable sedimentary strata (sands and carbonate deposits) and separated by layers of low permeability (clays and marls), that do, however, allow hydraulic communication to take place between the aquifers throughout the basin.

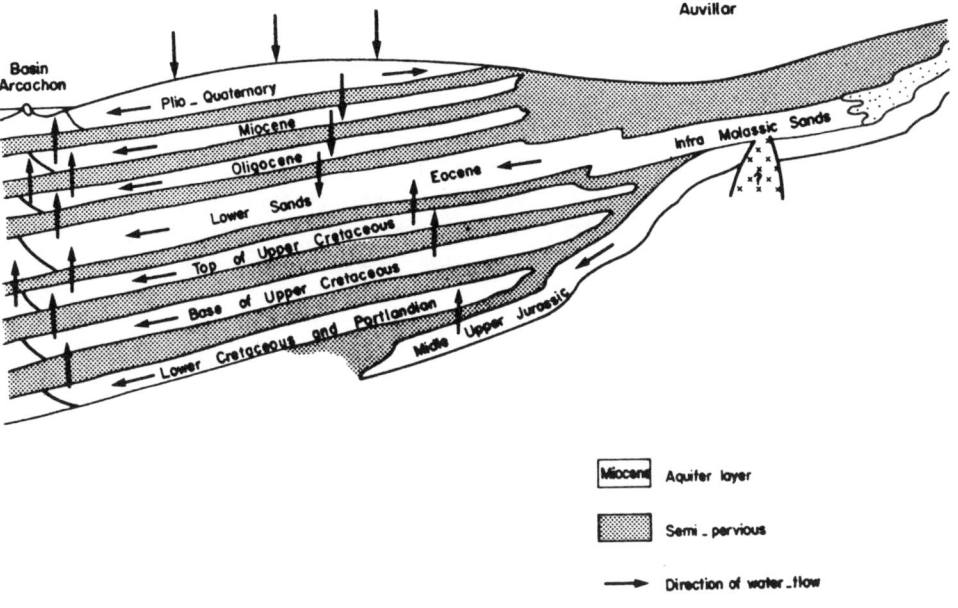

Fig. 1. Pattern of groundwater circulation in the Aquitaine Basin (from Besbes et al., 1974).

From top to bottom on Figure 1 and from north-west to south-east, we see the following:

(a) A plio-quaternary reservoir mainly consisting of sands from the Landes. This aquifer receives rainwater infiltration over its entire area; it is drained by streams and by the Atlantic Ocean.

(b) A miocene reservoir partly fed by the former and drained by the valleys and the sea.

(c) An oligocene aquifer of detrital carbonate character drained by the Garonne river, the region's main water course, and toward the sea.

(d) An eocene "infra-molassic" sandy aquifer which provides the major part of the water resources used in the Bordeaux region. It is recharged to the east and south through the outcrop zones as well as vertically by leakage and it is drained toward the Garonne and the Atlantic Ocean.

(e) A reservoir at the summit of the upper cretaceous in the chalk.

(f) A limestone reservoir at the bottom of the upper cretaceous.

(g) A lower-cretaceous and portlandian aquifer.

(h) An upper- and middle-jurassic aquifer of carbonate character, resting on a thick clay and marl stratum considered as the bedrock of the hydrogeological system.

The system as a whole has the shape of a halo with an incline of the layers toward the west in the direction of the sea. The general groundwater flow runs from the south-east to the north-west, the drainage is shared between the Atlantic Ocean, the Garonne and its main tributaries as shown by the piezometric maps of the different layers (Fig. 2).

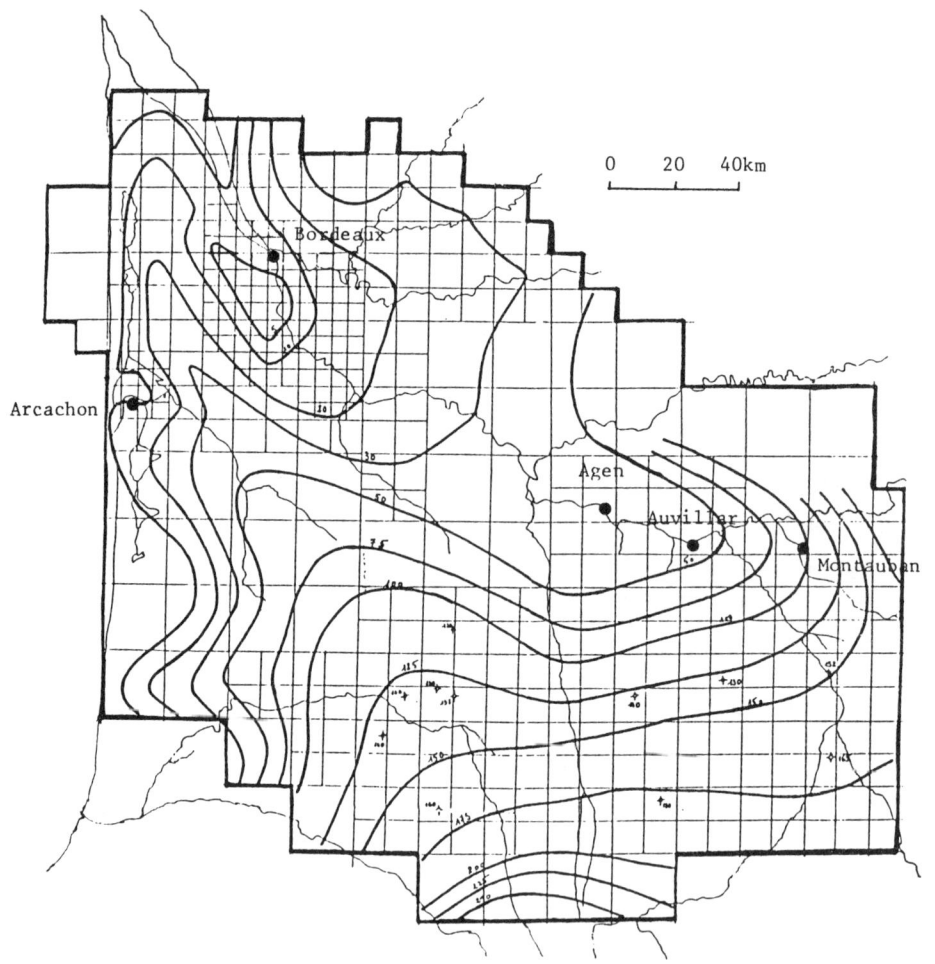

Fig. 2. Piezometric map of the aquifer in eocene and infra-molassic sand. The Auvillar anomaly. (from Besbes et al., 1974).

POSSIBILITIES OF ABNORMAL FLOW IN SEDIMENTARY BASINS

The flow shows a general downward tendency in the upper regions of the basin whereas in the vicinity of the coast the circulation is blocked by the salt water; the movement is then systematically upward.

A more detailed study of the piezometric map, in particular of that of the infra-molassic sands, reveals, however, certain deviations from the general pattern.

B. THE AUVILLAR ANOMALY

Between Montauban and Agen there is a large piezometric depression, the axis of which seems to coincide with the bed of the Garonne (Fig. 2).

The study of the lithology of the infra-molassic aquifer shows that this depression is taken up by a permeable lens enclosed in a formation of low permeability. Since no heavy pumping is done in the region, the origin of this piezometric depression which is 15 m below the drainage level of the river, can only be found in a vertical movement toward a deep aquifer.

There are two possible explanations for this movement: the eastern border of the depression coincides with the granitic dome of Castelsarrazin, which has been explored both geophysically and by boring. This formation probably contains large fracturing that provides communication with the Lias limestone which has a low piezometry. Another, probably stronger, possibility is that under the Garonne there exists a zone of high transmissivity in the carbonate upper jurassic that allows the drained flow to escape. In that case the flow would reappear downstream due to an ascending movement.

No matter which explanation is chosen, observations confirm the existence of a mechanism that generates a vertical movement over several hundreds of km2, disrupting the general pattern of flow in the upstream zone of the Aquitaine Basin.

C. THE ATLANTIC COAST AND THE ARCACHON BASIN

In the vicinity of the relatively straight coastline, the piezometric map shows an almost uniform flow perpendicular to the shore. There is, however, a disruption near the Arcachon Basin where the stream lines in all the 8 aquifers in question converge (Fig. 3).

This phenomenon indicates that the water percolates vertically upward toward a base level constituted by the Arcachon Basin itself.

POSSIBILITIES OF ABNORMAL FLOW IN SEDIMENTARY BASINS

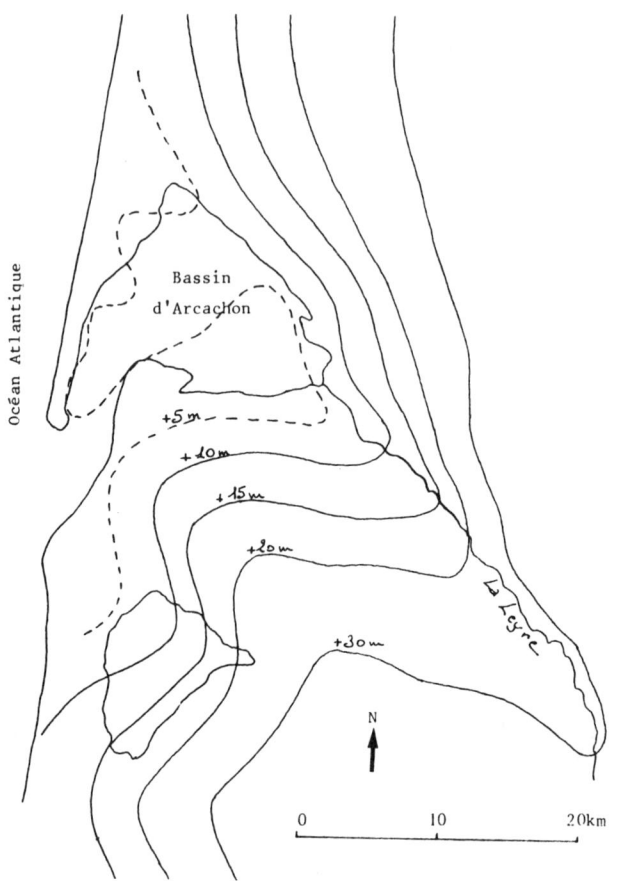

Fig. 3. Piezometric map of the miocene aquifer around the Arcachon Basin.

This is explained by the low permeability of the layers that separate the aquifers at this level and by the barrier to the flow created by the salt water emanating from the sea. The groundwater flow could also be blocked to the west by a variation in the rock characteristics together with tectonic accidents, known to affect this area and which would facilitate the vertical movement of the water.

In a study made on a mathematical model it was estimated that a quantity of water of around 2 m3/s is discharged by the Arcachon Basin from deep drainage. When this value is compared to that of 10 m3/s, which represents the remainder of the outflows on the Atlantic coast, the impact of the anomaly represented by this outlet on the hydrogeological system, becomes apparent.

POSSIBILITIES OF ABNORMAL FLOW IN SEDIMENTARY BASINS

The two preceding examples, taken in the overall context of the Aquitaine Basin, show that there are phenomena, probably of geologic origin, that cause significant disruption of the flow both in the upstream and the downstream areas of the system. If they are disregarded in the modelization of the system, it is probably still possible to give a general picture of the flow structure but one would then leave out flow movements which may strongly influence the formation of deposits.

II. DEPRESSED AQUIFERS IN WESTERN AFRICA

Since precise altimetric data have become available along the coastline of Western Africa, evidence has emerged of aquifers showing piezometric levels below that of the sea.

Among the many possible examples, we have chosen the cases of the Ferlo aquifer in Senegal and the Gondo aquifer in Mali. In each case we will endeavour to offer a plausible explanation of these anomalies based on the geological history of the area.

A. THE FERLO AQUIFER IN SENEGAL

The Ferlo aquifer is situated in north-eastern Senegal in deposits of limestone and marly limestone from the Middle Eocene as well as more or less clayey sands and sandstone from the Terminal Continental.

The piezometric surface is shaped like a large bowl the edges of which abut to the north on the Senegal river, to the west on Lake Guier and to the south and east on the borders of the sedimentary basin (Fig. 4). In the center of this region, one observes piezometric levels of 40 m less than the sealevel although the aquifer lies more than 90 m below the soil surface and the withdrawal through wells is insignificant. This hydrogeological situation is therefore in total disagreement with the normal pattern, which should present a general east-west flow toward the Atlantic Ocean and toward the main drainage axis represented by the Senegal river.

In the center of the bowl, there is a zone where the flows converge. We have seen that human intervention can be excluded as a reason for this phenomenon. The same is true for evaporation because of the depth of the aquifer. The hypothesis of downward drainage toward a deeper aquifer is equally unacceptable since the head in the underlying aquifer in Maestricht sands is higher than that of the Ferlo aquifer. If this were not the case, it would only change the position of the problem since one would then have to account for the depression in the deep aquifer.

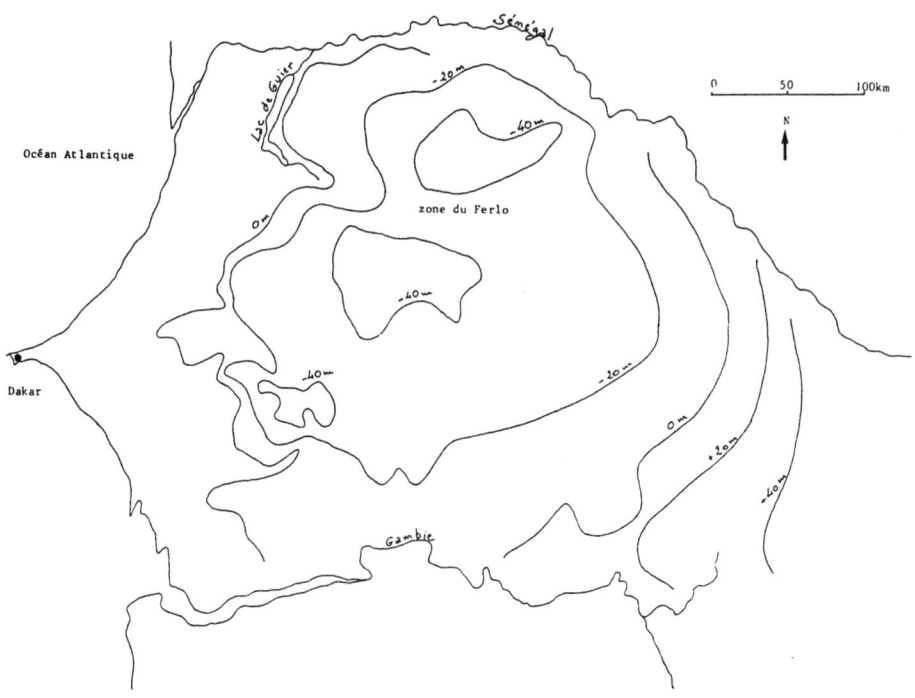

Fig. 4. Piezometric map of the Ferlo aquifer in Senegal.

The explanation of the present piezometric map seems to lie in the conjunction of two phenomena, one climatic, the other neotectonic. The current distribution of rainfall shows a strong gradient from south to north with annual figures of more than 1 200 mm in the south and less than 300 mm in the north, which prevents any recharge of the aquifer in the latter sector. It is therefore probable that the recharge of the aquifer has taken place in the past during the rainy periods of the Quaternary, about 8 000 years ago. It is, however, established that even during wet periods the northern zone must have remained in deficit compared to the southern one, which has made recharge to the aquifer precarious.

It has been found that simultaneously with this climatic evolution, between 20 000 and 5 000 years ago, the sealevel rose. Without recharge from the surface the return to equilibrium can only occur through lateral flow from the zones with high piezometric levels, i.e. the sea, the Senegal river and the areas to the south and the east, where recharge is high. The diffusivity of the aquifer is low (T/S, ratio between the transmissivity and the storage coefficient), which gives this transient regime a strong time constant that may explain why the central zone has not yet been filled up.

B. THE GONDO AQUIFER IN MALI

The Gondo depression on the border between Mali and Burkina Faso offers another example of a depression aquifer.

The aquifer is made up of fairly heterogeneous mio-pliocene continental deposits, bounded to the west by the silicious sandstones of Bandiagra and to the east by the crystalline formation of Burkina Faso. The aquifer has a good continuity and its piezometry shows a depression in its center (Fig. 5).

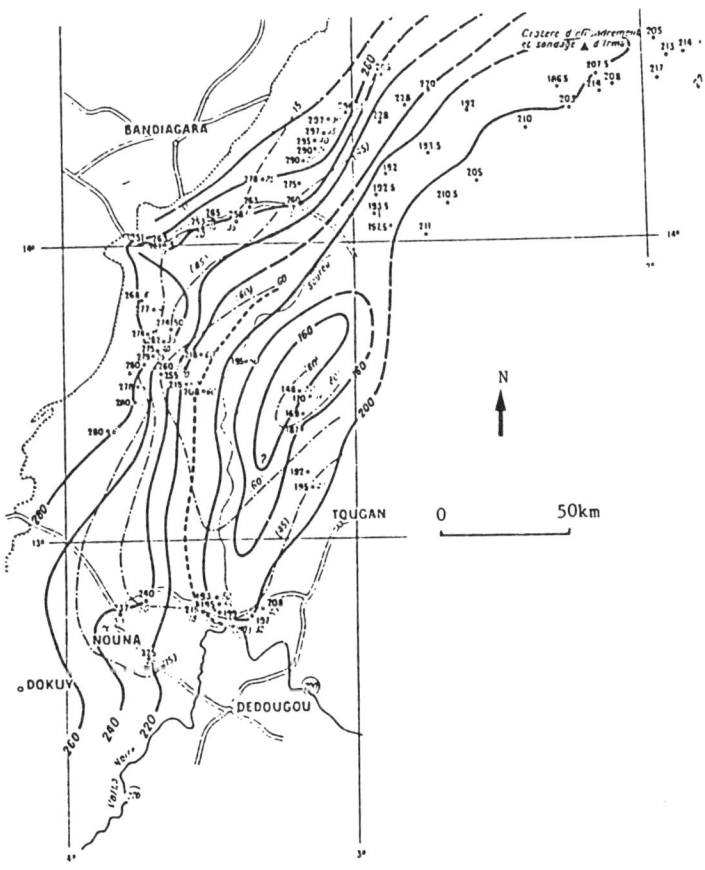

Fig. 5. Piezometric map of the Gondo aquifer in Mali.

The rainfall ranges between 500 and 800 mm which provides recharge by infiltration from the surface but the main source of recharge is the runoff from the sandstone cliff and the crystalline formation.

The depression cannot be attributed to excessive exploitation of the aquifer since the annual volume withdrawn does not exceed a total depth of 0.1 mm which should be more than sufficiently compensated for by infiltration. Similarly, evaporation can be discounted in an aquifer situated at a depth of 90 m below the soil surface. The possibility of the water emptying into an underlying aquifer must also be rejected since the lowest points of the depression are far below (around 80 m) the level of the hydrographic network of the Niger which drains the region.

The most likely hypothesis is that the Gondo depression is filling up with water after having caved in, probably during the Pliocene. A study of the depth of the water in the wells over 20 years has shown that the levels have a tendency to rise.

If this hypothesis is right, the depression is temporary and should gradually disappear while retaining a large time constant in view of the low infiltration rate and the weak transmissivity, which limits the lateral recharge.

CONCLUSION

Several phenomena may be responsible for abnormal groundwater flow which can last for a non negligible length of time on the geological time scale and have a regional impact on the scale of the basin.

Examples of anomalies provoked by geologic, tectonic or climatic causes have been described briefly. These examples were chosen among many others and do not by any means presume to give an exhaustive list of all the causes of anomalies since each basin is a special case.

The purpose of this article is to point out how ordinary, natural phenomena can disrupt a flow and make the task very complex for the person who is responsible for modelling this flow over periods on the geological time-scale.

REFERENCES

BESBES, M., MARSILY, G. de, EMSELLEM, Y., 1974 - Modèle hydrogéologique de l'ensemble du Bassin Aquitain. Ecole des Mines de Paris, Centre d'Informatique Géologique, rapport LHM/RD/74/7.

DEGALLIER, R., 1954 - Hydrogéologie du Ferlo septentrional (Sénégal). Mémoire du BRGM n° 2.

DEPAGNE, J., 1966 - Les nappes déprimées d'Afrique Occidentale. Bulletin du BRGM n° 2.

C. HERMANRUD[1]

ON THE IMPORTANCE TO THE PETROLEUM GENERATION OF HEATING EFFECTS FROM COMPACTION-DERIVED WATER: AN EXAMPLE FROM THE NORTHERN NORTH SEA

SUMMARY

Simple physical principles show that thermal effects of compaction-derived water as a rule can not influence hydrocarbon generation. The release of overpressure from deeply buried rocks may possibly give rise to short-lived, rapid fluid flow which influence the present day temperature. The thermal anomaly at the Troll Field, as described by Eggen (1984), can not be explained by the movement of compaction-derived water.

INTRODUCTION

It is generally assumed that the transport of heat within a sedimentary basin is mainly governed by two factors: Pure conduction and fluid convection. Heat flow equations thus usually contain both a conductive and a convective term (Stallman, 1963; Sharp & Domenico, 1976; Welte & Yükler, 1981; Smith & Chapman, 1983).

Bulk rock properties, such as heat capacity, density and thermal conductivity, are dependent on the water content of the rock. If the velocity of compaction-derived water (water squeezed out of the sediments, as a result of compaction of these

(1) *Regional Geology Department, Statoil, Stavanger, Norway.*

sediments) is slow, because the rock compacts slowly, more water stays in the rock compared to cases where the rocks compact more rapidly. In this respect, the movement of compaction-derived water has an impact on the thermal regime of a sedimentary basin. In the heat transfer equations of the authors cited above, this mentioned impact of compaction-derived water on the bulk rock properties are accounted for in the conductive terms.

The convective terms of the above mentioned heat transfer equations describe the heating effects of the moving water carrying heat from hot to colder regions of the sedimentary basin. The heating effects of such moving water will vary with the geological process which is operating. The forces governing the water velocity may be density contrasts (in the case of natural convection, (Bories & Combarnous, 1973)), differences in ground water level (when the sediments are exposed to ground water movement (Majorowiez el al., 1984)), or hydraulic gradients resulting from the compactional state of the sedimentary basin. Only the latter will be covered in this study.

To determine the thermal effects of comparition-derived water, it is necessary to take a closer look at the fluid flour patterns in compacting sedimentary basins. In calculating the fluid flow velocity and its thermal effects, the best estimates will be obtained using a 3-dimensional deterministic model. However, some simple rules can be employed without such computer modelling. One advantage of using such simple rules, is that the reader easily can evaluate the results, and transfer the methodology to related problems. Such use can seldom be got from results derived from computer modelling. The calculations will, in spite of their simplicity, prove to give valuable results.

HEATING EFFECTS FROM COMPACTION-DERIVED WATER

I. FLUID FLOW PATTERNS

I-A. VERTICAL WATER MOVEMENT

Concider a situation where all the water movement is vertical. If the sedimentary basin has a crystalline basement and contains no sediments when it starts to subside, the net flow of water during basin subsidence will be directed downwards relative to a fixed, non-subsiding point. On the other hand, if the basin contains an infinite sedimentary column, with a fixed porosity vs. depth relationship through time, the water will not move at all relative to this fixed, non-subsiding point (Bonham, 1980), see fig.1. This is the situation which gives the largest difference in velocity between the water and the sediments. At a spesific point in the sedimentary column, this relative velocity equals the subscidence rate of that point.

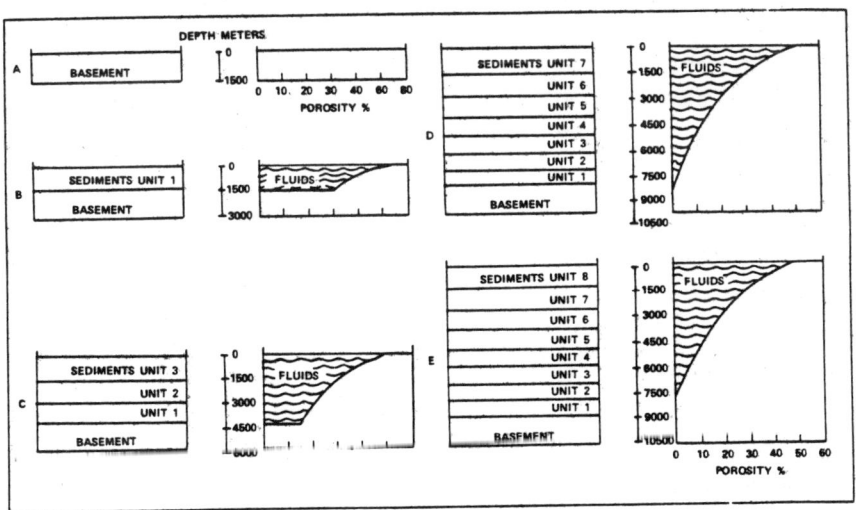

Fig.1. (From Bonham, 1980) Basin developement model. Schematic cross sections on the left represent sequential steps in basin subsidence, sedimentation, and compaction. Equivalent depth-porosity plot for each step is on right. Basin developes from A to E.

249

HEATING EFFECTS FROM COMPACTION-DERIVED WATER

As compaction is most significant at shallow depths of burial, the relative water velocity will here be at it highest. In the Northern North Sea, typical values for such water velocities in the past has been 5m/million year at 3km depth, and 30m/million years close to the sea beds. These numbers have been checked by Statiol's one-dimensional computer simulation program.

I-B. LATERAL MOVEMENT IN SLOPING LAYERS

If the horisontal permeability exceeds the vertical permeability in a sloping sedimentary layer, compaction-derived water may tend to move laterally within this layer. The magnitude of the lateral component is governed by the slope of the layer (and thereby the hydraulic gradient in this layer), and the contrast between vertical and lateral permeability. In shales, lateral permeability is about two orders og magnetude greater than vertical permeability, due to orientation of clay minerals (K. Magara, pers. communication). Such a large permeability contrast will force practically all of the water movement in lateral directions. This can be visualized as follows:

In a sloping layer (with a slope of α degrees), the vertical and lateral hydraulic gradients can be computed assuming a uniform pressure vs. depth relationship, see fig.2.

It is clear from fig.2 that the lateral (X) component of the hydraulic gradient is tan α. (the component of the hydraulic gradient in the normal (Z) direction). Using this relationship and the permeability ratio of Magara in Darcy's law, it turns out that water will prefer to move laterally in beds sloping more than 0,5 degree. If the ratio between the direction permeabilities is reduced with one order of magnitude (the lateral permeability is now ten times the vertical permeability), the water will tend to move laterally at slopes of 6 degrees and higher.

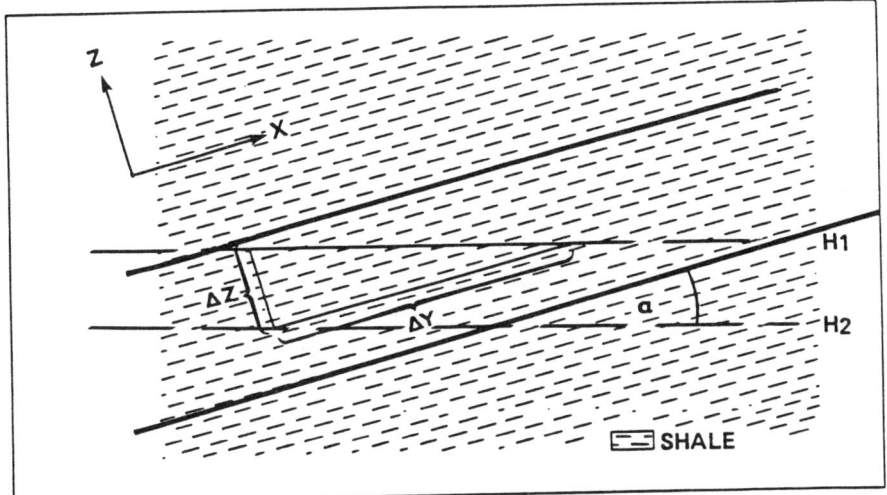

Fig.2 Hydraulic gradients in a sloping layer. The lines H_1 and H_2 run through points of equal hydraulic head. (Hydraulic head = Water pressure - hydrostatic pressure; Hydraulic gradient = Difference in hydraulic head/ distance. Hydraulic gradients govern flow flow through Darcy's law). The hydraulic gradients are $(H_2-H_1)/\Delta Z$ and $(H_2-H_1)/\Delta X$ in the Z and X directions, respectively. Note that $\Delta Z = \Delta X \tan \alpha$.

The contrast between lateral and vertical permeability is, however espesially large at the border between a sandstone bed and an overlying shale sequence. Here the vertical permeability is that of the shale, while the horisontal permeability is that of the sandstone. The water will, accordingly, be forced to move laterally below the border between these rocks (fig. 3).

The same situation will occur within a sandy layer when vertilcally moving water reaches a low permeability horizon (ex. a fine grained layer or a cemented or micaceous zone). The actual water velocity is dependent on the width of the layer with moving water, the subsidence and compaction rate of the sediments, and the mass of water available for movement.

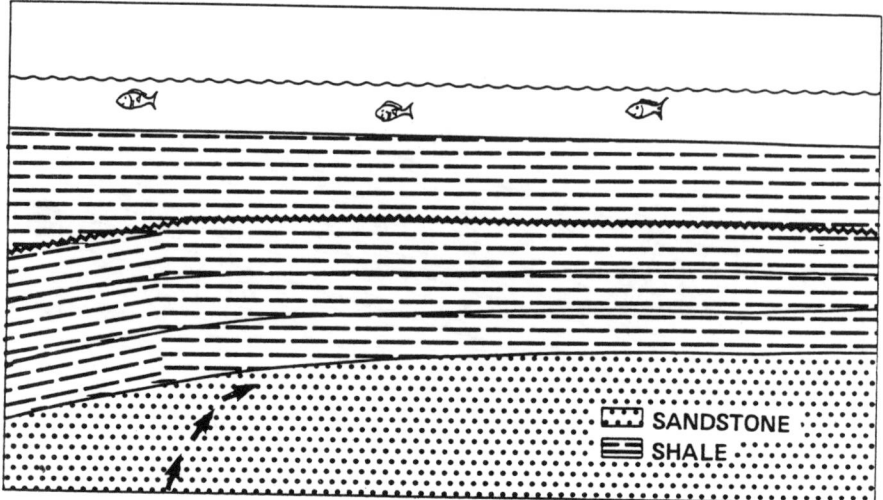

Fig.3. Fluid flow relative to the sedimentary particles. The water moves upwards through the sandstone. When it reaches the border between sandstone and shale, it moves laterally below this border.

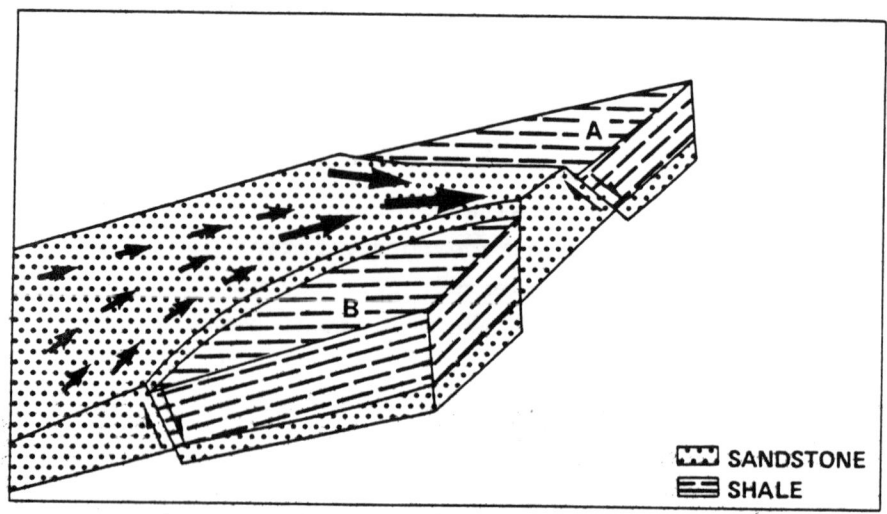

Fig.4. Lateral concentraion of fluid flow due to faults having changed the geometry of the sand body.

HEATING EFFECTS FROM COMPACTION-DERIVED WATER

This mass of water available for movement is dependent, amongst other things, on the stratigraphic separation between the lateral water conduits. If they are narrowly spaced, there will be less water available for movement below each low permeability horizon. Lateral concentraton og fluid flow can also increase the water velocity in a sedimentary basin, an example of this is shown in fig.4.

The uncertainties about the number, width, extent and efficiency of water conduits below low conducivity layers makes prediction of the water velocities during lateral water movement quite uncertain. A more extreme concentration of fluid flow is the situaton where fluids are moving through fractures (Fig.5).

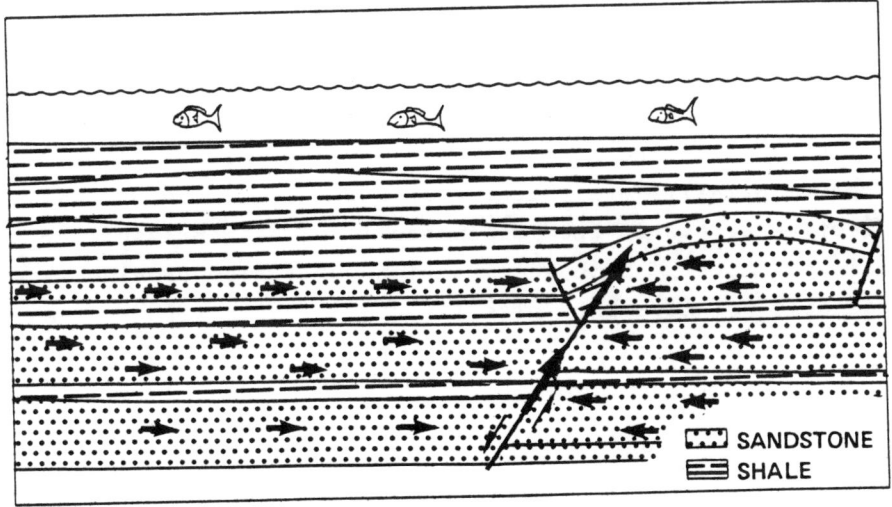

Fig.5. Concentration of fluid flow resulting from water movement through a fracture.

I-C. SHORT-TERM EFFECTS

The velocity of compacton-derived water in a sedimentary basin can for short periods of time be significantly higher than indicataed above. Very fast deposition of sediments can result in a compaction pulse (and following this a water pressure pulse), which may increase the water velocity quite substantially. An example of such rapid sedimentation is the deposition of several hundred meters of moraine material during the latest glaciation periods, in some cases resulting from one single glacier advance.

An other short-term effect which can increase the water velocity is the release of pressure from an overpressured, permeable section. The compressibility of water ($-1/v\, \delta v/\delta p$), is $2{,}2\ 10^{-4}$ bar^{-1} (Dake, 1978), and that of sandstone matrix is $1{,}1\ 10^{-4}$ bar^{-1} (Dickey, 1981). At 3 km depth of burial, a water volume with geostatic pressure would increase volumetrically by approximately 8% if all the overpressure was released, thus providing more fluid available for movement. The expansion increases with depth. If such a pressure release developed over a short time, e.g. due to a fault movement, then a short-lived relatively rapid fluid flow would result. If the sediments were poorly consolidated at the time of pressure relief, they might compact intially (grain rearrangement) due to the increased effective stress in the sediments (increased pressure at the grain contacts). Such a compaction would also provide water available for movement. The effect decreases with increased burial.

II THERMAL EFFECTS OF THE FLUID MOVEMENTS

II-A VERTICALLY MOVING WATER

The heat energy W involved in a temperature change ΔT of a water volume V is (Tipler, 1976):

$$W = \rho\, c\, V\, \Delta T \qquad (1)$$

Here ρ is the water density, and c is the specific heat of the water. This equation can be applied to estimate the heat flow contribution from compaction-derived water.

HEATING EFFECTS FROM COMPACTION-DERIVED WATER

Consider a rock volume V_h being heated by vertically moving fluids from below. For simplicity, assume that the fluids do not loose heat to the surrounding rocks when moving towards the rock volume V_h. On the other hand, it will be assumed that the fluids adjust to equlibrium conditions having reached V_h, emitting energy as described in equation 1. These simplifications will, of course, lead to a significant overestimating of the energy delivered to the rock volume V_h. It will, however, turn out that overestimatin the energy transported by compaction-derived water will not invaliate the conclusions of this paper.

The average temperature difference between the rock volume V_h and the fluids moving into it is ΔT. The water volume (V_w) supplied to V_h during the time t is

$$V_w = v \phi a t \tag{2}$$

where V is the actual fluid veloctiy, ϕ is the porosity and a is the base area of the rock volume V_h. The energy supplied during the time t is then

$$W = c \rho V_w \Delta T \tag{3}$$

If the system is in thermal equlibrium (i.e. not constantly being heated), this energy will flow further upwards, obeying

$$W = Q t a \tag{4}$$

with Q being the heat flow contribution from the compaction-derived moving water. Combining eqs. 2, 3 and 4 gives

$$Q = v \phi c \rho \Delta T \tag{5}$$

Assume a water veloucity of 10 m/million years, $\phi = 30\%$, $c = 4,2$ J $°C^{-1}$ g^{-1} $\rho = 1$ g cm^{-3} and $\Delta T = 25°C$. The T value is actually the largest temperature difference present between the water moving into the rock volume V_h and V_h

255

HEATING EFFECTS FROM COMPACTION-DERIVED WATER

itself, in a time span of 100 million years, with a geothermal gradient of 25°C/km. Equation 5 then gives $Q = 0,01$ mWm^{-2}. Compared with a gross heat flow of about 30 mWm^{-2}, this calculated contribution from compaction-derived water is 3000 times less than the contribution from conduction. Indeed, the severe overestimation of the energy delivered by moving fluids suggests that this energy hardly accounts for more than 1/10.000 of the total heat flow from the sediments. In other words, the flow of compaction-derived water must be concentrated at least 100 times to account for 1% of the total amount of energy transported into a rock volume.

II-B LATERALLY MOVING WATER

Estimates of the velocities of laterally moving water involve large uncertainties, for instance uncertainties concerning the three-dimensional permeability distribution in a sedimentary basin. To calculate the thermal effects of laterally moving compaction-derived water, an approach which does not necessitate determination of fluid flow velocities will be employed. As the volume of the compaction-derived water V_w having moved into a rock V_h is easier to estimate than the actual water velocity, this water volume will be compared to the rock volume V_h, of which heating is to be investigated. This heated rock body (V_h) has a vertical height and base area a,

$$V_h = a\ h \tag{6}$$

Equation 3 is still valid, although equation 2 cannot will be used to estimate V_w. The time span (t) in which the heated rock V_h emits the heat delivered from the compaction-derived water with an assumed even heat flow Q, equals the time necessary to supply the water volume V_w into the heated rock (V_h). Combining equations 3, 4 and 6 gives

$$Q\ t = (V_w/V_h)\ c\ \rho\ h\ \Delta T \tag{7}$$

HEATING EFFECTS FROM COMPACTION-DERIVED WATER

The best estimates of ΔT would be obtained by use of a differential equation analogous to that of Stallman (1963), but taking account for the compaction of the sediments. Computer simlulation is clearly necessary to solve such a problem.

However, it is clear that the heat transfer in a situation like fig. 5 is far more efficient than what is the case in fig. 6.

In the case of nearly vertical migration through fractures (fig. 5), the heat loss to the rocks in which the water migrates is minute compares to the case where long-distance movement in very gently sloping layers is postuated. Therefore, the ΔT value is chosen as: (the average temperature of the water when it enters the fractures) - (the assumed equlibrium temperature of the heated rocks V_h). If heating of the fracture zones did not occur, and no water remained in the fractures, equation 7 would describe the heat trasfer reasonably well. Using the above mentioned value for ΔT in a case like that described in figure 6 will, of course, largely overestimate the energy supply to V_h. Cross-stratal flux of water through fractures, transferring hot compaction-derived water to shallower, colder strata, has been suggested as a heat-transporting agent (Stegena, 1982; Galloway, 1984), in geological settings similar to that in fig. 5. To calculate the ratio (V_w/V_h) necessary to account for a heat flow contribution of 3 mWm^{-2} during a time span of 10 million years from a 100 m thick rock body, a geothermal gradient of 25°C/km will be assumed. As porosity is high right below V_h, (where the water is relativly cold) compared to the porosity at great depth (where the water is hotter) the ΔT value is clearly below the avarage temperature of the rock V_r sourcing the hot water. With $\Delta T = 50°C$ (implying water being sourced from a rock column extending about 6 km below V_h) the (V_w/V_h) ratio becomes 50.

HEATING EFFECTS FROM COMPACTION-DERIVED WATER

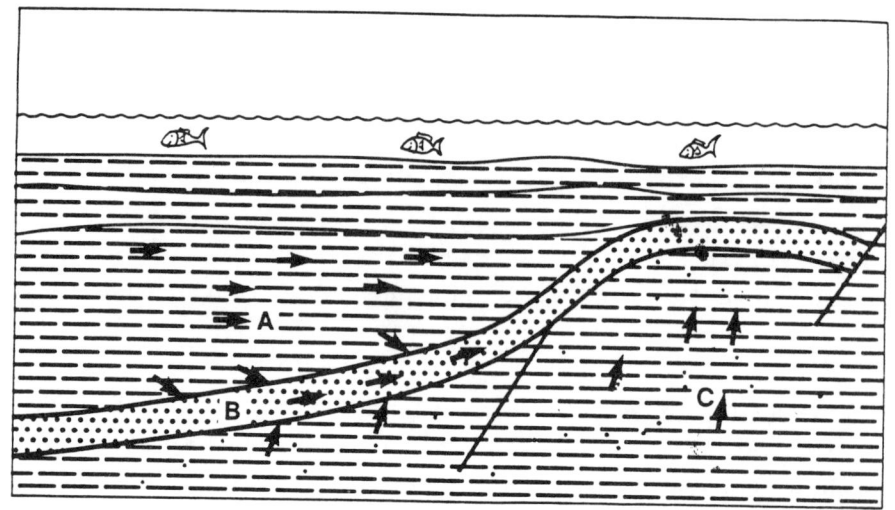

Fig. 6. A thermal situation showing an area (shaded) receiving heat energy from the hot water moving lateral in shales (A), in a carrier bed (B) and upwards (C).

Suppose that the rock from which the compaction-derived water is squeezed out (V_r) had its porosity reduced by 1% (relative to the bulk rock volume) due to compaction in this time span. The ratio (Vr/Vh) is then 5000. In a basin extending 6 km below V_h, with h = 100 m, this (V_r/V_h) ratio implies that the drainage area for water must exceed the area of the heated rock volume V_h at least 85 times.

This (V_r/V_h) ratio of 5000 is probably a dramatic underestimate, due to the simplifications made. Nevertheless, in a situation as shown in fig. 6, if V_h has the size of an oil field, water must migrate in a large rock volume before entering V_h. While moving, heating of the surrounding rocks will take place as much as heating of the rock V_h. If the (V_r/V_h) ratio is extended, so is the heat loss to the surrounding rocks.

It therefore seems clear that a steady state heat anomaly can not be maitained in an oilfield due to the supply of hot

HEATING EFFECTS FROM COMPACTION-DERIVED WATER

compaction-derived water moving in the subsurface sedimentary layers. Even in the case of water movement through fractures, the (V_r/V_h) ratio must have a number (5000) which in most cases is far larger than would be inferred from the geological interpretation, to account for a heat flow of 3 mWm^{-2}, which is about 10% of the total heat flow from the sediments.

II-C SHORT TERM WATER EXPULSIONS; EFFECTS ON TEMPERATURE.

1. Glacial sedimentation

The sedimentation rate due to glacial movements can be extremely high compared to normal sedimentation rates, in some cases perhaps as much as 1 million times higher. Such rates, however, only persist over very short time periods. This, of course, is inferred from the limited tickness of the glacial deposits. Typical thicknesses of Quaternary deposits offshore western Norway range from 50 to 300 meters (Rokoengen & Rønningsland, 1983). In the case of vertical fluid movement, with a geothermal gradient of 25°C/km, these thicknesses correspond to a maximum temperature increase of 0.2°C to 4°C of deeper stata (having taken account for the fact that compaction of the younger strata diminish the subsidence of deeply buried strata). These number are reasonable estimates, provided:

a. The deposition took place in an infinitely short time (thereby prevetning heat loss to the surrounding rocks during water migration).
b. The rocks initially adjust their porosity to the new depth of burial.
c. No heat loss from the rocks due to glacial cooling occured.

Failure in any of these statements will cause the water heating to diminish. Furthermore, as the hot water would almost immediately heat up the surrounding rocks V_h, the total temperature increase of the rocks would be approximately half of

259

the difference between rock and water temperature. Even if heat loss from the heated rocks after the deglatiation is not accounted for, any possible temperature increase due to rapid sedimentation in the glaciation periods of the Quaternary is seen to be negligible. In fact, most authors model a temperature decrease in rocks of glaciated areas, due to the low temperature of the glaciers and permafrost zones (Birch, 1948; Jessop, 1970; Beck, 1977; Hyndman et.al. 1977; Allis, 1978.)

2. Release of overpressure.

The release of overpressure has previously been mentioned as a possible cause for rapid displacement of water. If such an overpressure was released quite recently, it can be argued that the pressure release could create a heat pulse which is still detectable. In this case the water volume V_w expelled from highly permeable rocks (V_r) can be estimated from

$$V_w = V_r \phi \mu \qquad (8)$$

where V_r is the volume of the rocks from which the compaction derived water originate, ϕ is the average porosity of V_r and μ is the amount of water available for movement divided by the amount of water staying in the rock. The value of μ is determined by the rock and water compressibilities, the fluid pressure, and compaction induced by the pressure release. As μ hardly exceeds 0.1 this number will substitute μ in the calculations.

The excess heat flow transferred upwards from V_h, genereted by the energy transported into V_h by moving fluids can be modelled as being emitted at an even rate ($=Q_a$) from the time of water movement to the present time. This approach will never underestimate the present day heat flow contribution from the migrated water. In fact, Q_a can be substituded by $Q_a/2$ without violating this statement, as seen from fig. 7.

HEATING EFFECTS FROM COMPACTION-DERIVED WATER

Consider a situation where overpressure was relased 10000 years ago from deeply buried rocks, (average porosity = 10%), ΔT being as high as 100° C. Suppose that enough energy was transported into a 100 m thick rock body to allow for an even heat flow of 8 mWm^{-2} from this heated rock (implying $Q_a/2$ = 4 mWm^{-2}, which is the highest value that could be observed today under the given conditions). Equation 7 would imply a (V_w/V_h) ratio of 0.06, which implies (equation 8) a (V_r/V_h) ratio of 6.0.

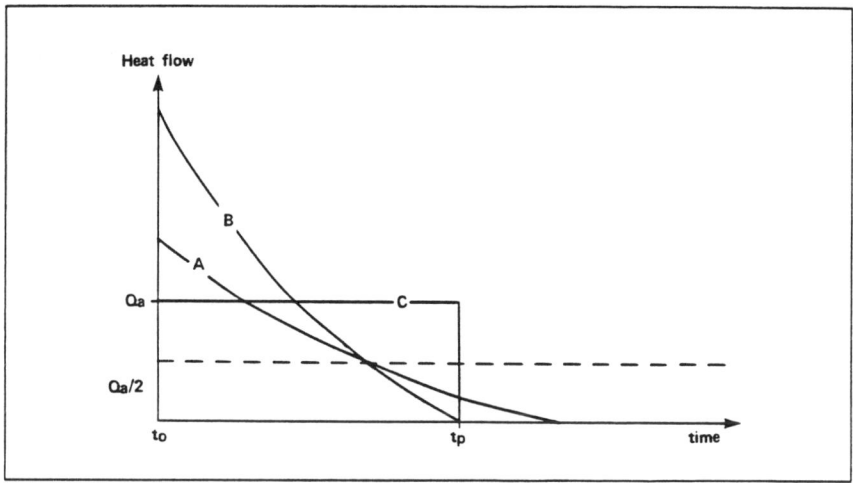

Fig. 7. The curves A, B and C describe different ways of modelling the heat flow (from water movement) vs. time relationships from the time of water expulsion (t_o) to the present time t_p. The area under the curves A, B and C are equal and describe emitted energy pr. area. The curves show that, as long as the energy emission is decreasing most rapidly immediately after the water expulsion (i.e. the curves A and B are not convex), the present day heat flow contribution from the expelled hot water can not exceed $Q_a/2$.

This calculated example will, of course, highly overestimate the the energy emission from the heated rock. Heat loss to the fractures in which the water migrates has been neglected. Furthermore, it is assumed that all the excess energy transported by the water is transferred to the rock V_h, thus neglecting the possibility of water moving through this rock volume V_h without being cooled to the prevailing equlibrium temperature. The magnitude of the errors depend heavily on the time needed for the pressure to adjust from overpressure to normal pressure. If such an adjustment takes place over a very short time span, for example resulting from a major tectonic movement, this will favour the validity of the calculations.

It can be concluded that, if there is reason to belive that overpressure was released quite recently from deeply buried rocks as a catastropic event, temperature anomalies might possibly be detectable today. The determination of recently released paleopressures will in general be quite uncertain. Indications of paleopressure release can come from thermal, geochemical or mineralogical evidence, from interpretation of the overall pressure regime, and from analysis of the tectonic development of the sedimentary basin.

The temperature increase of a rock volume resulting from a pulse of hot, moving water (due to the release of fluid overpressure) will only persist for a short time. The impact on hydrocarbon generation and source rock maturation will therefore be limited.

As an example, consider an increase in the surface heat flow from 43 mWm^{-2} to 47 mWm^{-2}. If the average thermal conductivity of the rocks is 1,4 $W°C^{-1} m^{-1}$ (somewhere between shale and sandstone), and the surface temperature is 7° C, the temperature 3 km below the sediment surface would increase from 99°C to 109° C, as can be seen from Fouriers law $Q = - K \frac{\delta T}{\delta z}$.

If the temperature increase persisted for 10000 years, the TTI value would increase with 0.5×10^{-2}, which is in the order of 0.1% of the total TTI-value one would expect for a 50 m. year old rock having suffered an even burial through time.

HEATING EFFECTS FROM COMPACTION-DERIVED WATER

It is therefore concluded that the effect of compaction - derived fluid movement to petroleum generation is in general of minor importance, even if the release of signigicant overpressure from large rock volumes is taken into account.

III. THE TROLL FIELD: AN EXAMPLE FROM THE NORTHERN NORTH SEA

The supergiant Troll gas field in the Norwegian North Sea (fig. 8) has a thermal anomaly which is assumed to be young (Eggen, 1984). Eggen (1984) also mentioned that compaction - derived fluid convection should be investigated in order to reveal the cause of this anomaly.

From the previous discussions, it is clear that moving compaction - derived water resulting from subsidence at an even rate can not account for the anomaly. The thickness of the heated section is in the order of 1 km, and the excess heat flow from the field is about 20 mWm^{-2} (Eggen, 1984). From equation 7, it follows that a (V_r/V_h) ratio of a least 30000 is necessary if even water expulsion through deep fractures is postulated. As seen from fig. 8, this ratio is quite far from the situation at the Troll field.

During the Quaternary, sediments above the Troll field were eroded. In the deglaciation periods, 200 m of moraine materiale was deposited. This is probably apprixmately equivalent to the eroded thickness and glacial cooling is likely to have occurred at times. As previouly shown, a subsurface temperature increase due to rapid sedimentation hardly occurred under the prevailing conditions.

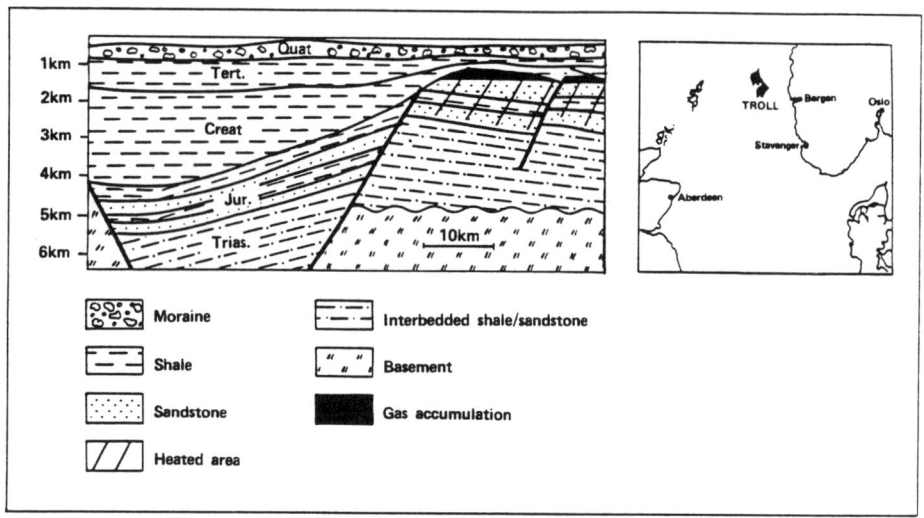

Fig. 8. Location of the Troll Field.

A release of overpressure from the rocks below the Troll field could possibly affect its temperature, provided it took place quite recently, and gave rise to a significant pressure drop. If all factors determining the size of the thermal energy transported into the field are given values which favour large energy transport, the time of the pressure release can be given a lower boundary.

Assume that the whole section of Triassic rocks, from the center of the Viking graben to the eastern part of the Troll field, had its perssure reduced from geostatic to hydrostatic pressure. About 50% of the Triassic is sandstone. Assume further that the average porosity is 10%, and low permeability is not obstructing the fluid flow (an optimistic assumption!). If additionally all of the excess fluid volume originating from the pressure release is transferred through steep fractures to the Jurassic sections of the Troll field, this could give a (V_v/V_h) ratio of 0.1. With $\Delta T = 90°$ C, $H = 1$ km and $Q = 20$ mWm^{-2}, equation 7 gives $t = 60000$ years.

This implies that if the thermal anomaly observed at the Troll field today was caused by moving hot water, expelled as a result of the release of overpressure from deeper strata, the pressure drop took place at the latest during the last deglaciation period in the Quaternary. However, no indications of such recent tectonic movements is known in the Troll area. A rapid fluid exchange in the Weichselian would, in addition, probably create a disequilibrium between the water chemistry and the rock forming minerals in the gas zone. As this is not the case at the Troll field, it is suggested that movement of compaction-derived water is not the cause of the present day thermal anomaly at the Troll field.

As this paper restricts itself to the discussion of heating effects from water mobilized by compaction, other possible causes for the thermal anomaly at the Troll field have not been examined. The topics listed below deserve attention in future studies of the thermal state of the Troll field.

Diagenetic or other geological evidence may suggest that circulation cells as described by Bories & Combarnous (1973) are active in the carrier beds west of the Troll field. If such circulation cells do exist they can promote the transport of thermal energy from the Viking graben to the Troll field significantly.

The fact that the Jurassic beds have been elevated above the sea level in the area east of the Troll field, implies that water flow driven by an elevated ground water level may have cooled the area in the past. This fact can not explain the thermal state of the Troll area alone, however, as the main problem here is to explain the high heat flow from the Troll field.

Finally, the three-dimensional distribution of the conductive heat transfer should be examined to predict the subsurface temperature distribution. Three-dimensional computer simulation should probably be employed to active the best description of this subsurface temperature distribution, but two-dimensional studies will certainly describe the influence of laterally varying thermal conductivities to temperature distributions quite well. This matter will be a subject of future research.

ACKNOWLEDGEMENTS

I thank Dr. Anthony M. Spencer for indispensable help with the preparation of this paper. Thanks are also due to Per Arne Bjørkum, Svein S. Eggen, Svein Inge Eide and Harald Hanche-Olsen for commenting an earlier version of the manuscript, and to Reidun Gabrielsen who cleverly typed this manuscript.

REFERENCES

ALLIS, R.G., 1978:   The effect of pleistocene climatic variations on the geothermal regime in Ontario: a reassessment.
Can. J. Earth Sci. Vol.15, p. 1875-1879, 1978

BECK, A.E., 1977:   Climatically perturbed temperature gradients and their effect on regional and continental heat-flow means.
Tectonophysics, Vol. 41, p. 17-39, 1977

BIRCH, F., 1948:   The effects of pleistocene climatic variations upon geothermal gradients.
Am. Journal of Science, Vol. 246, p. 729-760

BONHAM, L.C., 1980: Migration of hydrocarbons i compacting basins.
Am. Assoc. Petr. Geol.Bull., Vol. 64, No. 4, p. 549-567

BORIES, S.A and M.A. COMBARNEOUS:
Natural convection in a sloping porous layer.
J. Fluid Mech. 1973, Vol. 57, part 1, p. 63-79

DAKE, L.P., 1978: Fundamentals of reservoir engineering.
Elsevier Sci. Publishing Company

DICKEY, P. A.,1979: Petroleum Development Geology.
Penn Well Publishing Company

EGGEN, S., 1983: Modelling of subsidence and HC-generation on the Norwegian continental shelf.
in: B. Durand (e.): Thermal Penomena in sedimentary basins, international collquium Bordeaux, juni 7-10, 1983, p. 271-283.

GALLOWAY, W.E., 1984: Hydrogeologic regimes of sandstone diagenesis
in: McDonald & Surdan (eds): Clastic Diagenesis.
AAPG memoir 37, p. 3-13

HYNDMAN, R.D., A.M. JESSOP, A.S. JUDGE and D.S. RANKIN, 1979:
Heat flow in the maritime provinces of Canada.
Can. J. Earth Sei. Vol. 16, p. 1154-1165, 1969

JESSOP, A.M., 1971: The distribution of glacial pertubation of heat flow in Canada.
Can. J. Earth Sci.
Vol. 8, p. 162-166

MAJOROWICZ, J.A., F.W. JONES, H.L. LAM and A.M. JESSOP, 1984:
The variability of heat flow both regional and with depth in southern Alberta, Canada: Effect of groundwater flow?
Tectonphysics,
Vol. 106, p. 1-29, 1984

ROKOENGEN, K. and T.M. RØNNINGSLAND, 1983:
Shallow bedrock geology and quaternary thickness in the Norwegian sector of the North Sea between $60° 30'N$ and $62°N$.
Norsk geol. tidsskrift
Vol. 63, p. 83-102, 1983

SHARP, J.M. jr. and P.A. DOMENICO, 1976:
Energy transport in thick sequences of compacting sediment.
Geol. Soc. of Am. Bull.
Vol. 87, p. 390-460, 1976

SMITH, L. and D.S. CHAPMAN, 1983:
On the thermal effects of groundwater flow. I. Regional scale systems.
Jour. of geoph. research
Vol. 88, No. 131, p. 593-608, 1983

STALLMAN, R.W, 1963:   Computation of groundwater velocity from temperature data.
　　　　　　　　　　　in Bentall, R., ed: Methods of Collecting and Interpreting Ground-Water Data.
　　　　　　　　　　　U.S. Geol. Water Supply Paper 1544-H, p. 36-46, 1963

STEGENA, L., 1982:   Water migration influences on the geothermics of basins.
　　　　　　　　　　　Tectonophysics
　　　　　　　　　　　83(1982) p. 91-99

TIPLER, P.A. 1976:   Physics
　　　　　　　　　　　Worth Publishers Inc.

WELTE, D.H. and M.A. Yükler, 1981:
　　　　　　　　　　　Petroleum origin and accumulation in basin evolution - a quantitative model.
　　　　　　　　　　　Am. Assoc. Petr. Geol. Bull.
　　　　　　　　　　　Vol. 65, No. 8, 1981

M. QUINTARD, D. BERNARD[1]

FREE CONVECTION IN SEDIMENTS: NUMERICAL MODELLING AND TIME SPACE SCALING

I. INTRODUCTION

Modelling the thermal evolution of a sedimentary basin is of a great interest in the field of petrogenesis. Most of the existing models [1] only take into account conduction and forced convection. However we can insure that free convective motion probably occurs due to the heterogeneities in the porous sediments or the existence of sloping strata. This paper deals with the comparison between the scales (time and space) of the free convective phenomena and the compaction and conduction. It also deals with the possible occurence of a three dimensional motion.

Convective flow as a consequence of buyoancy forces in the gravity field is well known [2], [3]. If we only consider density gradient due to thermal gradient we can distinguish two kinds of problems :

a) the purely conductive gradient is not colinear to the gravity vector. Then natural convective motions do exist.

b) the purely conductive gradient is colinear to the gravity vector. Natural convective motions may exist depending on the thermal constraint.

Increasing the thermal constraint will lead to one or more bifurcating solutions from the conductive state.

---

(1) *Laboratoire Energétique et Phénomènes de Transfert, Centre National de la Recherche Scientifique, Talence, France.*

We can notice that even in case (a) increasing the constraint may cause the convective flow to bifurcate into another convective state.

How can we state the problem of modelling numerically the thermal evolution of a sedimentary basin ? It depends on the importance of the non linear-terms in the physical model of the transport process through the porous media.

In the case of a low thermal gradient we can expect conductive transport or convective flow of kind (a). If some properties (symmetry, ...) are known about the physical model we can then deduce the same properties for the solution of the model. For example if all the physical properties can be deduced by means of translation we can work on a two-dimensional model.

Nevertheless, if bifurcating solutions exist properties of the convective state explained above may not be true. Some general results as reported in the first part of this paper can be found for simple geometries. Unfortunately such results cannot be obtained in most situations : complex geometry, heterogeneity, anisotropy ... Furthermore, in the case of simple properties, for example translational properties, we can show that for a given constraint, 2D convective flow should change in a 3D convective motion. Depending on the non- linearities the convective flow may not be unique.

In this paper results obtained in very simple configurations are presented and discussed. They are not strictly representative of what occurs in sedimentary basins but provide, at leat, qualitative estimations of the effects that can be expected (heat Flux, water mass flux, ...). Soon more realistic results will be available using the finite elements model under development in the laboratory.

In the following paragraphs, first we discuss in the case of a paralelepipedic porous layer the existence of 2D or 3D stationary convective motion. Then we deal with effects involved by transient phenomena induced by property changes. For all of these aspects a comparison is made with time and space scales of the conduction and compaction phenomena.

II. STATIONARY NATURAL CONVECTION

A. MODEL

We consider the porous Layer shown in figure 1. Transport phenomena of a single fluid through the porous medium are described by a

FREE CONVECTION IN SEDIMENTS

simple model using the Darcy equation and the energy equation of a continuous onephase equivalent medium [3] :

(1) $$\nabla \cdot \vec{V} = 0$$

(2) $$\vec{V} = -\frac{k}{\mu}(\nabla p - \rho \vec{g})$$

(3) $$(\rho c)^* \frac{\partial T}{\partial t} = \nabla \cdot \lambda^* \nabla T - (\rho c)_f \vec{V} \cdot \nabla T$$

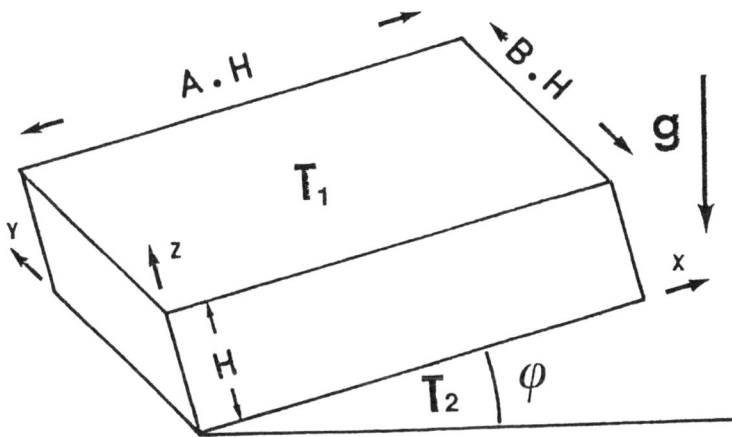

FIGURE 1 : Studied porous layer

In the first equation we assume that the porosity is constant and that the fluid dilatability and compressibility effects are negligible. However in the Darcy equation we admit that density changes linearly with temperature according to Boussinesq's hypothesis.

In our particular geometry we can seek for a solution of equations (1) to (3) in the form :

(4) $$\vec{V} = \frac{\lambda^*}{(\rho c)_f H}(v_x \vec{e}_x + v_y \vec{e}_y + v_z \vec{e}_z)$$

(5) $$\Theta = (T-T_1)/\Delta T - (1-z/H)$$

$\vec{V}$ and Θ can be represented as a linear combination of elementary functions such as :

FREE CONVECTION IN SEDIMENTS

(6) $v_x(l,m,n) = -b_{lmn}(t)\, A^2 l\, n\, \pi^2 \sin\dfrac{l\pi x}{AH}\cos\dfrac{m\pi y}{BH}\cos\dfrac{n\pi z}{H}$

$v_y(l,m,n) = -b_{lmn}(t)\, B^2 m\, n\, \pi^2 \cos\dfrac{l\pi x}{AH}\sin\dfrac{m\pi y}{BH}\cos\dfrac{n\pi z}{H}$

$v_z(l,m,n) = -b_{lmn}(t)(l^2+m^2)\,\pi^2 \cos\dfrac{l\pi x}{AH}\cos\dfrac{m\pi y}{BH}\sin\dfrac{n\pi z}{H}$

$\Theta(l,m,n) = a_{lmn}(t)\cos\dfrac{l\pi x}{AH}\cos\dfrac{m\pi y}{BH}\sin\dfrac{n\pi z}{H}$

The obvious solution $\vec{V} = 0$ and $\Theta = 0$ corresponds to the purely conductive state. Each elementary function defined by (6) is called a mode and is characterized by the numbers (l,m,n).

An important feature of the problem is the ratio between the total flux crossing the layer and the purely conductive flux : the Nusselt number Nu* (without convection Nu* = 1).

B. HORIZONTAL LAYER

If the layer is homogeneous, isotropic and horizontal ($\varphi = 0$) we obtain from (1) to (3) :

(7) $b_{lmn}(t) = \text{Ra}^* \dfrac{(l^2+m^2)}{\pi^2[A^2 l^2 n^2 + B^2 m^2 n^2 + (l^2+m^2)^2]} a_{lmn}(t)$

$\dfrac{d\, a_{lmn}}{dt} = -\left[\dfrac{l^2}{A^2} + \dfrac{m^2}{B^2} + n^2\right]\pi^2 a_{lmn} + (l^2+m^2)\pi^2 b_{lmn} + \mathcal{N}(a_{ijk} b_{pqr})$

Where $\mathcal{N}$ is representative of the non linear effects.

The factor Ra* appearing in the equation (7) is the Rayleigh number defined here as :

(8) $\text{Ra}^* = g\, \beta_{th}\, \rho_m\, \Delta T\, (\rho c)_f\, k\, H / \mu_m\, \lambda^*$

It expresses the importance of the thermal constraint applied to the porous layer. In the configuration we consider, the stability of the conductive state (i.e. the existence of free convection) is completely determined by the value of Ra*.

When studying the development of small disturbances of the conductive state the nonlinear term in (8) can be dropped. The time derivative of a_{lmn} will be positive, i.e. the conductive state is unstable for this mode, if the Rayleigh number is greater than a

FREE CONVECTION IN SEDIMENTS

critical value Ra^*_c :

$$(9) \quad Ra_c = \left[\pi^2 \left(\frac{l^2}{A^2} + \frac{m^2}{B^2} + n^2\right)(A^2 l^2 n^2 + B^2 m^2 n^2 + (l^2+m^2)^2)\right] / (l^2+m^2)^2$$

Equation (9) gives a means to estimate the effect of lateral confinement on the critical Rayleigh number [5]. In the case of infinite lateral extent the minimum of the cirtical Rayleigh number is the classical value $4\pi^2$. This result provides a simple way to determine if convection occurs. Unfortunately for this critical value it exists an infinite set of critical modes (l,m,n) and it is not possible in this way to predict the convective pattern that will take place.

Even if the aspect ratio is not large we can have such a dilemna. In the case of a cubic layer it can be shown that we have [6] :

- $4\pi^2 < Ra^* < 4.5\pi^2$: the only modes that exist are the 2D modes (1,0,0) and (0,0,1).

- $4.5\pi^2 < Ra^* < 1.5 \times 4.5 \pi^2$: the foregoing 2D modes or the 3D mode (1,1,1) can exist depending on the <u>initial conditions</u>.

For higher values of Ra^* more complexe convective pattern occur.

As a first conclusion 3 points must be noticed :

1 - in the case of the horizontal layer the criteria for the onset of convection are quite simple.

2 - stationary solution of the convective problem may not be unique.

3 - in that case the pattern depends clearly on the initial conditions.

C. SLOPPING LAYER

When the value of φ is not zero the purely conductive state does not longer exist for all Rayleigh numbers greater than zero (Ra^* defined by 8).

In the case of a layer with infinite lateral extent a solution to the convective problem is :

$$(10) \quad \Theta = 0 \qquad v_y = v_z = 0 \qquad v_x = Ra^* \sin\varphi \left(\frac{1}{2} - \frac{z}{H}\right)$$

275

FREE CONVECTION IN SEDIMENTS

This unicellular state is stable if the following condition is satisfied :

(11) $$Ra^* \cos \varphi < 4\pi^2$$

Otherwise two kinds of 3D convective motions appear depending on a certain critical angle φ_c [7] [8] (figure 2).

- $\varphi < \varphi_c$: pseudopolyedric cells characteristic of the horizontal problem

- $\varphi > \varphi_c$: helicoidal rolls with axis parallel to the x-axis

We now focus our attention on the case of a confined layer with Rayleigh number sufficiently low to ensure the stability of a unicellular state.

For aspect ratio A lower than 10 the convective state is sensitive to the effect of the lateral boundaries and the unicellular motion is slightly different from the one given by (10). The dependence on Ra^* and φ of Nu^* and $V_{\frac{1}{2}} = v_x$ ($x = \frac{AH}{2}$, y,z = 0) has been calculated for A = 2. The results are presented on Figures 3 and 4. For low Ra^* and φ the global transfer is poorly improved by convection as indicated by Nusselt numbers almost equal to one.

D. PHYSICAL ILLUSTRATION

The aim of this paragraph is to illustrate using realistic data the adimensional results given above.

In the following we assume that within the studied sedimentary basin we can isolate a paralelepipedic homogeneous porous layer limited by two isothermal and four adiabatic boundaries (figure 1). Then the Rayleigh number defined by the equation (8) characterizes the existence and the type of the convective motion.

Remarking that in the expression (8) of Ra^* all the parameters excepting $\Delta T, K$ and H are almost constant for the geological applications we consider and assuming that ΔT is given by the product of H by the geothermal gradient, the Rayleigh number is proportional to KH^2.

With the data of table one, the calculated Rayleigh number is about 12.5 so the conductive state in a horizontal layer is stable. With these physical properties the calculated critical depth H for which Ra^* is equal to $4\pi^2$ is 540 m. In that case the approximative wavelength of a convective roll will be 540 m.

FREE CONVECTION IN SEDIMENTS

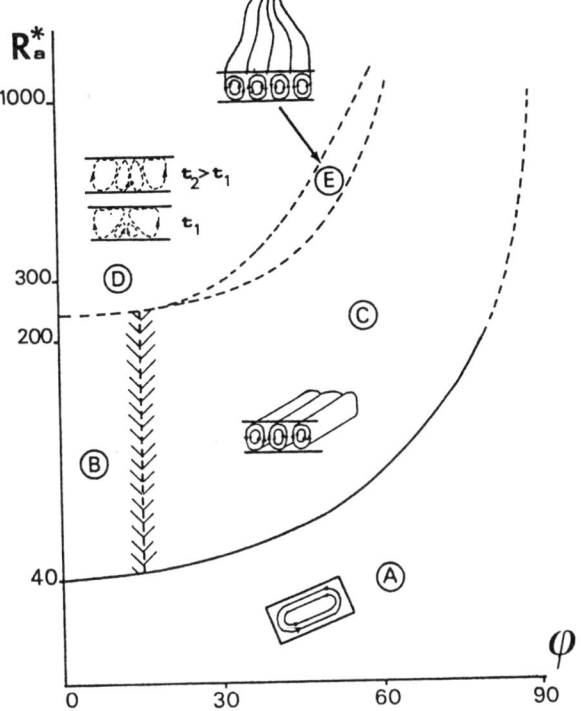

FIGURE 2 : The different convective motions in the diagram R_a^* - φ

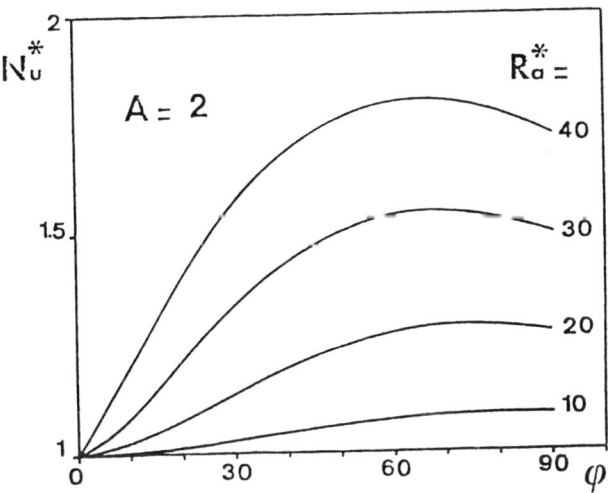

FIGURE 3. Variations of Nu* with Ra* and φ

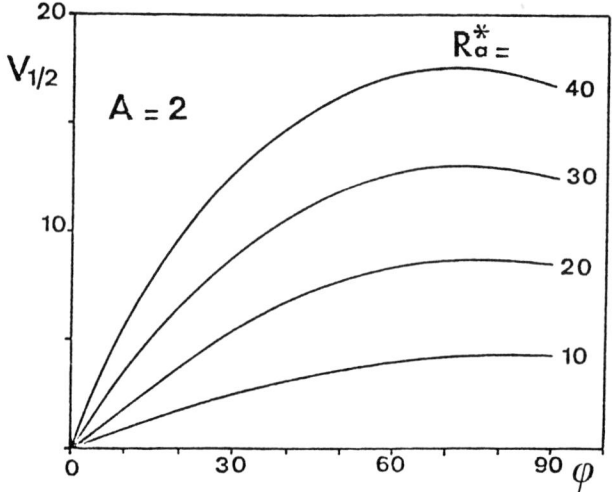

FIGURE 4. Variations of $V_{1/2}$ with Ra* and φ

ΔT	=	$0.03 * H$	(K)
H	=	300	(m)
K	=	$0.3\ 10^{-12}$	(m^2)
λ^*	=	1.5	($W\ m^{-1}\ K^{-1}$)
$(\rho c)^*$	=	$3.\ 10^6$	($J\ m^{-3}\ K^{-1}$)
⇓			
Ra^*	=	12.5	

Table 1 : Physical illustration . Numerical values

FREE CONVECTION IN SEDIMENTS

If we consider a sloping layer with equal to 20° the calculated values of $V_{1/2}$ and Nu* and their dimensional translation are given in table 2. The total flux does not greatly differ from the purely conductive one.

The total flux does not greatly differs from the purely conductive one, but the local flux presents large variations (figure 5). The temperature profile observed at the surface over the porous layer will be disturbed not because convection increases the heat transfer but because it reorganizes it. The shape of the Nusselt number curve presented figure 5 yields to a high temperature at the left side and a low one at the other side.

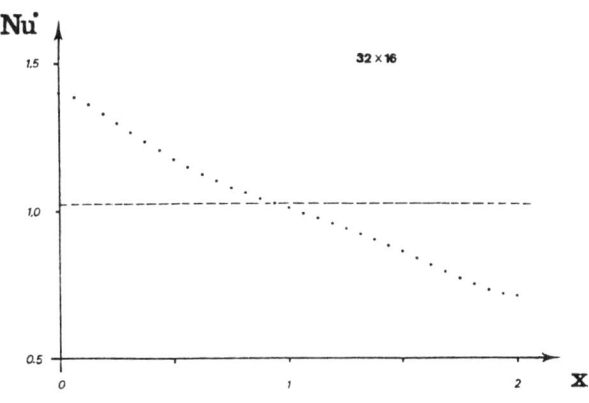

FIGURE 5. The local Nû (·) compared to the total Nû (--)

The water velocity induced by free convection is rather high compared to the velocities due to the compaction [9].

A	2	∞
Adimensional		
$V_{1/2}$ =	2.19	2.17
Nu* =	1.024	1.000
Dimensional		
$V_{1/2}$ = (m/year)	8.25 10^{-2}	8.18 10^{-2}
Total flux = (w/m²)	4.61 10^{-2}	4.50 10^{-2}

TABLE 2. Sloping layer with φ = 20° and Ra* = 12.5. Calculated values of $V_{1/2}$ and Nu* for A = 2 and A = ∞.

These values are only indicative, but these ideas may be applied to estimate roughly the relative magnitude of the various processes of transfer.

FREE CONVECTION IN SEDIMENTS

III. TRANSIENT EFFECT

This paragraph deals with transient effect induced by a sudden increase in the temperature at the lowest boundary of a horizontal layer. The initial conditions are :

(12) $\quad T = T_1 \quad\quad z \neq 0$
$\quad\quad T = T_2 \quad\quad z = 0$

Such a problem is a test problem of interest in our study of sedimentary basins in order to compare the time scales of the phenomena. This problem was first studied in the field of geophysics [10].

Free convection is a macroscopic phenomena produced by amplification of perturbations existing at the microscopic scale. The physical phenomena (thermal noise, microscopic defaults, ...) that initiate convection are not known. Numerically we simulate it by adding random values to the coefficients $a_{lmn}(t=0)$ appearing in the equations (6).

A. STABILITY

Previous results give some ideas about the stability. The methodology used to obtain these results is not discussed here and is available in [4], [11]. The stability domains are plotted in the Ra*-t plane on figure 6. For a given value of Ra*, say Ra*=100, we have :

- for $t < t_1$: any disturbances existing at t=0 or produced during the process are not amplified and will decrease monotically.

- for $t > t_1$: the condition described above is no longer valid, a disturbance may grow or decrease depending on its amplitude.

- for $t < t_2$: any disturbances of infinitesimal amplitude will increase.

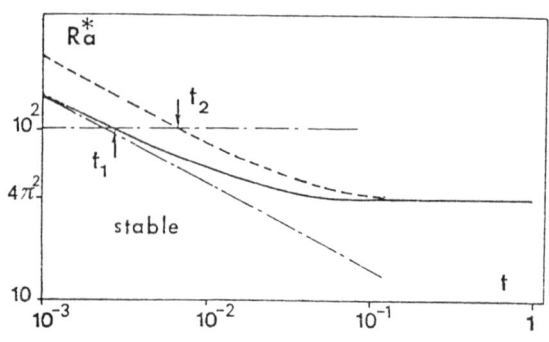

FIGURE 6.
Stability domains

FREE CONVECTION IN SEDIMENTS

After a certain amount of time, depending on the experiment, disturbances grow sufficiently to make the convective motion observable.

As we concerned with the wavenumber of the convective motion when time tends towards infinity we are in the case examined in § 2. However, the stability study shows that when time increases there is a continuous change in the critical wavenumber. This time evolution of the convective pattern should be studied with the whole nonlinear set of equations.

B. 3D SIMULATION

A spectral approximation of equations (7) can be computed [4] with (l,m,n) belonging to the set $[0,L] \times [0,M] \times [0,N]$. The motion is perturbed at t=0.

The absolute maximum of the perturbation is lower than a given value A_0. The system is solved numerically on a CRAY 01 computer. Figure 7 shows the time evolution of the Nusselt number Nu^* (z=0) for various values of Ra^* ($A_0 = 10^{-4}$, continuous line ; $A_0 = 10^{-5}$, discontinuous line).

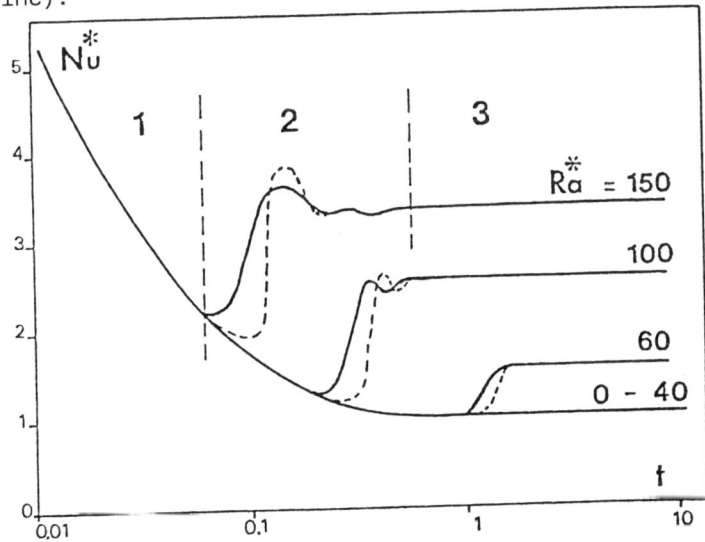

Figure 7 : Evolution of Nu* with time

When the Ra* number is lower than $4\pi^2$ Nu*(t) represent the evolution of the purely conductive thermal flux calculated at z=0.

When Ra* is greater than $4\pi^2$ the evolution of Nu* can be divided into three steps (as shown on the figure for Ra* = 150) :

- Step I : First $(t<t_1)$ all the absolute values of the coefficients a_{lmn} decrease. Then some of them begin to increase according to the critical times given by the stability study. At the end of this step the amplification is sufficient to produce a significant increase in the Nusselt number.

- Step II : this step is a short transient evolution of the motion characterized by a great increase in the flux. During this step there is a transition between an almost regular cellular motion (wavelength lower than H) and a motion representative of the steady state (wavelength about H).

- Step III : it corresponds to the setting up of the steady convective pattern. Surprisingly enough the real steady state is obtained for relatively high values of t (between 10 and 20). This fact is not sensitive on the value of the flux Nu* which seems nearly constant. The convective motion reorganizes itself by evolution between convective structures of nearly the same wavelength H.

This fact appears clearly on the figure 8 which represents isotherms in given horizontal planes for four values of time. Their heights z are chosen to correspond to the position of isotherm T_m (average temperature) in the purely conductive state at the same time.

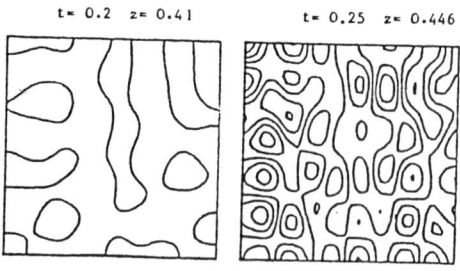

$Ra^* = 100$

$A = B = 4$

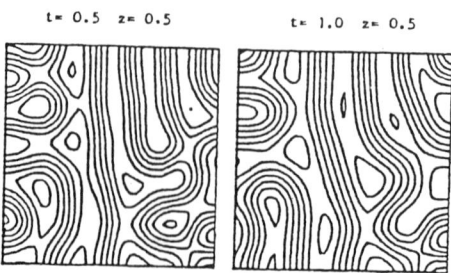

FIGURE 8. Isotherms in horizontal planes

FREE CONVECTION IN SEDIMENTS

In that case we observe changes in the convective structure for a time scale greater than the conductive one. In 2D simulations the elapsed time for the establishment of convection is generally lower than 1 [12], [13]. Our results should be attributed to the 3D character of the movements. This fact should be considered in the sedimentary basin simulation where there is a continuous change in the boundary conditions.

In the physical case examined in § 2.4 the real time corresponding to an adimensional time of t = 10 is about 57.000 years.

IV. PRACTICAL CONCLUDING REMARKS

In a sedimentary basin it is practically impossible to find a porous body where the conditions (heterogeneities, slope, thermal constraint, ...) permit the stability of the saturating fluid. Consequently free convection always exits.

The evaluation of its effects is difficult because no general results could be developed for realistic situations. Particularly in the case of inhomogeneous media with complex geometry and boundary conditions the significance of a Rayleigh number is casual compared to the simple case presented in § II.

Nevertheless the study of convection within a paralelepipedic homogeneous porous layer provides us some informations :

- the total heat transfer through the layer is generally only slightly improved by convection

- the local heat transfer is highly perturbated

- induced water flow should be important

- 3D motions should probably occur for Rayleigh numbers and aspect ratio large enough

- in some cases the time scale of the convective phenomena is of the same order as that of the purely conductive phenomena.

According to those results it is necessary to take natural convection into account when studying the evolution of sedimentary basin. To estimate its precise impact care must be taken when

283

defining the numerical model. Two difficulties arising in this field are greatly discussed, based on original works, in this paper :

- even if 2D approximation seems to be appropriate regarding to physical properties 3D motions can develop

- on large systems or with high thermal constraint the solution of the problem may not be unique and depends on the initial conditions.

ACKNOWLEDGEMENTS

part of this work was supported by the Institut Français du Pétrole.

NOTATIONS

A	Longitudinal aspect ratio
A_0	Maximum amplitude of the perturbations
$\vec{B}$	Lateral aspect ratio
$\vec{g}=g\,\vec{e}_z$	Gravity acceleration
H	Height
k	Permeability
Nu*	Nusselt number
T	Temperature
T_1	Temperature of the layer upper boundary
T_2	Temperature of the layer lower boundary
t	Time
$\vec{V}$	Velocity
$V_{1/2}$	Velocity at the point ($\frac{AH}{2}$, 0,0)
β_{th}	Coefficient of volumic thermal expansion for the fluid
$\Delta T = T_2 - T_1$	
Θ	Adimensional temperature
λ	Thermal conductivity
μ	Dynamic viscosity
ρ	Density
(ρc)	Volumic heat
φ	Sloping angle

Indices

c	Critical
f	Fluid
l m n	Refered to the mode (l,m,n)
m	At the temperature $T_m = (T_1 + T_2)/2$

Sub indices

*   Refered to the homogeneous media equivalent to the porous one.

REFERENCES

[1] WELTE D.H., YUKLER M.A., Petroleum origin and accumulation in basin evolution - A quantitative model
A.A.P.G. Bull, vol. 65, 8, 1387-1396, 1981

[2] COMBARNOUS M., Natural convection in porous media and geothermal systems
6th Int. Heat Transfer Confer., Toronto, 1978

[3] COMBARNOUS M., BORIES S., Hydrothermal convection in saturated porous media
Advances in Hydroscience, vol. 10, 231-307, Academic Press 1975

[4] QUINTARD M., GOUNOT J., CALTAGIRONE J.P., Structuration thermoconvective dans une couche poreuse brusquement chauffée par le bas : modélisation numérique 3D par une méthode spectrale.
Coll. Int. Informatique dans les Sciences de la Terre, Nancy Avril 1984

[5] BECK J.L., Convection in a box of porous material saturated with fluid
Phys. Fluids, vol. 15, 1377-1383, 1972

[6] STEEN P.H., Pattern selection for finite-amplitude convection states in boxes of porous media
J. Fluid Mech., vol. 136, 219-241, 1983

[7] BORIES S., COMBARBOUS M., Natural convection in a sloping porous layer
J. Fluid Mech., vol. 57, 63-79, 1973

[8] CALTAGIRONE J.P., BORIES S., Solutions and stability criteria of natural convective flow in an inclined porous layer
J. Fluid Mech., 155, 267-287, 1985

[9] JACQUIN C., POULET M., Essai de restitution des conditions hydrodynamiques régnant dans un bassin sédimentaire au cours de son évolution
Revue IFP, vol. XXVIII, 3, 269-297, 1973

[10] ELDER J., The bowels of the earth
Oxford University Press, 1976

[11] QUINTARD M., Convection naturelle en milieu poreux : systèmes non-stationnaires, déplacements
Annales des Mines, 5-6, 85-92, 1984

[12] POULIKAKOS D., BEJAN A., Unsteady natural convection in a porous layer
Phys. Fluids, vol. 26, 1183-1191, 1983

[13] PERRIN J.F., Modélisation du champ thermique dans les bassins sédimentaires - Application au bassin de la Mahakam (Indonésie)
Thèse docteur-ingénieur, Université de Bordeaux I, 1983

PART C

CONDUCTIVITIES, PERMEABILITIES, RADIOACTIVE HEAT GENERATION IN SEDIMENTS

A. J. SILVA[1], R. H. MORIN[2]

THE SENSIVITY OF SEDIMENT PHYSICAL PROPERTIES TO CHANGES IN TEMPERATURE, PRESSURE, AND POROSITY

ABSTRACT

Laboratory equipment has been designed and fabricated to determine the permeability and thermal conductivity of deep ocean sediments at in situ conditions of high temperature (up to 220°C), hydrostatic pressure (up to 62 MPa), and porosity. Results on two biogenic oozes and two clays show no discernible effects of hydrostatic pressure upon the permeability, while temperature effects are accounted for by applying a viscosity correction. The variation in permeability with porosity, however, is found to be substantial. Estimated values of bulk permeability for an entire sedimentary cover are shown to decrease by orders of magnitude across only a few hundred meters of layer thickness.

The effects of sizeable variations in hydrostatic pressure and temperature upon sediment thermal conductivity closely reflect the behavior of the conductivity of the liquid component alone. Unlike certain modulus parameters, such as shear strength, the thermal conductivity does not increase sharply as porosities decline into the transitional range between liquid-dominated and solid-dominated two-phase media. Estimated values of bulk conductivity increase by only about 30 percent across a section several hundred meters thick.

I. INTRODUCTION

To make accurate predictions of thermal and fluid flow processes in deep-sea sediments, it is necessary to know the relevant transport properties at in situ environmental conditions. Two approaches can be taken: (1) measure the properties in the natural state, or (2) test

(1) *University of Rhode Island, Kingston, Rhode Island, USA.*
(2) *United States Geological Survey, Denver, Colorado, USA.*

samples in the laboratory under simulated in situ conditions. The primary advantage of conducting laboratory measurements is that samples can be subjected to a variety of effects including consolidation and high temperature. However, there are obvious difficulties in reproducing deep-sea hydrostatic pressures of 60 MPa or more and in subjecting samples to temperatures in excess of 200°C. This paper briefly reviews the design of special instrumentation that enables testing under these severe conditions and presents the results of an extensive experimental program conducted on four different sediment types representing a wide range of mineralogies and porosities. The relative sensitivity of sediment-transport properties to environmental conditions are analyzed, and the results are interpreted in terms of their relevance to the study of the thermal evolution of sedimentary basins.

Motivation for the research came from two very different interests: one concerns hydrothermal activity and convective heat transfer mechanisms through the seafloor (Anderson and others, 1979, Langseth and Herman, 1981), and the other involves modeling of the coupled temperature and pore pressure fields near high level radioactive waste canisters that may be buried in deep-sea sediments (Hollister and others, 1981). Both of these applications require accurate information regarding the effects of high pressure and high temperature on the thermal conductivity and permeability of the sediment. The laboratory program described herein was directed toward these concerns.

II. BACKGROUND

Several attempts to study the effects of high pressures and temperatures on the physical properties of sediments have been reported in recent years. Laboratory tests measuring permeability need to be performed under pore water back pressures of at least 0.5 MPa to ensure dissolution of gases (Silva and others, 1981). Earlier studies have suggested that much higher pressures (over 30 MPa) might affect the double layer of fine grained sediments and change the permeability (Lang, 1967). It also has been suggested that thermal conductivity measurements made on samples at atmospheric pressure might be significantly different than at high in situ pressure conditions (Abbott and others, 1981).

Previous investigations concerning the effect of temperature on thermal conductivity were limited to a narrow range of 4 to 25°C (MacDonald and Simmons, 1972), while permeability studies on illitic clay exposed to temperatures of 80°C found that flow changes were accounted for entirely by those in pore water viscosity (Silva and others, 1981). Therefore, there was a need to investigate the effects of elevated temperatures and to combine the studies with the effects of high hydrostatic pressure. The apparatus that was developed enables systematic studies to be performed on various sediment types under high pressure and high temperature.

III. LABORATORY EQUIPMENT

The laboratory equipment is designed to measure permeability (or hydraulic conductivity) and thermal conductivity at hydrostatic pressures ranging from atmospheric (0.1 MPa) to 62 MPa and at temperatures from 22 to 220°C. Because of these extreme conditions, it was necessary to incorporate some novel methods and instrumentation. One of the challenges was to maintain a low, constant hydraulic gradient under high absolute pressures and to monitor very small volume flows. The existence of a highly corrosive environment (saltwater at high temperature) necessitated use of special materials. A schematic of the equipment is shown in Figure 1 and a photograph of the assembled apparatus is presented in Figure 2.

A. PERMEABILITY

Permeability is measured directly by producing a differential pressure across a specimen and measuring the corresponding volume of fluid that permeates through the sample. The new apparatus fabricated for this work is quite different from conventional geotechnical laboratory equipment. Stainless steel bellows equipped with linear variable displacement transformers (LVDT) provide the interfaces between hydraulic oil on the applied pressure side and the seawater pore fluid (Fig. 1). As seawater permeates the sediment sample, it flows out of one bellows and into the other; the resulting bellows movement is monitored by the LVDT. The capacity of each bellows is 72 cm^3, with a resolution of 0.01 cm^3.

The pressure gradient applied across a sediment sample during a test needs to be virtually constant to properly determine the permeability. This is accomplished using a unique pressure system composed of two dead-weight testers (DWT) and a hydraulic pump. Each DWT consists of a precision ground piston, on top of which is a pedestal and a series of weights. Oil pressure initially is generated by the pump, and pressure regulation is achieved in the same manner as a conventional dead-weight calibration device. A prescribed weight exerts a downward force on a piston, pressure from the pump is increased until the piston rises and floats in the oil reservoir, and the resulting hydrostatic pressure is proportional to the given weight divided by the piston area. Friction is reduced by designed oil leakage around the piston and is minimized further with an electric motor that continuously spins the piston and weight assembly. The differential pressure, produced by unequal loads on the DWT's, can be controlled within ± 3.5 kPa at absolute pressures to 62 MPa. A heat exchanger is incorporated into the oil line to cool the hydraulic fluid, and thus permeability tests can be made at substantial absolute pressure over several hours. Further details on the equipment and procedures can be found in Morin and Silva (1983), and Morin (1982).

B. THERMAL CONDUCTIVITY

A stainless steel hypodermic needle, 0.15 cm in diameter, is mounted vertically through the base of the pressure cell, and the sediment sample is pushed down axially over it (Fig. 1). The needle contains a thermistor and a nickel-chromium resistor wire potted in a high temperature epoxy resin, which enables tests to be performed at temperatures up to 220°C. Thermal conductivity is measured by the needle probe technique (Jaeger, 1958; Von Herzen and Maxwell, 1959; Chaney and others, 1983). The temperature response of the thermistor to a specified heat input from the resistance wire is monitored as a function of time and is directly related to the sediment thermal conductivity.

The pressure vessel, with base plate and piston, is made of Inconel 600 to resist the corrosive nature of high temperature seawater, and is designed to accommodate a sediment sample 5 cm in diameter by 10 cm long. The sample itself is confined in a thick-walled tube lined with teflon. The piston, which incorporates a double O-ring seal design, enters the pressure vessel through the top cap and permits one-dimensional loading and consolidation of the specimen. Using this loading system, the sample can be consolidated to change the porosity so that the compaction process in a sediment column can be simulated.

C. TEMPERATURE CONTROL

Three resistance heater bands, each with a 1500W capacity, surround the pressure vessel, heating the cell and its contents. Asbestos insulation is provided to reduce heat loss, and heat input is regulated with a proportional power controller. The sediment temperature is monitored from the thermistor that is inside the thermal-conductivity needle. Once the system has reached a steady state condition, sediment temperatures are maintained to within ±0.5°C.

IV. HIGH PRESSURE-HIGH TEMPERATURE TESTS

A. SEDIMENT SAMPLES

Except for the remolded smectite material, all samples were "undisturbed" subsamples obtained from large-diameter (10.2 cm) gravity cores. The samples represent a varied range of sediment types, locations and conditions (initial porosity), as shown in Table 1. The samples from each core were taken in close proximity to each other in order to minimize compositional variations among them.

Table 1. Sample Descriptions

Type of Sample	General location of sample	Water depth (m)	Grain size++ anal.,%		Init. poros., ϕ,	Mineralogical composition
			Silt	Clay		
Silic. ooze	Gulf of Calif.	1900	48	40	89	*54% SiO_2, 8%$CaCO_3$
Calcar. ooze	Galapagos Spreading Center	2700	42	42	85	*65% $CaCO_3$, 21% SiO_2
Illite	North Central Pacific	5900	34	65	73	**73% ill., 18% chlor., 6% kaol., 3% smect.
Smect.+	North Central Pacific	5900	27	72	80	**22% smect., 56% ill., 3% chlor., 19% kaol.

+Reconstituted, reconsolidated (remolded) *Percent of total.
++The balance is sand **Percent of clay fraction.

B. EXPERIMENTAL PROCEDURES

Hydraulic conductivity (or coefficient of permeability) was measured directly and flow was assumed to be governed by Darcy's law.

$$\frac{q}{A} = ki \qquad [1]$$

where q/A is the flow rate per unit area (cm/sec), k is the coefficient of permeability (cm/sec), and i is the hydraulic gradient (dimensionless). Applying a known pressure gradient across a sample and measuring the corresponding flow rate yields a value for the coefficient of permeability of the specimen. Once equalization of pressures throughout the sample was achieved, as determined from the steady-state flow response, the time for a typical permeability measurement was on the order of 1 hour.

Thermal conductivity data were obtained by the standard needle probe method (Jaeger, 1958; Von Herzen and Maxwell, 1959). For large times and a small radius, the temperature increase of the probe can be approximated by the following relationship:

$$T = \frac{Q}{4\pi j} \ln \frac{4\alpha t}{cr^2} \qquad [2]$$

where T = temperature of the needle;
    Q = heat input per unit length per unit time;
    j = thermal conductivity of sediment;
    α = thermal diffusivity of sediment;
    t = time;
    r = needle radius; and
    c = constant (1.781).

Each series of sediment samples from the four sediment types was tested at 22°, 80°, 150°, and 220°C. An initial experiment, which was performed at room temperature, involved the variable pressure studies. Tests at the higher temperatures were performed at a constant pressure equal to the in situ hydrostatic pressure of the specific sample. The samples were also consolidated under several incremental loads in order to study the effects of variable porosity. Due to frictional forces between the pressure-vessel piston and the O-ring seals, the exact magnitudes of the overburden stresses being transmitted to the samples were unknown. Therefore, the response of the sediments to effective stresses was not investigated in this laboratory program.

C. RESULTS

1. Pressure Effects

a. Hydraulic Conductivity (Permeability)

Evidence from previous scientific investigations, both analytical and empirical, indicates that the double layers present in clays actually deteriorate with increasing hydrostatic pressure (Owen and Brinkley, 1943; Hamann, 1957; Horne, 1964; Lang, 1967). Our first series of laboratory tests was designed to determine what effect the existence of this phenomenon would have on the permeability of ocean sediments. One sample of each of the four sediment types was exposed at room temperature to pressures ranging from 2 to 60 MPa; the porosity remained constant. Results showed negligible pressure effects on the hydraulic conductivity (±10%), and the average value for each sediment is shown in Table 2.

b. Thermal Conductivity

Results of the thermal conductivity versus hydrostatic pressure tests for all four sediment types are depicted in Figure 3. The data closely parallel the behavior of water alone as a function of pressure (Lawson and others, 1959) and the overall pressure effect for a specific sediment type is relatively small. This information can be translated into a thermal conductivity correction factor for hydrostatic pressure as a function of porosity. Based on these test results, a new relationship can be used to correct thermal conductivities to in situ pressure conditions for sediment porosities between 0.78 to 1.0 (Equation [3]).

$$\begin{bmatrix} \text{Press. Corr.} \\ \text{Therm. Cond.} \end{bmatrix} = \begin{bmatrix} \text{Therm. Cond.} \\ \text{@ 1 atm} \end{bmatrix} + \begin{bmatrix} 1.72 \times 10^{-5} \frac{\text{cal.}}{\text{cm. sec. }^\circ C} \end{bmatrix} \times \begin{bmatrix} \text{Wat. Depth,} \\ \text{(km)} \end{bmatrix} \quad [3]$$

The correction for sediments with lower porosities is less than this (see Morin and Silva, 1984). The correction factor suggested here is different than that recommended by Ratcliffe (1960); the differences between the two methods become significant for low-porosity sediments at large water depths.

Table 2. Average hydraulic conductivity for pressure range of 2 to 60 MPa

Sediment Type	Void* Ratio, e	Hydr. Cond. k, (cm/sec)
Sil. ooze	7.0	8.2×10^{-5}
Calc. ooze	3.0	4.3×10^{-6}
Illite	2.0	1.8×10^{-6}
Smectite	4.3	8.6×10^{-7}

* ϕ = Porosity = $\frac{e}{1+e}$; or e = $\frac{\phi}{1-\phi}$

2. Temperature Effects

a. Mineralogical Changes

Studies on chemical alteration of ocean sediments at high temperatures seem to be somewhat contradictory. The transformation of smectite to illite has been documented by Weaver (1979). For Gulf Coast clays, Burst (1969) also has recorded a gradual transformation of smectite into illite with increasing depth and temperature. This diagenesis, however, may occur only when there is a sufficient source of potassium in the sediment, because illite is produced when potassium becomes lodged between the contracting layers of smectite. Khitarov and Pugin (1966) confirmed this condition by exposing a number of montmorillonite samples to elevated temperatures.

Seyfried and Thornton (1981) have conducted experimental and theoretical investigations of thermally induced chemical alterations of

illitic clays. Analyses of seawater chemistry during both their 200°
and 300°C tests showed concentrations of magnesium in solution
decreased and concentrations of potassium increased. On the basis of
changes in solution chemistry, they concluded that illite, when exposed
to temperatures in excess of 200°C, is replaced in part, by smectite.

For siliceous and calcareous sediments, some diatoms and
foraminifera, go into solution as temperatures are increased, until the
pore fluid becomes saturated. This saturation level is very sensitive
to temperature variations. Adelseck and others (1973) have reported
that substantial precipitation of calcite occurred in a nannofossil
ooze when exposed to 300°C for 1 month. They found no significant
reaction, however, at 200°C.

A comparative study of compositional variations before and after
exposure to elevated temperatures showed no evidence of changes for any
of the four sediment types in our study. However, it appears from
previous work that increasing our maximum experimental testing
temperature from 220°C to 300°C might have produced substantial changes
in mineralogy. Testing at this temperature was beyond the scope of this
study; therefore, metamorphic changes were not considered.

b. **Hydraulic Conductivity (Permeability)**

One of the difficulties in comparing hydraulic conductivities at
different temperatures is that the relative viscosities of the
permeating fluid need to be taken into account. The absolute
permeability (K_{abs}) incorporates a temperature correction for the
viscosity of seawater and carries the units of length squared:

$$K_{abs} = k \frac{\nu}{g} \qquad [4]$$

where k is the sediment permeability, ν is the kinematic viscosity of
pore fluid, and g is the gravitational constant.

The effects of temperature on pore water chemistry, diffuse double
layer, mineralogical changes, and pH are very complex (Mitchell, 1976;
Bischoff and Seyfried, 1978; Lang, 1967; Clark, 1966). These effects
can result in competing chemical processes, and it is difficult to
predict the eventual or long-term changes in permeability due to
elevated temperatures.

Results of permeability tests for calcareous ooze as a function of
temperature are shown in Figure 4 (uncorrected for viscosity changes)
and in Figure 5 (absolute permeability). Similar data are presented by
Morin and Silva (1984) for the other three sediment types.

After initial pressurization and heating, each test specimen was
consolidated, by means of axial loads transmitted by the pressure-

vessel piston, through a series of decreasing porosities to simulate compaction through the sediment column. The results of the permeability tests performed on the two clays of interest closely approximate those of a study by Bryant and others, (1975) performed with unconsolidated marine sediments. Expressions for absolute permeability versus void ratio, which were derived from our data, are as follows,

$\log K_{abs} = 1.30e-14.57$, cm^2, (illite) [5a]
$\log K_{abs} = 0.72e-14.55$, cm^2, (smectite) [5b]
$\log K_{abs} = 0.62e-13.27$, cm^2 (calcareous ooze) [5c]
$\log K_{abs} = 0.35e-12.43$, cm^2 (siliceous ooze) [5d]

where e = void ratio of sediment.

c. **Thermal Conductivity**

Results of the thermal conductivity versus void ratio experiments for illite are presented in Figure 6, and others are shown in Morin and Silva, (1984). The expected trend of increasing conductivity with decreasing water content is consistent with previous work (Ratcliffe, 1960); the rate of variation is also consistent with several analytical models which describe the conductivity of two-phase media (Hashin and Shtrikman, 1962).

Regression analyses were performed on all the data to yield the following relationships for thermal conductivity as a function of temperature and void ratio.

$\log j = [(2.43 \times 10^{-4}) T - 0.084]e - 2.450$ (illite) [6a]
$\log j = [(1.31 \times 10^{-4}) T - 0.038]e - 2.591$ (smectite) [6b]
$\log j = [(5.03 \times 10^{-5}) T - 0.022]e - 2.605$ (calcareous ooze) [6c]
$\log j = [(3.57 \times 10^{-5}) T - 0.019]e - 2.618$ (siliceous ooze) [6d]

j in cal/cm sec $°C$, T in $°C$
($22°C \leq T \leq 150°C$; $1.5 \leq e \leq 8.8$)

The above equations were derived from our experimental data and are only applicable for a relatively limited range of porosities that were relevant to this study. Because the analysis assumes an increase in thermal conductivity with increasing temperature, the equations should not be used when sediment temperatures exceed approximately 150°C. This limit is suggested because the thermal conductivity of water increases with temperature until it reaches a maximum at about 150°C and then begins to decrease (Keenan and others, 1978). This effect is shown in Figure 7, which also presents a summary of thermal conductivity data for illite and smectite at constant void ratios (obtained from curves such as are shown in Figure 6). The same trends exist for the two oozes that were tested.

3. Long-Term Behavior

The X-ray diffraction analyses performed on the illite samples exposed to a temperature of 220°C did not show evidence of mineralogical alterations. However, the time of exposure (which was usually about 14 days, but only a few days at 220°C) was relatively short, and it is possible that small amounts of smectite would not be detected by the X-ray technique. To study further the effects of long-term exposure, we conducted a test on illite at a temperature of 140°C and a pressure of 13.8 MPa for 86 days. The results of this test, at a void ratio of 2.64, indicated only a small decrease in coefficient of permeability (not corrected for viscosity) after about 50 days of exposure at this temperature (from 8×10^{-6} cm/sec to 6×10^{-6} cm/sec). Unfortunately, it was not possible to maintain constant temperatures accurately above 140°C for long periods of time, and therefore, the possible effects at higher temperatures were not resolved adequately with this preliminary study. The slight decrease of permeability after 50 days may be indicative of some mineralogical alterations, but further testing at higher temperatures is needed.

V. BULK PHYSICAL PROPERTIES OF SEDIMENTARY LAYERS

In studying the thermal evolution of sedimentary basins, it is important to assign proper values of physical properties to the sediment strata. Bulk properties are affected by the thickness of the sediment column that, in turn, directly relates to overburden stress and associated consolidation behavior, as well as temperature and pressure. Laboratory results presented in the previous section can be combined with geotechnical information derived from other tests to formulate empirical expressions for the bulk physical properties of a sedimentary column as a function of its thickness.

This exercise begins by examining the consolidation behavior of marine sediments. The compression index C_c of a sediment sample is defined (Lambe and Whitman, 1969) as,

$$C_c = \frac{-\Delta e}{\Delta(\log \sigma'_v)} \qquad [7]$$

where e is the void ratio, and σ'_v is the vertical effective stress. A value for compression index is obtained through consolidation tests performed in the laboratory; the slope of the compression curve is C_c. From equation [7],

$$e = -C_c \log (A \sigma'_v) \qquad [8]$$

SENSITIVITY OF SEDIMENT PHYSICAL PROPERTIES

where A is a constant. The overburden stress σ'_v at any depth z is defined as the buoyant unit weight of the sediment integrated over the entire depth, assuming no excess pore pressures.

$$\sigma'_v = \int_0^z \gamma_b \, dz \qquad [9]$$

$$\gamma_b = \frac{(G-1)\gamma_w}{1+e} \qquad [10]$$

where G is the specific gravity, and γ_w is the unit weight of seawater. A boundary condition now is applied to equation [8]. It is assumed that, at a depth of 0.1 m, the void ratio has remained constant with depth and is equal to some initial value, e_o. The constant A now can be calculated.

$$e_o = -C_c \log \frac{A(.10\gamma_w)(G-1)}{1+e_o} \qquad [11]$$

$$A = \frac{(1+e_o) \, 10^{-e_o/C_c}}{(.10\gamma_w)(G-1)} \qquad [12]$$

It is assumed that e varies in some manner through the sediment column.

$$e = f(z) \qquad [13]$$

$$\sigma'_v = \int_0^z \gamma_b \, dz = \gamma_w (G-1) \int_0^z \frac{1}{1+f(z)} \, dz \qquad [14]$$

Substituting equation [14] into equation [8],

$$f(z) = -C_c \log \left[A\gamma_w (G-1) \int_0^z \frac{1}{1+f(z)} \, dz \right] \qquad [15]$$

$$10^{-f(z)/C_c} = A\gamma_w (G-1) \int_0^z \frac{1}{1+f(z)} \, dz \qquad [16]$$

Taking the derivative of both sides of equation [16],

$$10^{-f(z)/C_c} \left[\frac{-f'(z)}{C_c} \right] = A\gamma_w (G-1) \left[\frac{1}{1+f(z)} \right] \qquad [17]$$

299

$$f'(z) = \frac{(-D) \; 10^{f(z)/C_c}}{1 + f(z)}, \text{ where } D = C_c A \gamma_w (G-1) \qquad [18]$$

By definition,

$$f'(z) = \lim_{\Delta z \to 0} \frac{f(z+\Delta z) - f(z)}{\Delta z} \qquad [19]$$

$$\frac{f(z+\Delta z) - f(z)}{\Delta z} = \frac{(-D) \; 10^{f(z)/C_c}}{1+f(z)} \; ; \; f(.1) = e_o \text{ (boundary condition)} \qquad [20]$$

Equation [20] is solved numerically, and values of $f(z)$ for small increments of z are plotted in Figure 8 as solid lines. A linear fit to this semi-log function provides a good approximation for the dependence of void ratio upon depth up to some minimum value of e. It is assumed that, at sediment depths beyond about 400 m, these linear approximations are no longer valid and that the curves depicted in Figure 8 begin to flatten and approach the abscissa asymptotically. Therefore, this preliminary analysis of bulk physical properties is not directly applicable to investigations of thicker sediment sequences.

Values of C_c were estimated from laboratory consolidation tests, and values of e_o were determined from water content profiles of relevant piston and gravity cores for three of the four sediment types. Because the smectite samples were remolded, this clay type has not been included in the analysis of bulk properties.

The linear approximations shown in Figure 8 provide expressions for void ratio versus depth that take the following form,

$$e = -d_1 \log (d_2 z) \qquad [21]$$

where d_1 and d_2 are constant, and z is in meters. These expressions are substituted into the general equations for K_{abs} and a general relationship is derived.

$$K_{abs} = s_1 \; z^{-s_2} \qquad [22]$$

where s_1 and s_2 are constants, and K_{abs} is now in square meters.

Assuming vertical fluid flow, the bulk permeability of a number of sedimentary layers of varying permeability can be determined by employing a series model. This equivalent permeability is dominated by the least permeable layer and is described by the following expression (Freeze and Cherry, 1979),

$$K_{bulk} = \frac{\sum_0^n \Delta z_n}{\sum_0^n \frac{\Delta z}{K_{abs_n}}} \text{, where } \sum_0^n \Delta z_n = H \qquad [23]$$

Equation [11] now becomes,

$$K_{bulk} = \frac{H}{\int_0^H \frac{dz}{K_{abs}}} \qquad [24]$$

Substituting the values for K_{abs} from equation [22] into equation [24] yields the desired expressions for bulk permeability versus sediment cover thickness for the three sediment types of interest. These are listed below and are illustrated in Figure 9.

$$K_{bulk} = (8.366 \times 10^{-13}) \, H^{-1.55} \text{ (calc. ooze)} \qquad [25a]$$
$$K_{bulk} = (3.390 \times 10^{-14}) \, H^{-1.02} \text{ (silic. ooze)} \qquad [25b]$$
$$K_{bulk} = (3.978 \times 10^{-6}) \, H^{-1.04} \text{ (illite.)} \qquad [25c]$$

where K_{bulk} is in square meters and H is in meters.

Values of sediment thermal conductivity versus void ratio have been presented as functions of pressure and temperature. Although a small dependence on hydrostatic pressure was found, the magnitude of the thermal conductivity correction across a few hundred meters of sediment column is not regarded as significant to the application. Thus, for this analysis, thermal conductivity variations due to changes in pressure are considered negligible.

This same argument holds true for temperature variations. Because increases in temperature with depth usually are quite small (approximately 3°C per 100 m), this is a reasonable assumption when sediment thicknesses extend for only a few hundred meters. Even a temperature increase of 30°C through the sediment column will produce only approximately a 2% increase in thermal conductivity. The equations for thermal conductivity versus depth reduce to the following simplified form,

$$j = az^n \qquad [26]$$

where a and n are constants and z is in meters.

As in the previous instance concerning equivalent permeability, bulk thermal conductivity similarly can be determined again by using a series model,

$$j_{bulk} = \frac{H}{\int_o^H \frac{dz}{j}} = \frac{H}{\int_o^H \frac{dz}{az^n}} \qquad [27]$$

Appropriate values of the constants a and n for each sediment type were derived from the laboratory data. These are substituted into equation [27] and the required integration is performed. The following equations are the final expressions that relate the bulk thermal conductivity of a sediment layer to its thickness (also see Fig. 10),

$$j_{bulk} = (1.61 \times 10^{-3}) \, H^{.054} \text{ (calc. ooze)} \qquad [28a]$$
$$j_{bulk} = (1.66 \times 10^{-3}) \, H^{.054} \text{ (silic. ooze)} \qquad [28b]$$
$$j_{bulk} = (2.19 \times 10^{-3}) \, H^{.065} \text{ (illite)} \qquad [28c]$$

The laboratory results of permeability and thermal conductivity as functions of hydrostatic pressure, temperature, and void ratio can be combined with the sediment thickness analyses depicted by Figures 9 and 10 to derive a conceptual understanding of the relative sensitivities of transport properties to individual environmental factors. This information is summarized in Table 3.

Table 3. Relative dependence of transport properties to environmental conditions.

Transport Property	Pressure (0-62 MPa)	Temperature (22-220°C)	Porosity (78-100%)
Permeability	Negligible	Negligible after viscosity correction	Very large
Thermal conductivity	Small	Significant, but not large	Significant but not large

The sediment thermal conductivity is affected to a small extent by hydrostatic pressure and a pressure correction to thermal conductivity values has been indicated earlier (Equation [3]). This physical property also is dependent, to a more significant degree, upon temperature and porosity. Changes in hydrostatic pressure and temperature across a sedimentary section of a few hundred meters thickness are not large enough to produce significant adjustments to thermal conductivity. Variations in this transport property therefore are caused primarily by changes in porosity. The behavior of some physical properties, such as shear strength and compressional acoustic velocity, is linked to the bulk moduli of the material and the appearance of cementation. As a consequence, the magnitudes of these properties increase significantly during some stage of the consolidation process as sediments shift from liquid-dominated to solid-dominated two phase media. This behavior has been documented by Morin (1985) for calcareous sediments from the southwest Pacific. Unlike these modulus parameters, the thermal conductivity displays a very gradual change with decreasing porosity and does not exhibit an apparent transitional behavior. In addition, because the thermal conductivity values for the solid and fluid components are not substantially different ($j_{solid}/j_{fluid} \approx 6$), the effect of even large porosity changes is not very great. Consequently, compared to permeability, the values of thermal conductivity undergo relatively minor changes across hundreds of meters of sedimentary cover.

Hydrostatic pressure has a negligible effect upon absolute permeability, while temperature effects can be accounted for with a simple viscosity correction. However, permeability is found to be very sensitive to porosity. Decreases in porosity produced by lithostatic compaction can produce orders of magnitude reductions in permeability. Thus, values of absolute permeability can experience drastic variations across a few hundred meters of sediment thickness.

VI. SUMMARY

A novel, high pressure-high temperature laboratory apparatus was designed and successfully employed to conduct a test program on four different sediment types. Permeability and thermal conductivity were measured as functions of three variables: temperature, hydrostatic pressure, and porosity. Results show permeability to be relatively independent of hydrostatic pressure and temperature, but extremely sensitive to porosity, whereas thermal conductivity is found to exhibit noticeable, but not substantial, variations due to all three test variables. It is anticipated that the behavior of these transport properties would have been affected to a more significant degree had the temperature range of the experimental program been extended beyond 250°C, at which point mineralogical changes are expected to become prominent. Long-term permeability tests conducted at 140°C produced no clear variations in this property.

The laboratory results can be used to assign representative values of transport properties to sedimentary strata of variable thickness and thus, provide some insight into the relative sensitivity and effect of these properties when examining the thermal evolution of sedimentary basins.

The laboratory thermal conductivity data are combined with related consolidation information to arrive at a relationship for each of three sediment types for the bulk thermal conductivity of a sediment column versus its thickness. These expressions depict increases in the values of this transport property with increasing thickness, though the variation is only about 30% across a few hundred meters of the sediment column.

Similar expressions for bulk permeability versus sediment cover thickness also are developed from the laboratory results. Values of this property are shown to decrease by several orders of magnitude over only a few hundred meters of thickness. Indeed, bulk permeability is shown to be much more sensitive than its thermal counterpart to changes in the thickness of a sedimentary blanket. Because of this substantial range of values determined across relatively thin strata, permeability should play a major role in controlling various geologic processes that occur at the seafloor, such as consolidation and convective heat transfer. Consequently, this property needs to be considered carefully when attempting to model the evolution of sedimentary basins.

Acknowledgements

Support for this research was provided by the National Science Foundation, grant OCE 79119426; and the U.S. Department of Energy, Subseabed Disposal Project, Sandia Laboratories contracts 13-2561 and 13-9927.

REFERENCES

ABBOTT, D., W. MENKE, M. A. HOBART, and R. N. ANDERSON, Evid. for excess pore press. in southwest Indian Ocean sed., J. Geophys. Res., 86:3, 1813-1827, 1981.
ADELSECK, C. G., G. W. GEEHAN, and P. H. ROTH, Exper. evid. for the selective dissol. and overgrowth of calc. nannofossils during diagenesis, Geol. Soc. Am. Bull., 84, 2755-2762, 1973.
ANDERSON, R. N., M. A. HOBART, and M. G. LANGSETH, Geother. convect. through oceanic crust and sed. in the Ind. Ocean, Science, 204:4395 828-832, 1979.
BISCHOFF, J. L. and W. E. SEYFRIED, Hydrother. chem. of seawater from $25°$ to $350°C$, Am. J. Sci., 278:6, 838-860, 1978.
BRYANT, W. R., W. E. HOTTMAN, and P. K. TRABANT, Perm. of unconsol. and consol. marine sed., Gulf of Mex., Mar. Geotech., 1, 1-13, 1975
BURST, J. F., Diagen. of gulf coast clayey sed. and its poss. rel. to petroleum migration, Am. Assoc. Pet. Geol Bull., 53:1, 73-93, 1969.
CLARK, S. P. (Ed.), Handbook of Physical Constants, rev. ed. Geol. Soc. of Amer., Mem. 97, New York, 1966.
CHANEY, R. C., G. RAMANJANEYA, G. HENCEY, P. KANCHANASTIT, and H. Y. FANG, Sugg. test meth. for determin. of ther. conduct. of soil by ther.-needle procedure, Geotech. Testing J., 6, 220-225, 1983.
FREEZE, R. A. and J. A. CHERRY, Groundwater, Prent.-Hall, Englewood Cliffs, N. J., 1979.
HAMANN, S. D., Physico-Chem. Effects of Press., Butterworths Scient., London, 1957.

HASHIN, Z. and S. SHTRIKMAN, A variational approach to the theory of
the effective magnetic permeability of multiphase materials, J.
Appl. Phys., 33, 3125-3131, 1962.
HOLLISTER, C. D., D. R. ANDERSON, and G. R. HEATH, Subseabed disposal
of nuclear wastes, Science, 213:4514, 1321-1326, 1981.
HORNE, R. A., Structure makers and breakers in water: Press.-induced
changes in the hydration atmospheres of ions in sol., A. D. Little
Inc. Tech Rep. 3, Off. of Nav. Res., Wash., D. C., 1964.
JAEGER, J. C., The measure. of ther. conduct. and diffusivity with
cylindr. probes, Eos Trans. AGU, 13, 708-710, 1958.
KEENAN, J. H., F. G. KEYS, P. G. HILL, and J. G. MOORE, Steam Tables:
Ther. Prop. of Water; Vap., Liq., and Sol. Ph., Wiley, N. Y., 1978.
KHITAROV, N. I. and V. A. PUGIN, Behavior of montmor. under elev. temp.
and press., Geochem. Int., 3, 621-626, 1966.
LAMBE, T. W. and R. V. WHITMAN, Soil Mechanics, Wiley, N. Y., 1976.
LANG, W. J., The influence of pressure on the elect. resist. of clay-
water syst., Proc. Conf. Clays Clay Miner., 15th, 455-468, 1967.
LANGSETH, M. G. and B. M. HERMAN, Heat trans. in the oceanic crust of
the Brazil Basin, J. Geophys. Res., 86:11, 10805-10819, 1981.
LAWSON, A. W., R. LOWELL, and A. L. JAIN, Ther. conduct. of water
at high press., J. Chem. Phys., 30, 643-647, 1959.
MACDONALD, K. and G. SIMMONS, Temp. coeff. of the ther. conductiv. of
ocean sediments, Deep Sea Res., 19, 669-671, 1972.
MITCHELL, J. K., Fund. of Soil Beh., Wiley, N. Y., 1976.
MORIN, R. H., Thermophys. prop. of deep sea sed. and infl. upon oceanic
heat flow, Ph.D. Thes., U. Rhode Island, Kingston, R. I., 1982.
MORIN, R. H., Phys. prop. of calc. sed. from the southwest Pac., in
Kenneth, J. P., C. von der Borch, and others., Init. Repts. Deep Sea
Drilling Project, 90: Wash. (U.S. Govt. Print. Off.) 1985.
MORIN, R. and A. J. SILVA, High press.-high temp. lab. appar. for the
meas. of deep sea sed. phys. prop., Ocean Engrg., 10, 481-487, 1983.
MORIN, R. and A. J. SILVA, The effects of high press. and high temp.
on phys. prop. of ocean sed., J. Geophys. Res., 89:1, 511-526,
1984.
OWEN, B. B. and S. R. BRINKLEY, The effect of press. upon the
dielectric const. of liq., Phys. Rev., 64, 32-36, 1943.
RATCLIFFE, E. H., The ther. conductiv. of ocean sed., J.Geophys. Res.,
65:5, 1535-1541, 1960.
SEYFRIED, W. E. and E. C. THORNTON, Expor. and theor. modeling of
hydrother. proc. in near field environ., in Subseabed Disp.Ann.
Report, 80-86, Sandia Nat. Lab., Albuquerque, N.M., Jan.-Dec. 1981.
SILVA, A. J., J. R. HEATHERMAN, and D. I. CALNAN, Low-grad. permeab.
testing of fine-gr. mar. sed., ASTM., STP 746, 121-136,
1981.
VON HERZEN, R. P. and A. E. MAXWELL, The measur. of ther. conduct. of
deep-sea sed. by a needle-probe method, J. Geophys. Res., 64, 1557-
1563, 1959.
WEAVER, C. E., Geother. alter. of clay min. and shales: Diagen.,
Battelle Memor. Inst., Off. of Nucl. Waste Isol., Tech. Rep. ONWI-
21, Off. of Nucl. Waste Isol., Battelle, Columbus, OH, 1979.

SENSITIVITY OF SEDIMENT PHYSICAL PROPERTIES

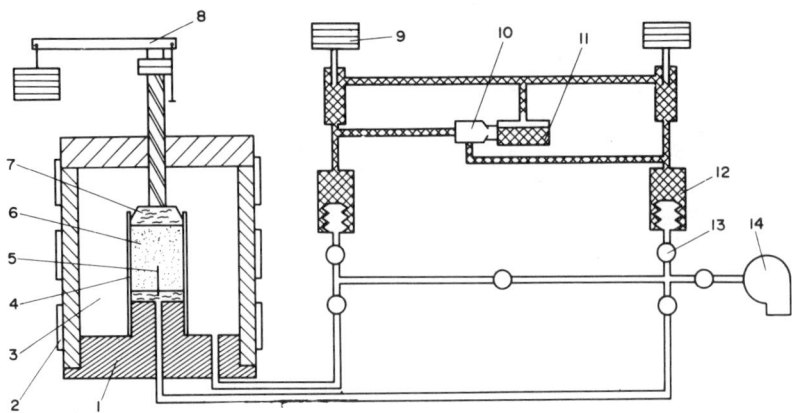

1. High pressure triaxial cell
2. Heating band
3. Sea water
4. Confining ring
5. Thermal conductivity probe
6. Sediment sample
7. Porous filter
8. Loading frame with weights
9. Dead weight tester
10. Recirculating hydraulic pump
11. Oil reservoir
12. Volume change device with bellows
13. High pressure valve
14. High pressure roughing pump

Figure 1. Schematic diagram of laboratory equipment.

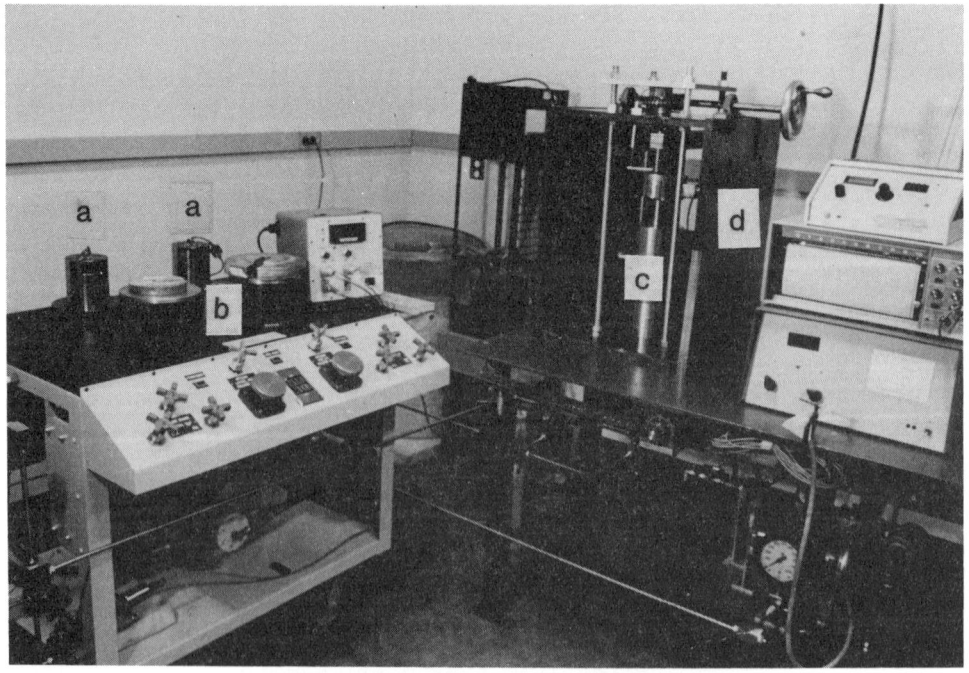

Figure 2. Photograph of equipment. Unit at left contains two volume change devices (a) behind the two dead-weight testers (b), a 5 HP motor and a hydraulic pump. Pressure cell (c) is shown inside loading frame (d) and electronic instrumentation controls are at right

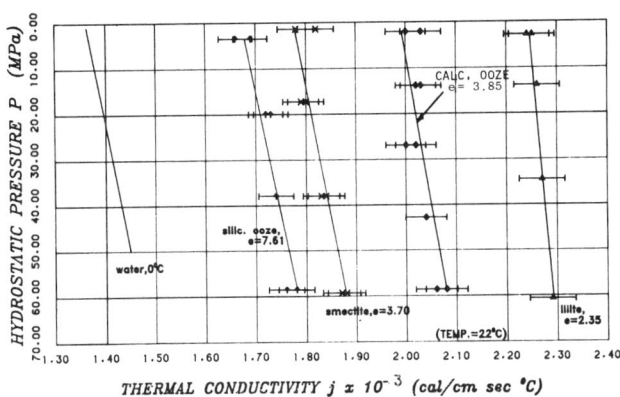

Figure 3. Thermal conductivity versus hydrostatic pressure for all four sediment types at room temperature and constant void ratio.

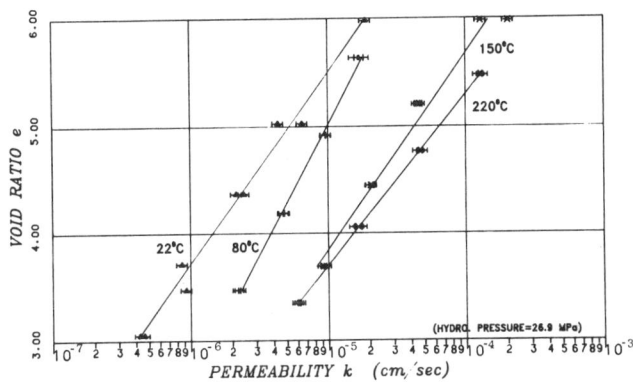

Figure 4. Coefficient of permeability versus void ratio at four temperature intervals for calcareous ooze. Hydrostatic pressure is constant at 26.9 MPa.

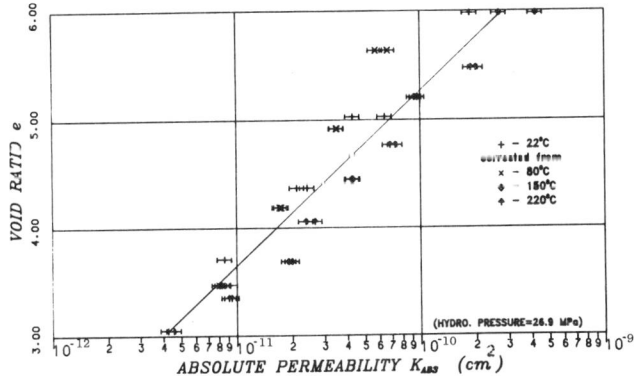

Figure 5. Absolute permeability (corrected for temperature) versus void ratio for calcareous ooze. Hydrostatic pressure is constant at 26.9 MPa.

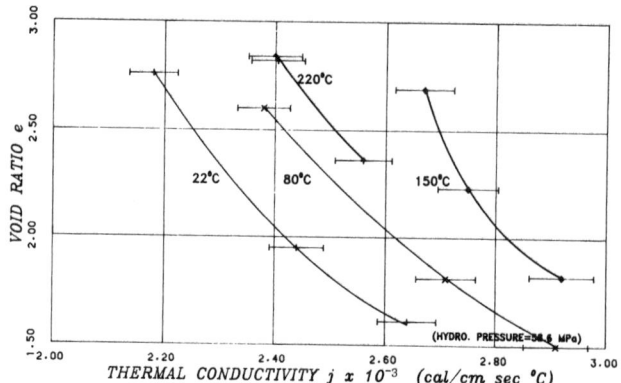

Figure 6. Thermal conductivity (j) versus void ratio at four temperature intervals for illite. Hydrostatic pressure is constant at 58.6 MPa.

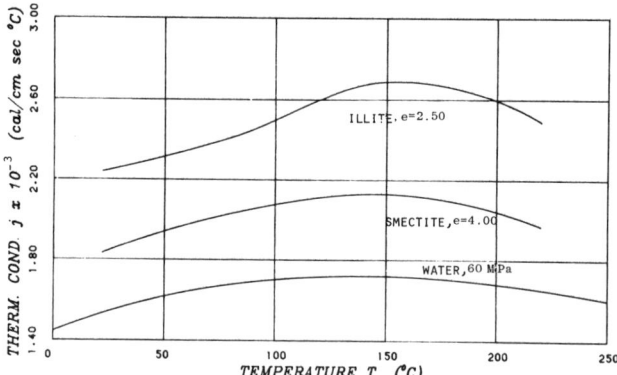

Figure 7. Thermal conductivity versus temperature for both clays at constant void ratio compared to water at 60 MPa hydrostatic pressure.

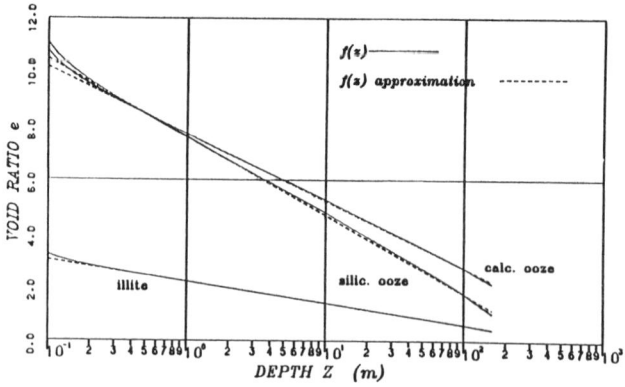

Figure 8. Approximation of void ratio versus depth for three sediment types.

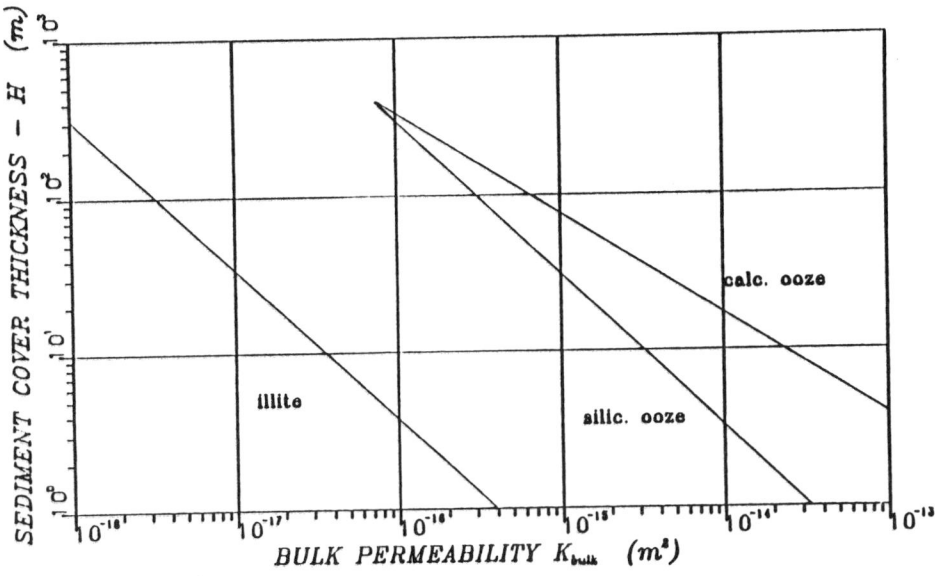

Figure 9. Estimation of bulk permeability versus sediment cover thickness for three sediment types.

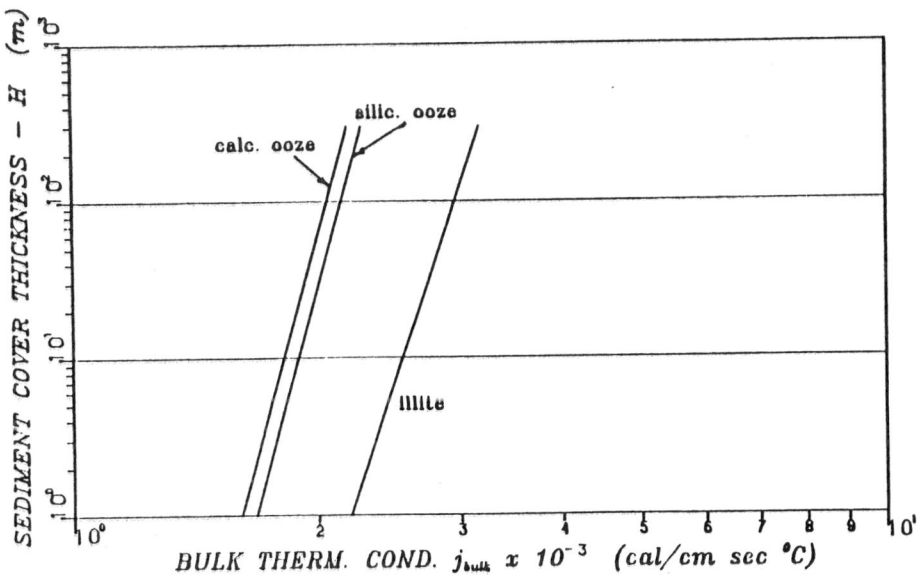

Figure 10. Estimation of bulk thermal conductivity versus sediment cover thickness for three sediment types.

L. RYBACH[1]

AMOUNT AND SIGNIFICANCE OF RADIOACTIVE HEAT SOURCES IN SEDIMENTS

Abstract

Heat generation by radioactive decay in sediments varies with lithology within a wide range (lowest in evaporites and carbonates, low to medium in sandstones, high in shales and siltstones and highest in black shales). There is still a great lack of reliable data for sedimentary environments.

Heat generation can be calculated from the uranium, thorium and potassium contents of the rock in question. These contents can be measured in sedimentary columns directly by borehole logging: natural gamma spectrometry (NGS) tools are now routinely available from various service companies. However, in the case of the vast majority of boreholes drilled and logged so far, the only information usually available on radioactivity is the natural gamma log (GR). An empirical relationship is presented which enables the determination of heat generation (A) from GR log readings (in API units):

$$A\ (\mu W/m^3) = 0.0145\ [GR(API) - 5.0]$$

Mean values of heat production for various lithologies are presented and the significance of contrasting heat generation on the geothermal field of sedimentary basins is demonstrated by simple model calculations.

---

(1) *Institute of Geophysics, ETH-Hoenggerberg, Zurich, Switzerland.*

RADIOACTIVE HEAT GENERATION IN SEDIMENTS

I. INTRODUCTION

In modelling the thermal evolution of sedimentary basins, especially with respect to maturation of organic matter and hydrocarbon generation, it is evident that crustal and lithospheric heat sources must be considered. As far as the heat generation in crystalline rocks of crust underlying sedimentary basins is concerned, its amount and distribution can be estimated from seismic velocity data (RYBACH & BUNTEBARTH 1982, 1984; see also ČERMÁK & BODRI, this volume).

Because even a small temperature change ($+10°C$) can have large effects in maturation, the question is often raised about the amount and significance of heat sources in sediments: How large are the effects of internal heat sources on the temperature field in sedimentary basins? Over what time scale is internal heating effective? It can be shown that radioactive heat generation is the main heat source owing to the long half-lives of the relevant heat producing radioelements; other internal heat sources, such as exothermal chemical reactions, occur over much shorter times.

In the following, after a short definition of radioactive heat generation and a brief review of the methods of determination, characteristic mean values will be given for common sedimentary lithologies. The effect of heat production on the geothermal field of sedimentary basins will be demonstrated by simple model calculations. Finally, since there is still a great lack of reliable data, it will be shown how heat generation in sediments can be determined from gamma ray (GR) well logs.

II. HEAT GENERATION IN SEDIMENTARY ROCKS

Radioactive decay converts mass into energy which in turn is converted into heast in the immediate vicinity of the decaying nucleus. All naturally occurring radioactive isotopes generate heat but the only significant contributions arise from the decay series of uranium and thorium, and from the isotope K^{40}.

The radioactive heat production A (in $\mu W/m^3$) of a given rock can be calculated by taking into account the heat generation constants (amount of heat released per unit mass of U, Th or K per unit time) and the uranium, thorium and potassium concentrations c_U, c_{Th} and

c_K of the rock:

$$A = 10^{-5} \rho (9.52 c_U + 2.56 c_{Th} + 3.48 c_K) \tag{1}$$

where ρ is the density of the rock (in kg/m^3); c_U and c_{Th} are in weight ppm, c_K in weight %. Owing to its high heat generation constant, 1 ppm uranium produces nearly four times more heat than 1 ppm thorium.

Heat production varies with lithology, unlike other sediment properties (density, porosity, thermal conductivity), over several orders of magnitude. Table I gives broad average figures for common sediment lithologies; it should be kept in mind that each rock type can exhibit a considerable width of variation. Heat production is lowest in evaporites and carbonates, shows low to medium values in sandstones, high values in shales and siltstones, and is highest in black shales. As a comparison, commercial grade uranium ore (with 2 kg U per ton) has a heat production of about 0.6 mW/m^3, average granite around 2.5 µW/m^3. The Th/U ratio, which can be indicative of the depositional environment, is low (< 1) in carbonates and black shales but more or less constant in other lithologies.

It is customary to determine hesat production, via eq. (1), by gamma spectrometric measurement of c_U, c_{Th} and c_K on rock samples like drillcores in the laboratory. Gamma spectrometry is a relatively simple and rapid analytical technique which enables simultaneous U, Th and K determinations. The spectrometric equipment, usually utilizing NaI(Tl) scintillation or Ge(Li) solid state detectors, must be calibrated using standards with known U, Th and K contents. The gamma ray spectra are evaluated by computer to yield c_U, c_{Th} and c_K and, in turn, A. Occasionally, x-ray fluorescence is used as an analytical technique (GALSON et al., 1983).

Only few sediment heat generation data have been reported in the literature and there is still a great lack of a reliable data base for specific sedimentary lithologies.

Table I:
Average contents of heat producing radioelements in sedimentary rocks (after RYBACH 1976, HAACK 1982 and RYBACH & ČERMÁK 1982)

Rock type	U (ppm)	Th (ppm)	K (%)	Th/U	Density* ($\cdot 10^3$ kg/m^3)	Heat generation (μW/m^3)
CARBONATES					2.6	
Limestone	2.0	1.5	0.3	0.75		0.62
Dolomite	1.0	0.8	0.7	0.80		0.36
EVAPORITES						
Salt	0.02	0.01	0.1	0.50	2.2	0.012
Anhydrite	0.1	0.3	0.4	3.0	2.9	0.090
SHALES & SILTSTONES	3.7	12.0	2.7	3.2	2.4	1.8
BLACK SHALES	20.2	10.9	2.6	0.54	2.4	5.5
SANDSTONES					2.4	
Quartzite	0.6	1.8	0.9	3.0		0.32
Arkose	1.5	5.0	2.3	3.3		0.84
Graywacke	2.0	7.0	1.3	3.5		0.99
DEEP SEA SEDIMENTS	2.1	11.0	2.5	5.2	1.3	0.74

*) Broad average since density strongly depends on porosity

III. THE EFFECT OF HEAT GENERATION ON THE GEOTHERMAL FIELD OF SEDIMENTARY BASINS

Crustal heat sources play a decisive role in shaping the temperature field of the crust. On continents, 20-60% of the surface heat flow originates from crustal radioactivity.

In sedimentary basins, the thermal structure of the rock pile can be influenced by many factors, both "internal" (i.e. properties of the sediment column) and "external" (controlled from outside the pile). Internal parameters include thermal conductivity and heat generation of the sediments while external factors include the heat flow into the column from below.

The effect of heat production on the geothermal field can be demonstrated by a simple, one-dimensional, purely conductive model in equilibrium (= steady state). The temperature at any depth z is

$$T(z) = T_o + \frac{q_b + A H}{K} z - \frac{A}{2K} z^2 \qquad (2)$$

where T_o is surface temperature, q_b the heat flow at the base of the column (at z = H, see Fig. 1), A the average heat production and K (in W/m °C) the average thermal conductivity of the sediments. The 1D-approximation is justified as long as the width of the sedimentary basin exceeds several times its depth, which is usually the case.

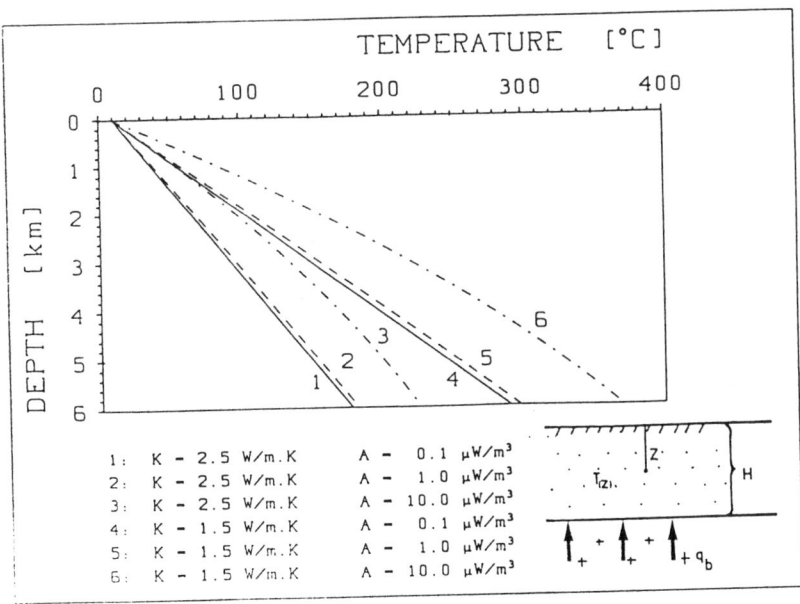

Fig. 1: The influence of sediment heat generation (A) and thermal conductivity (K) on the temperature-depth profile, T(z). The T(z) curves 1-6 have been calculated, with the K and A values given, from eq. (1) for H = 6 km with q_b = 70 mW/m² and T_o = 10°C.

Figure 1 shows the influence of A, if varied over the wide range from 0.1 to 10 µW/m³. The effect of changing K is also evident. For simplicity, K was taken constant over the entire sediment column; since K usually increases with depth (especially due to compaction) on one hand and decreases with increasing temperature (i.e. depth) on the other, the two effects might compensate each other. The third term in eq. (2) is responsible for the curvature of the temperature-depth profile and leads to a decrease of the geothermal gradient with depth. The effect of heat production is strongest at great depth and with high heat generation.

After considering the steady-state (equilibrium) conditions, the factor time shall be addressed. The internal radioactive heat sources of sediments deposited in a given area of accumulation, will raise the temperature of the sediment column. These heat sources have often been considered as the decisive one in the evolution of geosynclines (self-heating of continental-type sediments which would create, after some time, temperatures high enough to produce granitic melts). The temperature rise ΔT with time t has been usually estimated in these considerations by $\Delta T = (A/\rho c)t$, where A is heat production, ρ the density and c the heat capacity of the sediments. ΔT thus would rise continuously with time. However, if one accounts for the heat conducted away, the temperature rises much slower, as can be shown by the following simple model (see Fig. 2). We consider a radioactive layer (e.g. a shaly formation) embedded in a sequence with low radioactivity (e.g. clean sands). The heating by radioactivity will be highest in the mid-plane of the layer (GROSSLING, 1959) and is given by

$$\Delta T(t) = \frac{A t}{\rho c} [1 - 4i^2 \text{erfc} (\frac{D}{2\sqrt{\kappa t}}) + 2i^2 \text{erfc} (\frac{h+H+D}{2\sqrt{\kappa t}}) - 2i^2 \text{erfc} (\frac{h+H-D}{2\sqrt{\kappa t}})] = \Delta T_{max} \cdot \phi (t) \qquad (3)$$

where H and h are the depth to bottom and top of the radioactive layer, 2D is the layer thickness, κ is the sediment thermal diffusivity and $i^2\text{erfc}$ the repeated integral of the complementary error function. By assigning conservative thermal parameters to the model (see Fig. 2), the temperature rise (relative to the <u>in situ</u> sediment temperature without radioactive heating), remains as calculated from (2), still quite small even after 10 million years.

316

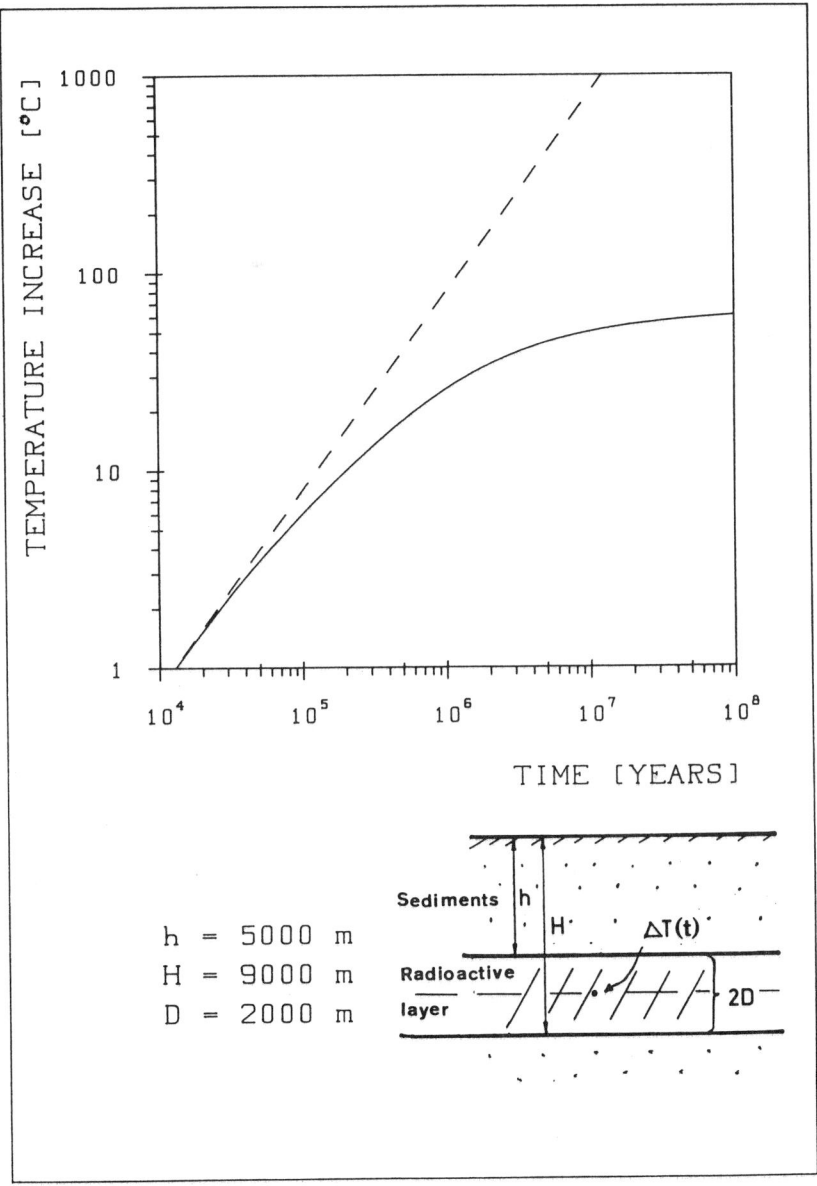

Fig. 2: Solid line: temperature increase with time in the mid-plane of a radioactive layer, $\Delta T(t)$, according to eq. (2). Dashed line: $\Delta T_{max} = (A/\rho c)t$. Model parameters as indicated in the lower left corner and $\rho = 2500$ kg/m^3, $c = 0.8$ kJ/kg$^\circ$C, $\kappa = 10^{-6}$ m^2/s with $A = 5$ µW/m^3.

The model is, admittedly, very simple; in reality the temperature rise will be somewhere inbetween the two lines of Figure 2. More sophisticated (2D, 3D) time-dependent calculations could be done on the basis of more detailed and realistic radioactivity distributions. Next it will be shown how such distributions can be obtained, for a given sedimentary sequence, from well logs.

IV. HOW TO DETERMINE HEAT GENERATION FROM WELL LOG DATA

According to equation (1), heat generation can be calculated from the uranium, thorium and potassium contents of the rock in question. In sedimentary columns these contents can be measured directly by borehole logging: natural gamma spectrometry (NGS) logging has recently become available from various service companies. The rock density which is also needed to calculate heat production, can be read from formation density (FDL, FDC) logs.

Not many drillholes have been logged yet by NGS tools since natural gamma spectrometry is a new logging technique. However, practically all logging programs include the natural gamma ray (GR) log; for the vast majority of boreholes logged so far, information about rock radioactivity is usually available in the form of GR logs.

The GR log measures "total" gamma activity, i.e. the sum of radioactive contributions from U (mainly gamma rays of Bi^{214}), from Th (mainly of Tl^{208}) and from K^{40}. On the other hand, uranium, thorium and potassium contribute in a similar way to heat production, cf. eq. (1). In fact, the sensitivities of GR logging tools for U, Th and K are in similar proportions as the corresponding heat generation constants of eq. (1). Therefore, it is sensible to establish an empirical relationship between GR log readings and heat production (A).

In order to establish such a relationship, the GR readings will be addressed first. Modern GR logs record rock radioactivity in API (=American Petroleum Institute) units. It is thus advisable to base the A - GR relationship on API units. In the earlier days of GR logging, the unit µg Raeq/t (microgram radium-equivalent per ton) was used. For most logging tools the conversion factor to convert readings in µg Raeq/t into API units is known, e.g. 1 µg Raeq/t = 16.5 API for the GNT-F tool of Schlumberger.

Figure 3 shows an empirical relationship between heat generation and GR log readings. This relationship has been established on the

RADIOACTIVE HEAT GENERATION IN SEDIMENTS

basis of NGS simultaneous NGS and GR readings (e.g. Fig. 8 on p. 88 in SCHLUMBERGER, 1982). Obviously, the GR readings must be corrected for borehole effects (sonde eccentricity, borehole diameter, and the presence of casing/cementation) before using Figure 3.

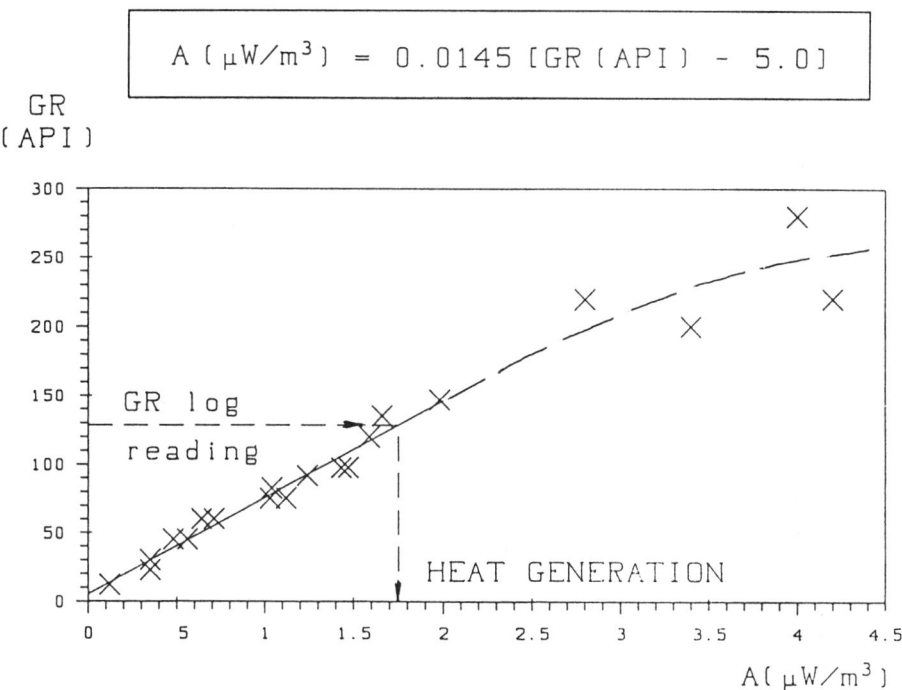

Fig. 3: Empirical relationship to determine heat generation (A) from gamma ray (GR) log readings in API units. Within the range 0 - 150 API, which encompasses practically all common sediments, the relationship is given by $A(\mu W/m^3) = 0.0145[GR(API)-5.0]$. Above 150 API the scatter of data points increases.

319

As far as heat generation is concerned, its value can be estimated from the empirical relationship

$$A\ (\mu W/m^3) = 0.0145[GR(API) - 5.0] \tag{4}$$

with an accuracy of about ± 0.1 $\mu W/m^3$. The reasonably good linearity of the GR - A relationship results from the fact that both heat generation and total gamma radioactivity (= the GR signal) are a sum of contributions from U, Th and K; as long as the Th/U and U/K ratios are fairly constant (cf. Table I) the relationship of eq. (4) holds in the range from 0 to about 150 API units, which is the range for "normal" sediments. Higher readings and higher heat production are usually due to elevated U contents associated with anomalously low Th/U ratios. The small intercept (about 5 API) of the empirical GR - A relationship can be attributed to mud and logging tool radioactivity.

V. CONCLUSIONS

Radioactive heat production in sediments, which is usually determined on rock samples (e.g. drillcores) in the laboratory, varies with lithology over orders of magnitude. Heat generation is lowest in evaporites and carbonates, low to medium in sandstones, high in shales and siltstones, and is highest in black shales.

Simple model calculations show that sediment heat production should be taken into account in thermal modelling of basin evolution if the basin is deep (>5 km) and if long time spans (>10 million years) have to be considered. For lithologies with low radioactivity (e.g. carbonates, evaporites), the effect of heat production in the sediments on the temperature field is negligible.

Heat generation can be directly calculated from the uranium, thorium and potassium contents which can be measured nowadays by the natural gamma spectrometry (NGS) log. For most drillholes, however, only natural gamma ray (GR) logs are available; an empirical relationship has therefore been established to convert GR log readings (in API units) into heat production (in $\mu W/m^3$). This relationship is valid in the range 0 - 150 API which covers practically all common sedimentary rock types.

RADIOACTIVE HEAT GENERATION IN SEDIMENTS

Acknowledgements

Thanks are due to Dr. J.J.H.C. Houbolt (Shell Research B.V., Rijswijk/Holland) who suggested the establishment of the gamma log - heat production relationship, to Dr. D. Galson for helpful comments on the manuscript, to Mr. W. Eugster and Mr. J. Dózsa for their help with the figures, and to Mrs. I. Siegel for typing the camera-ready manuscript.

VI. REFERENCES

GALSON, D.A., ATKIN, B.P. and HARVEY, P.K. (1983): The determination of low concentrations of U, Th and K by XRF spectrometry. Chem. Geol. 38, 225-237

GROSSLING, B.F. (1959): Temperature variations due to the formation of a geosyncline. Bull. Geol. Soc. Amer. 70, 1253-1282

HAACK, U. (182): Radioactivity of rocks. In: G. ANGENHEISTER (ed.): "Physical Properties of Rocks", Landolt & Börnstein, Group V/Vol. 1/Subvol. b, p. 433-481, Springer-Verlag, Berlin-Heidelberg-New York

RYBACH, L. (1976): Radioactive heat production: A physical property determined by the chemistry of rocks. In: R.G.J. STRENS (ed.): "The Physics and Chemistry of Minerals and Rocks", p. 309-318, Wiley & Sons, London

RYBACH, L. and BUNTEBARTH, G. (1982): Relationships between the petrophysical properties density, seismic velocity, heat generation and mineralogical constitution. Earth Planet. Sci. Lett. 57, 367-376

RYBACH, L. and ČERMÁK, V. (1982): Radioactive heat generation in rocks. In: G. ANGENHEISTER (ed.): "Physical Properties of Rocks", Landolt & Börnstein, Group V/Vol. 1/Subvol. b, p. 353-371, Springer Verlag, Berlin-Heidelberg-New York

RYBACH, L. and BUNTEBARTH, G. (1984): The variation of heat generation, density and seismic velocity with rock type in the continental lithosphere. In: V. ČERMÁK, L. RYBACH and D.S. CHAPMAN (eds.): "Terrestrial Heat Flow and the Structure of the Lithosphere", Tectonophysics 103, p. 335-344

SCHLUMBERGER (1982): Well Evaluation Developments - Continental Europe

Contribution no. 482, Institute of Geophysics ETH, CH-8093 Zurich/
Switzerland

V. V. PALCIAUSKAS[1]

MODELS FOR THERMAL CONDUCTIVITY AND PERMEABILITY IN NORMALLY COMPACTING BASINS

I. Introduction

Reconstructing the time-temperature history of a basin requires knowledge of the appropriate boundary conditions on the basin as well as the physical properties of the sediments. In this paper I will present results on the two coefficients that govern the transport of heat, i.e. thermal conductivity and permeability, and discuss how they influence the thermal evolution of a basin. The theoretical approach utilizes the effective medium theory in conjunction with a model for the micro-geometry of clastic sediments. This approach allows the computation of the effective thermal conductivity and permeability as functions of the properties of the individual components, thus illustrating their dependence on the mineral composition of the rock. The main objective is to identify the primary variables that control the magnitude of these coefficients and their expected variations with depth and lithology. In the following section I will present results for a normally compacting basin and the conditions under which these results can be expected to be attained. The subsequent sections contain various results on the permeability and thermal conductivity of clastic sediments, their dependence on the micro-geometry, and their variations with depth.

II. Normally Compacting Basins

Leaving temperature aside for the moment, the important variables describing the evolution of a basin are v_f (velocity of the fluid), v_s (velocity of the solids), P (fluid pressure), S (total vertical stress), and ϕ (porosity).

---
(1) *Chevron Oil Field Research Company, La Habra, California, USA.*

These variables are determined through the conservation equations of fluid and solid mass, Darcy's Law, stress equilibrium, and the constitutive equation describing the rheological behavior of the sediment. For clastic sediments the most widely accepted relation is that the porosity is only a function of the effective stress σ on the sediment, i.e. $\phi = F(\sigma)$ where $\sigma = S-P$, and we shall assume that it is applicable in what follows. This relation implies that all other parameters are held constant or are insignificant and that enough time elapses for transient fluid movements to take place. Thus the existence of the relation $\phi = F(\sigma)$ defines the drained compressibility coefficient $c = -d\ln\phi/d\sigma$ for a porous media. A discussion on how c is related to other elastic parameters of the sediments and how it varies with lithology is contained in Palciauskas and Domenico[1]. The porosity variations with depth z can be written as

$$- d\ln\phi/dz = c \, [\gamma - (\partial P_{ex}/\partial z)] \qquad (1)$$

where P_{ex} is the fluid pressure above the hydrostatic and γ is the weight of the submerged solids $\gamma = (\rho_s - \rho_f)(1-\phi)g$. Generally the solution of the conservation equations requires numerical methods, but for those cases where the excess pressure gradient $\partial P_{ex}/\partial z$ is small as compared to γ, the porosity will only be a function of depth, and a closed form solution can be easily found for a one-dimensional model. The associated geometry of the one dimensional model is illustrated in Figure 1a.

$$\phi = \phi_o \exp\{-c\gamma z\}$$
$$v_s(z,t) = v_s(h) \, [1-\phi(h)]/[1-\phi(z)] \qquad (2)$$
$$v_f(z,t) = v_f(h) \, \phi(h)/\phi(z)$$

We assume that the basin subsidence history h(t) has been reconstructed from the present day lithologic depths and ages. Thus the velocity of the solids and saturating fluid are completely determined by the porosity function and the subsidence rate h(t). Note that $v_s(h) = dh/dt$ is the rate of subsidence of the basement. The physical picture in this case is that the compaction process is driven by the subsidence and/or sedimentation and that the permeability is sufficiently large that the fluids can flow out of the sediment without producing "supernormal" pressures. This condition can be expressed quantitatively as

Hydraulic conductivity =
$$k\rho g/\mu \gg v_s(h) = \text{subsidence rate} . \qquad (3)$$

MODELS FOR THERMAL CONDUCTIVITY AND PERMEABILITY...

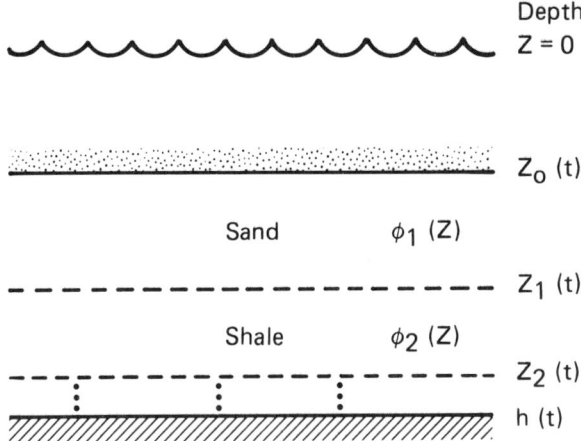

Figure 1a. Geometry of a one-dimensional compacting basin with varying lithologies.

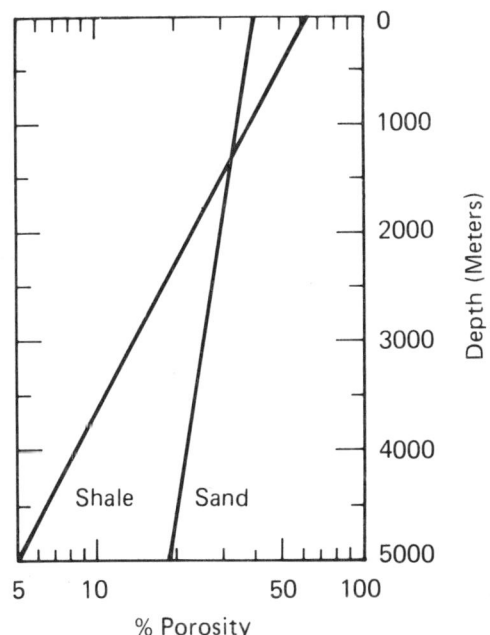

Figure 1b. Compaction curves for typical sand/shale lithologies.

Here k is the permeability, ρg is the unit weight of the saturating fluid, and μ is the viscosity. For subsidence rates of $v_s(h) = .1$ km/my $= .3 \times 10^{-10}$ cm/sec normal pressures are achieved for sediments whose particle size is an order of magnitude greater than 5×10^{-5} mm. Thus for this subsidence rate "supernormal" pressures will be developed in sediments containing fine clay particles as montmorillonite and illite or in sediments isolated from the surroundings by these types of sediments.

The energy transport equation for a normally compacting basin can then be written as

$$c^*(\partial T/\partial t) + c^*(h)v_s(h)(\partial T/\partial z) = \partial(\lambda \partial T/\partial z)/\partial z \qquad (4)$$

where c^* is the average specific heat of the sediment and λ is the effective thermal conductivity. The only different feature in this equation is that the solid and fluid velocities have been expressed in terms of the subsidence rate $v_s(h) = dh/dt$ using equation 2. This shows that the compactive effects are quantified by the inclusion of the rate of subsidence $v_s(h) = dh/dt$ in the convective term.

III. Effective Medium Theory

From equation (4) and the previous discussion, it easily seen that the two important parameters are the thermal and hydraulic conductivities of the sediment. In many practical situations in situ measurements are not available or perhaps not complete. In these cases it is useful to have a reliable theoretical model for predicting these transport coefficients. In addition, a theory is also required to extrapolate measurements to conditions different from those where the measurements were made. In previous work this has been achieved by empirical relations such as $\Lambda = \lambda_2(\lambda_1/\lambda_2)^\phi$ where ϕ is the volume fraction of medium 1 (i.e. fluid) and $1-\phi$ is the volume fraction of medium 2 (i.e. rock). This type of empirical relation is adequate only if the ratio λ_1/λ_2 is not very large or small. This relation would be inadequate if one of the components has zero conductivity. In this case the empirical relation incorrectly predicts that the effective conductivity Λ is zero for all values of the porosity.

The Effective Medium Theory first presented by Bruggerman[2] is very useful for describing the properties of a composite material if the volume fractions of the individual components and their properties are known. A very good recent review of the effective medium approach to various physical problems has been presented by Landauer[3]. The problem this theory attempts to solve can be stated

MODELS FOR THERMAL CONDUCTIVITY AND PERMEABILITY...

in the following way. For a randomly inhomogeneous medium where we are given the volume fractions c_i of the constituents and their individual conductivities λ_i, what is the effective conductivity of the n component mixture? Since the spatial distribution of the components is usually not known, normally only bounds on the conductivity can be established (Hashin and Shtrikman[4]). If the theoretical bounds are wide, as is the case for a mixture where the ratio λ_1/λ_2 is very small or large, then the bounds are not very useful. The effective medium theory on the other hand gives a unique prediction for the conductivity Λ by assuming that the distribution of the component "grains" is random. The basic result of this theory is

$$\Lambda^{-1} = \sum_{i=1}^{n} 3c_i (2\Lambda + \lambda_i)^{-1} \qquad (5)$$

where it is assumed that on a scale much larger than the "grain" size the composite is homogeneous and isotropic. This result is applicable for any number of components, any distribution of volume fractions, and all values of the individual conductivities. It has been tested many times experimentally as well as by comparison to numerical calculations for idealized inhomogeneous media. Only near a percolation threshold might a more accurate description be required (Kirkpatrick[5]).

IV. Micro-geometry of Clastic Sediments

As discussed in the previous section the effective coefficients for a composite medium can be predicted if the volume fractions of the individual conductivities are known. For certain properties, such as the thermal conductivity, this is quite sufficient since the differences in the individual conductivities is not very large and thus the detailed distribution of the components not very important. But for other properties, such as the permeability, the spatial distribution of the components is essential for an accurate prediction, as can be seen from most correlations (plots) of permeability versus porosity. The main reason is that the spatial distribution of the components is influenced by the particle size distribution, which must be taken into consideration. Particle size distributions and the packing of particles of different diameters has been studied for a long period of time. Figure 2 shows an example of the dense random packing of spheres of different diameters (Westman and Hugill[6]). Recently, Clarke[7] has discussed the importance of the bimodal particle size distribution, and how it can account for certain properties of sediments. In figure 3 we present this idealized picture of the bimodal model for clastic sediments. Here it is assumed that the sediment is composed of two distinct particle sizes, coarse and fine, whose diameters differ by more than an order of magnitude. $1-\phi_c$ represents the volume fraction of the coarse grained particles while X represents the total volume occupied by the fine grained particles (i.e. clays) plus their associated porosity ϕ_f. The actual porosity ϕ of a sediment can then be expressed in terms of the volume fraction of fines X, ϕ_c, and ϕ_f.

$$\phi = \phi_c - X(1-\phi_f) \qquad (6)$$

As X increases the fine particles must be packed into the pore space of the larger grain size particles which are carrying the overburden stress, thus decreasing the porosity. When $X > \phi_c$ all of the pore space of the coarse grain structure is filled and now all of the overburden is carried by the fines. In this figure it is assumed that the coarse grain porosity is $\phi_c = .4$ while the internal porosity of the fines is $\phi_f = .5$. Since the depositional environment affects the value of X, it can be expected that there will be large variations in the initial porosity of a sedimentary layer even before any diagenetic changes have occurred.

V. Thermal Conductivity

We will assume that for a typical clastic sediment the coarse grained particles are composed of a mixture of quartz and feldspar, the fine grained particles are clays, and the saturating fluid is water. The volume fractions of the individual components are as follows: coarse grains = $1-\phi_c$, fines = $X(1-\phi_f)$, and the saturating fluid = ϕ. Figure 4 contains the values of the individual components and the computed results for the effective thermal conductivity as a function of the volume fraction of fines using equation 5. For X=0 we have a saturated sediment of purely coarse grained particles with $\phi_c = .4$, while for X=1 we have a clay-water system with $\phi_f = .6$. The three curves represent three compositions of the coarse grained particles: the upper is for 100% quartz, the middle is for 50% quartz - 50% feldspar, and the lower represents 100% feldspar. If we define the region where $X < .4$ as "sandstone", then it is easily seen that the variability of conductivity of what we call "sands" can be significant.

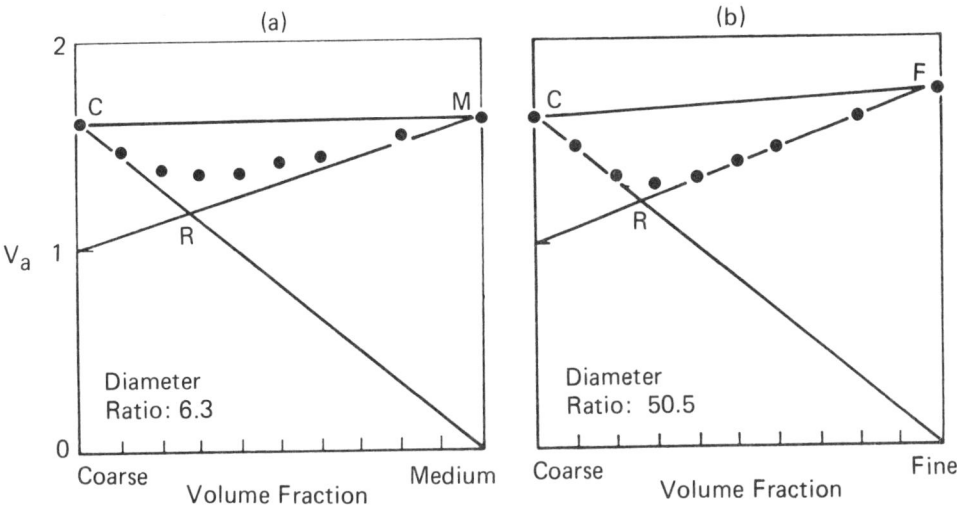

Figure 2. Observed volume for binary dense random packings of spheres with diameter ratios of (a) 6.3 and (b) 50.5. The volume is normalized by the total volume of the solid spheres, V_a = (Bulk Volume/Solid Volume).

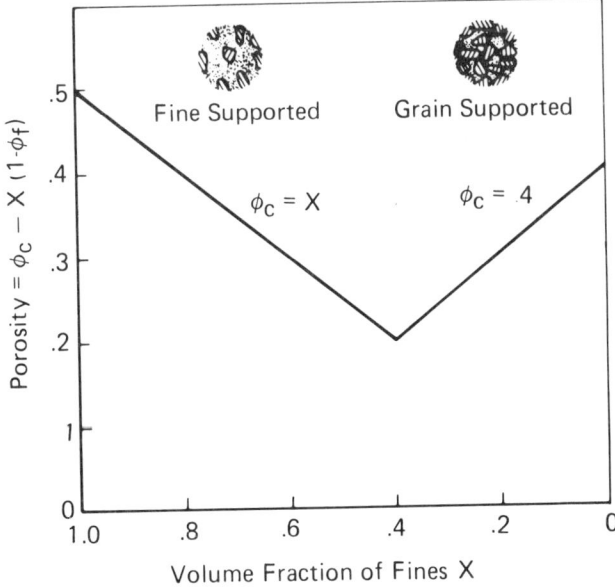

Figure 3. The porosity as a function of X, the volume fraction occupied by the fines.

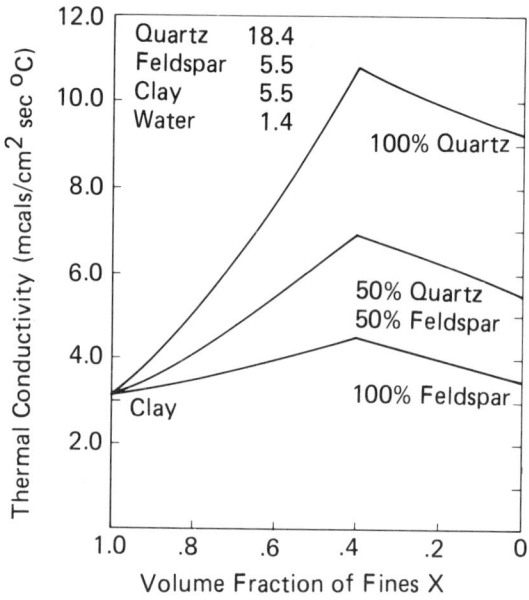

Figure 4. The effective thermal conductivity for clastic rock-water mixtures as a function of X, the volume occupied by the fines.

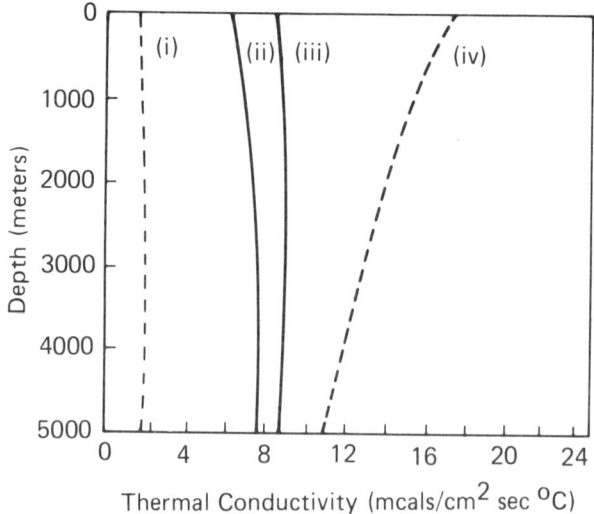

Figure 5. The thermal conductivity of a quartz-water mixture as a function of depth. Curves (i) and (iv) represent the conductivity of water (λw) and quartz grains (λq). Curve (iii) is the effective medium prediction. Curve (ii) is the prediction of the empirical relation $\lambda q (\lambda w/\lambda q) \phi$.

The variations of the effective conductivity with depth can easily be computed by including the effects of compaction on ϕ_c, ϕ_f, and ϕ and the effects of temperature on the λ_i. As an example we present the variation with depth of the effective conductivity for several clastic sediments. Figure 5 shows the effective thermal conductivity of a quartz-water mixture (curve iii) as a function of depth. The thermal conductivity (mcals/cm^2 sec deg C) of quartz (curve iv) has been assumed to vary with temperature as $\lambda_q = 18.4/[1+(4 \times 10^{-3})T]$ and the conductivity of water (curve i) as $\lambda_w = 2.23(1+T/273)^{3/2} - 0.88(1+T/273)^{5/2}$ where T is in degrees centigrade. For simplicity it was assumed that the temperature gradient was 30 deg C/km and the surface temperature was 20 deg C. The compaction curve has been taken as that shown for sand in Figure 1b. It is interesting to note that the effective conductivity is approximately constant with depth, implying that the decrease of the conductivity of the quartz grains due to increasing temperature compensates for the effects of compaction.

Feldspar and certain clays do not show such a marked temperature effect and thus the increase in the rock conductivity due to compaction can be significant. As an example we have modeled the clay and feldspar grains to have the same magnitude and temperature dependence, i.e. $\lambda_c = \lambda_f = 5.5/[5 \times 10^{-4})T]$. We consider two mixtures which represent the end members of the lowermost curve in Figure 4. The clay-water mixture is assumed to compact as shown in figure 1b (shale) while the feldspar-water mixture compacts as a sand. The results are shown in Figure 6. Due to the higher initial porosity, the shales (curve ii) have a somewhat smaller conductivity than the feldspar-water mixture (curve iii), but due to their larger compressibility the conductivity increases more rapidly with depth.

The thermal conductivity of sediments is predicted quite accurately by the effective medium theory *if* the mineralogy and the volume fractions are known. This is due to the fact that the micro-geometry of sediments is not a significant factor when the ratios of the individual conductivities are less than an order of magnitude. Unfortunately this is not true for other transport properties of a rock, such as the electrical conductivity and permeability, where the transport takes place primarily through one of the components and thus some of the details of the geometry have to be incorporated into the theory.

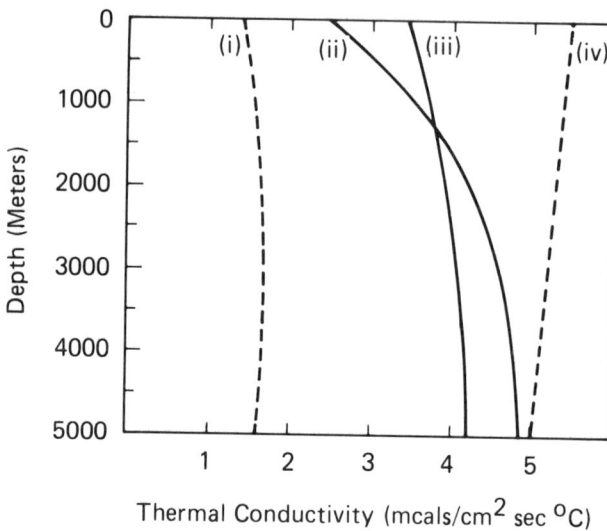

Figure 6. The thermal conductivity of a clay-water (ii) and feldspar-water (iii) mixtures as a function of depth. Curve (i) represents the conductivity of water and curve (iv) that of clay or feldspar grains.

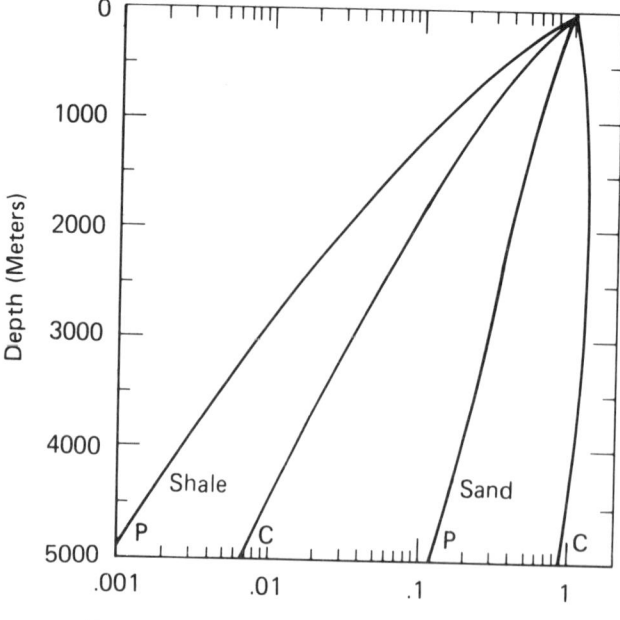

Figure 7. The permeability (P) and hydraulic conductivity (C) as a function of depth.

VI. Permeability

It has been known for a long time that the Carman-Kozeny equation adequately predicts the permeability of an isotropic rock of <u>uniform</u> grain size. This relation is

$$K = \phi^3/5S^2(1-\phi)^2 \ . \tag{7}$$

Here S is the surface area of the rock per unit solid volume and ϕ is the porosity. This equation is applicable for any rock of uniform particle diameter where the size of the grains is accounted for through the factor S. This equation can be derived by various methods, i.e. capillary tube or random network models for the pore space, but the assumption that the rock is composed of particles of approximately equal diameters must be always included. Unfortunately in practice this relation is often applied to rocks which contain a wide particle size distribution and then the validity of the relation is questioned. The permeability will change as the porosity and the specific surface S decrease with burial. Figure 7 shows the variation of the permeability and hydraulic conductivity of a "sand" and "shale" and their respective compaction curves. They can be considered to be end members of a bimodal sediment, X=1 and 0 respectively. All values are normalized by their surface values. The smaller decrease in the conductivity is due to the fact that the viscosity of water decreases by a factor of 10 over this temperature range.

The more interesting problem is of course how the permeability changes with X, the volume fraction occupied by the fine particles. In this case the geometry of the pore space has to be taken into account since the fines are "packed" into the pore space defined by the larger grains. A random network model which is defined by the packing of the coarse grains is very convenient for this description (Zallen[8]). Due to the uniform particle size the network contains elements (bonds) of approximately equal permeability k_c, while the connectivity of the network as defined by the coarse grains provides the additional geometric information on the pore space. For a uniform grain size sediment of porosity ϕ_c, the permeability by this model becomes

$$K = \phi_c k_c/\tau_c \quad \text{where} \quad \tau_c = 3/(1+2\phi_c) \ . \tag{8}$$

This result is equivalent to the Carman-Kozeny result if k_c is expressed in terms of ϕ_c and S by similar assumptions as those leading to equation (7). It is interesting to note that τ_c represents the tortuosity (geometry) of the isotropic network while k_c depends on the details of each of the components (bonds) of the

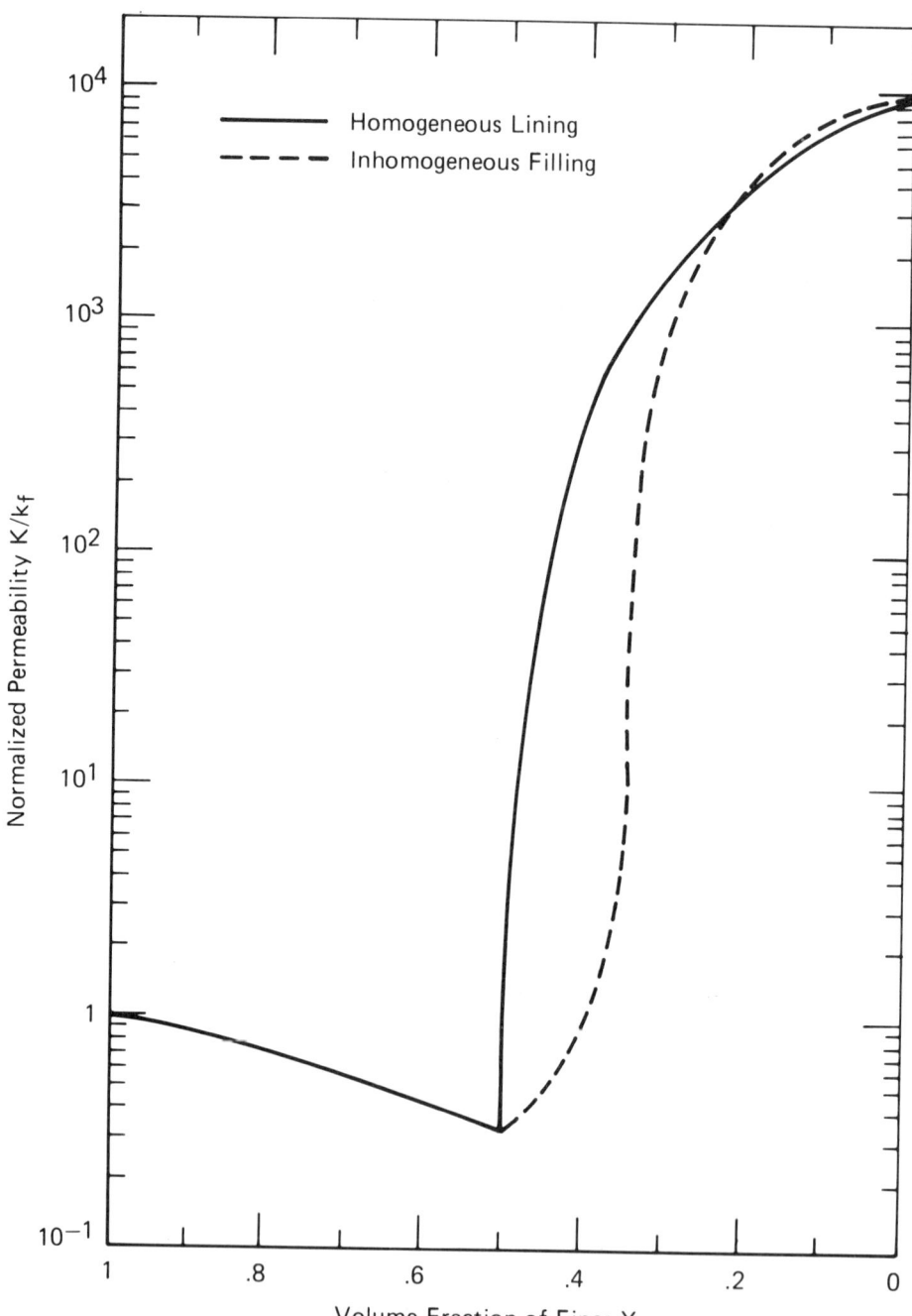

Figure 8. The permeability as a function of X, the volume fraction occupied by the clays.

network, such as size and shape. As the fine particles are now added, they must enter into the pore space defined by the network, and thus they can only alter the permeability of the individual bonds of the network. This fact implies that the permeability of sediment with a bimodal distribution of grain size will have the form

$$K = \phi_c \, k_{eff}(X)/\tau_c. \qquad (9)$$

Here $k_{eff}(X)$ is the effective medium value of the bond permeability for a given distribution of bond permeabilities. The distribution of bond permeabilities of course depends on the assumed distribution of fine particles within the rock. There are numerous ways of distributing the fine particles within the pore space ϕ_c, and normally this information will not be available. As an example we shall consider two extreme cases. First we shall assume that the fine particle volume X is distributed homogeneously on the surface of the larger grains, uniformally reducing the pore channels (solid line). The second case assumes that the fines are distributed at random completely filling some pores with probability X and leaving others totally empty with probability 1-X (Fig. 8 dashed line). In both cases, $k_{eff}(X)$ is computed through the effective medium theory and the results are presented in Figure 8. The computations are expressed in units of k_f, the permeability of the clay water system. As the clay particles are packed into the pore space the permeability decreases dramatically eventually reaching a minimum when ϕ_c (assumed to be equal to .5) is completely filled. With a further increase in the volume fraction of the fines (X>.5) the permeability increases since impermeable larger grains are replaced by a porous clay- water mixture of permeability k_f. This clearly illustrates how delicately the permeability depends on the volume fraction of clay particles and their distribution, an effect that is more significant than the permeability changes that occur due to compaction. The manner in which the clay particles are distributed in the pore space is very important only over a narrow range of values of X, in a range where the permeability varies dramatically from the coarse grain to the fine particle value which can be described by the Carman-Kozeny relation.

VII. Summary

The effective medium theory is extremely useful for describing the physical properties of rocks. The results for the thermal and hydraulic conductivities are presented as illustrations of the use of this theory as well as to point out the important

335

variables that effect these parameters. The permeability is primarily determined by the fine particle volume and grain size. As a rough rule "supernormal" pressures will be found in sections of basins were the volume fraction of fine particles is .4 or larger and the particle size is of the order of montmorillonite or smaller. The thermal conductivity on the other hand depends on the crystal structure of the individual grains as well as the volumes of the individual components. The pore space geometry has to be modeled appropriately for the prediction of the permeability but is not very important for the thermal conductivity. Generic descriptions such as "sand" and "shale" are not sufficient for an accurate quantification of these two parameters.

REFERENCES

1. V. V. PALCIAUSKAS and P. A. DOMENICO, Water Resources Research, 18, p. 281-290, 1982.

2. D. A. G. BRUGGERMAN, Ann. Phys. (Leipz.), 24, p. 636, 1935.

3. R. LANDAUER, Proc. Conf. On Electrical and Optical Properties Of Inhomogeneous Media (ETOPIM). The Ohio State University, Columbus, Ohio, September, 1977. That review includes all of the important historical references on this subject.

4. Z. HASHIN and S. SHTRIKMAN, J. Appl. Phys. 33, p. 3125, 1962.

5. S. KIRKPATRICK, Reviews of Modern Physics, 45, p. 574-588, 1973.

6. A. E. R. WESTMAN and H. R. HUGILL, J. Am. Ceram. Soc., 13, p. 767, 1930.

7. R. H. CLARKE, AAPG Bulletin, 63, p. 799-803, 1979.

8. R. ZALLEN, Fluctuation Phenomena, Ed. E. W. Montroll and J. L. Lebowitz (Amsterdam: North-Holland), Ch. 3, p. 117-228.

PART D

EXAMPLES OF THERMAL RECONSTRUCTIONS

F. HORVÁTH[1], Á. SZALAY[2], P. DÖVÉNYI[1],
J. RUMPLER[3]

STRUCTURAL AND THERMAL EVOLUTION OF THE PANNONIAN BASIN: AN OVERVIEW

ABSTRACT

The Pannonian basin formed by extension (17-10 Ma) and subsidence (17-0 Ma) of an Alpine orogenic area. A one-dimensional computer model has been developed to simulate its subsidence and thermal history. Model parameters were constrained by comparison of predicted and observed i) subsidence history, ii) present crustal thickness, iii) temperature vs. depth profiles and iv) heat flow.

Thermal maturation of organic matter was calculated by the use of the Lopatin method. Model results were compared with measured vitrinite reflectances (Ro) in eight "master wells" which provided an improved Time-Temperature Index vs. Ro relationship in the interval Ro $\leq$ 3.0%, and for temperatures up to 230°C. This relationship was used to reconstruct the maturation history of potential source rocks in the basin. The value of such modeling is shown by comparing the calculated maturity with the location of oil and gas fields in a well-known test area in the Great Hungarian Plain.

I. INTRODUCTION

The Pannonian basin in Central Europe, is one of

(1) *Geophysical Department, Eötvös University, Budapest, Hungary.*
(2) *Petroleum Exploration Company, Szolnok, Hungary.*
(3) *Geophysical Exploration Company, Budapest, Hungary.*

the largest young sedimentary basin in the Alpine folded belt. Extensive geophysical surveys and numerous deep drillings for hydrocarbon prospecting in the last decades have led to a good knowledge of the structural features and sedimentation history of the basin. This provides a suitable basis to attempt to model quantitatively the basin evolution and to predict future prospects for petroleum exploration.

In this paper we give a brief overview of our model calculations and show some implications. Further information and a similar model study of the Pannonian basin can be found in Royden et al. (1983a,b). More up-to-date and comprehensive data collection and interpretation will be published in a volume to be edited by Royden and Horváth (1985).

II. REGIONAL GEOLOGY OF THE PANNONIAN BASIN

Bally and Snelson (1980) have suggested that the Pannonian basin is the type example of continental backarc basins which are associated with continental collision and are located on concave side of an arc characterized by A-type subduction. Figure 1. shows the depth to basement map of the Pannonian basin, and the surrounding mountains of the Alps, Carpathians and Dinarids. The basin fill varies in age from early Miocene to Quaternary and locally can be as thick as 7000m. Basement morphology is characterized by a system of deep troughs which are separated by basement highs.

Interpretation of reflection seismic profiles and drillhole data suggests that the Pannonian basin formed by crustal extension during the middle Miocene (Horváth and Rumpler, 1984; Pogácsás, 1985). Large displacement along listric normal faults resulted in progressive tilting of originally horizontal strata and the formation of characteristic half grabens (Fig.2). Normal faults were connected to a system of coeval strike-slip faults. The most remarkable system is associated with a mid-Hungarian shear zone which crosses obliquely (from SW to NE) the whole Pannonian basin. Curvature, splaying and side-stepping of strike-slips segments occur frequently. This has given rise to contemporaneous formation of closely spaced zones of extension and local compression.

Evolution of the Pannonian basin

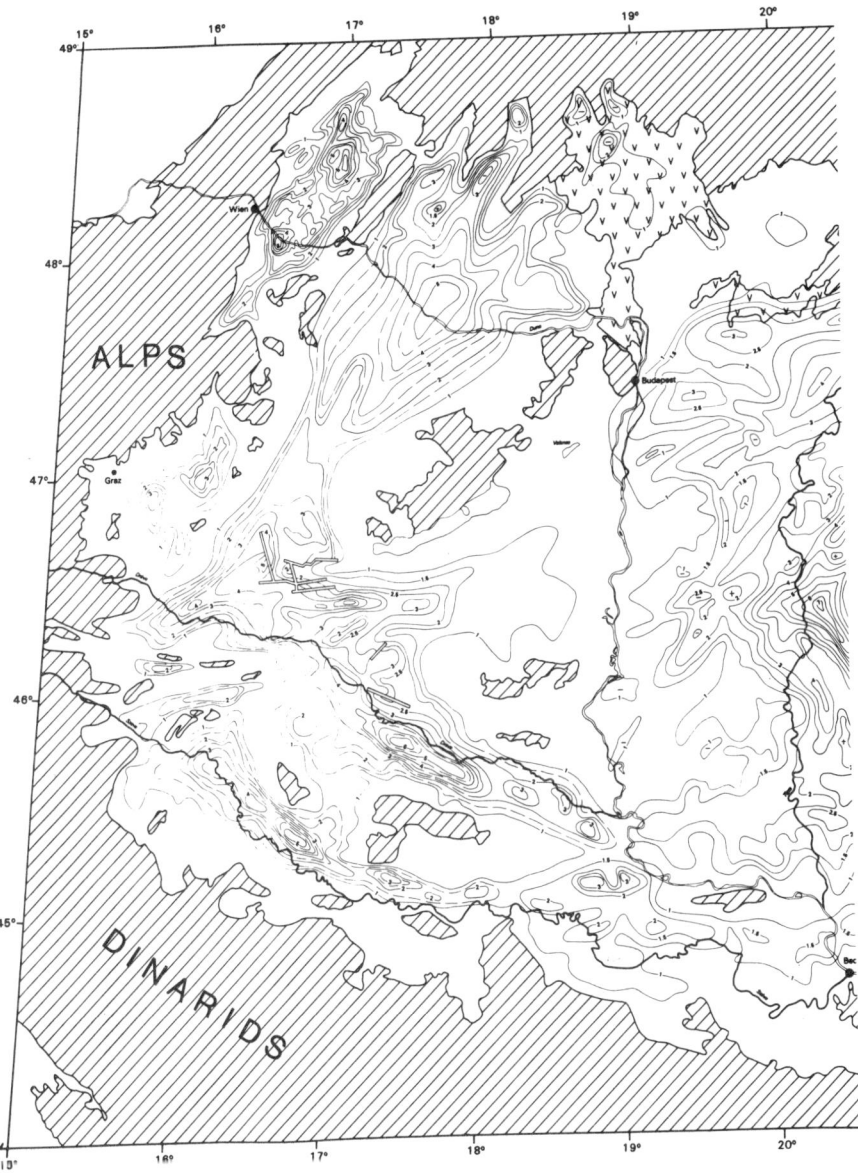

Figure 1. Isopach lines (in kilometers) of the Miocene through Quaternary basin fill of the Pannonian basin (Horváth and Royden, 1981). Hatchured areas indicate the pre-Miocene rocks of the Alpine, Carpathian and Dinaric mountains and the outcrops of the basement within the basin. Miocene to Pliocene calc-alkaline volcanic rocks on the surface are also shown (vvv).

Evolution of the Pannonian basin

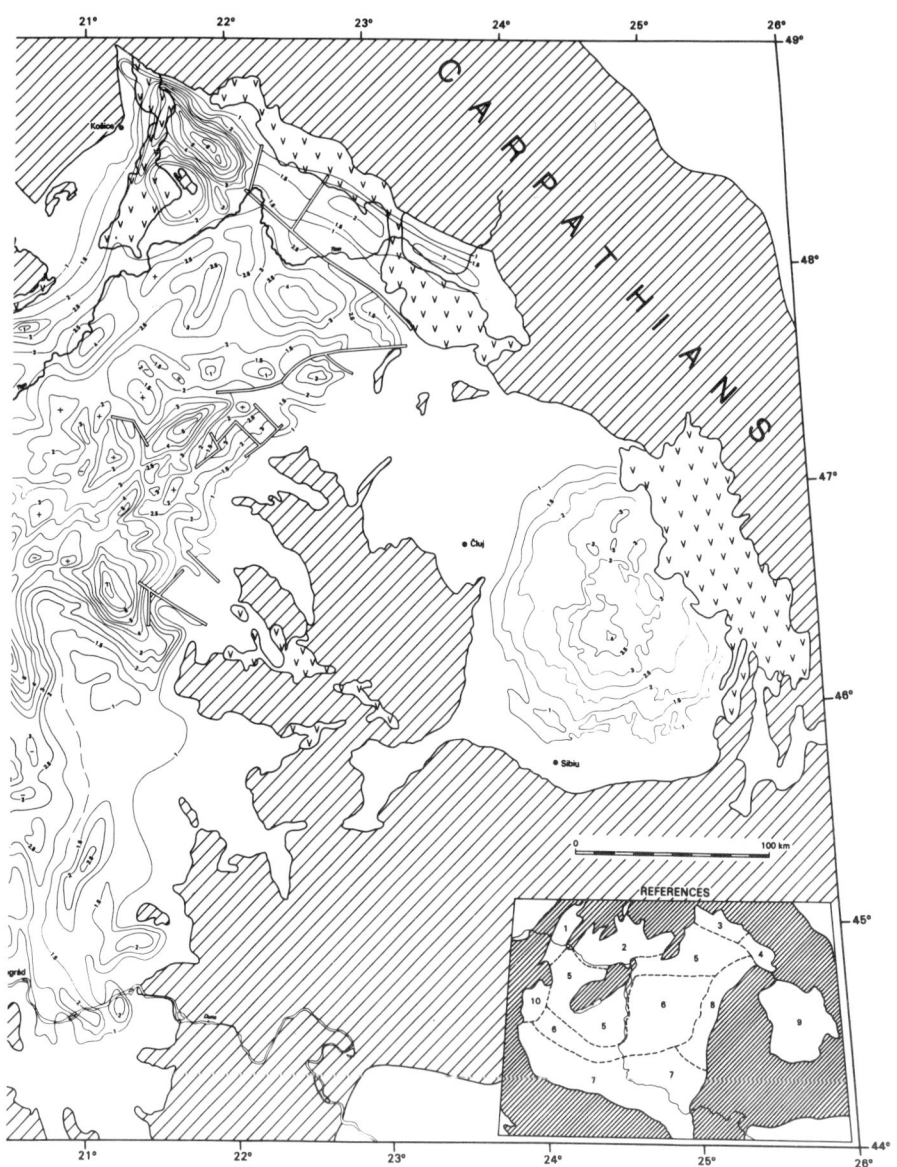

Figure 1 (cont'd). Note that three deep depressions close to the Carpathians are not considered part of the Pannonian basin. These are the Vienna basin, the Transcarpathian depression and the Transylvanian basin. They are indicated by 1, 3-4 and 9 respectively, on the small inset at the lower right corner. The Great Hungarian Plain lies to the east of the Duna river (see Fig.5).

Evolution of the Pannonian basin

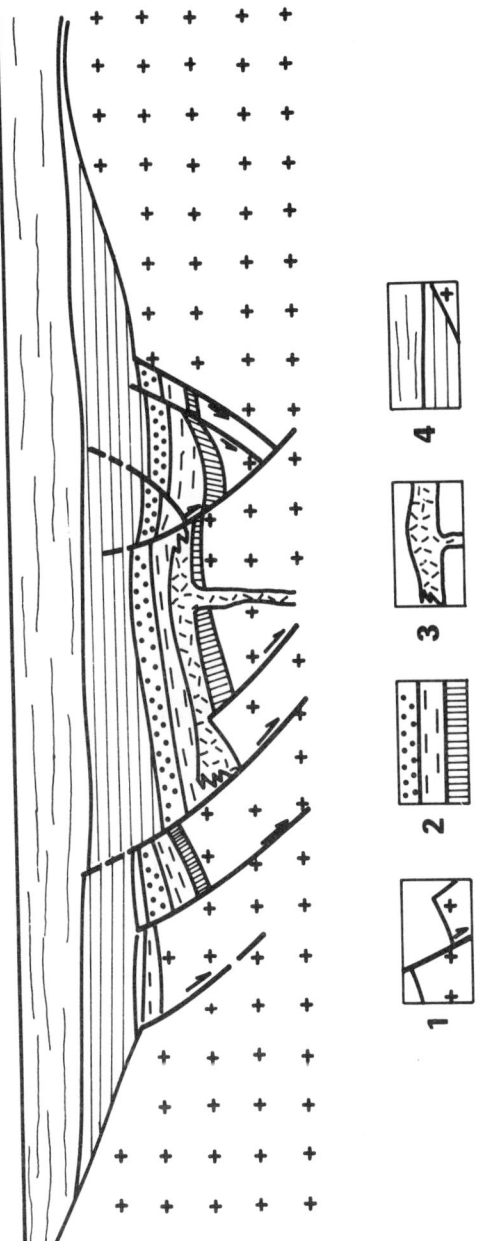

Figure 2. Sketch to show a characteristic extensional structure in the Pannonian basin. Legend: 1=Listric normal fault which involves the basement; 2=Synrift sedimentary rocks. They are predominantly deep water marls and turbidites of Middle Miocene age (Badenian and Sarmatian). 3=Synrift volcanic rocks. 4=Post-rift sedimentary rocks. They consist of Lower Pannonian prodelta to delta slope deposits which are overlain by Upper Pannonian and Quaternary delta plain and marsh deposits.

Evolution of the Pannonian basin

During the middle Miocene <u>initial phase</u> of basin evolution subsidence rates were usually higher than sedimentation rates. This brought about significant water depths (locally more than 1000m) in extensional troughs, which were separated by more stable and emergent blocks. In the troughs mostly fine-grained marls and clays were deposited with a relatively high organic content (Corg = 0.5% to 1.5%). At the beginning of this period the Pannonian basin belonged to the Paratethys sea. Marine connections finally ceased during the Sarmatian and the Pannonian basin became an isolated lake afterwards.

The late Miocene through Quaternary period is the <u>thermal phase</u> of subsidence. Little or no faulting involved the basement but frequently soft sediment deformation occurs. The sedimentation process has been controlled by upbuilding and southward progradation of a large fluvial-dominated delta system and by associated downslope mass transport. Distal turbidites are made up mostly of shales and contain an increased amount of organic matter of dominantly terrestrial origin. In general, subsidence rates were smaller than sediment accumulation rates during this period. Progressive infilling of the Pannonian basin led to shallow water conditions during the Pliocene, and eventually to the disappearance of the lake during the Quaternary.

III. RELEVANT GEOPHYSICAL FEATURES

Refraction and reflection seismic data indicate that the crust of the Pannonian basin is continental and the depth to Moho varies from 24km to 28km (Posgay et al., 1981; Horváth and Royden, 1981). Taking into account the sedimentary fill of the basin the minimum thickness of crystalline crust is about 18km. The pre-extensional crustal thickness is, of course, not known but it is reasonable to suppose that it was similar to the present thickness of the internal Eastern Alps and Western Carpathians (35-45km). This gives a constraint for the maximal value of crustal stretching: δ = 2 to 2.5.

The present thickness of the lithosphere is markedly less than normal. Deep seismic sounding, magnetotellurics and seismological data suggest that 40 to 80km thick lithosphere is underlain by an elevated, anoma-

Evolution of the Pannonian basin

lous mantle wedge (Posgay et al., 1981; Ádám and Wallner 1981). Significant mantle contribution to the thermal regime of the crust is shown by high temperature gradients, elevated heat flow and increased $3He/4He$ ratios (Oxburgh, this volume). Geothermal data including thermal conductivities of the basement and sedimentary rocks have been summarized recently by Dövényi et al. (1983). They also argue that convective thermal disturbances are either shallow or small in the Pannonian basin.

IV. COMPUTER MODELING

A one-dimensional computer model has been developed to simulate the subsidence, thermal and maturation history of the Pannonian basin. We call it MISS BASIN for Maturity Interpretation System for Sedimentary Basin. A simplified flow diagram is shown in Figure 3. The basic idea and fundemental mathematics is similar to that described by Royden et al. (1983 a.,b) for a two--layer stretching model. Differences from this model include the use of improved relationships to characterize:

i) the change of porosity with depth (normal compaction trend),

ii) the change of thermal conductivity of fine-grained (shales and marls) and coarse-grained (sandstones and conglomerates) sedimentary rocks with depth and temperature.

Pore-pressure vs. depth functions at different times were also estimated by using a simple method suggested by Magara (1978). During model calculations the same original crustal and lithospheric thicknesses were assumed for the whole Pannonian basin (36km and 126km, respectively). The time interval of active crustal stretching was also fixed (17Ma to 12Ma). Subcrustal thinning was assumed to occur during the same time, but in a few cases it was allowed to continue until 10Ma.

Particular attention was paid to constrain unfixed model parameters like stretching factors for the crust and mantle lithosphere, and heat production of the crust. This was done by comparing model predictions with actual data determined in or extrapolated for a borehole. Comparison of crustal thickness, temperature vs. depth

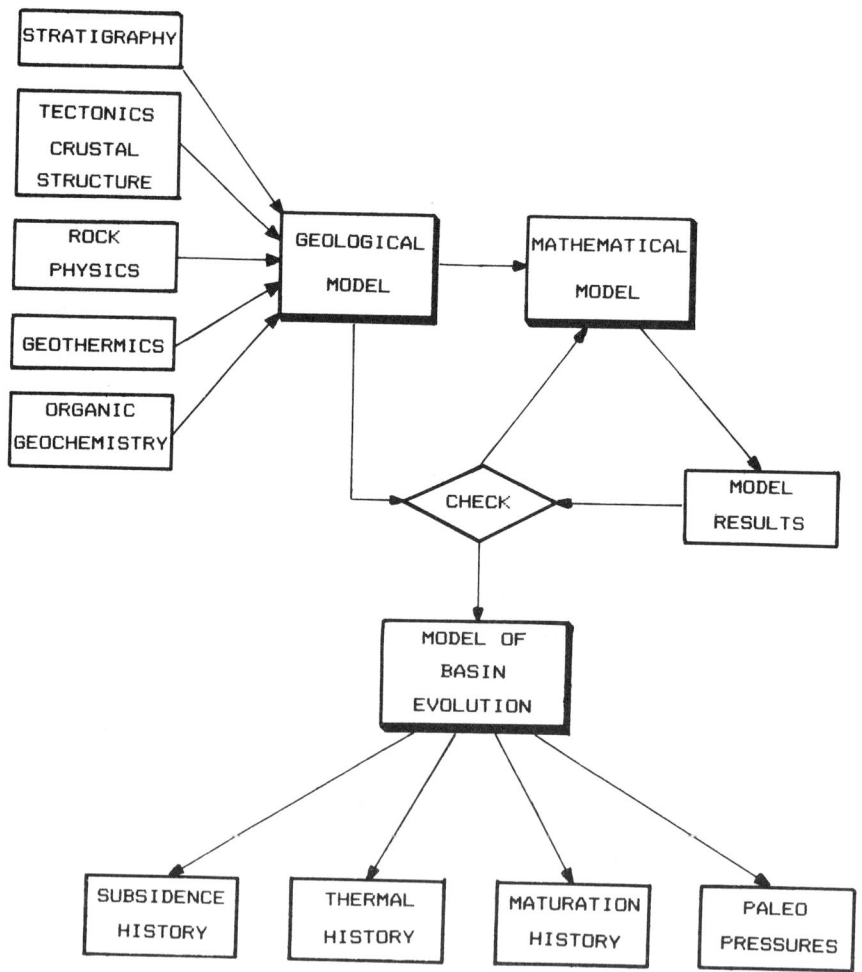

Figure 3. Simplified flow diagram to show the basic idea of the computer program (MISS BASIN) which is used to simulate basin evolution.

Evolution of the Pannonian basin

profiles, and heat flow is easy and straightforward. This is, however, not readily the case for the subsidence history. Decompacted thicknesses of chronostratigraphic units give only the sedimentation history. It should be combined with the variation of water depth in space and time to infer the "observed" subsidence history. The traditional view held that water depth was always shallow (0m to 100m) during the deposition of the Pannonian basin sediments. Recent stratigraphic analysis of seismic sections combined with drill core studies has led to a better understanding of facies conditions and paleo-depositional surfaces (Mattick et al., 1985). Figure 4 shows a generalized water depth pattern for deep basins (present basement depth more than 3000m), and it was used during the model calculations. Model parameters were accepted if the comparison of calculated subsidence history with the observed sedimentation history resulted in a reasonable water depth history.

According to our experience the present crustal thickness, temperature vs. depth profiles, heat flow and the water depth history are good constraints on appropriate solutions for the stretching parameters. We arrived at the conclusion that moderate crustal extension ($\delta \leq 2.2$) combined with major thinning of the mantle lithosphere ($\beta \leq 100$) offers an adequate mechanism to explain the formation of the Pannonian basin.

V. RESULTS OF MATURITY CALCULATIONS

Reconstruction of subsidence and thermal history of the sedimentary strata makes it possible to delineate the maturation history of the potential source rocks in the basin. The simple Lopatin-Waples relationship was used to calculate the change of Time-Temperature Index (TTI) of the basin fill through time (Waples, 1980). We did not accept, however, Waples' empirical relationship that related TTI and mean vitrinite reflectance (Ro). Instead, we recalibrated this relationship in order to arrive at a more appropriate relationship that is valid for young and hot basins like the Pannonian basin. We selected 8 "master" wells in the Great Hungarian Plain (Fig.5), which exhibited particularly good data sets, including vitrinite reflectance vs. depth profiles. The eight wells are characterized by rather different thermal and maturity

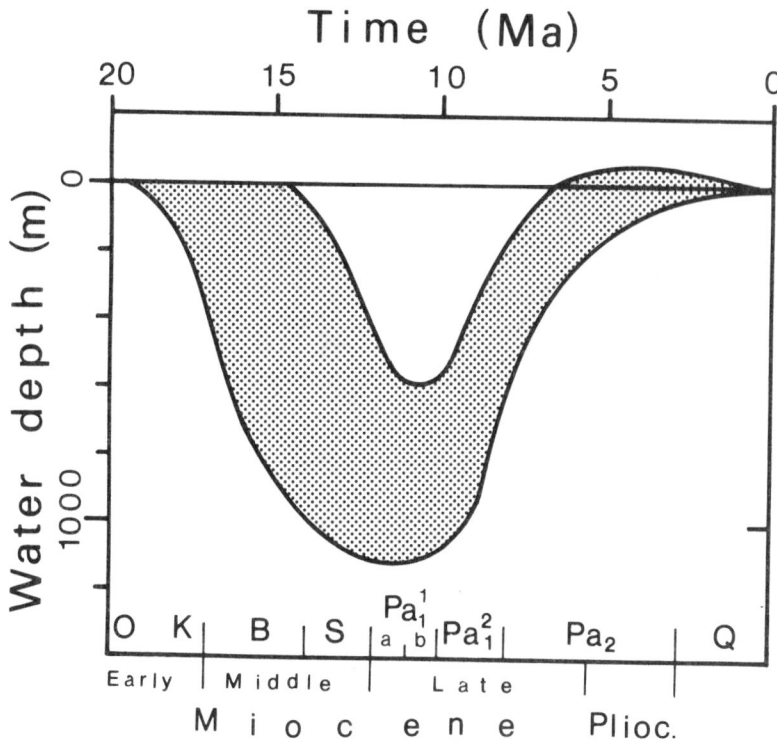

Figure 4. Generalized trend of water depth changes in the deep troughs of the Great Hungarian Plain during the deposition of early Miocene to Quaternary sediments. Actual water depth curve for a given locality should plot within the area dotted. Note that the time axis at the bottom shows the regional stage system: Ottnangian (O), Karpatian (K), Badenian (B), Sarmatian (S), early Lower Pannonian (Pa1,1a and Pa1,1b), late Lower Pannonian (Pa1,2), Upper Pannonian (Pa2) and Quaternary (Q).

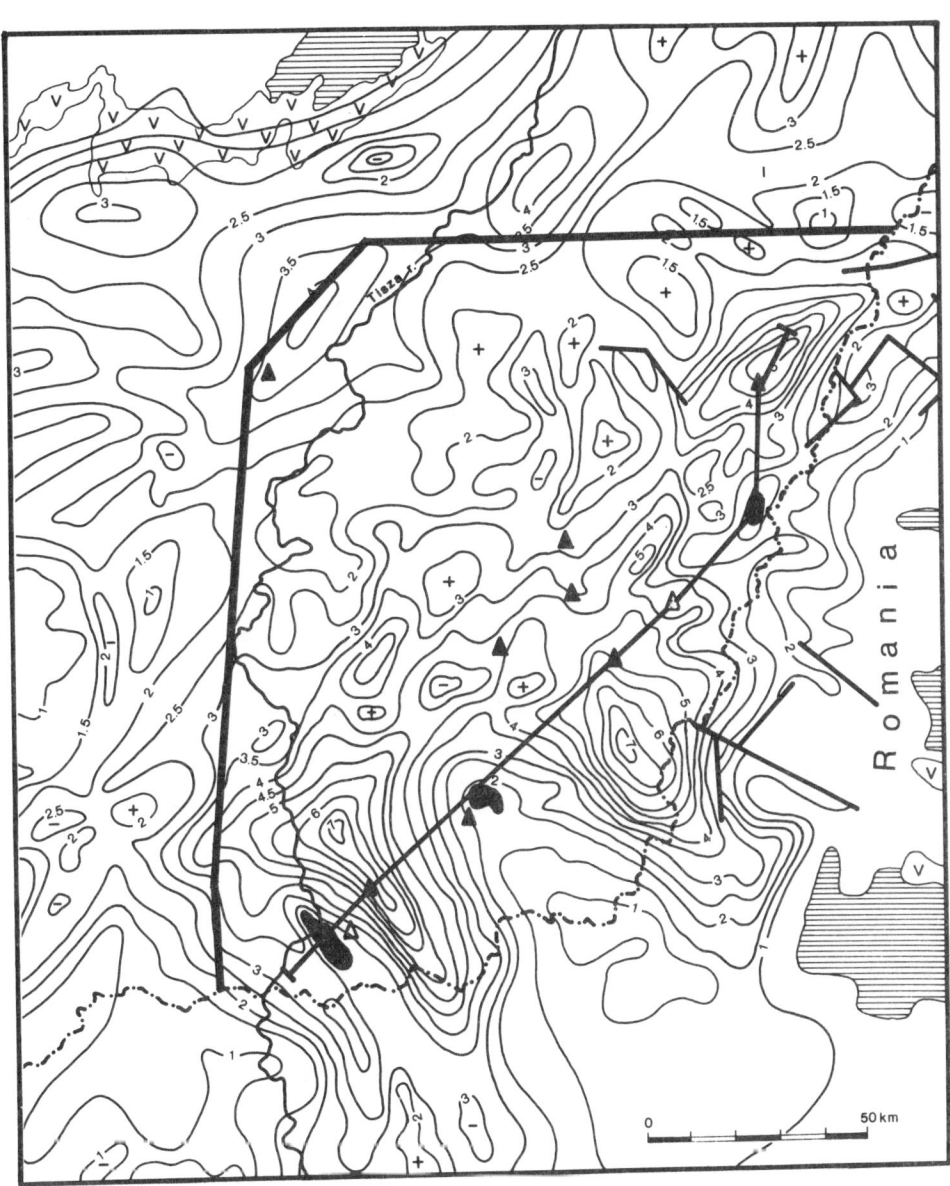

Figure 5. Isopach map of the Great Hungarian Plain showing the boundary of the test area, the location of the profile in Fig.8 and the master wells. Open triangles close to the profile indicate two additional wells which are also shown in Fig.8. Three oil fields discussed in the text are indicated by black spots.

conditions (Fig.6). For each sedimentary horizon that corresponded to measured value of vitrinite reflectance in the wells, a temperature history ($T(t)$) was determined by the MISS BASIN program, and then a Time-Temperature Index was calculated by the formula

$$TTI = \int_0^t 2^{\frac{T(t) - 105}{10}} dt$$

Current TTI values and measured vitrinite reflectances show a remarkably good correlation, and could be approximated with an empirical curve as is shown in Figure 7. This curve is significantly different from that of Waples for the Ro > 0.6%, and agrees perfectly in the 0.3% ≤ Ro ≤ 1% range with a new relationship derived by Issler (1984) for the Atlantic continental margin of North America and Europe. Our TTI-Ro conversion was checked for an other set of nine independent wells in the Great Hungarian Plain, which are again characterized by good vitrinite reflectance vs. depth profiles and other relevant data. We are fairly confident that this conversion is adequate and can be used in the Ro ≤ 3% interval and for temperatures up to 230°C.

We selected a test area of about 17 000km^2 in the eastern part of the Pannonian basin, which lies between the Tisza river and the Hungarian/Romanian national boundary (Fig.5). This area is remarkably well studied: the total length of multifold seismic sections is about 10 000km and some 200 wells deeper than 2000m have been drilled for hydrocarbon exploration. There are 53 productive oil and gas fields. The test area was covered by a rectangular grid system with a 5kmx5km spacing, and model calculations were performed at each grid point. This included the determination of the current maturity and its reconstruction at three different times (2.4Ma, 8Ma, 10Ma), along with the coeval temperatures and stratal thicknesses.

Figure 8 shows an approximately NE-SW oriented section to illustrate the current structural and maturity conditions. It can be seen that the oil-generation window (0.6% ≤ Ro ≤ 1.3%) starts at 2km to 3km and terminates at 3.5km to 5km. The M4,5 and Pal,1a units, which represent potential source rocks have already passed through this window. Reconstruction of the maturation

Evolution of the Pannonian basin

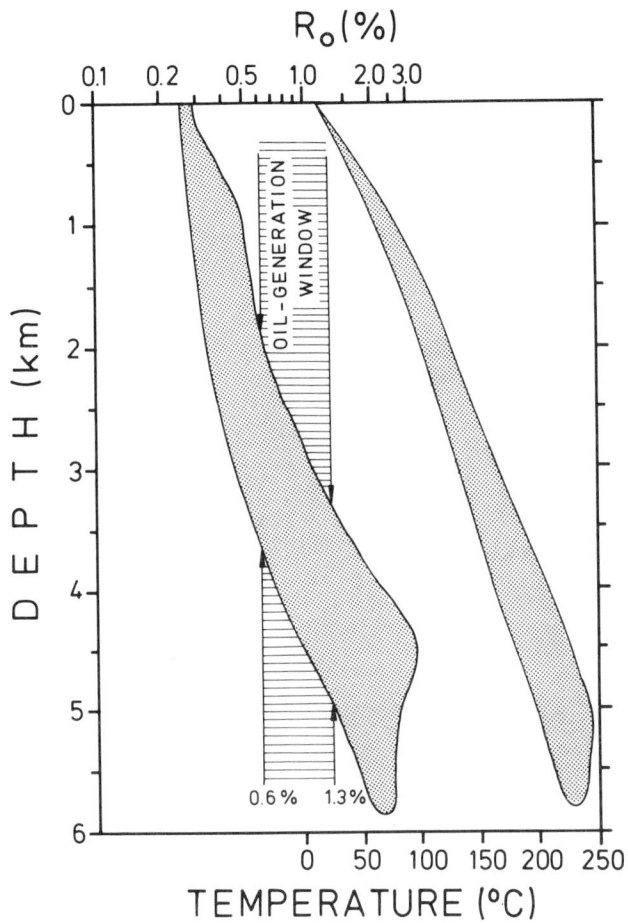

Figure 6. The range of mean vitrinite reflectances (Ro, on the left) and temperatures (on the right) as function of depth in the eight master wells of the Great Hungarian Plain. It is to show that maturity and thermal conditions are rather variable in the area under consideration. The onset of oil-generation window, for example, varies from a depth of 1.8 km to 3.6 km.

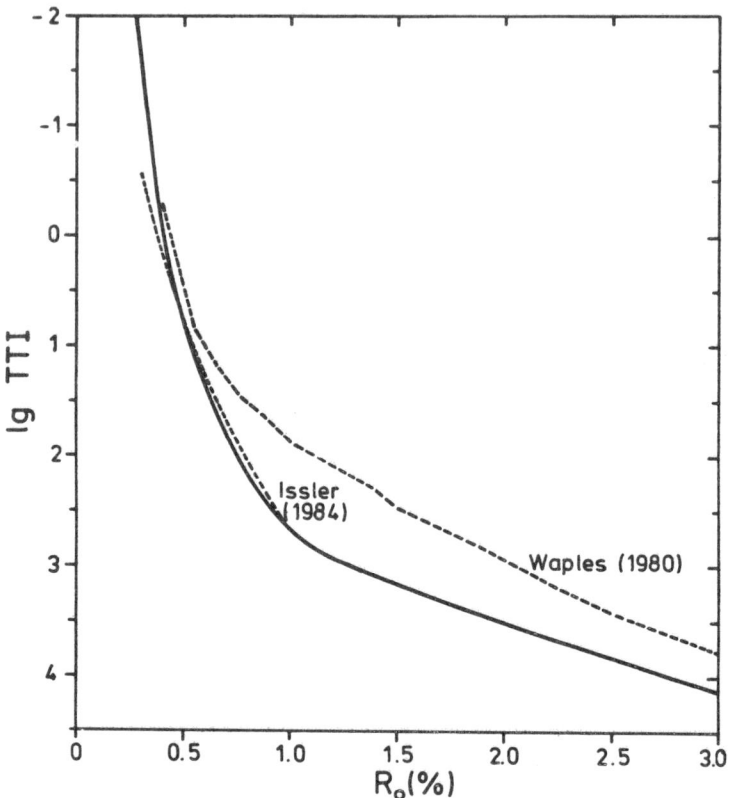

Figure 7. Relationships to convert Time-
-Temperature Index to vitrinite reflectance (Ro) after Waples (1980), Issler
(1984) and this paper.

and subsidence histories along the same section shows
that hydrocarbons have been generated in deep troughs
during the last 6 to 8Ma. 50 of the 53 known fields
in the test area were found outside of the oil genera-
tive troughs and above basement highs in immature reservoir rocks. Three fields are crossed by the section
in Figure 8. The Algyő field on the SW is the biggest
Hungarian oil field and is situated in a compaction
anticline of Pa1,2 and Pa2 strata above the elevated
basement. The Pusztaföldvár field was found in a similar
structural/lithological trap right above the basement.
Körösszegapáti is a small field along the section which
is particularly interesting because most of the oil was
accumulated in the Paleozoic basement. The basement here
exhibits high permeability due to fractures and fissures.

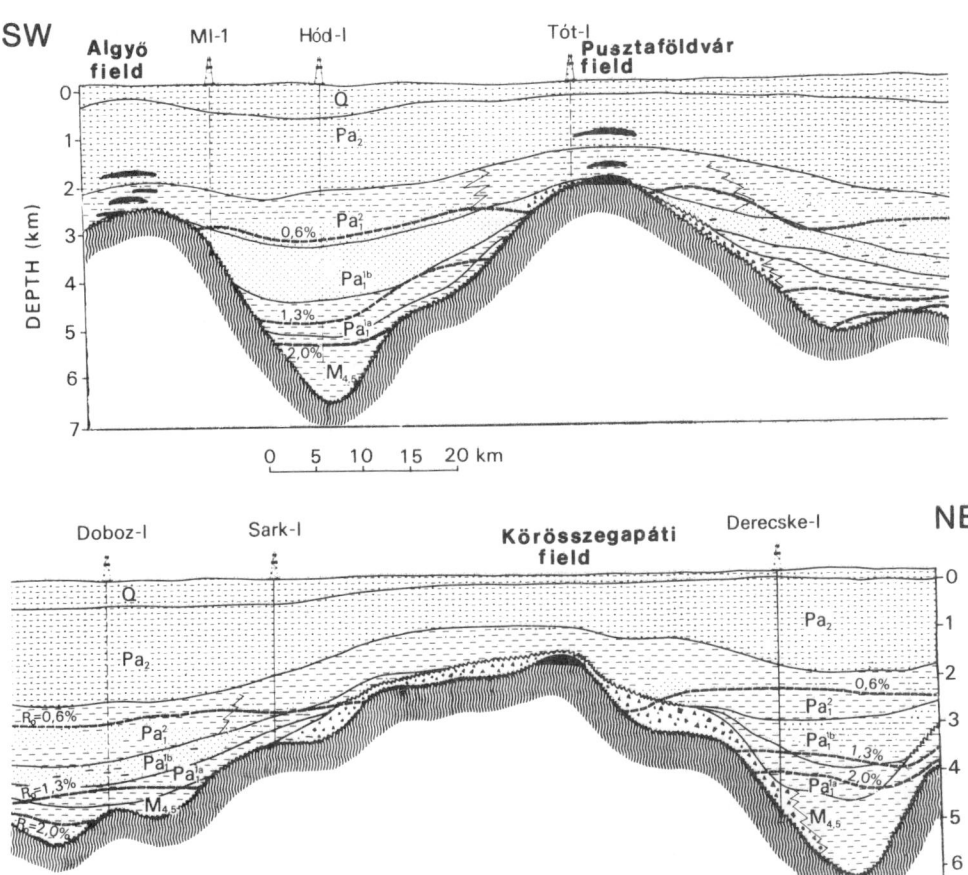

Figure 8. Profile to show the stratigraphy and maturity conditions in the Great Hungarian Plain. Three oil fields are indicated by black spots with vertical exaggeration. Legend: 1=Chronostratigraphic units: Middle Miocene (M4,5), Lower Pannonian (Pa1,1a, Pa1,1b, Pa1,2), Upper Pannonian (Pa2) and Quaternary (Q); 2= Marls and claystones; 3= Sandstones; 4= Conglomerates and tectonic breccias; 5= Mixed development consisting of alternating sand and clay beds; 6= Water; 7= Basement; 8= Isoreflectance lines (Ro= =0.6%, 1.3% and 2%). See location of the profile in Figure 5.

Hydrocarbons found in traps associated with basement highs were generated in the nearby deep troughs and then migrated updip until they were trapped in relatively shallow depths (1.5km to 2.5km). Although this seems to be a general mechanism in the Great Hungarian Plain, we think this is not the only type of fields which can be searched for. It is because pressure conditions have been such as to hamper a large part of generated hydrocarbons to migrate directly up towards basement highs. Reconstruction of pore pressures for different time epochs shows formation of remarkable overpressures in the shaley $M_{4,5}$ and $Pal,1a$ units at about the same time when they entered the oil generation window. Actual pressure data and calculations suggest that the largest excess pressure develops within the upper part of the thick shale complex. Such a "pressure seal" has a major control on the direction of fluid migration (Magara, 1978).

Figure 9. shows the calculated subsidence and maturation history for the Hód-I well (see location in Fig.8). The depth of the pressure seal and its change through time is also indicated. It can be seen that a lesser volume of potential source rocks (upper part of unit $Pal,1a$) has always remained above this pressure seal. Hydrocarbons generated over here could have migrated upward into the permeable $Pal,1b$ unit, and then laterally toward basement highs. However, a larger volume of the source rocks has been situated below the pressure seal and, hence, hydrocarbons generated in this volume should have been driven downward into permeable rocks at depth. Such rocks can be found:

i) in association with the unconformity between $M4,5$ and younger Pannonian strata,

ii) at the lower part of $M4,5$ strata (e.g. basal conglomerates),

iii) in the fractured basement.

These possible migration paths could have resulted in traps inside the basement and, more probably, subtle traps at the flanks of deep troughs. Exploration of these kinds of traps can be performed successfully by combining a reliable reconstruction of direction(s) of migration with an analysis of coeval trap formation.

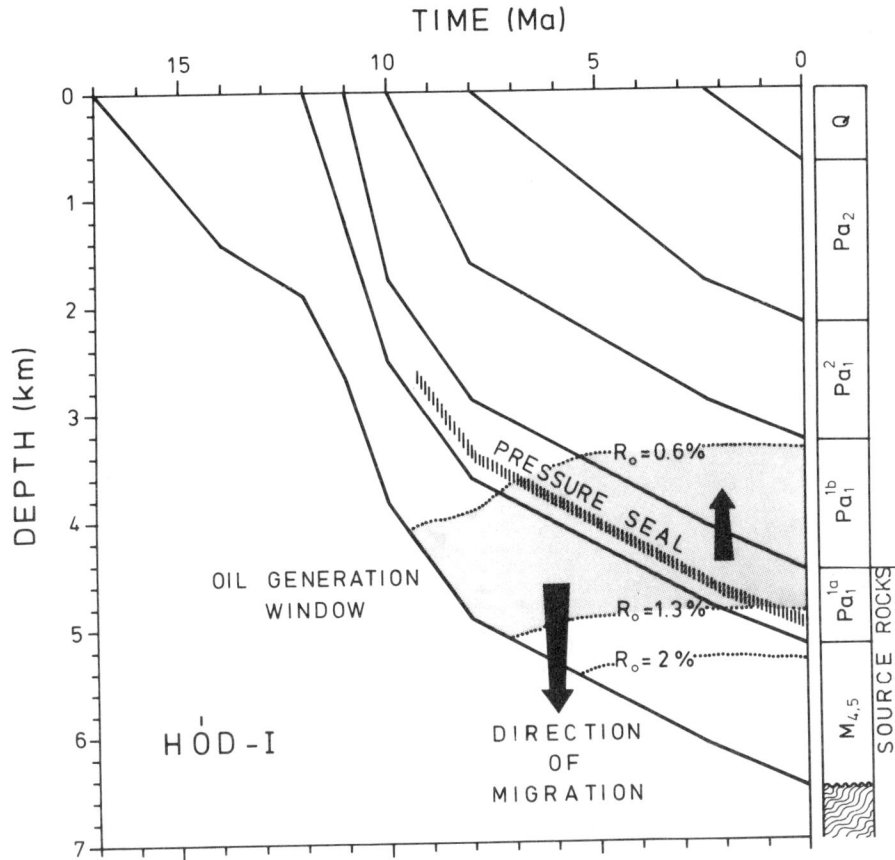

Figure 9. Subsidence and maturation history of the Hód-I well. The depth of the maximum overpressure is also shown through time. This pressure seal has had a major control on the direction of hydrocarbon migration out of the source rocks in units Pa1,1a and M4,5 (Szalay, 1985).

VI. EPILOGUE

The "state-of-art" of basin analysis in Hungary can be summarized as follows:

1) There has been a great progress in understanding of the evolution of the Pannonian basin. This includes i) seismic mapping of areas of major extension and

timing of extension, ii) implication of seismic stratigraphy to derive regional depositional models and paleo-water depth estimates and iii) new temperature, heat flow and maturity data in deep wells.

2) An improved computer model, called MISS BASIN, has been developed to simulate the subsidence, thermal and maturation history of an extensional basin.

3. Computer simulation in the Great Hungarian Plain confirmed the earlier geodynamic conclusion (Stegena et al., 1975; Sclater et al., 1980; Royden et al., 1983 a, b) that moderate crustal extension combined with major thinning of the subcrustal lithosphere has controlled the formation of the Pannonian basin.

4. It has been demonstrated that the simple Lopatin-Waples method works well for prediction of the maturity level of organic matter provided that it is calibrated properly for use in a young and hot sedimentary basin.

5) The known oil and gas fields in the Great Hungarian Plain have been found in immature reservoir rocks at relatively shallow depths (1.5km to 2.5km) above basement highs. These hydrocarbons derive from mature source rocks in nearby deep troughs ("hydrocarbon kitchen") and migrated updip a few tens of kilometers.

6) Model calculations suggest that a large volume of hydrocarbons should have migrated downward and/or laterally out of the source rocks, and could have been trapped in the basement and at the flanks of deep troughs at moderate depths (2km to 4km). Reliable modeling of basin evolution and studies of contemporaneous sealing conditions are thought to be a most powerful tool to search for these subtle traps in the Pannonian basin.

Acknowledgments

This project has been supported by the Petroleum Exploration Company, Szolnok. The work is still underway, and the authors are grateful to this company for the continuing interest and permission to publish the initial results.

The first author (FH) thanks the Institut Francais du Pétrole, the Hungarian Geological Survey and the Eötvös University, Budapest for help in defraying participation and travel expenses.

References

ÁDÁM, A. and Á. WALLNER, 1981, Information from electromagnetic induction data on Carpatho-Pannonian geodynamics: Earth Evol. Sci., v. 1, no. 3-4, p. 280-284.

BALLY, A.W. and S. SNELSON, 1980, Realms of subsidence: A.D. Miall (ed.), Facts and principles of world petroleum occurrence, Can. Soc. Petrol. Geol. Memoir 6, p. 9-94.

DÖVÉNYI, P., F. HORVÁTH, P. LIEBE, J. GÁLFI and I. ERKI, 1983, Geothermal conditions of Hungary: Geophys. Transactions, v. 29, no. 1, p. 3-114.

HORVÁTH, F. and L. ROYDEN, 1981, Mechanism for the formation of the intra-Carpathian basins: a review: Earth Evol. Sci., v. 1, no. 3-4, p. 307-316.

HORVÁTH, F. and J. RUMPLER, 1984, The Pannonian basement: extension and subsidence of an Alpine orogene: Acta Geol. Hung., v. 27, no. 3-4, in press.

ISSLER, D.R., 1984, Calculation of organic maturation levels for offshore eastern Canada - implications for general application of Lopatin's method: Canadian Journ. Earth Sci., v. 21, no. 4, p. 477-488.

MAGARA, K., 1978, Compaction and fluid migration: Elsevier Sci. Publ. Co., Amsterdam-Oxford-New York, p. 1-319.

MATTICK, R.E., J. RUMPLER and R.L. PHILLIPS, 1985, Seismic Stratigraphy and Depositional Framework of Sedimentary Rocks in the Pannonian Basin in Southeastern Hungary: AAPG Memoir, under publication.

POGÁCSÁS, Gy., 1985, Seismic stratigraphic features of Neogene sediments in the Pannonian basin: Geophys. Transaction, v. 30, no. 4, p. 373-410.

POSGAY, K., I. ALBU, I. PETROVICS and G. RÁNER, 1981, Character of the Earth's crust and upper mantle on the basis of seismic reflection measurements in Hungary: Earth Evol. Sci., v. 1, no. 3-4, p. 272-279.

ROYDEN, L., F. HORVÁTH and J. RUMPLER, 1983a, Evolution of the Pannonian basin system: 1. Tectonics: Tectonics, v. 2, no. 1, p. 63-90.

ROYDEN, L., F. HORVÁTH, A. NAGYMAROSY and L. STEGENA, 1983b, Evolution of Pannonian basin system: 2. Subsidence and thermal history: Tectonics, v. 2, no. 1, p. 91-137.

ROYDEN, L. and F. HORVÁTH (eds.), 1985, The Pannonian basin: A study in basin evolution: Amer. Assoc. Petrol. Geol. Memoir, under press.

SCLATER, J.G., L. ROYDEN, F. HORVÁTH, B.C. BURCHFIEL, S. SEMKEN and L. STEGENA, 1980, The formation of the intra-Carpathian basins as determined from subsidence data: Earth Planet. Sci. Lett., v. 51, p. 139-162.

STEGENA, L., B. GÉCZY and F. HORVÁTH, 1975, Late Cenozoic evolution of the Pannonian basin: Tectonophysics, v. 26, no. 1-2, p. 71-90.

SZALAY, Á., 1985, Maturation and migration of hydrocarbons in the southeastern Pannonian basin. AAPG Memoir, in press.

WAPLES, D.W., 1980, Time and temperature in petroleum formation: application of Lopatin's method to petroleum exploration. AAPG Bull., v. 64, p. 916--926.

I. HUTCHISON[1]

NUMERICAL MODELLING OF OCEANIC HEAT FLOW. A WESTERN MEDITERRANEAN CASE STUDY

1.0   INTRODUCTION

The Balearic and Tyrrhenian Seas in the Western Mediterranean form deep oceanic marginal basins within a tectonic regime dominated by the convergence of the European and African plates. In contrast to the Eastern Mediterranean, the basins are young, ranging from late Oligocene ages in the Balearic to Miocene-recent in the Tyrrhenian. Biju-Duval et al. (1978), and Boccaletti and Guazzone (1974) explain these basins as the result of extensional tectonics behind a southeastwardly migrating trench-arc system. The present expression of this subduction is seen in a Benioff zone which dips WNW from Calabria beneath the Central Tyrrhenian Sea and in the ring of recently active volcanic islands in the SE Tyrrhenian.

An important constraint on any lithospheric model of the Western Mediterranean is the sea floor heat flows of the deep basins. Amongst others, Sclater et al (1980) suggest that the thermal signature of deep marginal basins is the same as that of the major oceans. We also expect basins formed by large amounts of extension of continental lithosphere (eg. McKenzie, 1978) to show heat flow - age relationships similar to the oceanic cooling curve. If the Western Mediterranean has formed by extension and spreading which migrates to the southeast, we would expect the heat flow to increase through the Balearic and Tyrrhenian seas, and that the heat flow values should reflect those for Oligocene - Miocene aged oceanic lithosphere.

---

(1) *Geophysical Division, BP Exploration Company, Ltd, London, UK.*

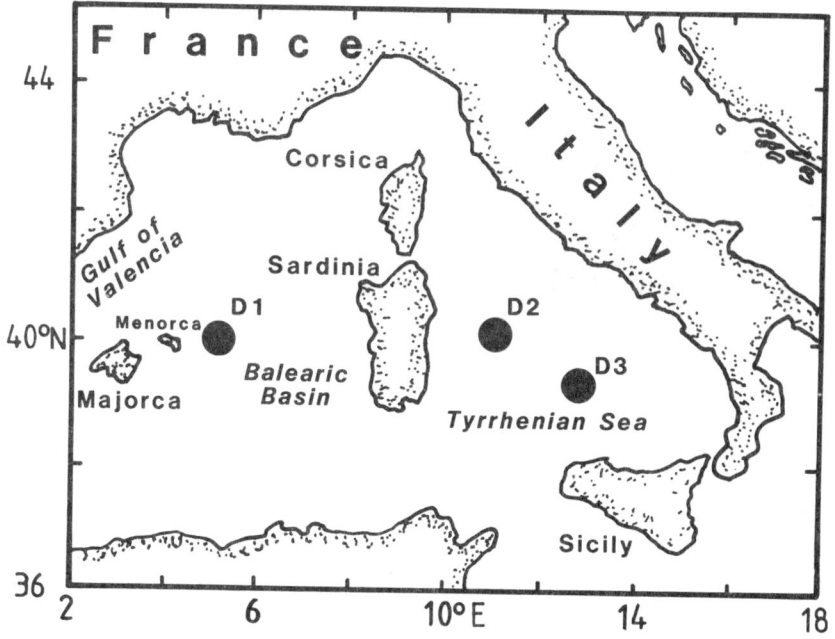

Fig 1 - Location map showing heat flow surveys D1 - D3

To test this model, three detailed heat flow surveys were made in December 1981 and January 1982 on the RRS Shackleton cruise 3/81. The first survey (Fig 1) consisted of 12 individual measurements centred in the Balearic abyssal plain at 40°01' N, 4° 55' E. The second and third, of 18 and 26 measurements respectively, were located around 40°16'N, 11°19'E and 37°17'N and 12°58'E in the western and southeast Tyrrhenian Sea. Within each survey, measurements are spaced at approximately 1km intervals along E-W and N-S profiles. The high measurement density allows the local variation in heat flow to be assesed and reliable area mean heat flows established.

2. DATA AQUISITION

Heat flow data were obtained using two different sets of oceanic heat flow probes - a 4m long microprocessor controlled instrument built by the University of Cambridge and a 5m probe from Woods Hole Oceanographic Institution (WHOI). Both probes had a nominal temperature resolution of 0.25 mK and the ability to measure thermal conductivity in-situ. In addition, piston core samples were taken in each survey area for ship-board conductivity

analysis using the needle probe technique. The instruments yield
temperature profiles in the first 4-5 m of of ocean floor sediment
which, with the measured thermal conductivity, allow calculation
of local sea floor heat flows. Fig 2 shows a sample station
record; in common with almost all the measurements, the
temperature gradient is constant and the thermal conductivity
shows only minor changes with depth. Correlation between in-situ
thermal conductivity and the needle probe values from nearby cores
was excellent.

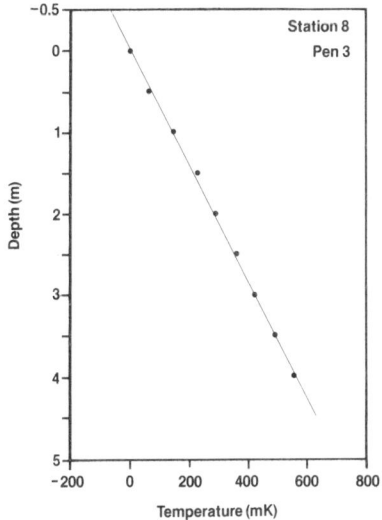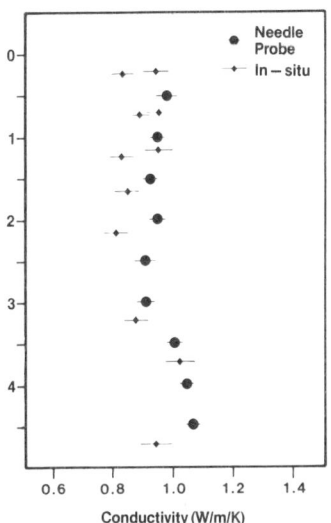

Fig 2 - Station record from survey D3, SE Tyrrhenian, showing
temperature profile and thermal conductivity

The calculated heat flow values are plotted in Fig 3. In all
three areas the variability between nearby stations is large, with
the differences in heat flow being much greater than the
measurement uncertainty. In particular, note the unusually high
values (up to 220 mWm-2) in the central part of survey area D3 in
the SE Tyrrhenian. The most obvious explanation of this
variability is thermal refraction through the high conductivity
Messinian salt and evaporite structures which are found throughout
much of the Mediterranean. For example, heat will flow
preferentially through a high conductivity salt diapir, giving a
local increase in flux over the crest of the structure and a
region of lower heat flow around the flanks. Another possible
reason is hydrothermal circulation, particularly in the SE
Tyrrhenian where the young oceanic crust is covered only by a thin
sediment layer and may locally come to outcrop.

WESTERN MEDITERRANEAN HEAT FLOW

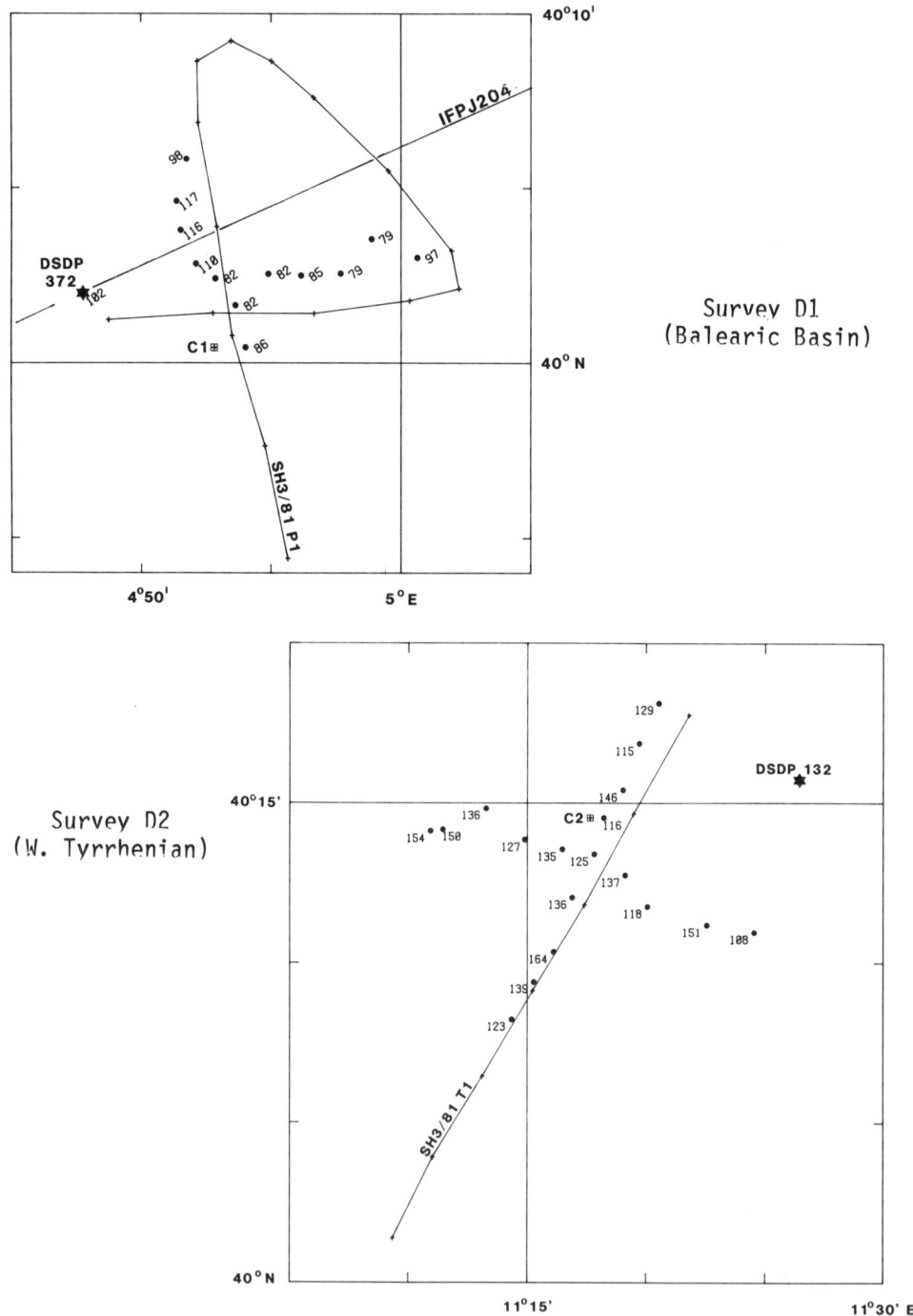

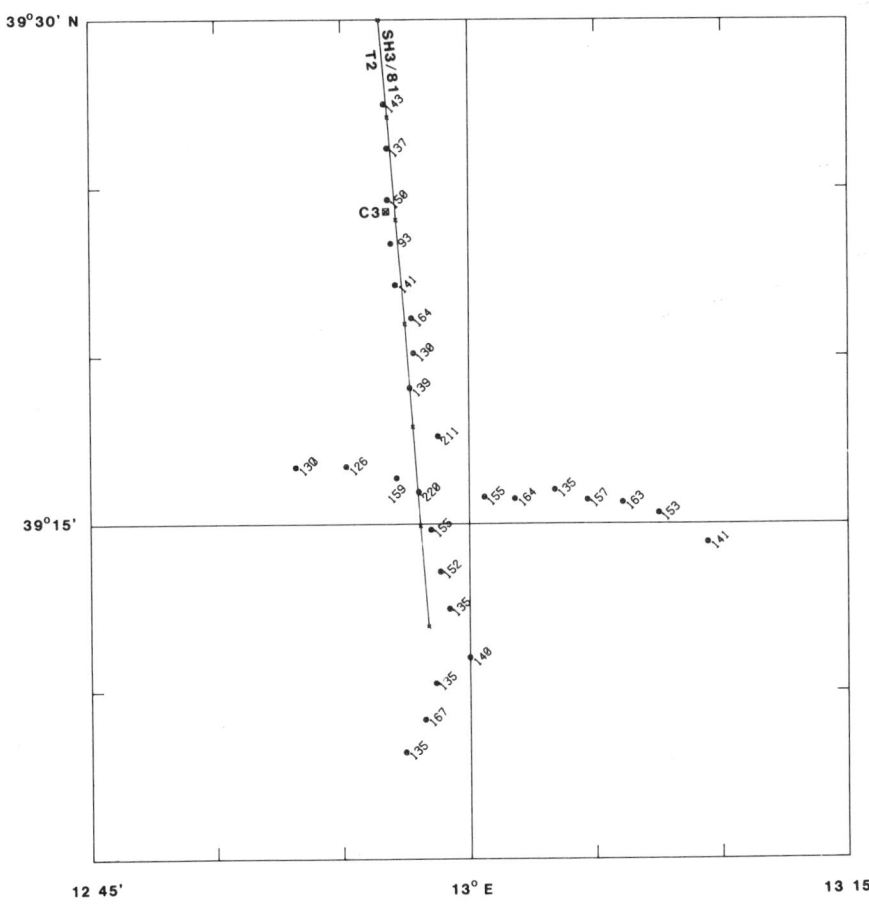

Survey D3 (SE Tyrrhenian)

Fig 3 - (This and facing page) Measured surface heat flows for survey areas D1 - D3.

Despite the variability, we see a general trend of increasing heat flow from survey area D1 to D3, with the mean sea floor heat flows in each area given below :

D1 (Balearic Basin) 92 ± 10 mWm-2
D2 (W Tyrrhenian)   134 ± 8 mWm-2
D3 (SE Tyrrhenian) 151 ± 10 mWm-2

(Errors in the mean quoted at 95% confidence level, based on Student's T test)

In contrast, averages of the in-situ and core thermal conductivities at each site shows a decrease from west to east, with values of 1.15 W/m/K at D1 through 1.01 W/m/K at D2 to 0.96 W/m/K at D3. This change is thought to result from regional variations in sediment type, with the high terrigenous input at D1 accounting for the unusually high conductivity.

3. NUMERICAL MODELLING

Before we can compare the mean heat flows with those predicted by any theoretical cooling model, we need to ensure (a) that the values are representative of the regional heat flow and (b) that the fluxes are corrected for any transient effects, such as rapid sedimentation or recent bottom water temperature changes.

3.1 Thermal Refraction

To model the variations in surface flux due to thermal refraction, we used numerical solutions to the 2D, steady state heat flow equation. The numerical model used a program developed by England (eq. England et al., 1980) to calculate the temperature distribution and surface heat flow for a 2D section. The boundary conditions applied are (1) constant surface temperature, (2) no lateral heat flow from the sides of the model and (3) a constant basal heat flow applied at a depth of 3-4 times greater than the deepest conductivity interface. The input depth sections for regions D1-D3 were derived from the seismic lines SH3/81P1, SH3/81T1 and SH3/81T2 respectively (Fig 3). The lines are interpreted in three units; sediments, salt/evaporite and basement. The thermal conductivity structure for each section was determined by estimated bulk lithology alone, with initial conductivity contrasts of 1:4:2 for sediment, salt and basement.

WESTERN MEDITERRANEAN HEAT FLOW

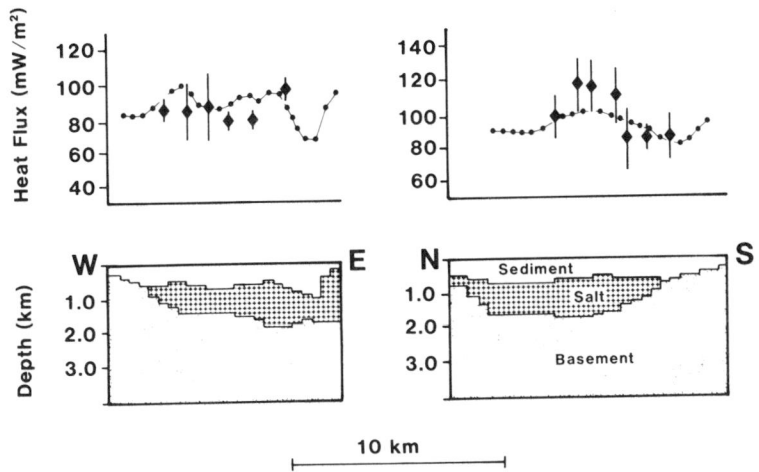

Fig 4 - 2D Conductive heat flow model for line SH3/81P1, survey D1 (Balearic Basin)

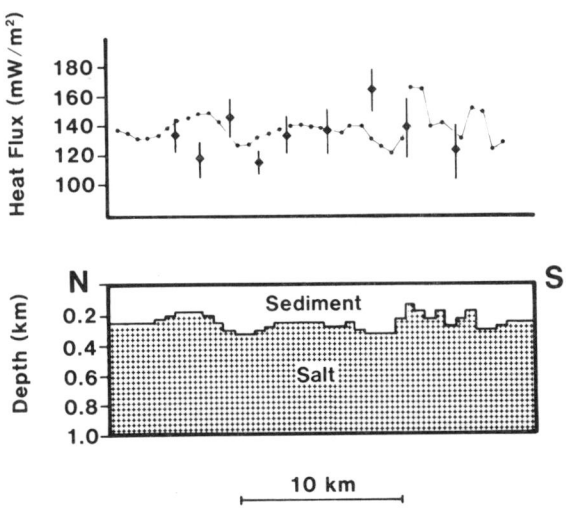

Fig 5 - 2D conductive heat flow model of line SH3/81T1, Survey D2 (Western Tyrrhenian)

In study areas D1 and D2 (Figs 4 and 5) we find that the overall size (c. ±20%) and shape of the observed heat flow anomalies are consistent with those modelled by thermal refraction. In detail, the correlation between individual measurements and the model values is occasionally poor, but this can be attributed to the effects of out-of section structure and the need to project heat flow measurements from distances up to 2km onto the line of section. Our models also show that the measurements cover areas of predicted high and low heat flows and so the calculated mean values should be reasonable estimates of the true heat loss.

The sediment lithology along line SH3/81T2 in survey D3 is uncertain; Malinverno et al. (1981) suggest that evaporites are absent, while Fabbri and Curzi (1979) claim the presence of an evaporite layer in the area. To account for this uncertainty, we combined the possible evaporite and basement layers and assigned conductivity contrasts from 1:2, 1:4 and 1:8 to the remaining two units, reflecting lithology contrasts from typical sediment/ basement to very low conductivity sediment/very high conductivity salt. Our results are shown in Fig 6; while refraction gives some variation in the heat flow, we see that the anomalously high and low values cannot be reproduced. The alternative explanation of convection of fluids in the possibly young and fractured oceanic basement beneath the low permeability sediment cover is advocated. The correlation of high heat flows with the topographic highs and the low heat flow at the base of the small scarp bounding the northern abyssal plain lends support to this model (c.f. Davies et al, 1980). In the well sedimented areas away from the topographic high, the heat flows are reasonably uniform, suggesting that any fluid flow is likely to be localised, associated with the lightly covered/outcropping basement. In this area, the calculated mean heat flow should give a reasonable estimate of the conductive heat loss only; if fluid circulation is taking place, we might also expect some additional, convective heat loss.

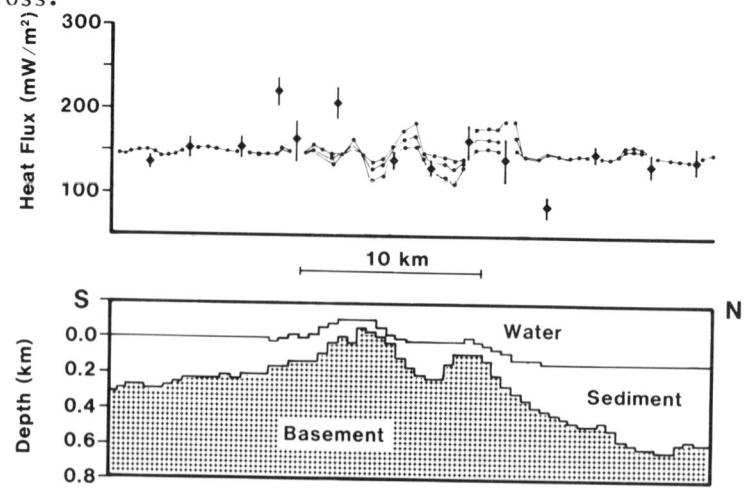

Fig 6. - 2D conductive heat flow models for line SH3/81T2, Survey D3 (S.E. Tyrrhenian)

3.2 Effects of Sedimentation

The second source of uncertainty in the mean heat flow values is the extent to which they have been reduced by recent rapid sedimentation. Analytical solutions for the sedimentation problem are only available for simple uniform half space models and take no account of compaction and pore water advection. To solve this problem, a physical model for the evolution of the sediment/basement system was developed and a numerical solution obtained for the resulting 1D time dependent heat flow equation. The model is outlined in Fig 7. For basement and salt/evaporite layers, we assume uniform thermal properties with no effective porosity. For the 'mudstone' component of the sediment, we assign an exponential porosity-depth variation of the form $\phi(z) = \phi_o e^{(-z/\lambda)}$. From this relationship, we can obtain analytical expressions for the sediment, pore water and basement advection rates with respect to a specified surface sedimentation rate. Thus, for a given depositional history, we can specify the advective and thermal properties at various depths in the 1D column through time. Coefficients for the 1D time dependent heat flow equation can then be defined. The equation is solved using an implicit finite difference scheme with boundary conditions of a given surface temperature history and the option of a fixed basal heat flow or basal temperature. With respect to the sea floor reference surface, sedimentation affects the temperature structure of the entire lithosphere and it is important that the base of the model is fixed at considerable depth. Analytical solutions for a uniform half space suggest that for reasonable geological time scales, a constant flux condition will only be valid when applied at depths greater than about 100km. The alternative constant temperature condition reflects the oceanic plate model, with the $1333^{\circ}C$ isotherm fixed at around 125km. Corresponding initial conditions are provided by the temperature structure of the lithosphere with a steady state, constant heat flow or from analytical description of a cooling plate of a specified age. Full details of the physical model and the numerical methods are given in Hutchison (1985).

The calculated alteration to the surface heat flow at each site is shown in Fig 8. Each diagram gives the results of a series of models with different boundary conditions and possible sedimentation histories. In survey D1, pre-Messinian sediment reduces the flux by up to 5%; rapid deposition of the Messinian evaporites gives a further reduction of 10% with Plio-Quaternary sedimentaion at 120 m/My giving a continuing decrease in heat flow to the present day. In area D2, the Messinian deposition leads to a reduction of 5-10% but the slower 38 m/My Plio-Quaternary sedimentation gives a slight recovery in the surface flux. D3, with only minimal (possibly no) Messinian sediment shows only a small, 5% reduction due to the Plio-Quaternary infill.

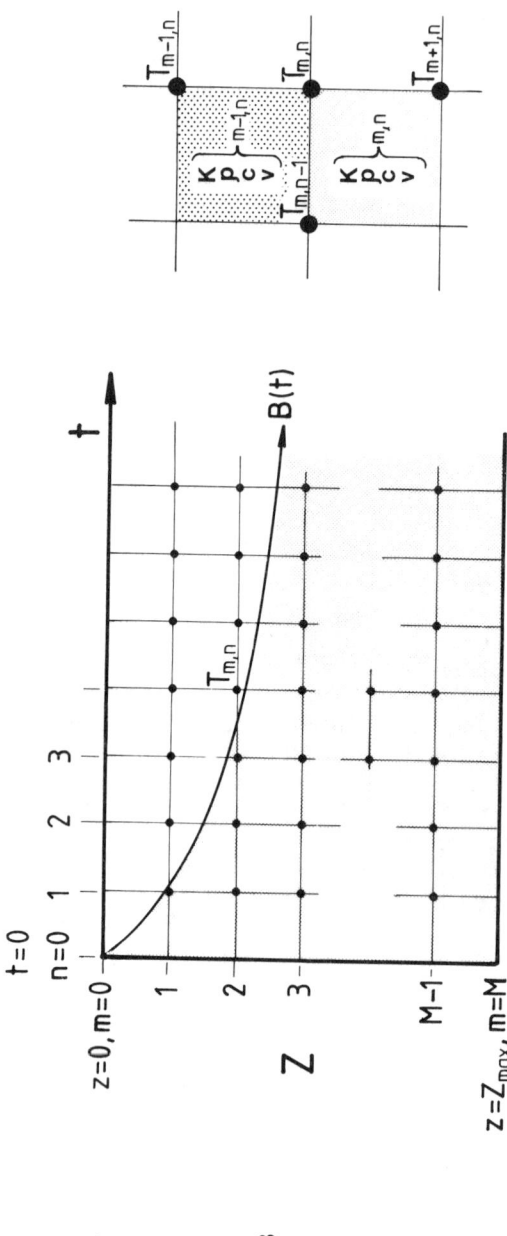

Fig 7 - Outline of numerical model of sedimentation

The analytical model describes the basement, sediment and pore water physical properties and advection rates. These are used to supply conductivity (K) density (ρ) velocity (v) and heat capacity (c) terms for the difference elements. Solution determines the Temperatures $T_{i,j}$ for various time steps Δt

WESTERN MEDITERRANEAN HEAT FLOW

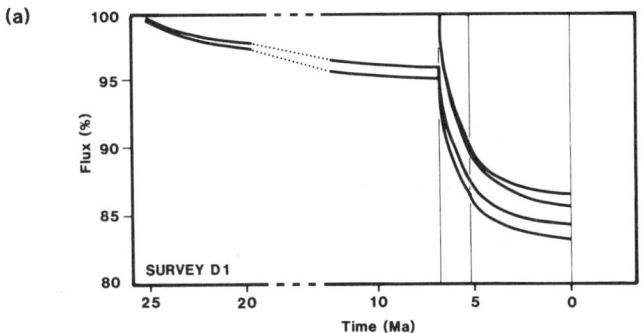

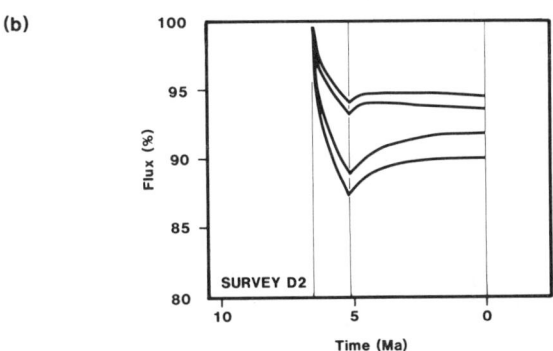

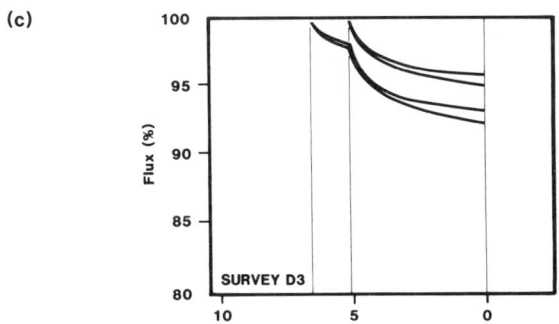

Fig 8 - Calculated percentage reductions due to sedimentation,
(A) - survey D1, Balearic Basin
(B) - Survey D2, W Tyrrhenian
(C) - Survey D3, SE Tyrrhenian

The final best estimates of the corrected heat flows are given below :

D1 (Balearic Basin) 109 ± 15 mWm-2
D2 (W Tyrrhenian)   145 ± 13 mWm-2
D3 (SE Tyrrhenian) 161 ± 14 mWm-2

Note that the corrections in Fig 8 do not include the variations in surface temperature which might have accompanied the 2-3km drop in effective sea level during the Messinian. In a separate test, this effect was modelled by an increase of 20°C for the period of evaporite deposition between 6.6 and 5.2 Ma. The results showed short lived transients at the beginning and end of the high temperature interval, but the alteration to the flux at the present day was not significantly different from the simpler, constant surface temperature case presented here.

DISCUSSION

To compare the corrected heat flow values with model derived lithospheric cooling curves, we need to assign ages to each survey area. Several lines of evidence constrain the Balearic Basin to late Oligocene/early Miocene ages, with the most likely range of 20-25 Ma. Ages for the Tyrrhenian are more speculative. The total sediment thickness at D2 is much less than in the Balearic, suggesting younger ages, while the occurence of Messinian evaporite implies existence prior to 7 Ma. Our best estimate for D2 is 7-12 Ma. Area D3 has little or no Messinian sediment and with oceanic basalts up to 8 Ma cored at the nearby DSDP site 373 we can form a possible age range of 5-8 Ma. In Fig 9 we compare our corrected heat flow/age points with theoretical cooling curves (a) from the oceanic plate model ($\beta = \infty$) and (b) from simple stretching models with β = 4, 6 and 10. The agreement with the plate model and high extension models is good. Remember that the D3 data is a minimum estimate and that the possible addition of a convective component would move the data further toward the plate model, $\beta = \infty$ line. Thus, after appropriate corrections have been made for sedimentation and account taken of possible convective heat loss, it appears that the thermal signature of the Western Mediterranean is consistent with that of oceanic crust of similar ages. Conversely, the systematic increase in heat flow from the Balearic through the Tyrrhenian agrees well with the model of the Western Mediterranean as the result of southeastwardly migrating extensional processes.

That earlier heat flow measurements do not show such a clear
E-W trend can be understood in the light of the present results.
The high variability seen within the detailed survey areas shows
that single measurement points selected at random would be
unlikely to give an accurate estimate of the regional heat loss.
Only by taking a large number of closely spaced measurements and
forming a reliable average do we obtain the true heat loss.

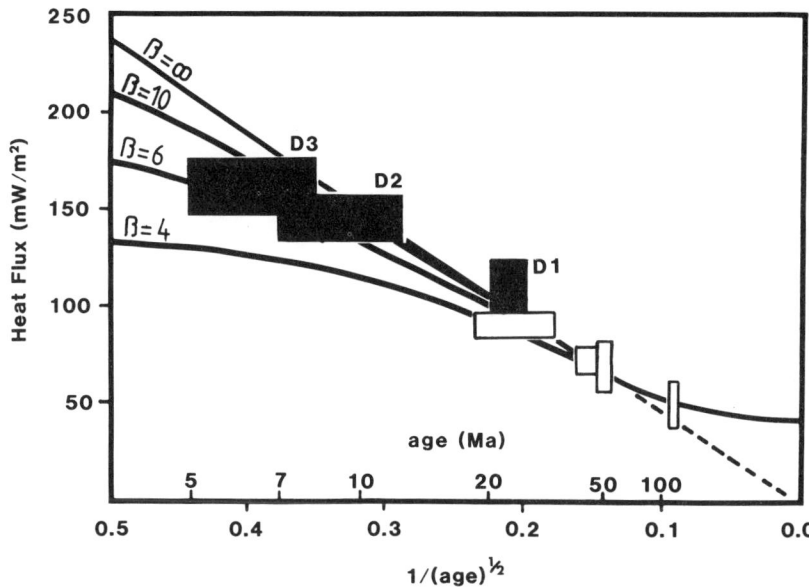

Fig 9 - Correlation of corrected mean heat flows for areas
D1-D3 with oceanic ($\beta = \infty$) and stretching model cooling
curves for $\beta = 4, 6$ and 10. (Unfilled boxes show data
from other marginal basins; after Sclater et al. 1980)

The importance of full numerical modelling must also be
stressed. The 2D steady state models show us the way to interpret
the data and offer an understanding of the scatter and, in this
study, the possibility of a systematic bias in area D3. The 1D
time dependent model gives a much more accurate representation of
the perturbation to the lithospheric temperature and heat flow
caused by sedimentation. It is interesting to compare our
calculated reductions of 10-15% with values 30-90% obtained by
Erickson and Von Herzen (1978) using only approximate analytical
solutions.

CONCLUSIONS

By using high accuracy, closely spaced heat flow measurements we have formed reliable estimates of the heat loss in three survey areas in the Western Mediterranean. Numerical modelling shows that the local variations in heat flow can be explained by thermal refraction for the two westernmost areas, but that hydrothermal proceess may influence the measurements in the third, SE Tyrrhenian survey. After correction for the effects of recent sedimentation, the area mean values agree well with heat flows predicted from oceanic and high extension cooling models, with a systematic increase from west to east reflecting the eastward decrease in age through the Western Mediterranean.

ACKNOWLEDGMENTS

I thank Shell International Oil Company for the support of a research studentship during the period of this work. Funding for instrumentation and ship time was provided by the United Kingdom Natural Environmental Research Council (NERC) awards GR3/3933 and ER3/4696 and U.S. National Science Foundation grants OCE-8025181 and OCE-8024287. BP Petroleum Development Limited provided valuable assistance in the production of this manuscript.

REFERENCES

BIJI-DUVAL. B., J. LETOUZEY, and L. MONTADERT, Structure and evolution of the Mediterranean basins, Initial Rep. Deep Sea Drill. Proj., 42(1), 951-984, 1978.

BOCALETTI, M., and G. GUAZZONE, Remnant arcs and marginal basins in the Cainozoic development of the Mediterranean, Nature, 252, 18-21, 1974.

DAVIS, E.E., C.R.B. LISTER, U.S. WADE, and R.D. HYNDMAN, Detailed heat flow measurements over the Juan de Fuca ridge system, J. Geophys. Res., 85, 299-310, 1980.

ENGLAND, P.C., E.R. OXBURGH, and S.W. RICHARDSON, Heat refraction and heat production in and around granite plutons in north-east England, Geophys. J.R. Astron. Soc., 62, 431-455, 1980.

ERICKSON, A.J., and R.P. VON HERZEN, Down-hole temperature measurements, Deep Sea Drilling Project, leg 42A, Initial Rep. Deep Sea Drill Proj., 42(1), 857-871, 1978.

FABBRI, A., and P. CURZI, The Messinian of the Tyrrhenian seismic evidence and dynamic implications,, G. Geol., 43(1), 215-248, 1979.

HUTCHISON, I., - Effects of Sedimentation and Compaction - Accepted for publication, Geophys. J.R.A.S. 1985.

MALINVERNO, A., M. CAFIERO, W.B.F. RYAN, and M.B. CITA, Distribution of Messinian erosional surfaces beneath the Tyrrhenian Sea: Geodynamic Implications, Oceanol. Acta 4, 489-495, 1981.

MCKENZIE, D.P., Some remarks on the development of sedimentary basins, Earth Planet, Sci. Lett., 40, 25-32, 1978.

SCLATER, J.G., C. JAUPART, and D. GALSON, The heat flow through oceanic and continental crust and the heat loss of the earth, Rev. Geophys. Space. Sphys., 18, 269-311, 1980.

F. LUCAZEAU, J. P. BARRIOT[1],
S. LE DOUARAN[2]

GRAVITY CONSTRAINTS ON THERMAL MODELS FOR EXTENSIONAL BASIN: EXAMPLE OF THE PROVENCAL BASIN

INTRODUCTION

Most of rifted zones evolve into deep sedimentary basins: a simple thermomechanical model of homogeneous extension (McKenzie, 1978) can explain the first order properties of these basins. The model assumes that the lithosphere (both crust and mantle) is thinned by a constant factor during an instantaneous phase of rifting. This results in an increase of the surface heat flow and an initial subsidence; following this stage, the lithosphere cools, the surface heat flow decreases to its initial value with a time constant of several tens of Ma, and the basin continues to subside due to this thermal cooling and the resulting contraction.

As "rifted basins" may contain significant hydrocarbon resources, this model has been extensively applied to study the temperature evolution of sedimentary layers, generally without accounting for heat transfers in the compacting sediment. We present in this paper a 2D model integrating the lithospheric and sediment thermal evolutions and their respective interactions: this model is also used to compute the density distribution which allows us to infer the bathymetry and the gravity field. In the case of young basins, the computation of gravity in connection with bathymetry provides a very useful constraint for isostatic and mass balances for the model since it is sensitive to both rifting and thermal evolution: the time dependent thermal component of the gravity anomaly, due to thermal expansion in the entire lithosphere, may represent about the same order of magnitude as the sum of the other anomalies (anomaly due to the basin and anomaly due to the crustal thinning) for young basins but decreases to 0 for basins older than 100 Ma (see figure 1 after Beaumont et al,1982).

---

(1) *Centre Géologique et Géophysique, Université des Sciences et Techniques du Languedoc, Montpellier, France.*
(2) *Société Nationale Elf Aquitaine (Production), Paris-la-Défense, France.*

GRAVITY CONSTRAINTS ON THERMAL MODELS FOR EXTENSIONAL BASINS

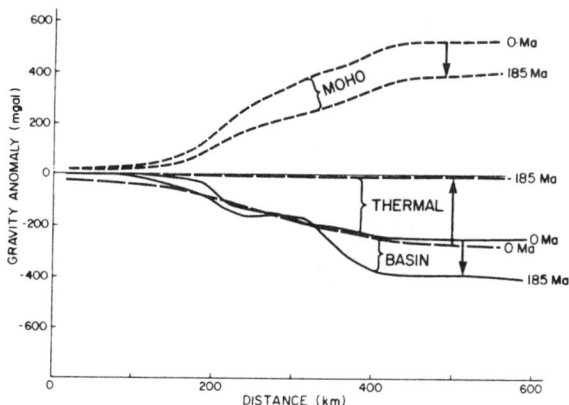

Figure 1: Component of the predicted free air gravity anomaly at 0 and 185 Ma after rifting (after Beaumont et al, 1982).

This model is applied to the study of the Provençal basin (Western Mediterranean), located between southern France and Corsica (Figure 2). There is no other basin as young as this one for which geological, geophysical and geodynamical studies give so many valuable constraints to its evolution. This basin was formed during Oligocene in the converging system of Africa and Europe, and is interpreted in terms of marginal basin (Biju Duval, 1978; Burrus, 1984). Schematically, the continental lithosphere was thinned from 30 to 23 Ma, then a sharp oceanic domain was formed from 23 to 19 Ma, and from 19 Ma to the present time, the lithosphere cools and subsides, with a very thick accumulation of Miocene Pliocene and Quaternary sediments. Although the model was elaborated to compute the history of temperature and burial of the sediment, in this paper we only emphasize the present geophysical constraints to the model (heat flow, bathymetry, gravity) on three N-S sections (figure 2) across the Provençal basin derived from refraction seismic profiles (Le Douaran et al, 1984).

MODEL

The main purpose of the model is to compute the temperature and burial history of the sediment deposited in a "rifted basin". There is extensive literature on this type of basins, essentially derived from the homogeneous stretching model (McKenzie, 1978; Royden and Keen, 1980; Beaumont et al., 1982). The model described here is a 2D finite elements model which uses the Lagrangian formulation: the mesh is deformed according to the geodynamic evolution. Therefore, there is no convective term in the heat transfer equation:

GRAVITY CONSTRAINTS ON THERMAL MODELS FOR EXTENSIONAL BASINS

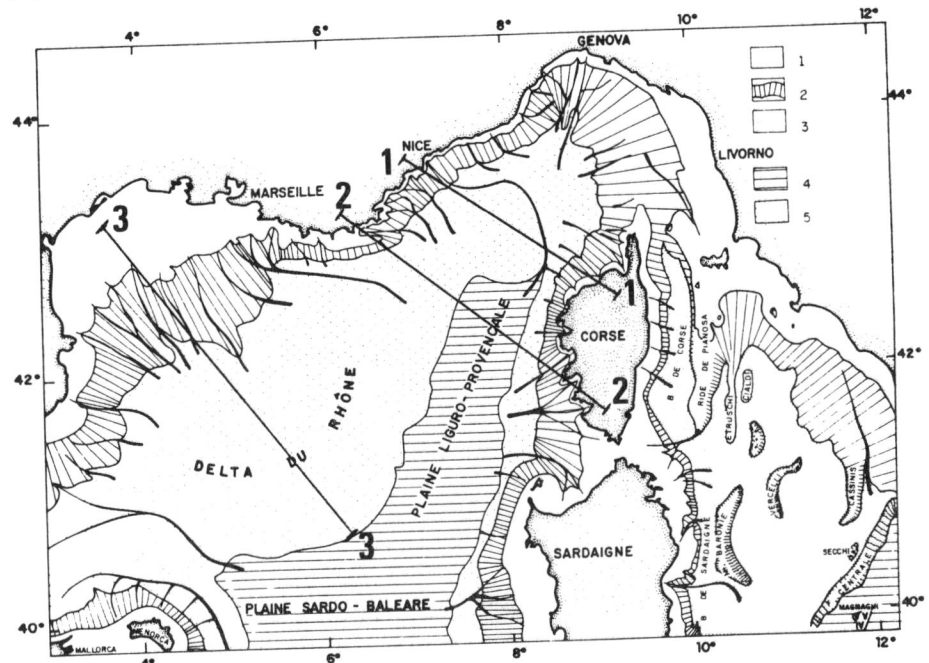

Figure 2: Location map of the 3 studied sections in the Provençal basin. 1: continental platform. 2: continental slope. 3: deltas. 4: bathyal plain. 5: northern Tyrrhenian sea sedimentary basins.

(1) $$\frac{\partial}{\partial z}(\lambda(X,Z,T)\frac{\partial T}{\partial z}) + \frac{\partial}{\partial X}(\lambda(X,Z,T)\frac{\partial T}{\partial X}) + A(X,Z) = \rho c(X,Z)\frac{\partial T}{\partial t}$$

where $\lambda(X,Z,T)$ is the thermal conductivity supposed to be equal in both X and Z directions. It varies with temperature (Houbolt and Wells, 1982) according to :

(2) $$\lambda(T) = \lambda_0(1 + \alpha T)^{-1}$$

is the conductivity in laboratory conditions and $\alpha = 5.10^{-3} \, °C^{-1}$. In the sediment, also varies due to compaction effects and resulting reduction of the porosity (Woodside and Messner, 1961):

(3) $$\lambda = \lambda_s^{(1-\phi)} \lambda_w^{\phi}$$

377

λ_s is the conductivity of the matrix, λ_w the conductivity of water and ϕ the porosity. $A(X,Z)$ is the distribution of heat production which is assumed to be constant for a given crustal layer and for a given sedimentary layer, but can vary along the X and Z axes according to thinning processes in the crust and compaction in the sediment. The density may vary with different materials, temperature and, in sediment, with compaction. The variation in density due to the thermal effect is given by:

$$(4) \qquad \rho = \rho_0 (1 - \alpha_e \Delta T)$$

where α_e is the coefficient of thermal expansion ($3 \cdot 10^{-5}\ {}^\circ C^{-1}$) and ΔT is the temperature difference between computed and initial states. The variation in density due to compaction is given by:

$$(5) \qquad \rho = \rho_R (1 - \phi) + \rho_w \phi$$

where ρ_R is the density of rock matrix, ρ_w is the density of water and ϕ the porosity at depth Z.
The specific heat capacity ρc follows the same law.

Initial conditions

Thermal equilibrium is computed assuming a given surface heat flow, a given distribution of heat production in the crust and a fixed temperature at the surface boundary condition.

Rifting

Rifting has a finite duration. The initial state of the lithosphere is computed from the final state (present state) assuming that the horizontal extension exactly corresponds to the vertical thinning for a given vertical column, and that crustal and lithospheric thicknesses were originally constant along the X axis. If l_i is the final width of the i th column and β_i the estimated thinning factor, the initial width of column L_i is:

$$(6) \qquad L_i = l_i / \beta_i$$

The deformation rate g_i is assumed to be constant during rifting:

$$(7) \qquad \dot{\varepsilon}_x = -\dot{\varepsilon}_z = g_i$$

and the integrated deformation is given by the thinning factor β_i^t (t = duration of rifting)

$$(8) \qquad \beta_i^t = \exp(g_i t)$$

The boundary conditions are fixed temperatures at the top and bottom of the equilibrium lithosphere, and no lateral heat flow is assumed to cross the vertical boundaries. Therefore, as the base of the lithosphere is continously deformed, it is necessary to create new elements at the bottom of the grid to ensure the lower boundary condition.

Oceanic accretion

Oceanic accretion velocities are inferred from kinematics and, from a numerical point of view, advection of hot oceanic material is simulated by creating new elements on a vertical side of the mesh (where continental crust is already thinned) and fixing a constant temperature along the ridge axis.

Sedimentation and compaction

As the first goal of the model is the calculation of temperature history of the sediment, it is necessary to pay special attention to sediment properties. The precise knowledge of the compaction processes, that is the way in which porosity decreases with depth, is very important as thermal conductivity and density depend strongly on this parameter. Commonly used exponential laws are assumed (but one can also use discrete laws in more complex cases) (Ruley and Hubbert, 1960) :

(9) $$\Phi(Z) = \Phi_o \exp(-C Z)$$

where Φ_o is the porosity at $Z = 0$ and C the compaction parameter. Values of Φ_o and C can be found in papers of Sclater and Christie (1980) ; Beaumont et al. (1980).

Temperature field

The model gives as output the temperature, for a given time of the evolution, at each node of the mesh belonging to sediment or to the lithosphere. It is therefore easy to compute the surface heat flow as the distribution of the thermal conductivity is known.

Subsidence

Subsidence is computed using local isostatic assumption. For a given column, the initial mass is given by:

(10) $$M_o = \int_o^L \rho_o(z) \, (1 - \alpha_e T_o(Z)) \, dZ$$

where ρ_o and T_o are respectively the original density and temperature distribution, L the base of the equilibrium lithosphere (considered as the reference level).

For a given time of the evolution, the mass of the column is:

(11) $$M(t) = \int_o^L \rho(z) (1 - \alpha_e T(z,t)) \, dz$$

The resulting subsidence, compensated at the asthenospheric level, is :

(12) $$S(t) = (M(t) - M_o) / (\rho_a - \rho_w)$$

where ρ_a is the density of asthenosphere and $\rho_w = 1.03$ g/cm^3 the density of water.

Gravity

For each step of the computation, the density distribution is calculated. Then, the gravity field at sea water level can be computed.

Free Air anomalies are calculated using classical analytical solutions for the first derivative of the gravitational potential for 2D polygons. Bouguer anomalies are computed in the same way, but water is replaced by sediment using a density reduction of 2.67 g/cm3. The geoid can also be computed for 2D bodies, provided that the local isostasy is ensured (Chapman, 1979). The geoid is numerically computed by integration of the horizontal derivative of the potential along the X axis.

STRUCTURE OF THE PROVENCAL BASIN

The deep structure of the Provençal basin was investigated by recent seismic refraction surveys (Le Douaran et al., 1984). The center of the basin is formed by a typical two layers oceanic crust, but there is a progressive transition between the thinned continental crust to the pure oceanic domain. The continental crustal thickness is typically 30 Km in Corsica or in the Southern Alps (Hirn et al., 1977; Thouvenot, 1985), and can be thinned to 5-6 Km close to the oceanic domain. This deformation of the crust is given as input parameters to the model.

Heat flow measurements have been performed both on land and at sea, and regional trends for the Western Mediterranean have been proposed (Lucazeau et al, 1985). In the Maure-Esterel massif (outcrop of Hercynian basement in the southern part of France), heat flow measurements give a low value $Q_o = 60$ 5 mWm^{-2} whereas it is 76 ± 15 mWm^{-2} in Corsica. This discrepancy between the 2 domains can be explained by the high erosion which is known to exist in Corsica and could enhance surface heat flow.

Bathymetry (Rehault,1981) shows that the water depth presently varies from 2000 to 2850 m in the central part of the basin, with a strong gradient on the Corsican margin.

Gravity anomalies are derived from Morelli et al (1977). Bouguer anomalies roughly reproduce the shape of the Moho discontinuity with an amplitude of +200 mgals (Le Douaran et al, 1984). Geoid anomalies have been computed from SEASAT altimetric data (Bernard et al, 1983), and filtered to remove long wavelengths. The short wavelengths present negative anomalies (-3 to -4 m) with an uncertainty of about 0.8 m.

MODEL RESULTS

In this study, we have chosen the parameters which seemed the most realistic to us, but we have (generally) not tried to match the observations. We start from an initial equilibrium corresponding to a 60 mWm^{-2} surface heat flow (as observed in Maures-Esterel), with a component of 27 mWm^{-2} from the mantle and the remaining part from crustal radiogenic heat sources. The crust is initially 30 Km thick. The evolution of the upper part of the grid (up to the Moho level) is represented in Figure 3 from initial state at 30 Ma to the present state, and the corresponding isotherms are represented on the right part of the figure. The section represented corresponds to section 2 of Figure 2 (Toulon-Ajaccio). It is obvious that the current thermal regime is still transient as isotherms at the Moho discontinuity are shifted up in the central part by about 10 Km. It may therefore be expected that the effect of thermal expansion on the density distribution is very important for this basin, and that gravity and bathymetry might currently constrain the integrated thermal structure. In order to show the influence of deep thermal anomalies on geoid, we computed 2 cases for the Gulf of Lion (section 3), one corresponding to a lithospheric thickness L = 125 Km with $T_{(z=L)}$ = 1333 °C and one corresponding to L = 100 Km with $T_{(z=L)}$ = 1090 °C. For the other section, only the first case was studied.

Heat flow (Figure 4)

There is a good agreement between observed and computed values for the gulf of Lion (section 3) (for both the cases corresponding to 100 and 125 Km lithospheric thickness), but without taking into account the thermal blanketing effect of the sediment, the computation cannot explain the observations: this clearly shows the importance of this phenomenon for heat transferts in the sediment (De Bremaeker, 1983; Lucazeau and Le Douaran, 1985). On the other hand, the agreement is not so good for the 2 other sections in the Ligurian

GRAVITY CONSTRAINTS ON THERMAL MODELS FOR EXTENSIONAL BASINS

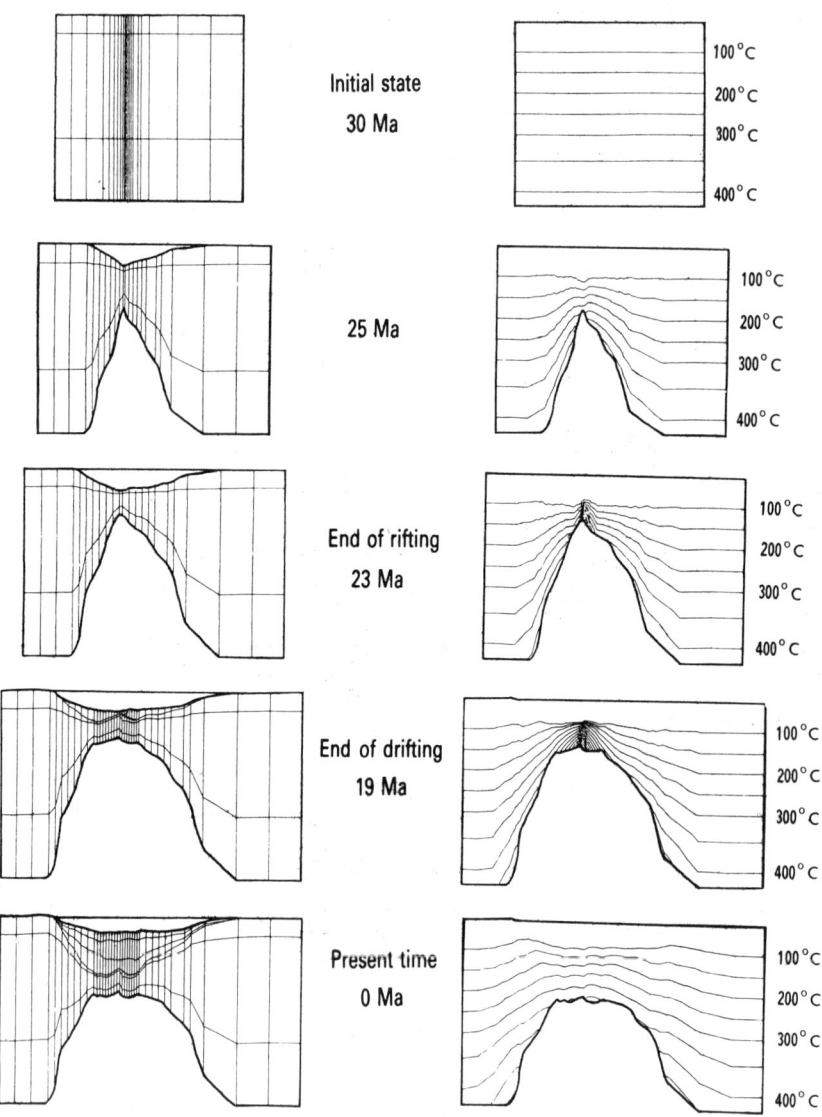

FIGURES 3 : SECTION 2 - EVOLUTION OF THE UPPER PART OF THE MESH (CRUSTAL PART) AND CORRESPONDING THERMAL FIELD FROM INITIAL SITUATION (30 Ma) TO THE CURRENT SITUATION (0 Ma).

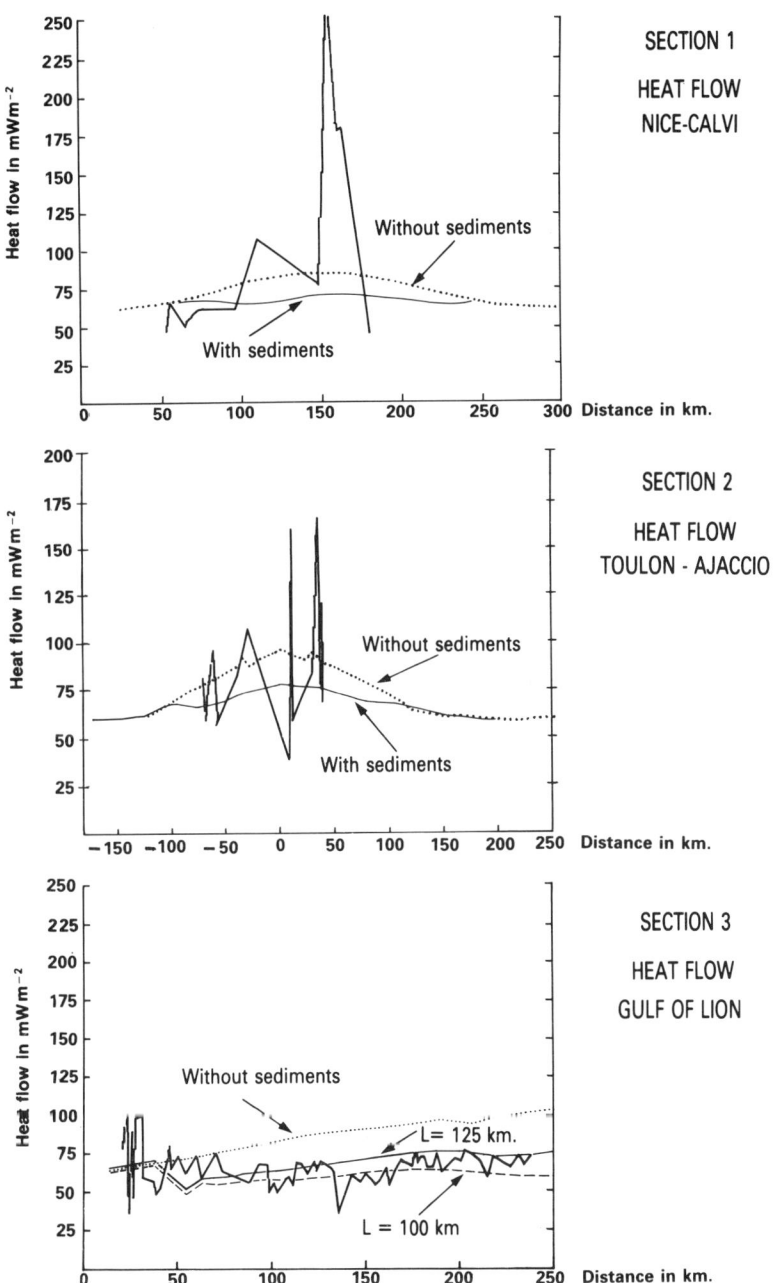

FIGURE 4 : HEAT FLOW VALUES FOR THE CURRENT SITUATION. THE OBSERVED PROFILES ARE PRESENTED BY HEAVY LINES. THE COMPUTED PROFILES ARE REPRESENTED BY CONTINUOUS LINES (LITHOSPHERIC THICKNESS : 125 km) OR DASHED LINES (LITHOSPHERIC THICKNESS : 100 km.) THE DOTTED LINES REPRESENT THE HEAT FLOW WITHOUT SEDIMENTS

GRAVITY CONSTRAINTS ON THERMAL MODELS FOR EXTENSIONAL BASINS

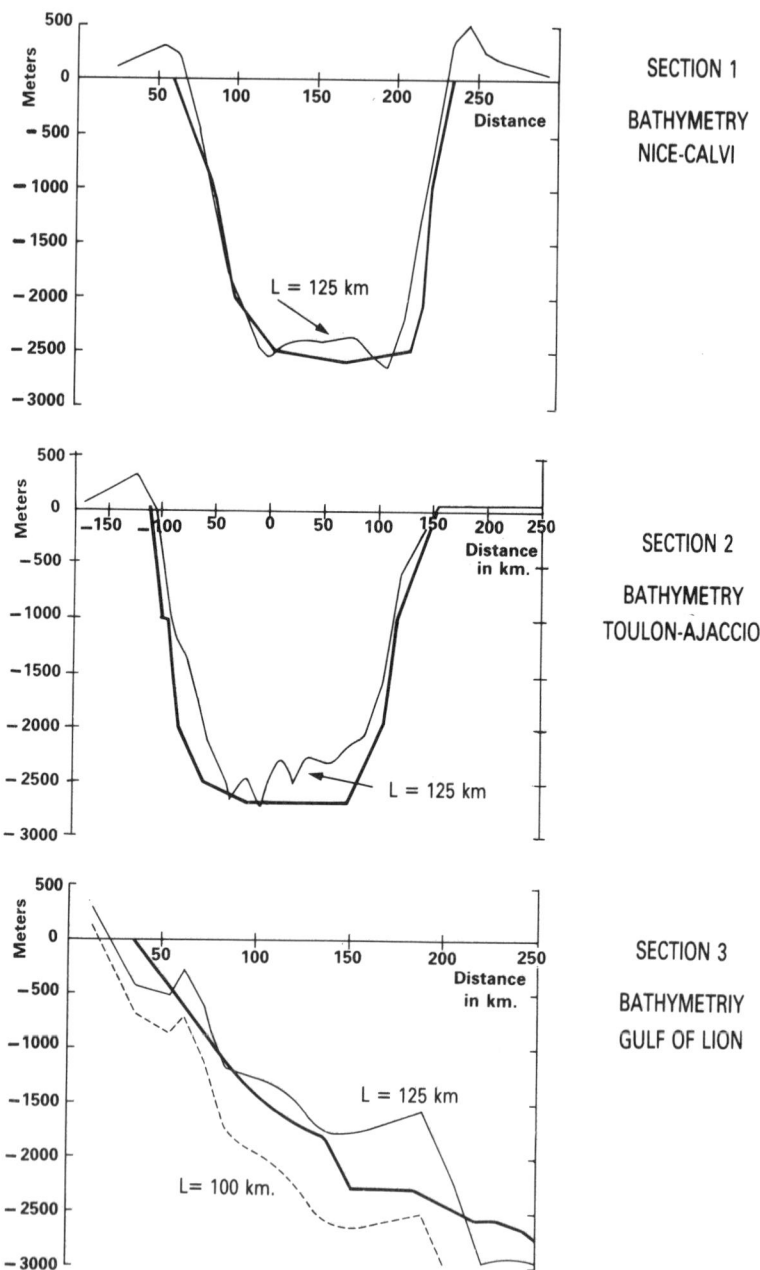

FIGURE 5 : BATHYMETRY PROFILES FOR THE CURRENT SITUATION.
HEAVY LINES REPRESENT OBSERVED PROFILES. CONTINUOUS LINES (LITHOSPHERIC THICKNESS : 125 km) AND DASHED LINES (LITHOSPHERIC THICKNESS : 100 km) REPRESENT COMPUTED BATHYMETRY USING AIRY ASSUMPTION.

sea: the observed heat flow varies between 40 and 260 mWm^{-2}, mainly on the Corsican side of the basin; as there is no correlation between the occurrence of salt domes and heat flow anomalies (Burrus, 1985), this is probably due to water circulations. However, the long wavelength component of heat flow can be explained by the model, but this is a very poor constraint to the thermal evolution of the basin.

Bathymetry (Figure 5)

The agreement between observations and computations is more or less good for the 3 sections. The long wavelengths of the bathymetry and the amplitude can be explained, but the small ondulations computed by a local isostatic model are not observed, suggesting that isostasy occurs at a regional scale.

Free Air anomalies (Figure 6)

The effect of the temperature field on the free air anomaly can be estimated on figure 6b: it represents about +150 mgals more than the expected value, whereas the computation including these thermal effects of expansion can match the observation quite well except for section 3 (Gulf of Lion) where 3D effects are probably important due to a disturbed bathymetry, as free air anomalies are very sensitive to the surface sources (sediment-water and sediment-basement interfaces).

Bouguer anomalies (Figure 7)

As for Free Air anomalies, the thermal effects of expansion can be estimated on Figure 7b: the anomaly is shifted up by 150 mgals and its wavelength is wider. There is general accord between the results of computations and observations, suggesting that both crustal thinning and deep thermal structure are well modelized.

Geoid anomalies (figure 8)

Geoid is very sensitive to deep structures. On the three computed sections, we may often observe a general negative trend due to crustal thinning and an opposite positive trend due to the lithospheric thinning. On the sections observed, the negative anomaly is dominant suggesting that the deep thermal stucture is not so well modelized as expected by the fit of Bouguer anomalies. Therefore, for section 3 for which the agreement is the worst, we changed the lithospheric thickness (100 Km instead of 125 Km) and the temperature of the asthenosphere in order to keep the same initial heat flow (60 mWm-2): consequently the thermal anomalies are concentrated at shallower levels in the deep lithosphere. This assumption is also in agreement with S waves studies (Panza and Mueller,1981) predicting

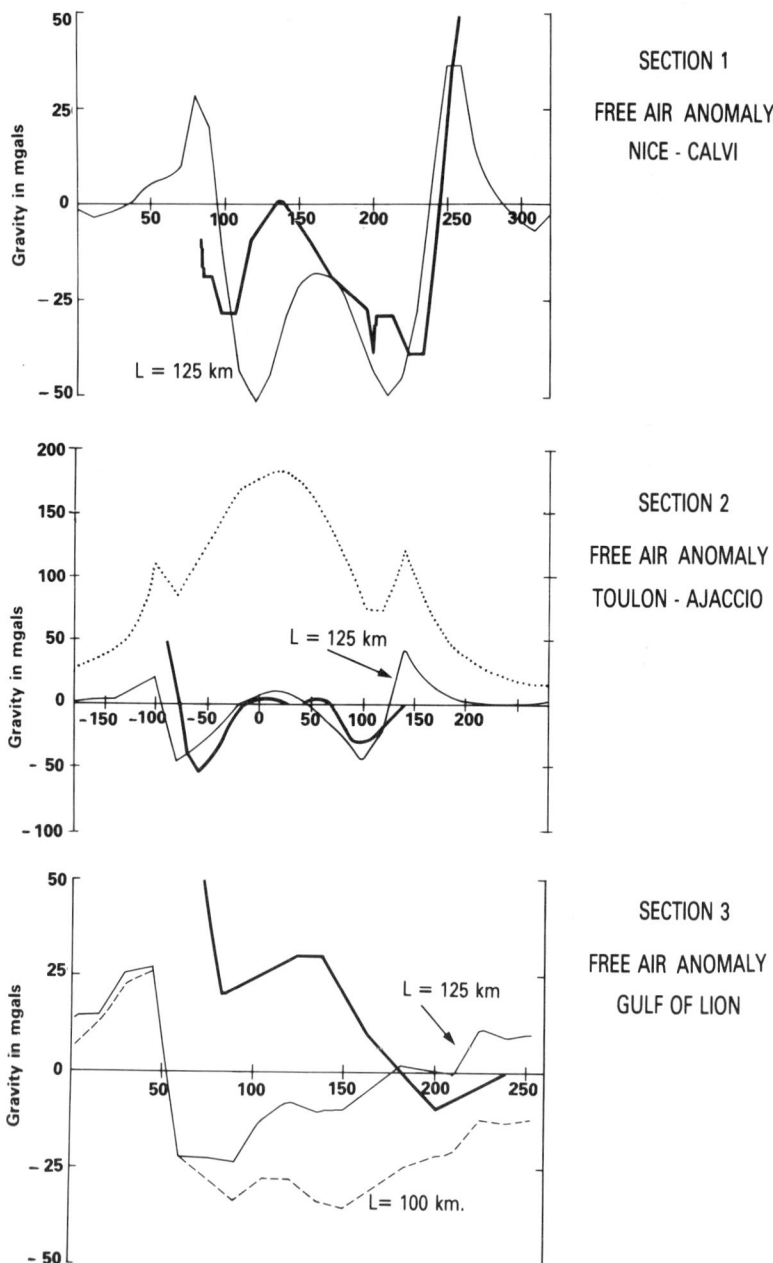

FIGURE 6 : FREE AIR GRAVITY ANOMALY FOR THE CURRENT SITUATION -
HEAVY LINES CORRESPOND TO OBSERVED ANOMALIES; CONTINUOUS LINES (LITHOSPHERIC THICKNESS : 125 km)
AND DASHED LINES (LITHOSPHERIC THICKNESS : 100 km) CORRESPOND TO COMPUTED VALUES.
FOR SECTION 2, THE DOTTED CURVE CORRESPONDS TO THE PREDICTED ANOMALY IN THE ABSENCE
OF THERMAL EFFECTS.

GRAVITY CONSTRAINTS ON THERMAL MODELS FOR EXTENSIONAL BASINS

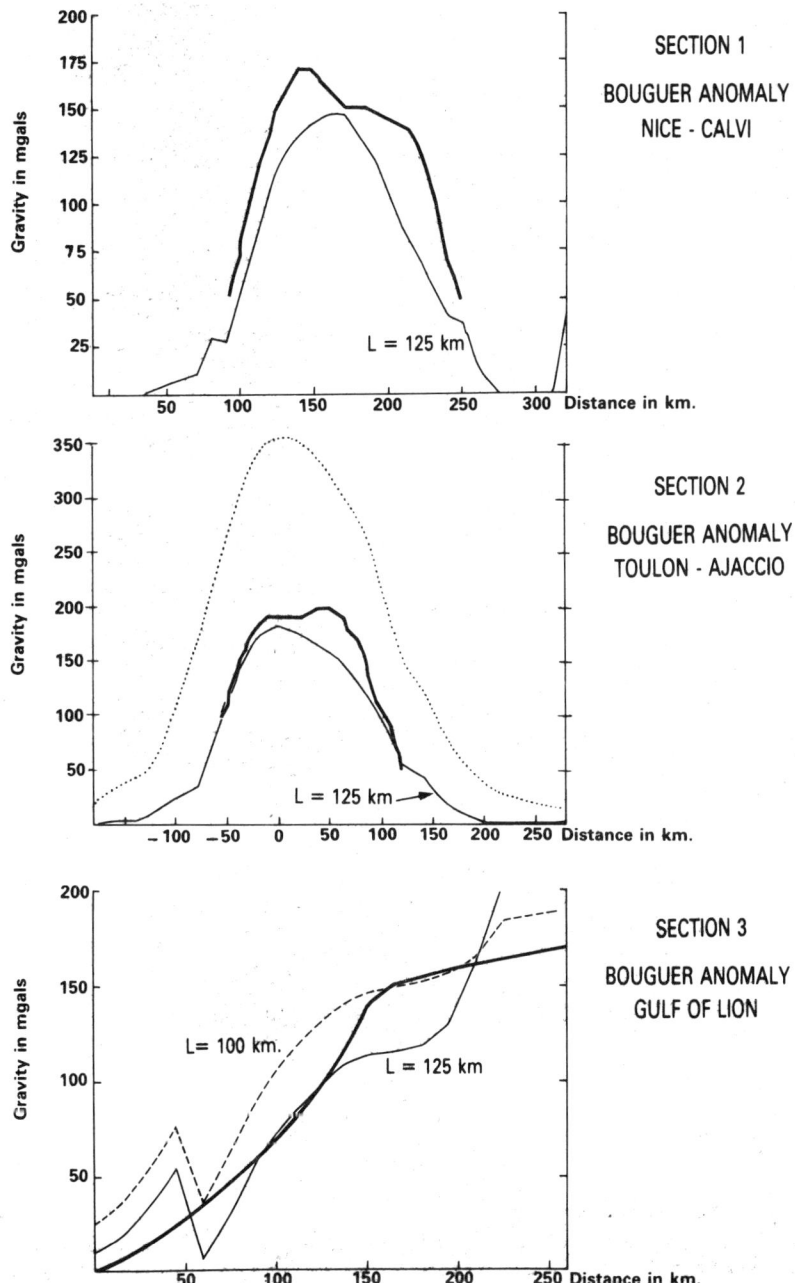

FIGURE 7 : BOUGUER ANOMALY FOR THE CURRENT SITUATION.
FOR SECTION 2, THE DOTTED CURVE CORRESPONDS TO THE PREDITED ANOMALY, IN THE ABSENCE OF THERMAL EFFECTS.

GRAVITY CONSTRAINTS ON THERMAL MODELS FOR EXTENSIONAL BASINS

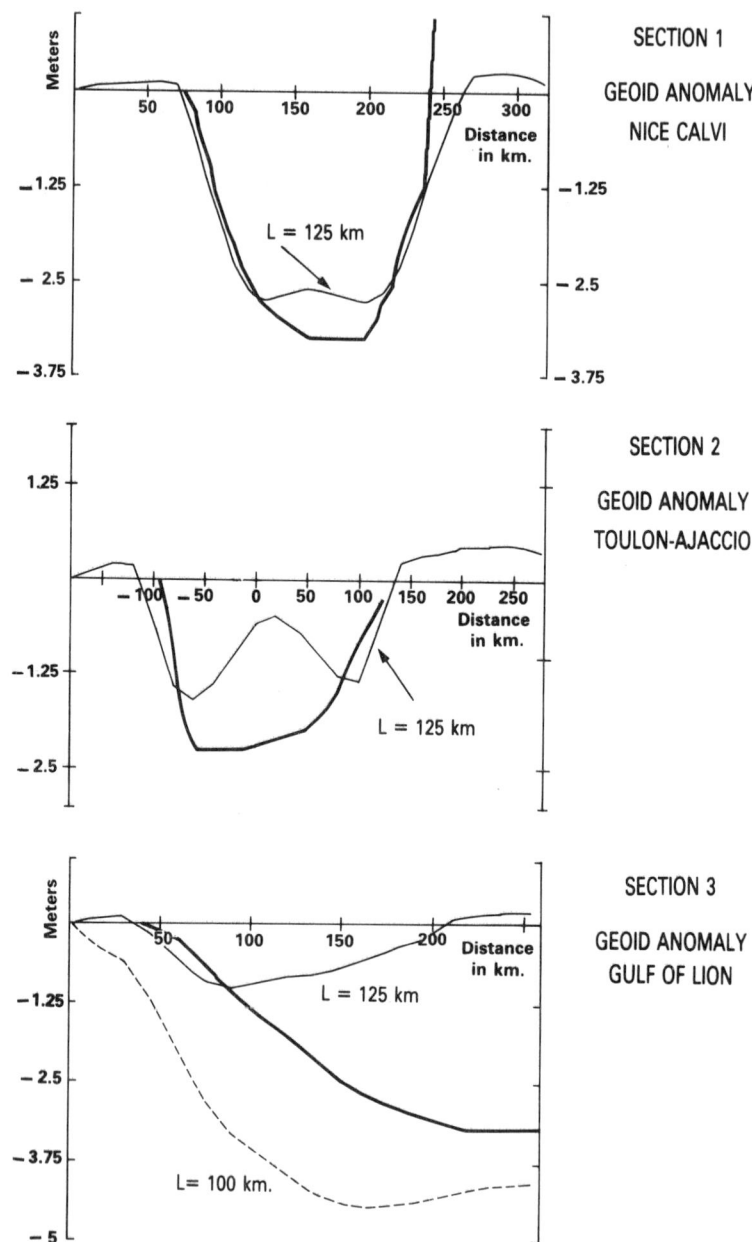

FIGURE 8 : GEOID ANOMALY FOR THE CURRENT SITUATION.
FOR SECTION 3 THE TWO COMPUTED CURVES,
CORRESPOND TO LITHOSPHERIC THICKNESSES OF 100 AND 125 KM THE BEST FIT IS FOR 100 km.

that the normal thickness for the seismic lithosphere is 90-100 Km in Western Europe. Then, the agreement between observations and computations for gulf of Lion becomes acceptable.

Therefore, the geoid can provide a very sensitive constraint to the deep thermal regime of young stretched basins such as the Provençal basin.

CONCLUSIONS

Thermal models for the evolution of sedimentary basins are used in oil research to predict the maturation of organic matter. These models become more sophisticated and therefore require more parameters over which we have no control. Thus it is necessary to use more powerful constraints to test the validity of these models. Three types of constraints are often used:
- Tectonic subsidence curves and current bathymetry. These constraints are very useful as they integrate the total thermal structure but they depend on the assumed mechanical model.
- Present day heat flow and temperatures or paleogeothermometers. They represent the less useful constraints to the global dynamic of the lithosphere because local thermal effects are sometimes more important in sedimentary basins than the background thermal field (fluid circulations, salt domes,...).
- Crustal structure inferred from refraction seismic profiles. It gives a very useful constraint on the deformation of the lithosphere but is seldom available.

In the case of young basins, gravity data can be used to constrain the thermal modelisation. We develop a 2D numerical model to predict the temperature and the burial history of the sediment deposited in a rifted basin. This model allows us to calculate the temperature field and surface heat flow, the subsidence and the current bathymetry, and also the gravity (free air anomaly, Bouguer anomaly and geoid anomaly). For the Provençal basin formed during Oligocene, we have shown that gravity data can give valuable constraints to the modelization as they integrate the structure of the basin, the deformation of the crust and the thermal regime of the entire lithosphere. Moreover, they are less perturbed and more abundant than surface heat flow data. Free air anomalies are sensitive to superficial structures, Bouguer anomalies are a good indicator of Moho ondulations, and geoid anomalies are very sensitive to deep structures and can characterize the deep thermal structure of the lithosphere. As the model can match different and independent sets of information, we may be confident of the validity of the global predicted history of temperatures in sediment: this is the first condition for the success of the thermal model and therefore gravity data can be used as any other constraints for young basins.

REFERENCES

C. BEAUMONT, C.E. KEEN and R. BOUTILIER (1982) - On the evolution of rifted continental margins : comparison of models and observations for the Nova Scotian margin. Geophys. J. R. astr. Soc., 70, 667-715.

J. BERNARD, F. BARLIER, J.P. BETHOUX and M. SOURIAU (1983) - First SEASAT altimetry data analysis on the Western Mediterranean Sea. J. Geophys. Res., 88, 1581-1588.

B. BIJU-DUVAL, J. LETOUZEY and L. MONTADERT (1978) - Structure and evolution of the Mediterranean basins. Initial Rep. Deep. Sea Dril. Proj., 42, 951-984.

J. BURRUS (1984) - Contribution to a geodynamic synthesis of the Provencal basin. (NW Mediterranean). Mar. Geol., 55, 247-269.

J. BURRUS and F. BESSIS (1985) - Thermal evolution of the Provencal basin, Western Mediterranean. Internal Research Conference Modelling evolution of sedimentary basins. Carcans, Maubuisson, 3-7 June 1985.

J. BURRUS, J.P. FOUCHER, F. AVEDIK and S. LE DOUARAN (1985) - Deep structure, thermicity and geodynamics of the Provençal basin. 2nd EGT workshop - Venice meeting, 7-9 Feb.

M.E. CHAPMAN (1979) - Technics for interpretation of geoid anomalies. J. Geophys. Res., 84, 3793-3802.

J.C. DE BREMAECKER (1983) - Temperature, subsidence and hydrocarbon maturation in extensional basins, a finite element model. AAGP, 67, 1410-1414.

J.P. FOUCHER (1983) - Thermal regime of Atlantic type of continental margins : the examples of the Biscay and Gulf of Lion margins. International Colloquium "Thermal Phenomena in Sedimentary Basins", St Médard en Jalles, June 1983.

A. HIRN and M. SAPIN (1976) - La croûte terrestre sous la Corse : données Sismiques. Bull. Soc. Geol. France, 7, 1192-1195.

J.C. HOUBOLT and P.R.A. WELLS (1980) - Estimation of heat flow in oil wells based on a relation between heat conductivity and sound velocity. Geol. En. Mijnbouw, 59, 215-224.

S. LE DOUARAN, J. BURRUS and F. AVEDIK (1984). Deep structure of the Western Mediterranean basins : results of a two ship seismic survey. Mar. Geol., 55, 325-345.

F. LUCAZEAU and S. LE DOUARAN (1985) - The blanketing effect of sediments in basins formed by extension : a numerical model. Application to the Gulf of Lion and Viking Graben. Earth Planet. Sci. Lett., 74, 92-102.

F. LUCAZEAU, G. VASSEUR, J.P. FOUCHER and F. MONGELLI (1985) - Heat flow along the southern segment of EGT. Second EGT Workshop : The southern segment. D. Galson and S. Mueller ed., European Science Foundation.

K. MAGARA (1985) - Porosity - depth relationship during compaction in hydrostatic and non hydrostatic cases. Internal Research Conference. Modelling thermal evolution of sedimentary basins. Carcans, Maubuisson, 3-7 June 1985.

D. McKENZIE (1978) - Some remarks on the development of sedimentary basins. Earth Planet. Sci. Lett., 40, 25-32.

C. MORELLI, P. GIESE, M.T. CARROZO, B. COLOMBI, I. GUERRA, A. HIRN, A. LETZ, R. NICOLICH, C. PRODHEL, C. REICHERT, M. ROWER, M. SAPIN, S. SCARASCIA and P. WIGGER (1977) - Crustal and upper mantle structure of Northern Apennines, the Ligurian Sea and Corsica derived from seismic and gravimetric data. Boll. Geof. Teor. Appl., 75-76, 199-260.

V.V. PALCIAUSKAS (1985) - Heat transfert in normally compacting basins. Theoretical models for conductivity and permeability. International Research Conference - Modelling thermal evolution of sedimentary basins - Carcans - Maubuisson, 3-7 June 1985.

G.F. PANZA, S. MUELLER and G. CALCAGNILE (1980) - The gross features of the lithosphere - asthenosphere system in Europe from seismic surface waves and body waves. Pure and Applied Geophysics, 118, 1209-1213.

J.P. REHAULT (1981) - Evolution tectonique et sédimentaire du bassin de Ligure (Méditerranée Occidentale). State Thesis, Univ. Paris VI.

L. ROYDEN and C.E. KEEN (1980) - Rifting process and thermal evolution of the continental margin of eastern Canada determined from subsidence curves. Earth Planet. Sci. Lett., 51, 343-361.

W.W. RUBEY and M.K. HUBBERT (1960) - Role of fluid pressure in mechanics of overthrust faulting. II - Overthrust belt in geosynclinal area of Western Wyoming in loight of fluid pressure hypothesis. Bull. Geol. Soc. Am., 60, 167-205.

J.G. SCLATER and P.A.F. CHRISTIE (1980) - Continental stretching : an explanation of the post Mid-Cretaceous subsidence of Central North basin. J. Geophys. Res., 85, 3711-3739.

F. THOUVENOT, J. ANSORGE, C. EVA and A. HIRN (1985) - The EGT-S 1983 seismic experiment in Western Alps. First results. E.G.U. Biennal Meeting, Strasbourg, 1-4 April 1985.

W. WOODSIDE and J.H. MESSMER (1961) - Thermal conductivity in porous media. J. Appl. Phys., 39, 1688-1706.

J. BURRUS, F. BESSIS[1]

THERMAL MODELING IN THE PROVENCAL BASIN (NW-MEDITERRANEAN)

INTRODUCTION

Ideas about the geodynamic phenomena pertaining to the thermal evolution of extensional sedimentary basins have been widely developed in academic circles these five last years, encountering an increasing interest among people involved in industrial operations, mainly in the oil exploration. Nevertheless, these theoretical concepts are generally well ahead of our knowledge of the set of data necessary to constrain thermal reconstructions: chronostratigraphy, deep structure of the crust, thermal gradients, physical properties (conductivities, densities, porosities, permeabilities) of rocks, geodynamic context, geochemical controls on the thermal history, etc. In most published studies, some of these data are not available, and assumptions have to be made. Another situation is encountered when a quite complete set of data has been collected, but the simplified numerical models used do not account for physical phenomena that might be essential for the thermal evolution of the basin: for example, water circulations, abnormal compaction of sediments, two-dimensional effects, time-transient effects.

The example of the Provencal Basin we present in this paper is not perfect (there is a lack of maturation data and the industrial deep boreholes recently drilled in the Gulf of Lions have not been published), but the Provencal is one of the young basins in the world in which the most complete set of seismic and thermal data has been collected.

---

(1) *Institut Français du Pétrole, Rueil-Malmaison, France.*

In the first part of this paper, we review these data and discuss qualitatively the major trends of the heat-flow distribution, in connection with the geological and geodynamic context. In the second part of the paper, the thermal evolution of the NW margin of the Provencal Basin (the Gulf of Lions) is studied using various numerical models. The sedimentary cover is backstripped and the tectonic subsidence is evaluated. A one-dimensional model is applied for a first-order approximation of the subsidence and of the thermal history of the margin. Then a true two-dimensional model is run, allowing heat- and water-circulations, and sedimentation; the geodynamic evolution of the margin, described in terms of non-uniform extension during rifting, is discussed.

Seismic and thermal data; geodynamic context:

1) Geological framework

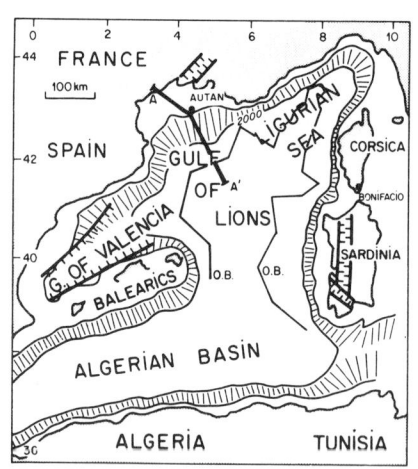

Fig. 1 - Location map of the Western Mediterranean
(O.B.: Oceanic Boundary;
AA': section modelized in chapter 2).

The Provencal Basin is a small oceanic basin created after a rifting stage during Oligo-Miocene times in relation with the eastwards rotation of Sardinia and Corsica initialy closer to the Provence (see reviews by Biju-Duval et al., 1978; Rehault et al, 1984; Burrus, 1984); oceanic accretion ended during Burdigalian times. The basin includes (Fig. 1) the Ligurian Sea, between Southern France and Corsica, and the Gulf of Lions, between France, Balearics and Sardinia. In the deepest part, it is covered by 2.8 km of water and 8.0 km of sediments, which are generally divided into four main sequences: (Cravatte et al., 1974; Montadert et al., 1978; Hsü et al., 1978; Mauffret et al., 1981). -1. Synrift

MODELING THE PROVENCAL BASIN

sediments were deposited in half-grabens during Rupelian-Chattian and Aquitanian times (30-24 My). -2. Miocene sediments are mainly deltaïc shales. -3. During the Messinian "crisis", the whole Mediterranean was isolated from the Atlantic and underwent a regression; the margins were eroded, an evaporitic sequence was deposited in the center of the basin, including a thick salt layer, affected by halokinetics during Pliocene. -4. Thick Plioquaternary deltaïc shales were then deposited after a Pliocene transgression.

2) Seismic and thermal data

Supported by French Government funding (through the CEPM, Comité d'Etudes Pétrolières Marines) a significant effort was made these last years by a joint group including IFP, IFREMER, and SNEA(P); a various set or original data were collected, in addition to numerous previous investigations:

- The seismic structure of the ante-Messinian sediments has been precisely described in the Gulf of Lions (unpublished Ligo 1 and 2 surveys), despite the presence of Messinian halokinetics.

- The deep crustal structure of the Basin has been investigated by the ESP technique; improving previous investigations (Fahlquist and Hersey, 1969; Hinz, 1972; Leenhard et al., 1972; Recq et al., 1979), Le Douaran et al. (1984) have shown that the center of the basin was oceanic and have documented the thinning of the continental margins around the Basin.

- Heat-flow measurements have been carried out in 1981 and 1982; 200 geothermal gradients have been measured and 120 acceptable determinations obtained, covering both Ligurian Sea and Gulf of Lions (Burrus et Foucher, 1986), in addition to rare previous measurements (Foucher et al., 1976; Rehault, 1981; Jemsek et al., 1985).

Figures 2, 3, 4 represent the deep crustal structure and present heat-flow along three transects of the Provencal Basin. Depth-sections are derived from previously mentionned ESP. Heat-Flow values have been obtained from thermal gradients and conductivity determinations, gradients being determined on a bottom-penetrating probe (5-10m in lengh) and conductivities measured on core-sample using a needle-probe technique, or during in-situ determinations.

. Gulf of Lions (Fig. 2): At the West, the post-rift sediments (Vp = 2.9 to 5.0 km/s) overlie a crystalline type substratum, which thins towards the center of the basin. The attenuation of the lower

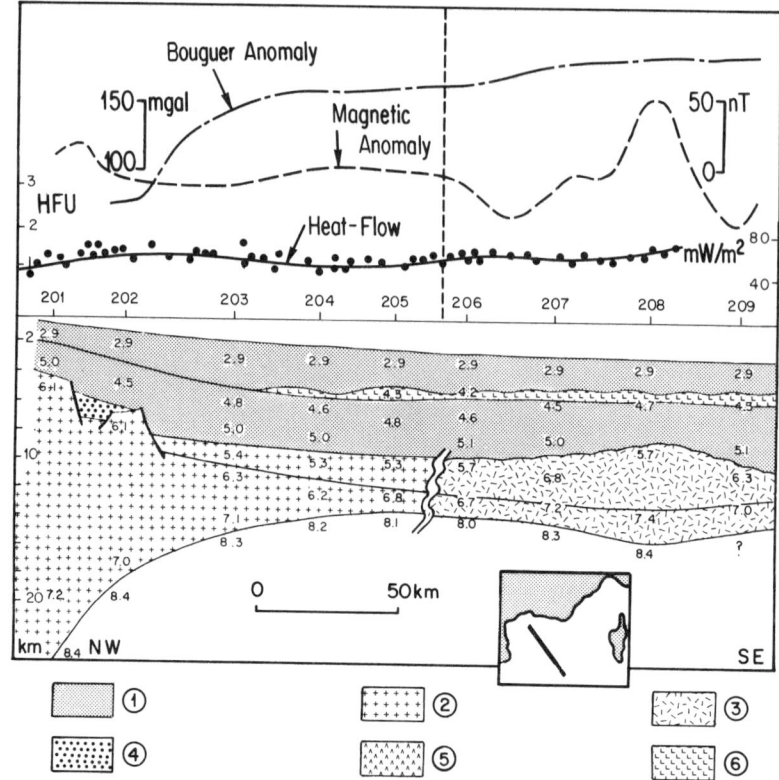

Fig. 2 - Simplified deep structure in the Gulf of Lions according to CROC ESP, surface Heat-Flow, Magnetics (after Galdeano) and Gravity (after Morelli) P-Velocities are in Km s^{-1}. 1.: Postrift sediments; 2.: Anterift substratum; 3.: Oceanic substratum; 4.: Synrift sediments; 5.: Volcano; 6.: Messinian salt or evaporites.

crust (Vp = 7.0 km/s) appears more rapid than for the upper crust (Vp = 6.2 km/s). Between ESP 205-206, a major discontinuity corresponds to the oceanic transition: crustal velocities (5.8 and 6.8 km/s) are similar to those of oceanic layers 2 and 3, contrasting magnetic signatures (Galdeano and Rossignol, 1977) do also characterize this transition. Below ESP 208, the center of the basin fits with an uplift of the oceanic substratum and with an axial magnetic anomaly possibly corresponding to the former oceanic-ridge, as discussed by Burrus (1984). The heat-flow increases regularly from 45-50 to 60-65 mW/m^2 towards the center of the basin. The Bouguer profile (Morelli et al., 1977) is very similar to the Moho profile, supporting the seismic interpretation.

. Southern Ligurian Sea (Fig. 3): Between ESP 220 and 230, a narrow oceanic domain is surrounded by the steep margins of Provence and Corsica. The oceanic crust appears here very thin, and no indication for any paleo-oceanic ridge is found. The magnetic profile appears very asymmetrical in the oceanic domain. The heat-flow increases from 80-90 mW/m^2 on the Provencal Margin to a maximum of about 110-120 mW/M^2 on the Eastern Ligurian oceanic domain, where the maximum extension of Messinian salt structures is observed. The Bouguer profile is here again very similar to the Moho profile.

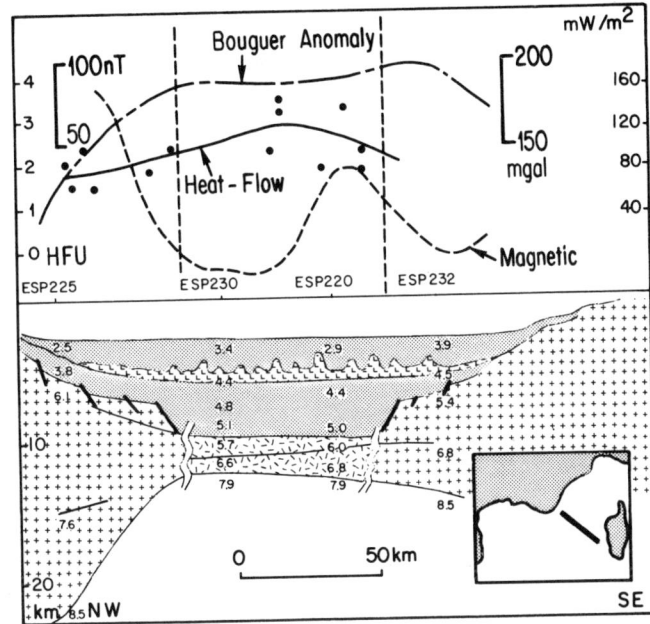

Fig. 3 - Simplified deep structure, heat-flow, magnetics and gravity across the Southern Ligurian Sea. Legend in Fig. 2.

. Northern Ligurian Sea (Fig. 4): According to ESP 229 and 222/223, the conjugated margins of Provence and Corsica have a very similar deep structure; nevertheless the Provencal margin appears steeper than the Corsican margin, this asymmetrical thinning will be discussed below. Below ESP 224, the crust is less rapid and thicker than elsewhere in the oceanic domain, and could be of intermediate nature. Below ESP 223, a synrift volcano is associated with a high, circular magnetic anomaly. The geometry of the Moho is supported by the shape of the Bouguer profile. The heat flow exhibits a dramatic increase from the Provencal margin (60-70 mW/m^2) towards Corsica (60-80 mW/m^2), with several maximum values of 160 mW/m2 eastern of ESP 223. These values are very similar to the one obtained by Jemsek et al (1985) on the same transect. We do not believe that this thermal anomaly is only relevant to salt structures, as far as no domes are observed in this area, where salt is replaced by evaporites and detritics.

MODELING THE PROVENCAL BASIN

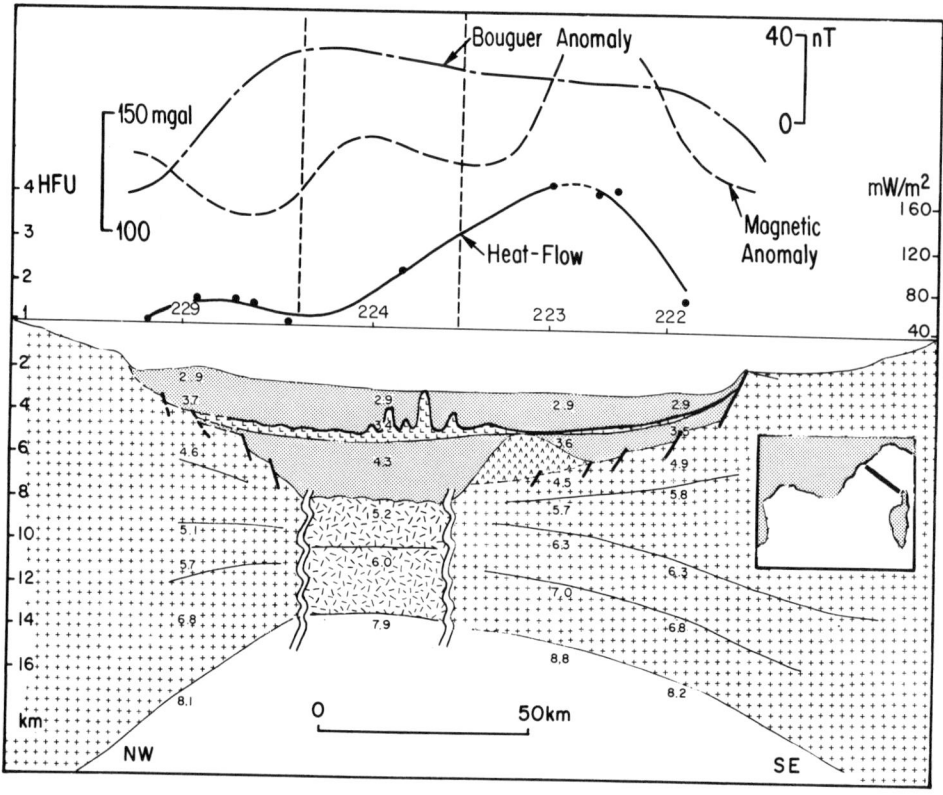

Fig. 4 - Simplified deep structure, heat-flow, magnetics and gravity across the Northern Ligurian Sea. Legend as in Fig. 2. High heat-flows are recorded in the absence of salt structures.

3) Qualitative discussion of the main trends of the heat-flow pattern:

Fig. 5 is a tentative map of the heat-flow in the Provencal Basin: despite considerable scattering of the data, it suggests that the Provencal margin is rather cold (average 50-80 mW/m^2). It seems that the "warmest" area extends from NW-Corsica to W-Sardinia slightly displaced towards the eastern oceanic boundary, and covers an area of about 60 x 300 km. Furthermore, corrected heat flow from Hercynian Corsica (76 $\pm$ 5 mW/m^2, Lesquer et al., 1983) is also higher than heat-flow on Hercynian Provence (60 $\pm$ 5 mW/m^2, Lucazeau et al., 1985). This thermal asymmetry might be a recent and superficial feature due for instance to water circulations in the Plio-quaternary sediments; the lack of data on the hydrodynamical regime in the area does not make it possible to argue this hypothesis and to quantify it. Nevertheless the extension and location of the thermal anomaly does not support this kind of interpretation. The asymmetry might also be due to the blanketing effect due to sedimentation: sediments are probably thicker on the Provencal margin, the surface heat-flow could be smaller there. In fact, numerical modeling, as discussed in the second part of this paper, shows that the difference of thickness is too small to explain the thermal contrast. On the other hand, radioactive contribution seems higher on the Provence (3 $\mu W/m^3$, Lucazeau et al., 1985) than in Corsica (1.8 $\mu W/m^3$, Lesquer et al., 1983). Consequently, we favor the following explanation; the asymmetrical distribution of the present heat-flow could be the consequence of asymmetrical geodynamic processes that might have been active during the formation of the basin, considering the following: 1.- Conjugated margins of Provence and Corsica underwent unequal attenuation during rifting (Fig. 4); 2.- According to figure 6, the magnetic pattern in the oceanic domain is asymmetrical, suggesting that asymmetrical accretion processes could also have characterized the drifting phase.

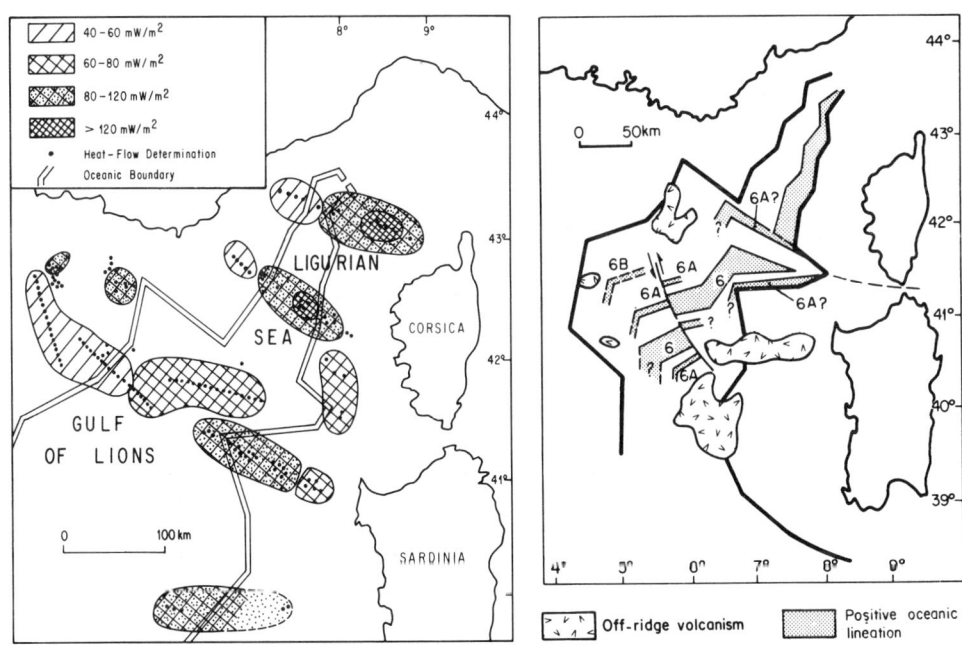

Fig. 5 - Tentative map of the heat-flow according to flumed data.

Fig. 6 - Magnetic interpretation in the Provencal Basin (after Burrus, 1984)

If these indications are correct, these mechanisms, active during rifting and drifting, might still be active today and would be the cause of the present asymmetrical heat-flow distribution, which would be due to a deep subcrustal abnormal high-flow.

As a preliminary conclusion, an important set of seismic and thermal measurements has been collected through out the Provencal Basin; the hight density of measurements, several times greater than in most conventional studies, makes it possible to observe second order features (as the asymmetrical crustal attenuation, the asymmetrical magnetic and heat-flow pattern) that are tentatively interpreted in terms of asymmetrical basin formation processes.

Modeling the thermal evolution of the Gulf of Lions margin

We reconstruct here the thermal evolution of a cross section of the Gulf of Lions represented on Fig. 1, . This section is characterized by a succession of horst and grabens, in particular the horst of Autan drilled at the Autan well (Cravatte et al., 1974); the tectonic subsidence and the thermal history have been investigated by the use of one-dimensional and two-dimensional thermal models.

1) Seismic backstripping:

Using seismic information (Fig. 7B) several tens of fictive wells were considered across the margin, and simultaneously back-stripped using a discrete porosity depth relationship derived from the Autan well (Bessis, 1986). The paleogeographic reconstruction, particularly the evaluation of the paleobathymetry and of the eroded thickness during the Messinian regression, can be precisely constrained by such a kind of two-dimensional approach, more reliably than with the conventional technique of backstripping of single wells, as discussed by Bessis (1986). Qualitatively, (Fig. 7C) the water-loaded tectonic subsidence exhibits a very classical behaviour: a rapid initial synrift subsidence is followed by an exponentially decreasing long-term subsidence. Nevertheless, a different style of tectonic subsidence appears in the deep basin: the post-rift tectonic subsidence (Fig. 7C) appears linear and is even accelerated in the last five million years, and no indication for exponential decay is observed. Quantitatively, knowing the amounts of synrift- and present post-rift tectonic subsidence, it is possible to adjust a synrift and a post-rift lithospheric attenuation parameter, β_I and β_∞, following the uniform extension model (Mac Kenzie, 1978) and

MODELING THE PROVENCAL BASIN

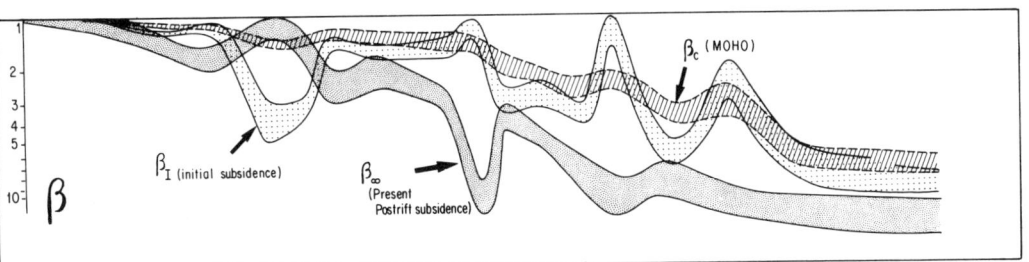

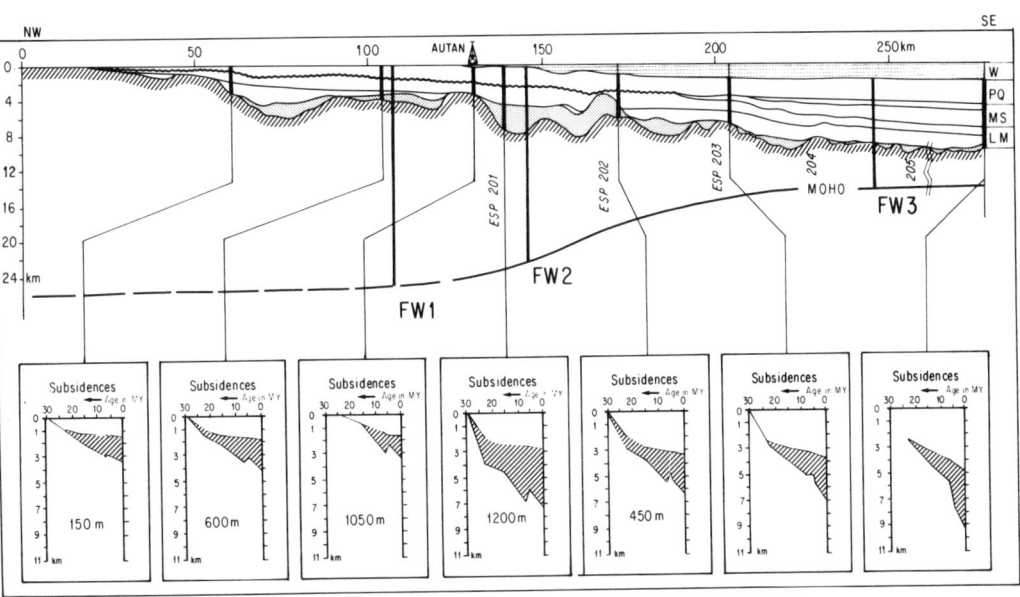

Fig. 7 - Cross-section in the Gulf of Lions:

A - Lithospheric thinning factor derived from synrift (β_I) and post-rift (β_∞) subsidence data, compared with measured (β_{Moho}) crustal thinning.

B - Deep structure of the margin as determined from seismics. Several tens of fictive wells have been backstripped. FW1, FW2, FW3 are fictive wells studied with the MONOCLE thermal model. This 2D section has been studied using the THEMIS model (Fig. 10).

C - Examples of total and tectonic (water-loaded) subsidence curves across the margin: after the rapid initial subsidence, and exponential decay is observed, except in the deep basin, where subsidence rate remains high.

compare them to the crustal thinning factor β_c derived from the present depth of the Moho; parameters used in this calculation are indicated in Annexe 1. According to Fig. 7 we note the following:

- the β_I and β_∞ curves are highly scattered, which reflects the structuration of the margin in horsts and grabens, locally controled by the brittle behaviour of the upper crust, and not by the deep thermal evolution; only smoothed curves would reflect these internal processes, but smoothing the scattered curves observed would be very unprecise. It is suggested that backstripping an unique well, often located on a particular high point, would probably not permit to draw any significant conclusion concerning the geodynamic evolution. For instance, the Autan well (Fig. 7B) was backstripped in the past (Steckler and Watts, 1980), but the calculated crustal thinning factor obtained, β = 10, did not make any physical sense.

- the fit between the β_c and β_I curves is not too bad (Fig. 7A), but the β_∞ curves departs significantly from these curves; the uniform extension model, in which the crust and the subcrustal lithosphere are thinned by the same factor, does not explain the post-rift subsidence history, which appears to be abnormally important at the East of Autan. This feature has not been commonly reported; in many previous subsidence studies (Royden, Keen, 1980; Sclater et al., 1980; Beaumont et al., 1982,; Royden et al., 1983), the calculated synrift subsidence was too large while present post-rift subsidence and heat-flow could be correctly estimated: the concepts of "subcrustal attenuation" and "non-uniform extension" were introduced to account for these features. We examine below wether these concepts apply to the Gulf of Lions margin.

2) One-D thermal modeling

After calculation of the tectonic subsidence, and following the previous remarks, we have applied a one-dimensional thermal model MONOCLE to three fictive wells FW1, FW2 and FW3 (Fig. 7B) selected in order to avoid the peaks and troughs (due to synrift structuration) of the β curves (Fig. 7A).

The principles of the MONOCLE model developped by IFP are presented in Annexe 2 with some details. The lithosphere is represented by a column with a basal temperature of 1330°C. The rifting phase is described by a finite-duration (Jarvis et Mc Kenzie, 1980) or instantaneous non-uniform lithospheric extension, by factors β_{cr} (in the crust) and β_{ma} (in the mantle). During rifting and/or after rifting, sediments can be deposited and compact at each time step; the sedimentary history can include non-depositional and erosional periods. Undercompaction is not taken into account in "MONOCLE". Other parameters (conductivity density, calorific capacity, radiogenic production, etc.) depend on lithologic changes during sedimentation (see conductivity profile on Fig. 8 for instance).

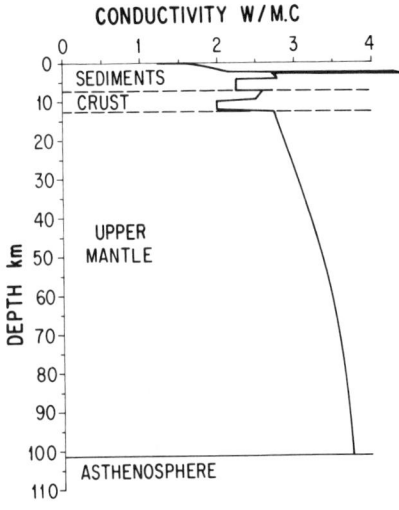

Fig. 8 - Conductivity is assumed to depend on lithology and porosity in the sediments (the peak corresponds to salt), and on temperature in the lithosphere.

The time transient heat equation is solved for conduction, forced convection due to compaction (negligible in fact in most realistic cases), and radiogenic production in the crust and sediments. Such a model permits a first order approximation of the thermal history of the sediments and, coupled with a maturation model (Tissot and Welte, 1978), gives an approximation of the oil and gas-windows. This later aspect will not be developed in this paper (Chénet, 1984; Burrus et al., 1986). The subsidence history can be reconstructed assuming local isostasy.

The chronostratigraphic characteristics of the three wells FW1, FW2, FW3 are indicated on Table 1; Table 2 contains the petrophysical parameters used. Fig. 9 (A,B,C) represents the calculated and observed tectonic subsidence and heat-flow on these wells for two different types of rifting: Type 1 is uniform rifting ($\beta_{cr} = \beta_{ma} = \beta_{Moho}$) of finite duration (30-23 My), including synrift sedimentation; Type 2 is non-uniform rifting, with greater mantellic attenuation ($\beta_{ma} = 3\beta_{cr}$), supposed to occur instantaneously at 23 My. The following results were obtained:

403

MODELING THE PROVENCAL BASIN

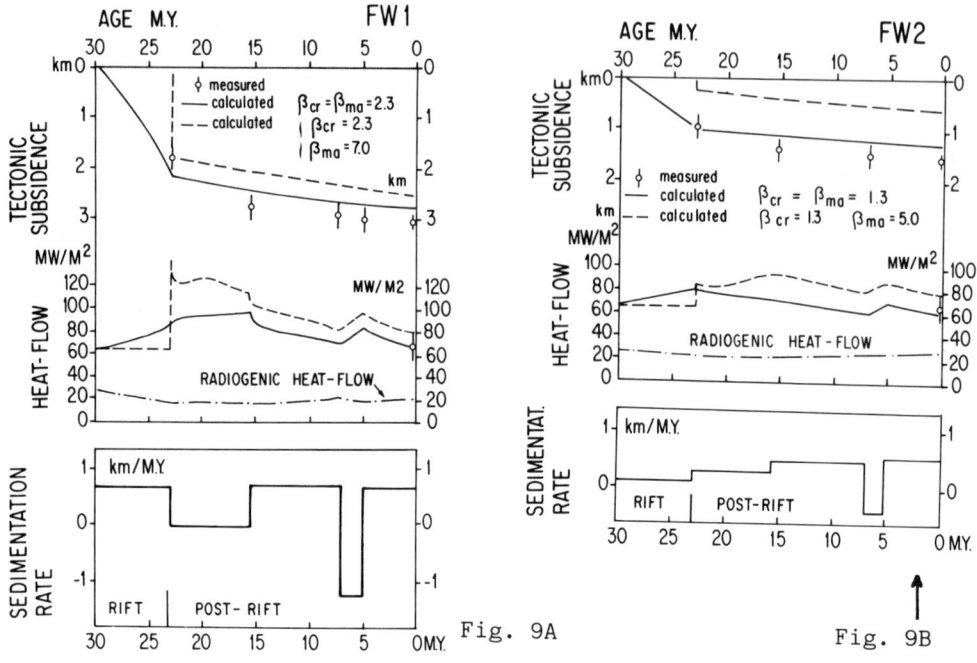

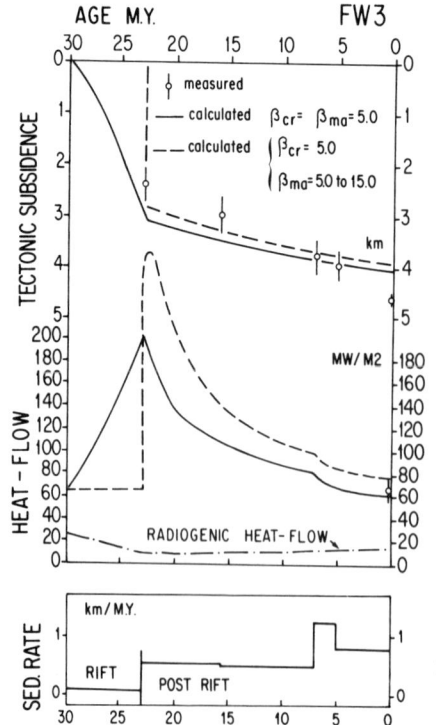

Fig. 9 - Calculated and observed tectonic subsidence and heat-flow for the three wells FW1(A), FW2(B), FW3(C).
Two types of rifting are compared: uniform and finite duration (full lines); non-uniform and instantaneous (dotted lines). The history of sedimentation is indicated (decompacted sedimentation rate).
Observed values are indicated by circles.

- Fig. 9A shows that the non-uniform rifting cannot account for the observed water-loaded tectonic subsidence of well FW1. In contrast, the uniform-extension model ($\beta_{cr} = \beta_{ma} = 1.3$) explains satisfactorily the synrift subsidence and the trend of the post-rift subsidence, even if the calculated value of the present tectonic subsidence appears 300 m too large: this difference is of the order of magnitude of the uncertitude on paleobathymetries. The observed present day heat-flow is also better accounted for by the uniform rifting than the non-uniform rifting, which results in exceeding heat-flow (Fig. 9A).

- Fig. 9B shows similar results for well FW2: the calculated tectonic subsidence and the present heat-flow are best accounted for by the uniform rifting model ($\beta_{cr} = \beta_{ma} = 2.3$). Nevertheless, the observed post-rift subsidence appears 300 m greater than the predicted one; it is the opposite for the end-rift subsidence; this discrepancy is also about the order of the uncertitude on the paleo-bathymetries.

- Fig. 9C indicates different conclusions for well FW3. For such high values of crustal attenuation as $\beta_{cr} = 5.0$, the uniform and the non-uniform rifting are not very much differenciated considering the subsidence history. None of these model can explain the observed high present subsidence, which exceeds by about 600 m the prediction, and the linear trend of tectonic subsidence since 23 My. Interpretation of this discrepancy can be two-fold:
1.- one can consider that, due the uncertitudes observed between 23 and 10 My, only the present subsidence does not fit with predictions; the discrepancy between model and observations would thus be a recent feature, that has appeared during the last five million years. 2.- one can also argue that the linear shape of the observed post-rift subsidence is not an artefact due to uncertitudes, but has a geological signification.

The authors favor the last interpretation. Such a linear subsidence curve cannot be accounted for by any passive cooling model, whatever the initial thermal perturbation is, neither can it be accounted for by still active heating after 23 My. Two kinds of explanation can be put forward: first, effect of regional compressive stress, accelerating continuously the subsidence in the center of the basin; second, continuous density changes at depth, which are not easy to imagine (subcrustal metamorphism, subcrustal erosion, etc.) and which would have no visible effects on the unstructured sedimentary cover. The MONOCLE code has been used to evaluate the thermal blanketing effect due to the sedimentation on the cooling of the lithosphere. Even if considerable effect on the surface heat-flow was obtained (up to 30 % in the deep abyssal plain confirming the discussion by Lucazeau and Le Douaran, 1985), nevertheless the likely difference of sedimentation rate during the

Plio-quaternary does not make it possible to explain a difference of heat-flow greater than 5 % between the "thick" Provencal margin and the "thinner" Corsico-Sardinian margin. The thermal asymmetry mentionned in the first part cannot be explained by a differential thermal blanketing effect.

3) Two D modeling, discussion

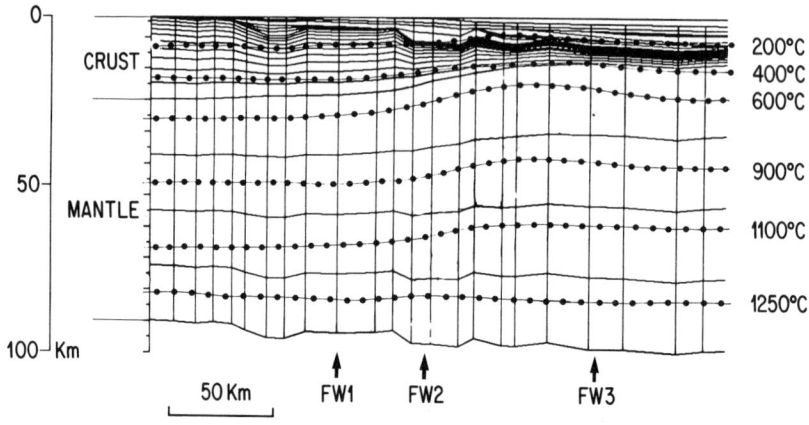

Fig. 10 - 2D reconstruction of the isotherms at present time during the THEMIS model. The grid used in the calculations is also represented.

The previous thermal reconstruction has been extended in two dimensions using the THEMIS model developped by IFP (see Doligez et al., this volume). In this case, we assumed that compaction was normal; the effect of impervious Messinian salt and evaporites as a possible cause of overpressuring have not been considered. The same petrophysical parameters (conductivities, porosities, radioactivities, densities, etc) than in the 1-D study have been considered. Uniform, finite duration (30-23 My) rifting, as in the previous 1-D study, has been introduced. The only difference is that heat can be conducted laterally, as well as water-flow driven by compaction. Fig. 10 shows the calculated thermal structure obtained at present time. Isotherms are still deformed at depth in the eastern part of the margin, as a consequence of the rifting, as could have been expected for a young neogene margin. The temperature at the Moho decreases from 500°C to about 400° C towards the center of the basin, while temperatures of more than 200°C are observed at the base of the deepest sediments. Comparing with the results obtained by MONOCLE reveals excellent agreement (deviation $\simeq$ 3 %), suggesting that the lateral thermal effect are negligible for such a 200 km wide margin, as indicated by Alvarez et al. (1985).

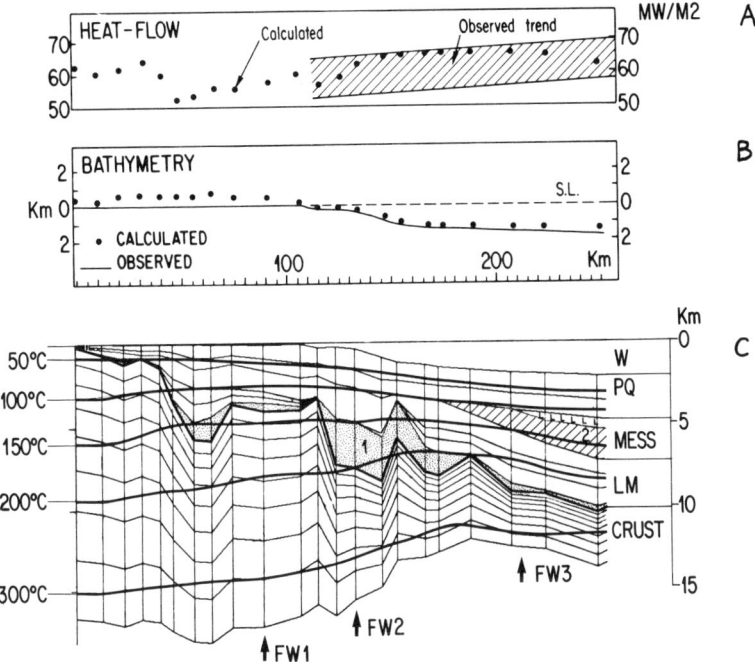

Fig. 11 - Same distribution of the isotherms than on Fig. 10; note details of the 2D grid which includes rectangular and triangular elements (pinch-outs of the Messinian); comparison of the calculated and observed heat-flow (11 A), calculated and observed present bathymetry, assuming local isostasy (11 B).

Such a 2-D thermal reconstruction can be constrained by the calculation of the surface heat-flow (Fig. 11A), which shows a good agreement with the measured trend. Considering subsidence response provides another kind of control: the present day observed bathymetric trend fits in a first approximation with the calculated trend (Fig. 11B), assuming local isostasy. Nevertheless, as noted in the discussion of the 1-D results, the observed top of sediments is up to 500 m too deep in the abyssal plain, where as the continental platform is 500-600 m below its predicted elevation which could be a recent feature as indicated above. A significant part of this discrepancy could be do to the flexural effect induced by the loading of the Plio-quaternary sediments on the slope and on the abyssal plain.

Table 1 - Chronostratigraphy of wells FW1, FW2, FW3

FW1
- 0My / 0 m
- PQ
- 5My / 1370 m
- LM
- 15.5My / 2800 m
- LM
- 23My / 3500 m
- SR
- 30My / 3850 m

FW2
- 0My / 0 m
- PQ
- 5My / 2400 m
- LM
- 15.5My / 4430 m
- 23My / 4440 m
- SR
- 30My / 7300 m

FW3
- 0My / 0 m
- PQ
- 5My / 2400 m
- SA
- 5.5My / 2900 m
- EV
- 7My / 4300 m
- LM
- 15.5My / 5400 m
- LM
- 23My / 7200 m
- SR
- 30My / 7400 m

Eroded thickness between 7 My and 5 My: FW1 = 500 m; FW2 = 1650 m; FW3 = 0 m
(PQ = Plioquaternary; SA = Messinian Salt; EV = Messinian Evaporites;
LM = Lower Miocene; SR = Synrift).

Table 2 - Petrophysical parameters

	Conductivity $Wm^{-1}C^{-1}$	Calorific Capacity pc $10^6 Jm^{-3}C^{-1}$	Radiog. Production Q_o μWm^{-3}	Density $Kg\ m-3$	Temp. Corr. on Cond C^{-1}
Water	0.6	4.18	0	1030	0
Upper Crust	3.09	3.88	3.0	2800	0.75×10^{-3}
Lower Crust	2.0	3.88	0	2800	0
Mantle	2.5	3.88	0	3300	-2.5×10^{-4}
Sediments (Matrix):					
Plio-Quaternary	2.57	2.43	1.0	2670	
Salt	5.85	1.85	0.0	2160	
Evaporites	3.37	2.35	0.75	2570	2.0×10^{-3}
Lower Miocene	2.57	2.43	1.0	2670	
Synrift	3.26	2.63	1.0	2670	

Asthenospheric temperature: 1330°C
Thermal Expansion Coefficient: $3.25 \times 10^{-4}\ C^{-1}$
Crustal Radiog. Contribut. Thickness: 10 km
(after Beaumont et al., 1982 and Cermak, 1982)

CONCLUSION

The deep seismic structure of the Provencal Basin is typical of a small oceanic basin surrounded by passive margins, showing surface extensional structuration as well as considerable crustal attenuation. Detailed thermal investigation suggests an asymmetrical distribution of the heat-flow, which is about 50-80 mW/m^2 on the Provencal side, and 60-110 mW/m^2 on the Corsico-Sardinian margin. The cause of this asymmetry is believed to be related to asymmetrical subcrustal heat-flow, first because no convincing or simple explanation can be mentioned in the sediments (influence of salt, regional water-flow), second because asymmetrical processes have been evidenced during rifting (thinning of the crust) and drifting (magnetic pattern).

Classical concepts of uniform and non-uniform lithospheric attenuation are applied to the study of the evolution of the NW margin of the Basin, the Gulf of Lions. It is suggested, provided methodological restrictions accounting for vertical movements due to the brittle structuration of the upper crust, that the first order magnitude of heat-flow and tectonic subsidence are satisfactorily enough explained by the uniform extension concept consistently with the observed depth to the Moho. Nevertheless, significant discrepancy is observed between observations and predictions:

- the observed bathymetry in the abyssal plain is about 400-500 m too deep, compared with predictions accounting for the blanketing effect due to sedimentation; the linear subsidence curve is not compatible with a simple cooling model.

- the observed bathymetry on the platform is about 50 m, where as 500 m elevation above sea-level is predicted.

The measured heat-flow is satisfactorily explained by the cooling of the perturbated lithosphere, but a significant (15-25 mW/m^2) radiogenic component has to be introduced, as indicated by radiogenic measurements.

The discrepancy between observed subsidence and calculated subsidence mentionned above cannot be reduced by increasing the mantellic attenuation factor with respect to the crustal factor: the main effect would be an increase of the heat-flow and a decrease of the synrift tectonic subsidence, which are not observed. Some recent flexural effects due to the load of the Plioquaternary sediments could explain part of the misfit on the platform. Nevertheless, the misfit in the abyssal plain (observed all over the basin, Rehault et al., 1984; Jemsek et al., 1985) can probably not be explained by the same cause. The linear trend of tectonic subsidence is not easy to discuss. Has it to be explained by the regional compression trend? by recent internal processes leading to density changes in the crust? Is it related to the asymmetrical pattern of the heat-flow? Further investigations involving improved petrophysical, seismic and thermal data and focusing on the nature of lithospheric response to the sedimentary load are still necessary. Nevertheless, right now, realistic thermal predictions can be made, which can be constrained, and the precision of which can be evaluated.

BIBLIOGRAPHY

Adam, A., 1978 Geothermal effects in the formation of electrically conducting zones and temperature distribution in the earth. Phys. Earth Planet. Int. 17, p. 2128.

Alvarez, F., Virieux, J., Le Pichon, X., 1984 - Thermal consequences of lithosphere extension over continental margins: the initial stretching phase. Geophys. J.R. astr. Soc., 78, p. 389-411.

Beaumont, C., Keen, C.E., Boutilier, R., 1982 - On the evolution of rifted continental margins: comparison of models and observations from the Nova Scottian margin. J.R. astr. Soc., 70, p. 667-715.

Bessis, F., 1986 - Some remarks on subsidence study of sedimentary basins: application to the Gulf of Lions margin (W-Mediterranean), Marine petroleum Geology (in press).

Biju-Duval, B., Letouzey, J., Montadert, L., 1978 - Structure and evolution of the Mediterranean basins. Initial Report of the DSDP, 42, p. 951-984.

Burrus, J., 1984 - Contribution to a geodynamic synthesis of the Provencal Basin, Marine Geology, 55, p. 247-269.

Burrus, J., Foucher, J.P., 1986 - Crustal and thermal structure of the Provencal Basin (NW Mediterranean), submitted to Tectonophysics.

Burrus, J. Bessis, F., Chénet, P.Y., 1986 - Thermal reconstruction and petroleum occurrence in the Western Mediterranean, In: K. Louden (Editor), Handbook of Sea Floor Heat-Flow, CRC Series in Marine Science, (in preparation).

Cermak, V., 1982 - Regional pattern of the lithosphere thickness in Europe. In: V. Cermak and R. Haenel eds., Geothermics and Geothermal Energy. E. Schweizerbartsche Verlagsbuchhandlung, Stuttgart, p. 1-10.

Chénet, P.Y., 1984 - Thermal transfer in sedimentary basins, paleotemperature reconstruction and maturation studies in the Gulf of Lion margin, In: B. Durand (Editor), Thermal phenomena in sedimentary basins, Ed. Technip, p. 257-269.

Cravatte, J., Dufaure, P., Prim, M., Rouaix, S., 1974 - Les sondages du Golfe du Lion, stratigraphie et sédimentologie, Notes Mém. CFP, 2, p. 209-274.

Fahlquist, D.A., Hersey, J.B., 1969 - Seismic refraction measurements in the Western Mediterranean Sea, Bull. Inst. Oceanogr. Monaco, 67, p. 1-52.

Foucher, J.P., Auzende, J.M., Rehault, J.P., Olivet, J.L., 1976 - Nouvelles données du flux géothermique en Méditerranée occidentale, 4° RAST, Paris, 174.

Foucher, J.P., Tisseau, C., 1984 - Thermal regime of Atlantic type continental margins: Bays of Biscay and Gulf of Lion, In: B. Durand (Editor) Thermal phenomana in sedimentary basins, Ed. Technip, p. 221-225.

Galdeano, A., Rossignol, J.C., 1977 - Assemblage à altitude constante de cartes d'anomalies magnétiques couvrant l'ensemble du bassin occidental de la Méditerranée, Bull. Soc. Géol. de France, 7, XIX, p. 461-468.

Hinz, K., 1972 - Results of seismic refraction investigations (Projet Anna) in the Western Mediterranean Sea South and North of the Island of Mallorca. Bull. Cent. Rech. Pau, SNPA, 6(2), p. 405-426.

Hsü, K.J., Montadert, L. et al., 1978 - Site 372 Minorca Rise. In: Report DSDP, vol. 42, Part 1, US Govt. Printing Office, Washington DC, p. 59-150.

Jarvis, G.T., Mc Kenzie, D., 1980 - Sedimentary basin formation with finite extension rates, Earth Planet Sci. Lett., 48, p. 42-52.

Jemsek, J., Von Herzen, R., Rehault, J.P., Williams, D.L., Sclater, J., 1985 - Heatflow and lithospheric thinning in the Ligurian Basin (NW Mediterranean), Geophys. Res. Lett., vol. 12, n° 10, p. 693-696.

Le Douaran, S., Burrus, J., Avedik, F., 1984 - Deep structure of the North Western Mediterranean: a two ships seismic survey, Marine Geology, 55, p. 325-345.

Leenhardt, O. et al., 1972 - Results of the Anna Cruise. Three North South Seismic Profiles through the Western mediterranean Sea, Bull. Centre Rech. Pau, SNPA, 6, p. 365-452.

Le Pichon, X., Sibuet, J.C., 1981 - Passive margins: a model of formation, J. Geophys. research, 86, p. 3708-3720.

Lesquer, A., Pagel, M., Orsini, J.B., Bonin, B., 1983 - Premières déterminations du flux de chaleur et de la production de chaleur en Corse, C.R. Acad. Sci. Paris, 297, p. 491-494.

Lucazeau, F., Le DOuaran, S., 1985 - The blanketing effect of sediments in basins formed by extension: a numerical model; application to the Gulf of Lions and Viking Graben, Earth Planet. Sci. Lett., 74, p. 92-102.

Lucazeau, F., Vasseur, G., Foucher, J.P., Mongelli, F., 1985 - Heat-Flow along the southern segment of EGT, in "Second EGT worshop: the southern segment", D.A. Galson et St. Mueller (Eds.) European Science Foundation, p. 59-63.

Mauffret, A., Rehault, J.P., Genesseaux, M., Bellaiche, G., Labarbarie, M., Lefebvre, D., 1981 - Western Mediterranean Evolution: from a distensive to a compressive regime. Coll. Urbino, October 1980. In: Sedimentary basins of Mediterranean margins, 61-81, CNR Italian Project on Oceanography, Tectoprint, Bologna.

Mc Kenzie, D., 1978 - Some remarks on the development of sedimentary basins, Earth Planet. Sci. Lett., 40, p. 25-32.

Montadert, L., Letouzey, J., Mauffret, A., 1978 - Messinian event: seismic evidence. In: K.J. Hsü, L. Montadert et al., In: Report of the DSDP, vol. XLII, part I, U.S. Govt. Printing Office, Washington D.C., p. 1037-1050.

Morelli, C., Giese, P. et al., 1977 - Crustal and upper mantle structure of the Northern Apennines, Ligurian Sea and COrsica, derived from seismic and gravimetric data. Bull. Geol. Teor. Appl. 75-76, p. 199-260.

Mueller, S., Panza, G.F., 1983 - The lithosphere-asthenosphere system in Europe, in first EGT Workshop, J. Galson, S. Mueller Editor, European Science Foundation, p. 23-26.

Recq, M., Bellaiche G., Rehault, J.P., 1979 - Interpretation de quelques profils de sismique réfraction en mer Ligure, Mar.

Geol., 32, p. 39-52.

Rehault, J.B., 1981 - Evolution tectonique et sédimentaire du bassin Ligure, Thèse, Paris VI.

Rehault, J.P., Boillot, G., Mauffret, A., 1984 - The Western Mediterranean basin geological evolution, Marine Geology, 55, p. 447-477.

Royden, L., Keen, C.E., 1980 - Rifting process and thermal evolution of the continental margin of eastern Canada determined from subsidence curnes, Earth Planet. Sci. Lett., 51, p. 343-361.

Royden, L., Horvath, F., Rumpler, J., 1983 - Evolution of the Pannonian Basin System: Subsidence and thermal history, Tectonics, vol. 2, N°1, p. 91-137

Sclater, J.G., Royden, L., Horvath, F., Burchfiel, B.C., Semben, S., Stegena, L., 1980 - The formation of the Intra Carpathian Basins as determined from subsidence data, Earth Planet. Sci. lett., 51, p. 132-162.

Steckler, M.S., Watts, A.B., 1980 - The Gulf of Lions: subsidence of a young continental margin, Nature, 287, 5781, p. 425-430.

Tissot, B., Welte, D., 1978 - Petroleum formation and occurrence. A new approach to oil and gaz exploration, Springer Verlag, Berlin.

ANNEXE 1
LITHOSPHERIC ATTENUATION, INITIAL SUBSIDENCE, LONG TERM SUBSIDENCE

Following Mac Kenzie (1978) and Le Pichon and Sibuet (1981) the initial water-loaded subsidence following instantaneous rifting by factor β_I is:

$$Z_I = \frac{hl(\rho_m - \rho_c) \, hc/hl(1 - \alpha_d/2 \, Ta \, hc/hl) - \rho_m \, \alpha_d/2 \, Ta}{\rho_m (1 - \alpha_d \, Ta) - \rho_w} (1 - 1/\beta_I) \quad (1)$$

where :
- h_c anterift crustal thickness — 25 km
- h_l anterift lithospheric thickness — 90 km
- ρ_m mantelic density at 0°C — 3.3 g/cm^3
- ρ_c crustal density at 0°C — 2.8 g/cm^3
- ρ_w water density (constant) — 1.03 g/cm^3
- α_d coefficient of thermal expansion — 3.28 10^{-5} C^{-1}
- Ta asthenospheric temperature — 1330°C

hc and hl cannot be directly evaluated; nevertheless, the present thickness of the crust of the upper margin of the Gulf of Lions is 25 to 30 km (Hirn, 1980); taking into account the possibility of a Mesozoic crustal thinning in the Gulf of Lions, we choose the lower value of 25 km. The regional lithospheric thickness has been evaluated by various techniques: seismological data (Mueller-Panza 1983), calculation of regional heat-flow pattern (Cermak, 1982), magneto-tellurics (Adam, 1978). Even if based on very different principles, all theses studies suggest that the lithospheric thickness in France is less than the standard value of 125 km generaly adopted in thermal modeling: measured values range between 75 and 100 km, with an average value of 90 km which was adopted. Moreover the calculated anterift steady-state heat-flow 65 mW/m^2 (given some realistic assumptions on the conductivities and heat-production in the lithosphere, (Tab. 2) compares with the regional heat-flow (55-70 mW/m^2) measured in the Hercynian substratum (Lucazeau et al., 1985). The agreement is much better with a thin lithosphere, than with a 125 km thick lithosphere (calculated heat-flow 45 mW/m^2). Equation (1) thus gives:

$$Z_I \text{ (km)} = 4.11 \, (1 - 1/\beta_I)$$

The post-rift subsidence is compared to calculated subsidence curves from Mac Kenzie (1978) and a thinning parameter β_∞ is determined. It is assumed that the instantaneous rifting occured 30 My ago.

MODELING THE PROVENCAL BASIN

ANNEXE 2
THE 1D MONOCLE (MONO dimensional CoupLEd) THERMAL MODEL

The physical model

In a fixed referential with upward z axis, the compaction of sediments is supposed to be :

$$\phi = \frac{1}{L(H + A - Z) + Lo}$$

ϕ = porosity
H = sedimentary thickness
A = anterift lithospheric thickness
L, Lo = parameters

This relation is used to reconstruct the sedimentation rate S(T) at the well, by decompaction of the different layers. Writing mass conservation for water and matrix, makes it possible to calculate at each time t:

the Water filtration velocity U(z)
the grain velocity V(z)
the instantaneous sedimentary thickness H

The heat equation :

$$\lambda \frac{\partial^2 T}{\partial z^2} + \frac{\partial T}{\partial z}\left(\lambda \log\left(\frac{\lambda_w}{\lambda_s}\right) L \phi^2 - \rho c_s (1-\phi) V - \rho c_w U\right) + Q = \rho c \frac{\partial T}{\partial t}$$

includes a conductive term and a convective term due to compaction. The conductivities are described as follows:

sediments: λ bulk conductivity $\lambda = \lambda_w^{\phi} \lambda_s^{1-\phi} \cdot 1/(1+\alpha T)$
λ_w water conductivity
λ_s grain conductivity
α temperature correction

crust: $\lambda = \lambda_{uc}/(1 + \alpha_{uc} T)$ in the upper crust
$\lambda = \lambda_{lc}$ in the lower crust

mantle: $\lambda = \lambda_m/(1 - \alpha_m T)$

(see discussion in Cermak, 1982)

The heat-flow is supposed to be continuous at each lithologic interface (substratum/sediment or sand/shale, etc.). The radioactive production decreases exponentialy in the upper crust:

$$Q(z) = Q_o \exp^{-z/z_o}$$

A small constant production is introduced in the mantle:

$$Q(z) = Q_m$$

The rifting event is described in terms of lithospheric attenuation. Two kinds of evolution are described:

- the first case is an uniform non-instantaneous thinning of the lithosphere by a final factor β ; ΔT is the duration of the rifting. The heat-equation in the lithosphere is then (Jarvis, Mac Kenzie, 1978):

$$\lambda \frac{\partial^2 T}{\partial z^2} - \frac{\log \beta}{\Delta T}(A-z)\rho c \frac{\partial T}{\partial z} + Q = \rho c \frac{\partial T}{\partial t}$$

Synrift sedimentation is taken into account, but synrift sediments are not extended, which is geologically correct but physically consistent only if thinning is not too important.

- the second case is an instantaneous non-uniform stretching of the lithosphere by factors β_{cr}, for the crust, and β_{ma} for the mantle, following Royden and Keen (1980).

The numerical model:

Adimensional variables T^*, z^*, t^*, are calculated, and the adimensional heat-equation is solved with a Crank-Nicholson implicit finite-difference method.

Due to the adimensional transformation, despite sedimentation or rifting, the heat-flow equation is solved in a non-moving medium of length 1. Sediments are represented by a constant number of knots (N = 50), and the lithosphere by 200 knots. Typical time-step is 0.05 to 0.5 My.

J. A. NUNN[1]

SUBSIDENCE AND THERMAL HISTORY OF THE MICHIGAN BASIN

I. ABSTRACT

Mechanical evolution of the Michigan basin is modeled as flexure of the lithosphere caused by thermal contraction. Results from both elastic and viscoelastic lithosphere models are consistent with the subsidence record of the sediments, structural contours and gravity anomalies across the basin. Model values for effective elastic thickness (20-120 km) and relaxation time (1-10 m.y.) of the lithosphere are compatible with previously estimated rheological parameters. Deviations in subsidence curves from exponentials associated with thermal contraction can be explained by changes in sediment supply.

The magnitude and spatial distribution of the computed thermal contraction load are consistent with the presumed mechanism. The driving load, presumably from subsurface processes including crustal stretching and emplacement of dense rocks into the crust, is roughly equivalent to the load generated by 10-20% crustal extension. Theoretical gravity results indicate that the driving load is centered at a depth of approximately 15 km.

Temperature histories for selected horizons in the Michigan basin are determined from excess temperature due to the thermal anomaly plus burial temperature predicted from subsidence curves. For an equilibrium temperature gradient of 22°C/km and surface temperature of 20°C, paleotemperatures do not exceed 110°C. This estimate is consistent with upper limits set by paleomagnetic studies. The low value for maximum paleotemperature is caused by

---

(1) *Department of Geology, Louisiana State University, Baton Rouge, Louisiana, USA.*

the concentration of the thermal anomaly below 15 km, in agreement with gravity results. The great depth of the thermal anomaly can explain the lack of evidence for an initial heating event prior to subsidence.

Once the thermal history of the sediments is specified, the oil potential of the basin can be determined from empirically derived kinetic equations for degradation of kerogen to petroleum. For the Michigan basin, predicted temperature conditions are sufficient for source rock (greater than 25%) conversion of type II kerogen only in Ordovician and older rocks in the southern peninsula of Michigan. By this model, petroleum found in rocks younger than Ordovician would have had to migrate upward from the older rocks. Geochemical studies of Dundee (Devonian) and Trenton (Ordovician) crude oils in Michigan are compatible with this interpretation. Niagaran (Silurian) crude oils appear to come from a different source than the Dundee and Trenton oils. If their source is Silurian rocks on the flanks of the basin, either the scheme used to calculate maturity is not applicable to these oils or the temperature of the source rocks was significantly higher in the past. More likely, the model underestimates the depth of burial of Silurian source rocks near the center of the basin, especially in local regions of downfaulting.

II. INTRODUCTION

This study quantitatively examines the subsidence history, thermal evolution and petroleum generation of sedimentary basins formed by surface cooling and subsequent thermal contraction of continental lithosphere. Although model results are constructed to explicitly resemble the Michigan basin in Middle Ordovician and later times, the techniques will apply to the development of any sedimentary basin produced by a thermal disturbance to the lithosphere.

A solution for subsidence of a sedimentary basin caused by thermal contraction of continental lithosphere is given by Nunn and Sleep (1984). A brief summary of the important observations and techniques is given below.

The Michigan basin is several kilometers of gently dipping, shallow water sediments on the North American continent. The outcrop pattern (Fig. 1) and cross-section (Fig. 2) indicate a nearly circular geometry and a decrease in basin width with time. The oldest strata which exhibit the present characteristic shape

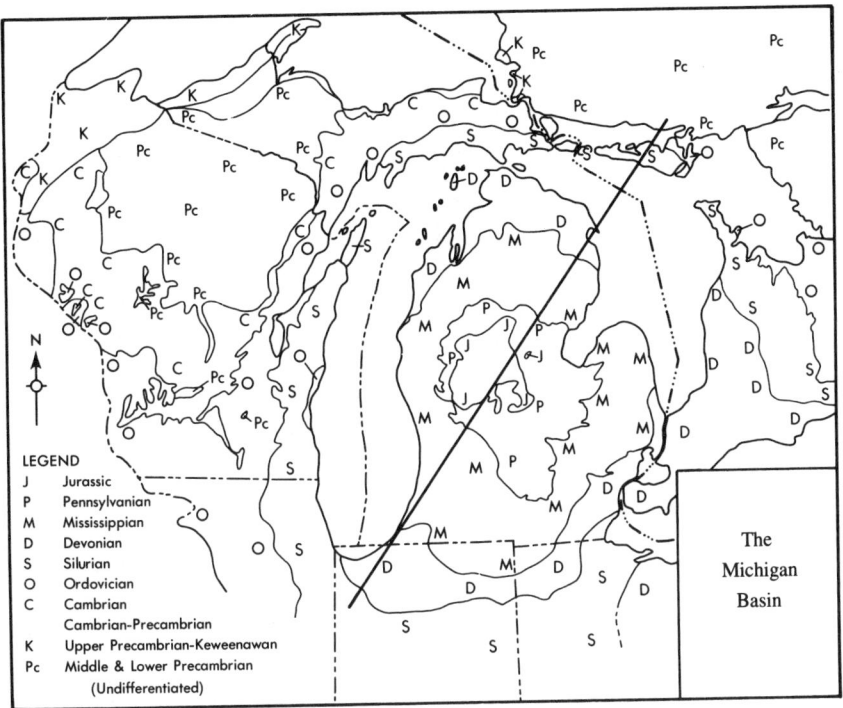

Figure 1. Geological map of the Michigan basin after Stonehouse (1969). The northeast striking path of the cross section in Figure 2 is shown as a heavy line.

of the Michigan basin are of Middle Ordovician age (462 Ma). The youngest sediments that record the subsidence are of Pennsylvanian age (300 Ma). However, subsidence probably continued throughout the Paleozoic and perhaps later (Haxby et al., 1976). Michigan basin strata are predominately limestone and dolomite or interbedded limestone, dolomite and shale.

Studies of the geometry of sedimentary basins with horizontal dimensions of a few hundred kilometers suggests the lithosphere responds to loads by regional flexure of a strong elastic or viscoelastic lithosphere overlying an inviscid fluid asthenosphere (Sleep and Snell, 1976; Turcotte, 1980).

Thermal contraction of the lithosphere is a probable cause of the gradual subsidence indicated by basin sediments (Sleep and Snell, 1976; Turcotte, 1980). Assuming horizontal conduction of heat is minor in the case of interest, the thermal contraction load

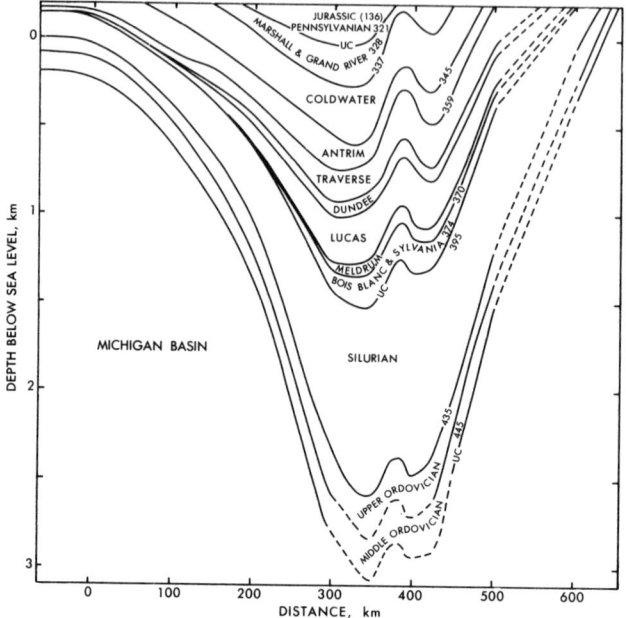

Figure 2. Post-Sauk cross section of the Michigan basin after Sleep and Snell (1976). Major unconformities are marked with 'UC'. No attempt is made to show minor structures, the distribution of post-Paleozoic sediment, or the present position of the land surface.

caused by surface cooling of the lithosphere increases exponentially as

$$P = P_o[1 - \exp(-t/t_p)] \qquad (1)$$

where t_p = thermal decay time of the lithosphere and P_o = load at the end of cooling.

Accumulation of sediments in an initial depression imposes a load on the lithosphere which sags owing to its weight. Substantial thicknesses of shallow and deep-water sediments may accumulate simply as a result of sediment loading (Walcott, 1972). However, for an average sediment density of 2.6 g/cm^3 (Hinze et al., 1978), an improbably deep initial depression of 1 km is needed to produce the 3-4 km of sediment in the Michigan basin.

III. SUBSIDENCE HISTORY

The present thickness of strata for both elastic and viscoelastic lithosphere models were determined using different values of effective flexural rigidity, thermal decay time $t_p = 50$ m.y., and viscoelastic decay times between 1 and 10 m.y. Eustatic sea-level changes, sediment compaction and variations in sediment density were ignored. Results are compared with observed position of strata (Figs. 1 and 2).

A. RHEOLOGY OF THE LITHOSPHERE AND ESTIMATED DRIVING LOAD

Two-dimensional flexural models for the Michigan basin are compatible with previously estimated rheological parameters. For an elastic lithosphere model, an effective mechanical thickness of 20 km can explain observed gravity anomalies and structural contours across the Michigan basin. If the lithosphere is modelled as a viscoelastic slab with viscosity between 1 and 10 m.y., effective elastic thickness can be as high as 120 km. For a three-dimensional elastic or viscoelastic lithosphere model, effective elastic thickness must be approximately a factor of two smaller: 8 km and 56 km, respectively. Larger values leave an excessive remaining load at the center of the basin, which is not indicated by observed gravity anomalies (Nunn and Sleep, 1984). The decrease in effective flexural rigidity is attributed to greater difficulty in equidimensional bending of a 3-D plate or shorter average flexural wavelength in a 3-D basin.

The magnitude of the thermal contraction load, P_o, is computed from the observed deflection for each value of effective elastic thickness (e.g. it is the load necessary to produce the observed deflection at $t = $ infinity). Gravity anomalies associated with platform basins constrain the depth and position of the load that drives basin subsidence. The best fit between theoretical and observed gravity anomalies is for a depth between 15 and 20 km, in agreement with previous studies (Haxby et al., 1976; Sleep and Snell, 1976). The great depth of the thermal anomaly may explain the lack of evidence for igneous activity and/or other thermal activity immediately prior to formation of the Michigan basin. For all models compatible with observed gravity anomalies, the thermal contraction load constitutes approximately 25% of the theoretical driving load. The remaining load is due to the weight of accumulated sediments. The magnitude of the thermal contraction load, approximately 280 million dyne-cm^3, roughly corresponds to the load generated by 10-20% uniform extension of the lithosphere.

B. PRESENT POSITION OF STRATA

Computed subsidence curves for basin models of various lithospheric rheologies are illustrated in Figure 3. When sediment loading is passive (continually filled basin), model subsidence curves follow exponentials defined by thermal contraction load and rheology of the lithosphere (Fig. 3a). Consequently, model results are poor when observed subsidence deviates from an exponential curve. For the Michigan basin, continuously filled basin models cannot simultaneously explain: slow subsidence during the Ordovician period; unconformities in the Early Devonian and Carboniferous; and rapid subsidence from Middle Devonian to Mississippian (Fig. 3a).

As mentioned above, sediment overburden constitutes approximately 75% of the load driving model subsidence. Consequentially, temporal and/or spatial variations in sediment supply may have a significant effect on computed subsidence. For example, periods of non-deposition and/or erosion marked by unconformities in Early Devonian (415-399 Ma) and Carboniferous (356-346 Ma) times show pronounced decreases in predicted subsidence (Fig. 3b). The correspondence between observed and predicted subsidence curves could be further improved by including effects of sediment compaction, paleobathymetry and sea level changes. For example, a modest eustatic sea-level drop of 100 meters can explain the observed unconformity during Early Middle Devonian times, in agreement with Sleep (1976).

The best fit to observations for both 2-D and 3-D models is a viscoelastic lithosphere with a low effective flexural rigidity (Nunn and Sleep, 1984). However, it is likely that these models are simply more forgiving of errors in lithostratigraphic data and/or other initial assumptions. With presently available data, both elastic and viscoelastic lithosphere models give acceptable results for the Michigan basin.

SUBSIDENCE AND THERMAL HISTORY OF MICHIGAN BASIN

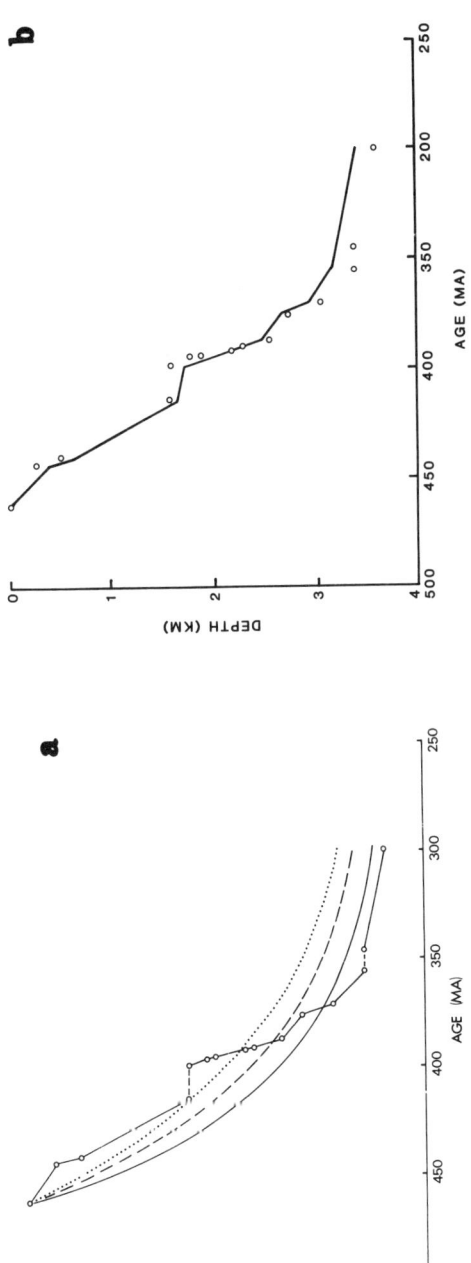

Figure 3. Cumulative subsidence at the center of the Michigan basin, as compiled by Sleep and Snell (1976), is plotted as a function of absolute age (open circles). The observed subsidence is not corrected for sediment compaction or eustatic sea level changes. (left) Theoretical subsidence curves are superimposed: solid line - elastic lithosphere model with mechanical thickness of 7 km; dashed line- viscoelastic lithosphere model with effective mechanical thickness of 19 km; and dotted line - viscoelastic lithosphere model with effective mechanical thickness of 52 km. A continuous sediment supply is assumed. (right) Theoretical subsidence curve (solid line) is computed from viscoelastic model with an effective mechanical thickness of 52 km. The weight of accumulated sediments is explicitly included. Relaxation time for all viscoelastic models is 1 m.y.

423

IV. THERMAL EVOLUTION

Using the above results for subsidence of a 3-D, continually filled sedimentary basin (fig. 3a), the thermal evolution of the Michigan basin is examined. The temperature distribution within the subsiding basin is composed of two parts: excess temperature due to the postulated thermal anomaly and equilibruim temperature after the lithosphere has cooled. A detailed derivation of the equations governing the thermal evolution of a continually filled sedimentary basin is given by Nunn et al., 1984.

Paleotemperature for sediments in the basin is the sum of excess temperature and equilibrium temperature. Ignoring higher vertical harmonics, the excess temperature in a three-dimensional plate having isothermal boundaries is

$$T_e(x,y,z,t) = \pi \exp(-t/t_p)\gamma P(x,y)\sin(\pi z/h)/(2h\rho g) \qquad (2)$$

where P = thermal contraction load, h = thickness of lithosphere, γ = coefficient of thermal expansion and ρg = specfic weight of lithosphere. In this study, equilibrium temperature is assumed to be a linear function of depth

$$T_o(x,y,z) = \Delta_T Z(x,y) + T_s \qquad (3)$$

where Δ_T = temperature gradient, and T_s = surface temperature. Because subsidence at the Michigan basin ceased in Jurassic (or earlier) times, the observed temperature distribution (temperature gradient = 22°C/km and surface temperature = 20°C) is used as the equilibrium value (Pollack and Watts, 1976).

Excess temperature in a plate is a function of both depth and time (eqn. 2). It increases with depth (distance from the isothermal boundary at the surface) and decreases with time as the thermal anomaly decays. During rapid subsidence, excess temperature increases along a time-stratigraphic horizon because deflection away from the free surface is faster than upward migration of temperature isotherms due to heat loss. As the thermal anomaly continues to decay, the subsidence rate decreases and excess temperature eventually begins a monotonic decline to zero. For the Middle Ordovician time-stratigraphic horizon (Fig. 4) excess temperature reaches a maximum of 15°C after 30 m.y. of subsidence (432 Ma, Early Silurian) and dwindles to zero within 250 m.y. (212 Ma, Jurassic). The paleotemperature along the Middle Ordovician time-stratigraphic horizon increases throughout geological time because depth of burial is the dominant thermal effect (Fig. 4).

SUBSIDENCE AND THERMAL HISTORY OF MICHIGAN BASIN

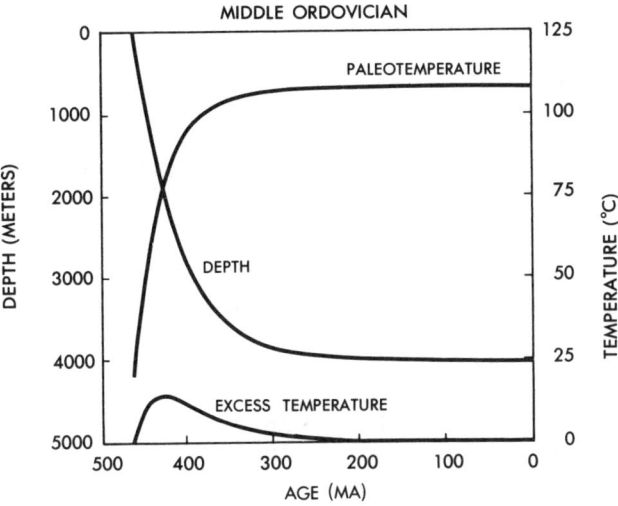

Figure 4. Model values for depth, excess temperature and paleotemperature are plotted as a function of absolute age. Calculations are for the Middle Ordovician time-stratigraphic horizon (462 Ma) at the center of the basin.

V. PETROLEUM GENERATION

Once the thermal history of the sediments is specified, the oil potential of a basin can be determined from laboratory derived kinetic equations for the degradation of kerogen to petroleum (Tissot and Espitalié, 1975). Using subsidence and temperature models developed above, the hydrocarbon potential of the Michigan basin is computed. Results are compared with the observed spatial distribution and stratigraphic location of oil fields in Michigan. Because initial results were not entirely satisfactory, thermal effects of post-Pennsylvanian burial and subsequent erosion are also considered. Finally, computed maturation curves are compared with results from geochemical studies.

At any time, the evolution of kerogen is measured by the transformation ratio, which is the ratio of hydrocarbons generated to the genetic potential (total amount of hydrocarbons which can be produced by a certain kerogen). A transformation ratio of 0.1

corresponds to the onset of oil generation and a transformation ratio of 0.4 corresponds to the oil/wet gas boundary.

Tissot and Espitalié (1975) define three kerogen types based on chemical composition. Types I and II are derived from algal matter, whereas type III contains terrestrial plants, which did not evolve until later in the history of the Michigan basin. The chemical composition of Michigan basin kerogen in unavailable. It is probably either type I or II.

Oil production in the Michigan basin is primarily restricted to the central portion of the southern peninsula of Michigan. In general, peripheral oil fields are in older stratigraphic units: Salina-Niagara (Silurian) in the Northwest and Trenton-Black River (Ordovician) in the south central section of the peninsula. The largest cumulative oil production is in rocks of Middle Devonian age. However, much of the oil is derived from stratigraphic traps (reefs and fault zones) and therefore it may have migrated upward from older rock units.

A. INITIAL RESULTS

Cumulative hydrocarbon generation for types I and II kerogen is illustrated in Figure 5. Because of the great age of the Michigan basin, conversion of labile organic matter to petroleum is in equilibrium with temperature for both types of kerogen and thus independent of the details of basin history. Type I has a maximum transformation ratio of approximately 0.1, well below the 0.25 minimum used to denote source rocks. Analytic solutions at constant temperature indicate a minimum temperature of 115°C for 200-400 m.y. is required for larger conversion of type I kerogen (Nunn et al., 1984).

Transformation ratios and maximum paleotemperature versus distance from the center of the basin are computed for the Middle and Upper Ordovician (462 and 444 Ma, respectively) time-stratigraphic horizons (Fig. 6). As expected, degradation of type I kerogen to petroleum is less than 15 percent everywhere in the Michigan basin. Temperature conditions sufficient for source rock generation of hydrocarbons from type II kerogen only exist in sediments of Middle Ordovician (and older) age. Source rock generation of type II kerogen in Michigan basin sediments (Middle Ordovician and younger) is confined to within 100 km of the point of maximum deflection (Fig. 6). Consequently, extensive lateral and upward migration of petroleum is required for this model to explain the observed distribution of oil fields in Michigan.

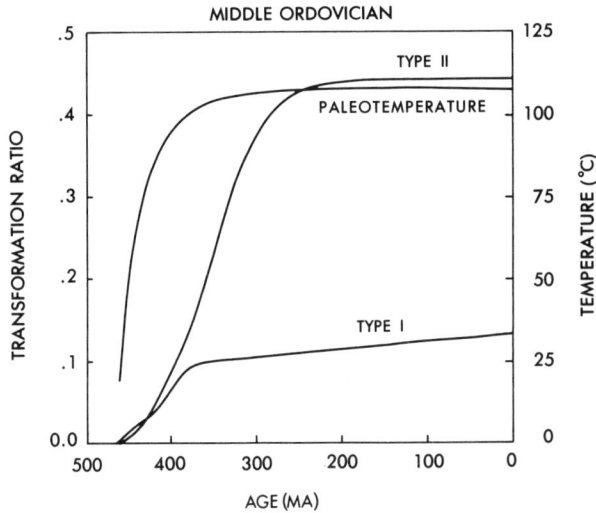

Figure 5. Model values for paleotemperature and transformation ratios for types I and II kerogen are plotted as a function of absolute age. Calculations are for the Middle Ordovician time-stratigraphic horizon (462 Ma) at the center of the basin.

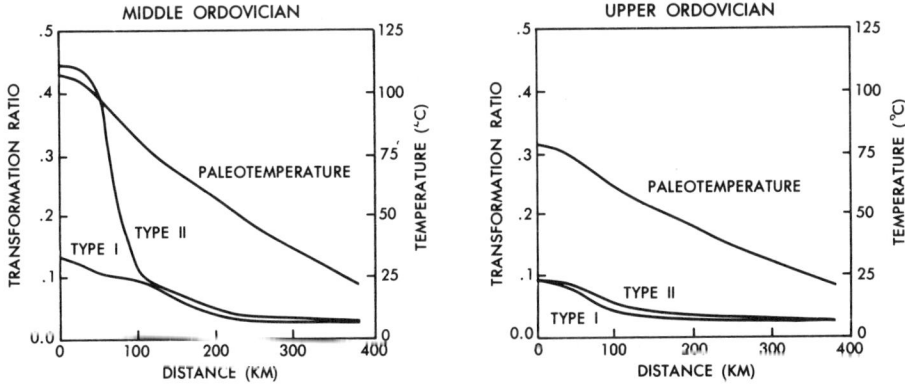

Figure 6. Model values for maximum paleotemperature and transformation ratios for types I and II kerogen on various time-stratigraphic surfaces plotted as a function of distance from the center of the basin. (left) Middle Ordovician (462 Ma) and (right) Upper Ordovician (444 Ma).

427

B. RESULTS WITH POST-PENNSYLVANIAN OVERBURDEN

Subsidence at the Michigan basin continued throughout the Paleozoic and perhaps later (Haxby et al., 1976). Most of the sediments deposited after the Pennsylvanian were subsequently eroded. A minimum of 300 meters of erosion for the interior of Michigan was estimated from the degree of organic metamorphism and the rank of coals in the Pennsylvanian section (Sleep et al., 1980). Assuming an equilibrium temperature gradient of 22°C/km, a uniform sediment overburden of 500 meters from Permian (300 Ma) to Jurassic (180 Ma) times raises the maximum temperature by 11°C for 120 m.y.

Addition of post-Pennsylvanian overburden approximately doubles the maximum transformaton ratio for type I kerogen to just under 0.2 (Fig. 7). If elevated temperatures are maintained until the present time, the transformation ratio would approach 0.25. However, temperatures in excess of 150°C are necessary for source rock degradation of type I kerogen to occur on a regional scale (Nunn et al., 1984). Burial by post-Pennsylvanian sediment increases the conversion rate for type II kerogen, but does not increase cumulative hydrocarbon generation (Fig. 7). For a time interval of 120 m.y., maximum temperature must exceed 130°C to significantly increase petroleum generation from type II kerogen (Nunn et al., 1984).

Predicted hydrocarbon generation in Upper Ordovician (444 Ma) and younger sediments is unaffected by post-Pennsylvanian temperature changes. Maximum paleotemperatures are less than the 100°C required for source rock conversion of either kerogen type. As a result, petroleum generation is insignificant.

Further burial dramatically increases the lateral extent of type II kerogen degradation in Middle Ordovician and older rocks. Source rock conversion of type II kerogen in Middle Ordovician sediments now covers most of the central portion of the southern peninsula of Michigan (Fig. 8).

C. GEOCHEMISTRY OF MICHIGAN BASIN CRUDE OILS

Model results from the preceding sections suggest two testable inferences: 1) oil in the Michigan basin has migrated from Ordovician (and older) age rocks in the center of the basin and 2) the oil is at only an intermediate level of maturation. Upward migration of oil from Middle Ordovician to Devonian age rocks in the Michigan basin presents some difficulties: up to 2 kilometers of

SUBSIDENCE AND THERMAL HISTORY OF MICHIGAN BASIN

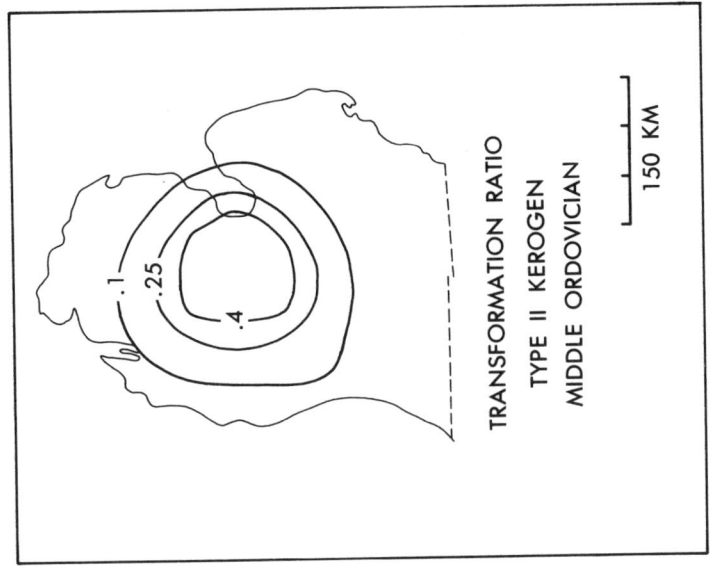

Figure 8. The transformation ratio for type II kerogen on the Middle Ordovician time-stratigraphic horizon (462 Ma) is computed with 500 m of Permian to Jurassic overburden.

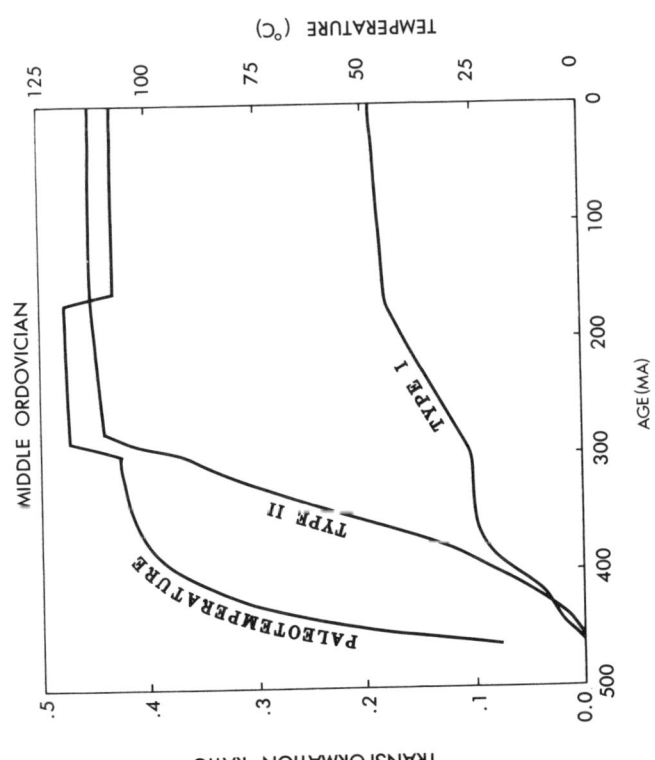

Figure 7. Model values for paleotemperature and transformation ratios for types I and II kerogen plotted as in Figure 5. A 500 m layer of sedimentary overburden is assumed to exist from Permian to Jurassic time.

429

vertical migration and traversal of relatively impermeable salt and evaporite layers in Silurian strata.

Analysis of Michigan basin crude oils, including carbon isotope ratios, nickel-vanadium analyses, capillary gas chronmatograph and GC-MS analyses indicates that Dundee (Devonian) petroleum is chemically indistinguishable from Trenton-Black River (Ordovician) petroleum (Vogler et al., 1981). Chemical similarity between Dundee and Trenton-Black River oils was also noted in a recent study by Gardner and Bray (1983). The similarity between Dundee and Trenton-Black River crude oils is especially interesting because the intervening Salina-Niagaran (Silurian) petroleum is chemically distinct.

Trenton-Black River crude oil has the following distinctive characteristics (Volger et al., 1981): 1) A strong odd predominance of n-paraffins, 2) An abrupt decline in n-paraffin after C-17, 3) Pristane predominance over phytane, and 4) A tendency for C-11 n-paraffins to be the predominant n-paraffin. These characteristics infer a source composed of marine (algal) organic material (type I or possibly type II kerogen), convertion to petroleum by decarboxylation, and a relatively mild maturation (Welte and Waples, 1973). Dundee petroleum, being like Trenton-Black River oil, would require a similar source material.

Silurian petroleum produced from Salina-Niagaran reservoirs has these characteristics (Volger et al., 1981): 1) A tendency towards even n-paraffin predominance below n-C15 followed by an odd predominance beyond n-C-15, 2) Phytane is prominent over pristane, 3) Only a gradual decline in n-C15 to n-C30, and 4) Abundant branched, cyclic and aromatic hydrocarbons are present, especially below n-C12. One and three above suggest a mixture of marine and terrestrial organic matter (type II kerogen), dominated by marine organic material converted to petroleum by reduction (Welte and Waples, 1973). The abundance of branched, cyclic and aromatic hydrocarbons suggests a moderate level of maturation. Higher levels of maturation are noted in basinward fields where a higher gas content is found.

VI. DISCUSSION

Temperature in the Michigan basin can be represented by a simple depth of burial curve. An uneventful thermal history is consistent with temperature restrictions calculated from paleomagnetic data (van der Voo and Watts, 1978). Palynamorph studies also fail to indicate much higher temperatures in the past (Gardner and Bray, 1983). Excess temperature in basin sediments is small and transient because the bulk of the thermal anomaly is deep in the lithosphere.

Temperature conditions predicted from the thermal-mechanical model are sufficient for source rock (greater than 25%) conversion of type II kerogen to petroleum only in Middle Ordovician and older sediments in the central section of the southern peninsula of Michigan. The spatial distribution of oil fields is consistent with this conclusion. Chemical composition of Michigan basin crude oils is also consistent with the moderate thermal environment and consequently incomplete degradation of kerogen determined from the thermal-mechanical model. Extensive upward migration is required for this model to explain the presence of petroleum in Silurian and Devonian rock units. Upward migration is plausible in light of the similarity in chemical composition of oil in Dundee (Devonian) and Trenton-Black River (Ordovician) age rocks. The differential pattern of oil and gas entrapment in Silurian pinnacle reefs also indicates migration from the center of the basin (Gill, 1982). However, many workers favor in situ formation of petroleum in Silurian and Devonian age rocks. For example, Gardner and Bray (1983) have suggested the Salina A-1 carbonate unit as a principle source rock for Silurian oils using carbon isotope data. It would be extremely difficult for our model to explain generation of hydrocarbons in Devonian strata.

A possible explanation for the anomalous Silurian oils is that the observed temperature gradient and/or depth of burial was greater in the past. A recent study by Cercone (1984) proposed a Paleozoic thermal gradient of 45°C/km plus 1 km of post-Pennsylvanian deposition and subsequent erosion. This much higher Paleozoic temperature gradient was required to explain in situ generation of Silurian and Devonian oils. Although postulating a higher thermal gradient in the past is an attractive solution to the problem of high levels of organic maturity in the Michigan Basin, a few simple calculations show that such a large increase in temperature gradient cannot be achieved by any reasonable mechanism (e.g. lower thermal conductivity, heating event or hydrothermal circulation). For example, assuming a thermal conductivity of 0.005 cal/cm-sec-°C, an additional 20°C/km in the thermal gradient throughout the Paleozoic development of the Michigan basin (~160 m.y.), requires 5×10^9 calories of heat per square centimeter of surface area (Nunn et al., 1984). This is equivalent to the heat per unit area generated by the cooling of a 50 km thick "batholith" beneath the basin. A thermal event of this magnitude would have certainly left other evidence behind.

Another difficulty with this explanation is that a higher geothermal gradient, especially one high enough for in situ petroleum generation in the Silurian reefs on the flanks of the basin and upper Devonian rocks anywhere in the basin, would imply much higher paleotemperatures deeper in the basin. For example, Amoco's Letts Unit #2-36 well in Gladwin County, Michigan has produced 4.4 mcf/d gas plus 72 bbl 60° API condensate from a Cambrian-Ordovician age unit at a depth of approximately 3.4 km (W. E. Moore, personal communication, 1985). Assuming constant temperature and a cooking time of at least 450 My, inversion of Tissot and Espitalié's method estimates formation temperatures between 115 and 122°C. This is clearly incompatible with a formation temperature of approximately 200°C estimated from Cerone's model (45°C/km goethermal gradient plus 1 km of additional burial). Such an elevated temperature regime would have completely burned out the unit. In addition, paleomagnetic studies of a sample from a depth of 5.3 km indicate a maximum paleotemperature of 200°C (van der Voo and Watts, 1978).

It is also possible that kerogen in the Michigan basin has a different kinetic reaction rate than those derived by Tissot and Espitalié (1975). Time may be a more important maturation parameter over hundreds of millions of years. For example, Daly and Lilly (1985) have suggested that Lopatin's method (Waples, 1980) predicts significantly higer levels of organic maturity for older sedimentary basins. However, results from both methods for the Michigan basin are remarkably consistent in both extent and timing of hydrocarbon maturation (Fig. 9). Not only do both methods predict a present maturation level in the wet gas zone, but they predict the occurrence of oil generation between Middle Devonian and Early Permian (390 to 280 Ma). Similarly, results from Lopatin's method for the Silurian time-stratigraphic horizon plot in the immature zone (Fig. 10), in agreement with results from Tissot and Espitalié's method.

Some Michigan basin kerogen might have higher Arrhenius constants than either type I or II. This is a plausible explanation for Silurian age source rocks because they were deposited in a different depositional environment (mesosaline) from the Ordovician rocks which the model indicated are the most probable source beds. Because of the restricted number of biological species that can survive in a mesosaline environment, the chemical composition of the resulting kerogen is probably different from kerogen deposited in a normal marine environment (Evans and Kirkland, 1981).

A final possibility is that the model underestimates the depth of current burial of some Silurian rocks near the center of the basin and hence their paleotemperature. From Figure 3a, the lowermost Silurian in the basin is about 600 m deeper than in the models. In addition, downfaulted portions of various structures near the center of the basin may add another few hundred meters of burial depth to Silurian rocks. Recomputing the temperature history for the Silurian time-stratigraphic horizon assuming that the model underestimated depth of burial by 600 meters, predicted organic

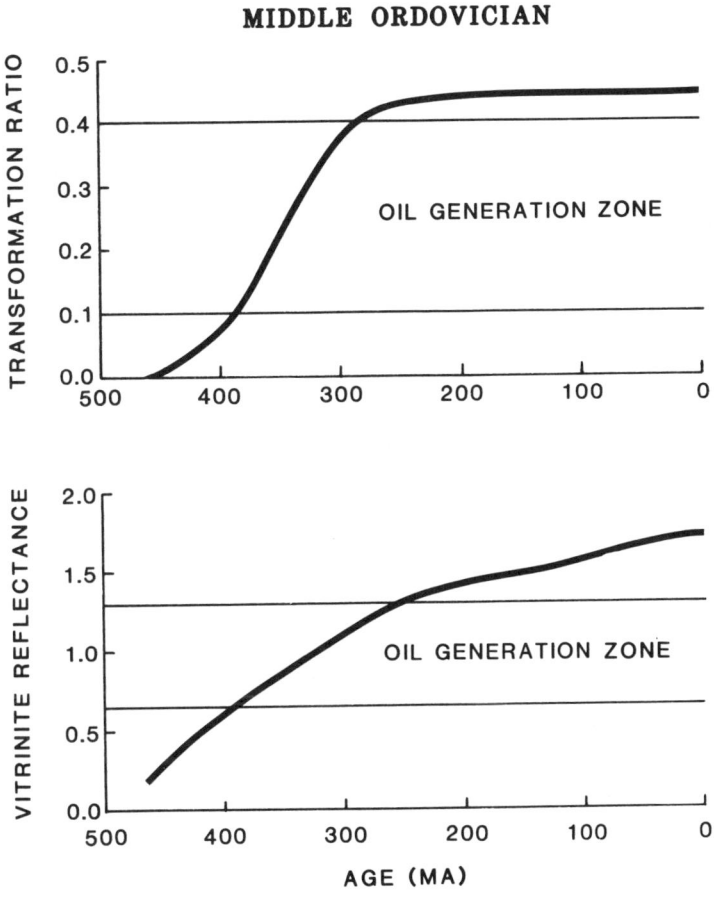

Figure 9. Model values for organic maturity on Middle Ordovician time-stratigraphic horizon (462 Ma) are plotted as in figure 5. (top) Organic maturity (transformation ratio) computed from Tissot and Espitalie's method. (bottom) Organic maturity (vitrinite reflectance) computed from Lopatin's method. The oil generation zone is delimited by straight lines. No attempt is made to scale transformation ratio to vitrinite reflectance as the relationship is not well defined.

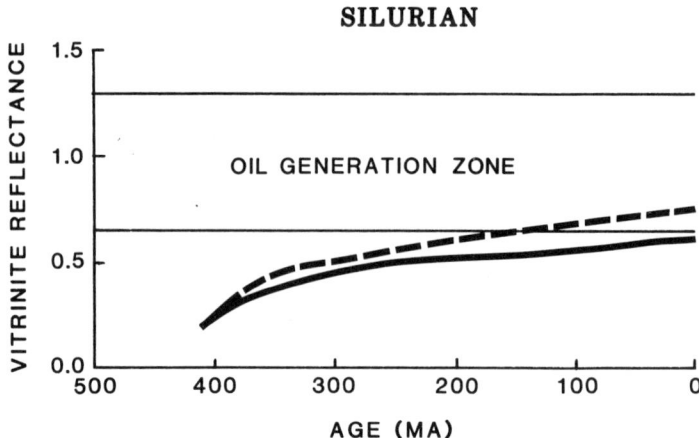

Figure 10. Model values for organic maturity on Silurian time-stratigraphic horizon (441 Ma) are plotted as a function of absolute age. Organic Maturity (estimated vitrinite reflectance) is determined from Lopatin's method. (solid line) Thermal history estimated from original model at center of basin. (dashed line) Thermal history computed assuming original model underestimated depth of burial of the Silurian by 600 meters.

maturity is within the oil generation zone (Fig. 10). This explanation would give the different source of the Silurian oils implied by the geochemical data.

Of the alternatives discussed above, easier conversion of kerogen and generation of hydrocarbons in locally downdropped regions at the center of the basin are considered most likely. It is even conceivable that the Silurian oils come from various of the sources which we have discussed.

VI. REFERENCES

CERCONE, K. R., 1983, Thermal history of the Michigan Basin: AAPG Bulletin, v.6, p. 130-136.

DALY, A. R. and D. H. LILLY, 1985, Comment on "Thermal subsidence and generation of hydrocarbons in Michigan basin" by Nunn, Sleep and Moore: AAPG Bulletin, in press.

EVANS, R., and D. W. KIRKLAND, 1981, Potential of evaporitic environment as source of petroleum: AAPG Bulletin, V. 65, p. 1357.

GARDNER, W. C., and E. E. BRAY, 1983, Oils and source rocks of the Niagaran reefs in the Michigan basin: GSA Special Paper, in press.

GILL, D., 1979, Differential entrapment of oil and gas in the Niagaran pinnacle-reef belt of northern Michigan: AAPG Bulletin, V. 63, p. 608-620.

HAXBY, W. F., D. L. TURCOTTE, and J. M. BIRD, 1976, Thermal and mechanical evolution of the Michigan basin: Tectonophysics, v. 36, p. 57-75.

HINZE, W. J., J. W. BRADLEY, and A. R. BROWN, 1978, Gravimeter survey in the Michigan basin deep borehole: Journal of Geophysical Research, v. 83, p. 5864-5868.

NUNN, J. A., and N. H. SLEEP, 1984, Thermal contraction and flexure of intracratonal basins: a three-dimensional study of the Michigan basin: Geophysical Journal Royal Astronomical Society, v. 76, 587-635.

NUNN, J. A., N. H. Sleep, and W. E. MOORE, 1984, Thermal subsidence and generation of hydrocarbons in Michigan basin: AAPG Bulletin, v.68, p. 296-315.

POLLOCK, H. N., and D. WATTS, 1976, Thermal profile of the Michigan basin: Transactions American Geophysical Union, v. 57, p. 595.

SLEEP, N. H., 1976, Platform subsidence mechanisms and "eustatic" sealevel changes: Tectonophysics, v. 36, p. 45-56.

SLEEP, N. H., J. A. NUNN, and L. CHOU, 1980, Platform basins: Annual Reviews of Earth and Planetary Science, v. 8, p. 17-34.

SLEEP, N. H., and N. S. SNELL, 1976, Thermal contraction and flexure of midcontinent and Atlantic marginal basins: Geophysical Journal Royal Astronomical Society, v. 45, p. 125-154.

TISSOT, B. P. and J. ESPITALIÉ, 1975, L'évolution thermique de la matiére organique des sediments: applications d'une simulation mathematique: Revue de l'Institut Francais du Pétrole, v. 30, p. 743-777.

TURCOTTE, D. L., 1980, Models for the evolution of sedimentary basins: in Dynamics of Plate Interiors, Geodynamics Series v. 1, American Geophysical Union, Washington D.C., p. 21-26.

VAN DER VOO, R., and D. WATTS, 1978, Paleomagnetic results from igneous and sedimentary rocks from the Michigan basin borehole: Journal of Geophysical Research v. 83, p. 5844-5848.

VOGLER, E. A., P. A. MEYERS, and W. E. MOORE, 1981, Comparison of Michigan basin crude oils: Geochimica Cosmochimica Acta, v. 45, p. 2287-2293.

WAPLES, D. W., 1980, Time and temperature in petroleum formation: application of Lopatin's method to petroleum exploration: AAPG Bulletin, v. 64, p. 916-926.

WELTE, D. H. and D. WAPLES, 1973, Uber die Bevorzuganggeradzahlimger n-Alkane in Sedimentgesteinen: Naturwissenschafter, v. 60, p. 516-517.

CHAPTER 3
CONTROL BY GEOCHEMICAL METHODS

PART A

ORGANIC METHODS

B. DURAND[1], B. ALPERN[2],
J. L. PITTION[3], B. PRADIER[2]

REFLECTANCE OF VITRINITE AS A CONTROL OF THERMAL HISTORY OF SEDIMENTS

ABSTRACT

Vitrinite is a maceral family of humic coals, and more generally of sedimentary organic matter of terrestrial origin, composed of gels and gelified tissues produced by the decay of ligno-cellulose parts of higher plants.

In early stages of coalification, peat and brown coals, there is no vitrinite sensu stricto. The components which will produce vitrinite further on are called huminite.

During coalification, reflectance of huminite, then vitrinite, increases from 0.25% at the peat stage to more than 4% at the metaanthracite stage. Above 1.5%, an anisotropy begins to develop, which increases strongly above 2.0%.

Practically, measuring reflectance of vitrinite consists of realizing histograms of measurements. At least 100 measurements are desirable; the mean value, R_m is taken as the significant parameter. The use of R_m to follow coalification is examplified here by a study of a Wesphalian coal series from the Gironville 101 well, Lorraine France· broadly speaking, the method is satifactory. Difficulties due to the occurrence of different vitrinite families and the influence of the preparation methods are discussed: they result in incertitudes which cannot be neglected, in comparison of the small range of variation of R_m during coalification.

---

(1) *Institut Français du Pétrole, Rueil-Malmaison, France.*
(2) *Groupe d'Etudes des Combustibles Fossiles, Université d'Orléans, Orléans, France.*
(3) *Total-Compagnie Française des Pétroles, Pessac, France.*

REFLECTANCE OF VITRINITE AS A CONTROL OF THERMAL HISTORY OF SEDIMENTS

In coal series, and more generally in terrestrial series even in the absence of coals, R_m patterns may be used to compare roughly thermal regimes of different basins. However, R_m values result from both temperature and time. Therefore a thorough reconstruction of the thermal regime and of its variation through time requires on the one hand a good reconstruction of burial history and on the other hand a physical model of R_m variations with temperature and time. No such models exist up to now. Only empirical models are available. A possibility is to use tabulations of R_m versus transformation ratios of terrestrial organic matter calculated by kinetic models of kerogen degradation. But the method is limited to the range of petroleum and wet gas formation, i.e. 0.6%-2%. A drawback of most models is that they have to be calibrated on natural series whose thermal history cannot be very precisely reconstructed.

In spite of these inconveniences, given the long experience of coal petrographers and petroleum geochemists, vitrinite reflectance is probably, when work is done with care, one of the best controls of thermal history of terrestrial series availables for the present.

The concept of vitrinite has been extended to marine and lacustrine series. Based on the example of a well in the Paris, it is shown that this extension is hazardous, for the following reasons:

- the particles whose reflectance can be measured are rarely true vitrinite, i.e. material derived from ligno-cellulose parts of higher plants. They are mostly bituminites which are derived from a planctonic material and may obten be mistaken for vitrinites. Their physico-chemical properties, hence their responses to thermal stresses are not the same than for true vitrinite.

- there is a variety of bituminites populations, even within a single formation and no population has been found yet, which is present in all series and could be used safely for a calibration.

Therefore in such series it is highly advisable to not rely on reflectance and to use other properties of organic matter. The same problems are also encountered in pre-Devonian series, even terrestrial ones, which cannot contain true vitrinite, just because higher plants did not exist yet.

Vitrinite reflectance studies are also hazardous even in terrestrial series, in case of lean samples and particularly in case of sandstones, because most of the organic matter is likely to be made of altered or reworked material.

REFLECTANCE OF VITRINITE AS A CONTROL OF THERMAL HISTORY OF SEDIMENTS

1. INTRODUCTION

Reflectance of vitrinite is presently the most popular among the possible indicators of thermal phenomena in sedimentary basins.

This is due more to ancientness, easy use and very good normalization of measurements by International Committee of Coal Petrography (ICCP) than to real mastery of its applications. Introduced by coal petrographers as a measure of coal rank, it was transposed, thanks mainly to M. Teichmüller, B. Alpern and P. Robert in Europe, N. Bostick in the United States, I. Ammosov and N. Lopatin in the Soviet Union, to the assessment of maturation, i.e. the stage reached by sedimentary organic matter in the process of petroleum formation.

The clear demonstration of B. Tissot (1969) that maturation was ruled by thermal history of sediments opened the door to a large utilization of organic matter maturation stages as a record of this thermal history.

Vitrinite reflectance was for long the only available maturation indicator and therefore was first to be used for this kind of work.

However in recent years it was progressively realized that if measurements of vitrinite reflectance were easy and very precise thanks to very good normalization and equipments, their interpretation in terms of thermal history were not simple, because vitrinite was not as simple a material as it was thought at first.

This paper is not a complete review of the problems posed. It more modestly aims to bring before the non-specialists who are interested in the utilization of vitrinite reflectance, and more particularly the structural geologists who might be interested in controlling the thermal histories they deduce from their geodynamical models, the conditions which should be realized when using vitrinite reflectance as a control of thermal history of sediments.

2. GENERALITIES ON REFLECTANCE OF SEDIMENTARY ORGANIC MATTER

Reflectance of a material is the ratio of reflected light intensity to incident light intensity. For solid organic matter (S.O.M.) it is measured with a microscope on carefully polished block-sections. In case of an organic rich sediment (coal, oil shale etc...) the polished block-section is made directly from the sediment, but in most cases, the S.O.M. has first to be concentrated by a physical method (flotation or sink-float after crushing of the sample) or a chemical method (most often demineralization by HCl or

HF); then it is included in a matrix (generally a synthetic resin) from which is made the polished block-section.

Practically, reflectance cannot be safely measured on organic particles or areas less than 5 µm in size (Alpern, 1980). Therefore reflectance measurements on S.O.M. will consist of measuring the reflectance on as many convenient particles or areas as possible and realizing histograms such as those of figure 1.

The reflectance is measured by means of a photometer and is expressed as for any material, when the incident light is perpendicular to the polished section (this is of course the usual procedure), by the Fresnel-Beer's formula:

$$R = \frac{(n - N)^2 + n^2 k^2}{(n + N)^2 + n^2 k^2}$$

where n and k are respectively the refractive and absorption indices of the material and N is the refractive index of the immersion medium, at the wavelength of the incident light. Since N, n and k vary with the wavelength, normalized measurements are made with a monochromatic light at 546 nm (green). In Western countries, microscopes with oil immersion objectives are used. The refraction index of the oil is $N = 1.517$. Such measurements should be indicated as R_o (O for oil), in contrast with R_a when the measurement is made in air (dry objective), as it is frequently the case in the Soviet Union.

S.O.M. may be optically anisotropic. This means that n and k vary with the orientation of the polished section with respect to "optical axes" of the S.O.M.; of course such a phenomenon can occur only when S.O.M. contains organized structures, which are preferentially orientated within the sample. This results in an anisotropy of S.O.M. reflectance, with a minimum reflectance R_{min} and a maximum reflectance R_{max}.

Reflectance of carbonaceous materials like S.O.M. is a weak phenomenon. The maximum possible value, about 16% R_o, is encountered for R_{max} of the graphite crystal, which is measured when the light is perpendicular to graphitic layers. Graphite is a strongly anisotropic negative uniaxial crystal, R_{min} being about 0.8% (Ergun 1967). Figures for S.O.M. are mostly between 0.1% and 4% R_o. Precision of measurement is very good with modern equipment and the second decimal is considered as significant.

As for any substance, reflectance of S.O.M. is a function of three basic factors (Rouzaud and Oberlin 1984, Rouzaud 1984):

- the nature of the atoms composing the substance, mainly C, H, O in case of S.O.M.;

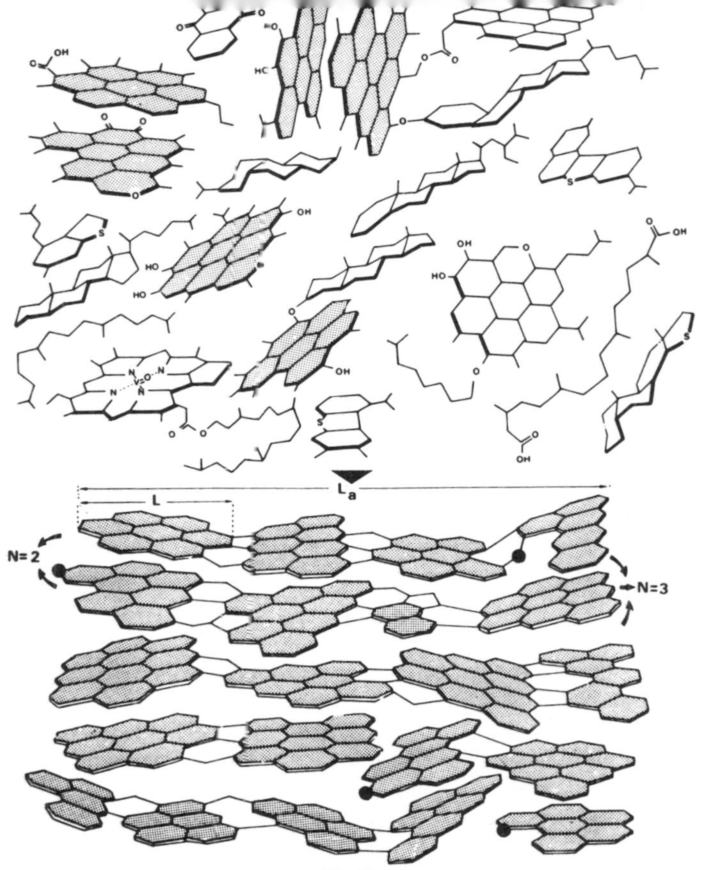

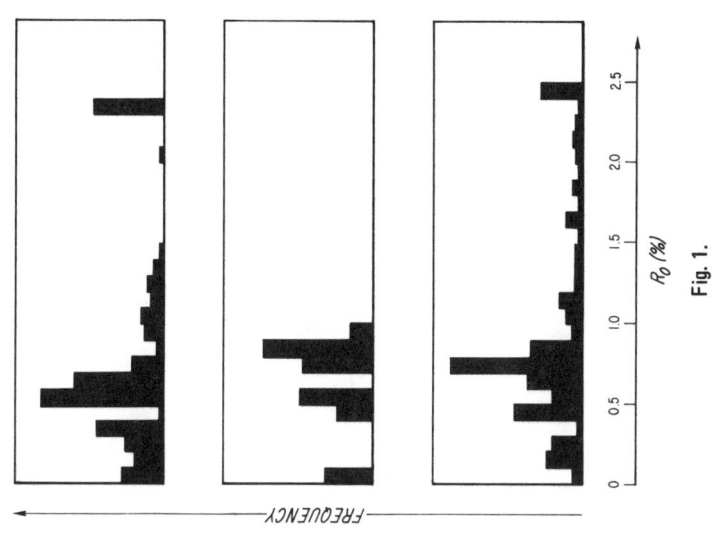

Fig. 1.

Examples of histrograms of reflectances of S.O.M. particles, realized after crushing of samples, concentration by a physical method, inclusion in a matrix, then polishing.

Fig. 2.

Model of S.O.M. structure (S.O.M. from the Lower Toarcian Shales of the Paris Basin and Posidonienschiefer of Western Germany), after Oberlin et al. 1980.
2a: End of diagenesis.   2b: Beginning of metagenesis.

Note the existence of BSUs, who have a strong intrinsic reflectivity and are anisotropic because they behave like negative uniaxial crystals. At the beginning of metagenesis S.O.M. is mainly made of BSUs, which have close orientations within small domains.

- the way the atoms are linked together, i.e. the chemical structures;

- the way the chemical structures are spacially disposed, i.e. the microtexture. This last factor rules mostly the anisotropy of reflectance.

It may also depend on the compaction of the sample, which depends on the mechanical stresses in the sedimentary column.

Broadly speaking, reflectance of S.O.M. increases when O/C and/or H/C decrease. A very important feature is the existence in S.O.M. of very small polyaromatic layers less than 10 Å in size, made of a few fused aromatic rings (Oberlin et al 1980). They are called basic structural units (B.S.U.s)(fig. 2). Such structures have a strong intrinsic reflectance because their bear delocalised electrons like metals do and are the main responsible for anisotropy of reflectance, when they are not randomly orientated, because they behave like very small negative uniaxial crystals.

When maturation proceeds, S.O.M. of a given elemental composition follows a maturation track (Durand and Espitalié 1973, Tissot et al 1974).

In a first step, called diagenesis, S.O.M. looses first preferentially O to produce CO_2 and H_2O (see fig. 3). In the same time, BSUs begin to build up. They are randomly distributed (fig. 2a) (Oberlin et al 1980), so there is no anisotropy of reflectance at the scale of a S.O.M. particle.

The overall effect is nearly no increase of reflectance if S.O.M. particles were initially H rich (and therefore O poor) and a weak increase of reflectance if S.O.M. particles were initially H poor (and therefore O rich).

In a second step, S.O.M. looses preferentially H to produce hydrocarbons. This is catagenesis. B.S.U.s go on building up while their concentration increases in the organic matter, owing to the departure of mobile products, mainly hydrocarbons. They are still randomly distributed. However S.O.M. may be plastic at this stage, and in response to mechanical stress due to sediment loading, B.S.U.s may be preferentially orientated and therefore a weak anisotropy of reflectance (flow anisotropy) may exist. The overall effect of catagenesis is mainly a strong increase of reflectance, which is as the strongest and the latest as S.O.M. was initially H rich.

At the end of catagenesis, S.O.M has become a H and O poor carbonaceous residue mainly made of B.S.U.s. These B.S.U.s reorganize themselves in small clusters (50 to 500 Å in size according to the initial elemental composition of S.O.M.), where their $\vec{C}$ axis have similar orientation (fig. 2b). However, there is no fusion of B.S.U.s together, i.e. there is no crystal growth. In the same time S.O.M. goes on loosing slowly H and O. Orientated pressures (lithostatic pressure) result in an orientation of most clusters of BS.U.s which is parallel to the bedding planes of the rock. The overall effect will be the development of reflectance anisotropy, which progressively becomes very strong, and an increase of the mean reflectance.

As a whole, reflectance of S.O.M. in the sedimentary column increase continuously with maturation, from values less than 0.5% to values about 4% for non metamorphised sediments. Higher values are generally found in metamorphism.

It is of paramount importance, when using this increase to calibrate maturation or thermicity of sedimentary basins, to realize that the values of reflectances and the shape of the reflectance increase vary considerably with the maturation track of S.O.M. (however, reflectances of the different maturation tracks will more or less converge as maturation proceeds, as elemental analysis do in the H/C O/C diagram of figure 3).

Therefore:

1 - Any calibration curve has to be made with particles which followed the same maturation pathway. This means primarily that they had initially (at the recent sediment stage) close elemental composition (Rouzaud 1984).

2 - Using a calibration curve in series makes sense only if the particles examined in this series had also the same initial elemental composition than those used to establish the calibration curve. Also it should be noted that increase of reflectance is not linear with depth, even if the geothermal gradient was constant and linear through the whole geological history. It is made of three successive parts, corresponding to departure of oxygen (diagenesis) departure of hydrogen (catagenesis) and microtextural reorganization of BSUs (metagenesis).

During catagenesis, reflectance values are mostly a function of the H/C ratio of S.O.M. Indeed H and C will be in most cases the two important atoms quantitatively and H/C expresses in a first approximation the aromaticity of S.O.M., which relates closely to concentration of BSUs in the chemical structures. Moreover, anisotropy is still not developed except possibly a weak flow anisotropy (see above). Therefore at a same stage of catagenesis, the scattering of reflectances of S.O.M. particles such as those recorded on figure 1 reflect primarily the dispersion of the H/C ratios of these particles: the lower the reflectance, the higher is the H/C ratio. These particles of course belong to a large variety of maturation tracks.

REFLECTANCE OF VITRINITE AS A CONTROL OF THERMAL HISTORY OF SEDIMENTS

Calculating a mean reflectance value for such histograms, which represent measurements done on different populations of particles located on a variety of maturation tracks makes no sense. Using such histograms requires the capacity of selecting particles belonging to a well defined maturation track, i.e. to a well defined submaceral (see chapter 3 for the definition of macerals). This may be impossible. Diagnosis may be considerably improved by special care given to the preparation of samples (R. Bertrand et al. 1985). Such facts make hazardous in many cases the use of reflectance for maturation or thermometric studies, particularly when made by unexperienced people.

Indeed we will see hereunder that good conditions can be reasonably realized for the moment for humic coals and more generally for S.O.M. derived from higher plants only.

Difficulties are even greater when a strong anisotropy exists, i.e. for samples having reached the metagenesis zone. Indeed:

– diagnosis of particles will be made more difficult due to vanishing of criteria of recogniscence;

– it will be hard to evaluate the respective part of elemental analysis and anisotropy in the dispersions of measurements;

– it will become, even for particles which have been proven to belong to a definite maturation track, more difficult to calculate a mean reflectance value. Hevia (1977) have calculated what this mean reflectance should be theoretically. It is expressed by:

$$R_m = \frac{2 R_{max} + R_{min}}{3}$$

R_{max} and R_{min} being measured under polarized light with a circular stage.

Therefore reflectance measurements, when anisotropy is strongly developed, i.e. for $R_o > 2\%$, will be more difficult to use for maturation or thermometric studies. This is the case of deep samples. But there may be in shallow samples particles reworked by the erosion of strata having previously been deeply buried. The risk is all the greater as the sediment is the poorer in S.O.M. and lean samples should not been taken for such studies. In particular, sandstones should be avoided, because it is very likely that their organic content is mostly reworked. Moreover, in sandstones, the organic material is also very likely to be weathered by water circulation at the time of deposit and sometimes during burial.

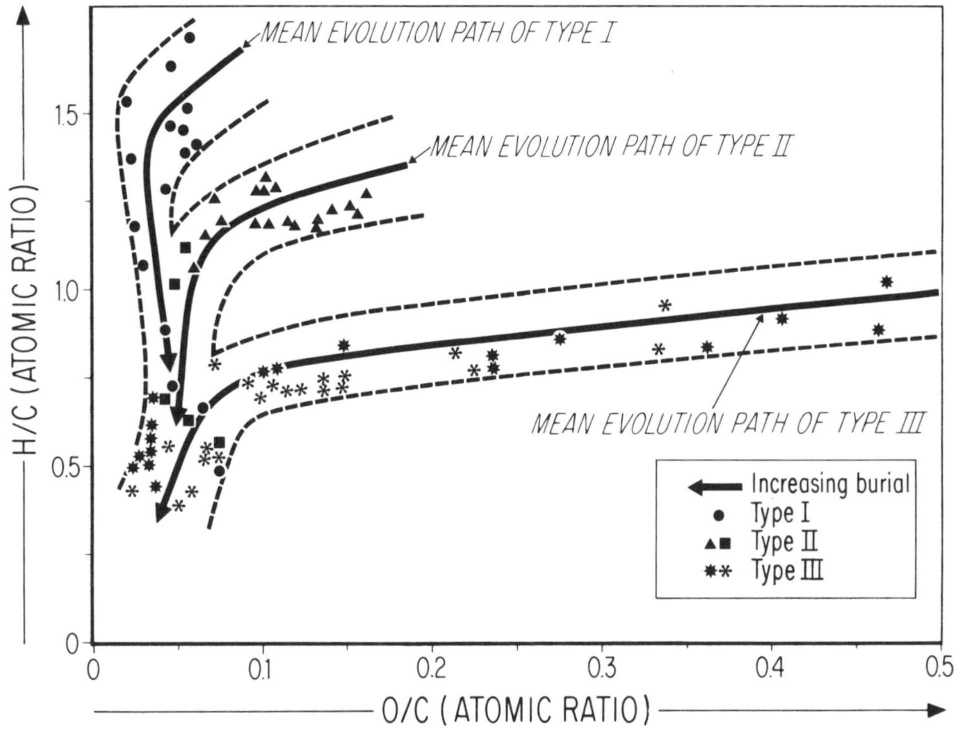

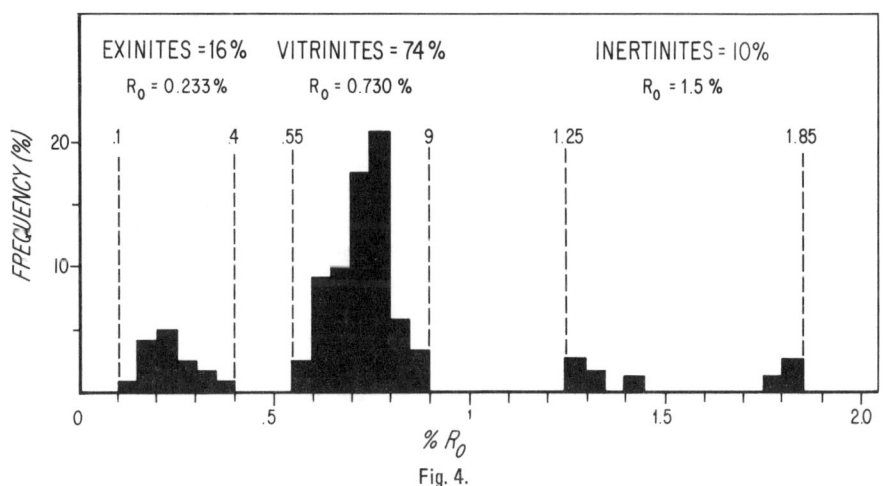

Fig. 3.

Maturation tracks in a H/C versus O/C diagram of three types of S.O.M. having initially very different elemental compositions, after Durand and Espitalié 1973 and Tissot et al. 1974. Note the convergence of the maturation pathways towards pure carbon.

Fig. 4.

Histogram of reflectances realized on a paleozoic humic coal at a low maturation stage (beginning of catagenesis).

- anisotropy is largely a result of mechanical stresses, thus reflectance will no longer depend on temperature and time only.

3. SEDIMENTARY ORGANIC MATTER DERIVED FROM HIGHER PLANTS

S.O.M. derived from higher i.e. terrestrial plants rich in ligno-cellulose tissues, are often called terrestrial or continental S.O.M. They occur in a concentrated form (humic coals) or a dispersed form (in clays). Typical series containing this S.O.M. are deltaïc and parallic series.

3.1. Humic coals

Humic coals are nearly pure S.O.M. Polished block sections can be made directly and easily and their is no problem in making histograms of reflectances on optically homogeneous areas such as the one of figure 4.

This histogram, which corresponds to a maturation stage at the beginning of catagenesis, is neatly broken down in three populations which do not overlap each other and which can also been distinguished by a variety of criteria. These populations are said to be maceral families (see table I).

- The low reflectances belong to the maceral family of liptinites. Liptinites are made of the following macerals (table I):

. alginite, derived from recognizable unicellular algae;

. exinites (the only one to be present on fig. 4) which can be recognized, mainly on the basis of morphological criteria, to be derived from outer parts of higher plants: spores, pollens, cork cell walls, leaves cuticles;

. unfigured liptinites (bituminites), which are amorphous;

. resinites, derived from resins of trees.

Liptinites make generally 5-15% of humic coal samples.

- The medium reflectances correspond to the maceral family of vitrinites which make up 60 to 80% of humic coal samples of the Northern Hemisphere. Vitrinites are gels or gelified tissues which are mostly derived from the decay of ligno-cellulose parts of higher plants. But this can rarely be deduced from morphological criteria because vitrinites looks mostly amorphous. Some vitrinites might be derived from material other than ligno-cellulose, at least partly (saprovitrinite and desmocollinite, see table I). At the recent sediment stage (peat stage), precursors of vitrinites are called huminites.

- The high reflectances correspond to the maceral family of inertinites. It consists of reworked, oxidised or carbonized particles whose precise origin is generally not recognizable. Oxidation may result for instance from dessication or from forest fires.

Inertinites are particularly abundant in coals from the Gondwana (Australia, India etc...), which therefore are less rich in other macerals.

The three families are not always clearly separated as on figure 4. Problems of diagnosis may also occur, particularly as concerns the distinction between vitrinites and inertinites, because of the frequent lack of morphological criteria in these two families. The difficulty increases when maturation proceeds, because of the merging of the optical properties of the different families. Distinction of liptinites is made easier by using UV illumination, which results in fluorescence. Inertinites do not fluoresce and some vitrinites fluoresce weakly. However, fluorescence vanishes after mid-catagenesis.

Table I

Main macerals of coal (after ICCP)

Decreasing H/C, increasing reflectance				
	LIPTINITES		Alginites	
			Exinites (spores, cuticles, cell walls of cork etc...)	
			Bituminites	
			Resinites*	---Transition L-V
	VITRINITES	Mostly Fluorescent gels	Saprovitrinites*	
			Desmocollinite (vitrinite B)	Vitrinite sensu stricto
		Mostly non-fluorescent tissues	Telocollinite (vitrinite A)	
			Telinite	
			Pseudovitrinite	---Transition I-V
	INERTINITES		Reactive semifusinite	
			Inert semifusinite	inertinite sensu stricto
			Degradofusinite	
			Pyrofusinite	

* Resinites and saprovitrinites are still under discussion in the ICCP working groups.

Dispersion of reflectances within the vitrinites family is always mostly present. This reflects, as explained in chapter 2, a variability of their H/C ratio and also anisotropy when metagenesis is reached. Such a variability of H/C has several causes, for instance:

- variations in the kind of components whose decay produced vitrinite;

- variations in the environmental conditions during coal formation, and particularly redox potential, bacterial activity, and length of transportation of vegetal debris.

Coal Petrographers have recognized many categories of vitrinite (see Stach's textbook of coal petrology, Alpern 1980 and table I), but make basically the distinction between to main kinds: Vitrinite A (also called telocollinite or homocollinite), which is a cryptotissue completely gelified, and vitrinite B (desmocollinite), which derives from a micropudding of very small particles embedded in a true gel or matrix.

Reflectance of vitrinite B is lower than reflectance of vitrinite A at the same maturity.

Differences in the parent vegetation may also result in different vitrinites. For instance Sittler (1979) found, for the coal series he studied, lower reflectances in upper Cretaceous and Tertiary coals derived from resinous trees than in Paleozoïc coals, for a same maturation stage.

Such differences result in a variety of maturation pathways. Nevertheless, vitrinites of humic coals had initially relatively close elemental analysis and the petrographer has the capacity to select them in the histograms of reflectance. Therefore they are a priori a suitable material for a calibration of maturation in many cases.

However the distinctions between the different kinds of vitrinites are not straitghforward, specially when the preparation is made from powdered samples where the spacial relations between the macerals can no longer be examined. Moreover the vitrinite histogram is generally considered as a whole and a mean value calculated, noted $\bar{R}_o$ or R_m. But because the variability of vitrinites is not negligible this results in fairly large uncertainties on the maturation stage. Petrographers are now conscious of the interest in making distinctions between the different kinds of vitrinite, so as to have a better accuracy in maturation studies.

Bad preparation of the samples or differences of preparation from one laboratory to another will also contribute to increase the uncertainty. Figure 5 shows for instance the modification in reflectances due to differences in the quality of polishing.

REFLECTANCE OF VITRINITE AS A CONTROL OF THERMAL HISTORY OF SEDIMENTS

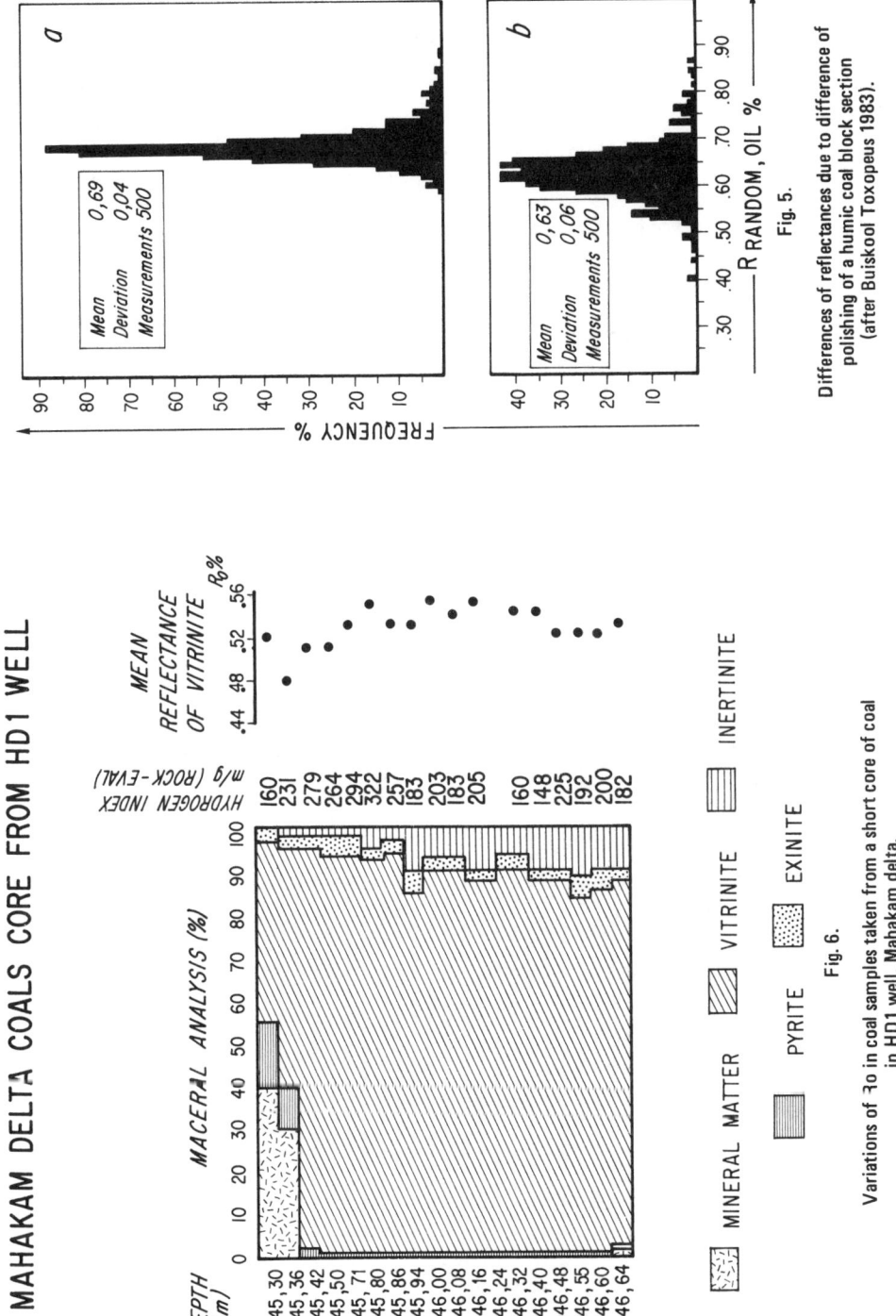

Fig. 5. Differences of reflectances due to difference of polishing of a humic coal block section (after Buiskool Toxopeus 1983).

Fig. 6. Variations of R_o in coal samples taken from a short core of coal in HD1 well, Mahakam delta.

An interesting discussion of these problems (differences in the vitrinites of humic coals and effects of preparation) can be found in Buiskool Toxopeus (1983) and Price and Barker (1985), together with a more general criticism of the use of vitrinites.

Figures 6 and 7 show the example of the Miocene coals of the Mahakam delta (Indonesia). These coals originate in an equatorial forest of resinous trees. Measurements made regularly on a core (fig. 6), therefore at a same maturation stage, show variations of R_o from 0.48 to 0.56%, i.e. 15% of the mean value. This variation is likely to be due to slight differences in sedimentation conditions. Figure 7 shows the variation of R_o with depth for a well of the Mahakam delta. Histograms of vitrinites are narrow. The large number of available coals compensate the scattering of R_o mentioned above. A precise trend of increase of maturation at depth is recorded.

Figure 8 shows another example which is the Gironville 101 well in the Paleozoïc of Lorraine (France). Histograms are larger. Nevertheless a fairly good curve can be established of the variation of R_o with depth, although the number of samples is insufficient to make a very good work.

In this case, two kinds of vitrinite, A and B, may sometimes be distinguished. It can be seen that their maturation tracks are slightly different. It is likely that vitrinites from the Mahakam delta coals, because they derive from resinous trees, are more hydrogenated than those from Gironville 101 well and therefore have a slightly lower reflectance at the same maturation stage, if we follow Sittler.

Gironville 101 is a deep well where the increase of reflectance with maturation can be followed on a large scale. Over 2% of reflectance, a strong anisotropy is observed, with Rmax reaching nearly 6.5% at bottom well (5875 m).

3.2. Dispersed organic matter

Terrestrial S.O.M. (derived from higher plants) occurs also in clays from coal measures and from deltaïc and parallic series. The S.O.M. contents are often low and S.O.M. is concentrated by physical or chemical methods. Physical methods result in partial recovery of S.O.M. and chemical methods might modify slightly the optical properties, although this was never clearly demonstrated. The recovered S.O.M. is included in a matrix, and carefully polished.

Figure 9 shows the variations of R_o at depth in such clays for the Gironville 101 well, compared to that of coals. S.O.M. of clays was concentrated by a sink-float method.

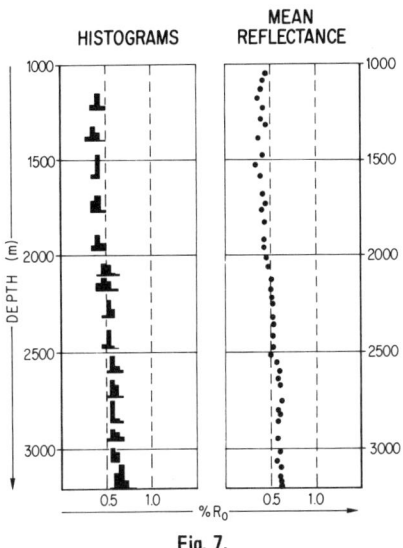

Fig. 7.

Variations of Ro at depth in coals from a well of the Mahakam delta. After Oudin 1984.

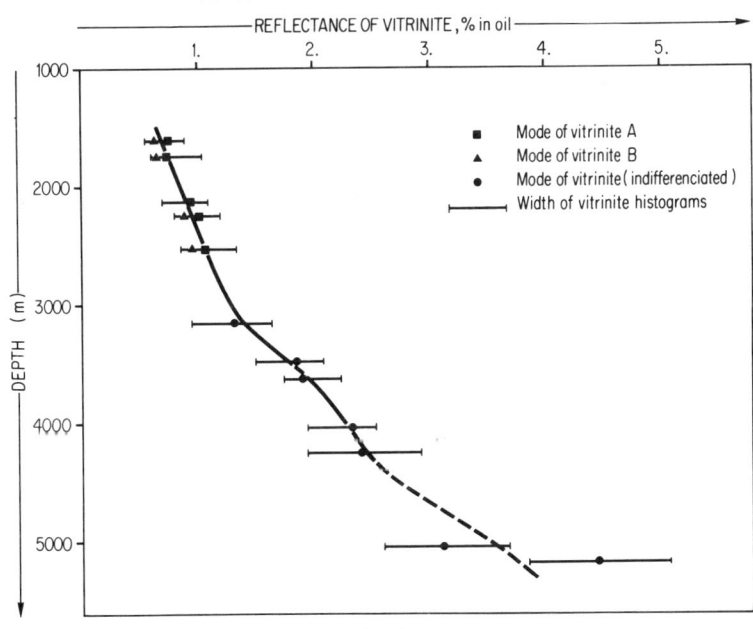

Fig. 8.

Variations of Ro at depth for the coals of the Gironville 101 well, Paleozoïc, Lorraine, France.

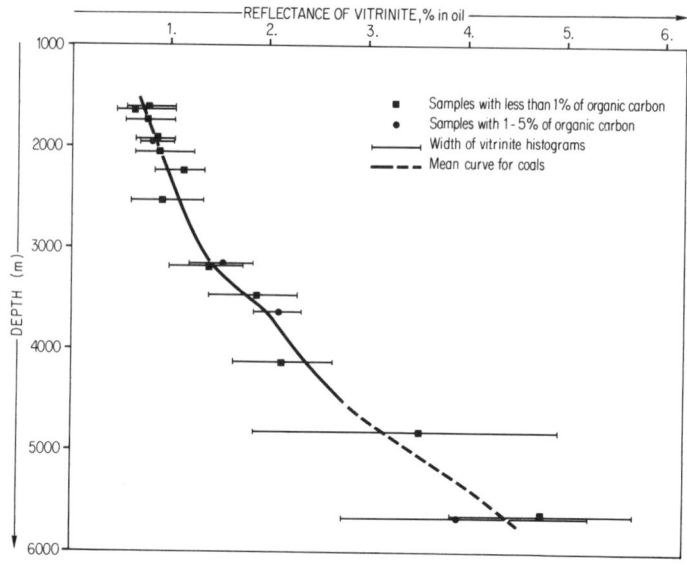

Fig. 9.

Variations of Ro at depth for SOM from clays of the Gironville 101 well.

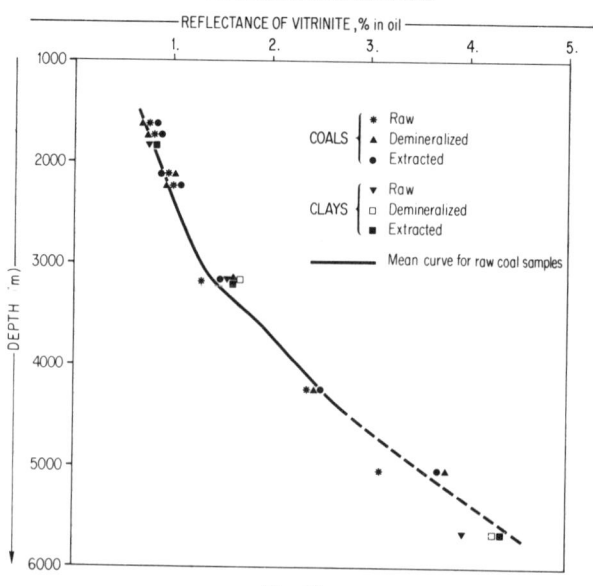

Fig. 10.

Comparison of Ro for raw, CHCl₃ extracted and demineralized samples in the Gironville 101 well.

A good agreement is observed here between clays and coals. However histograms are larger and mean reflectance values of vitrinites are more scattered for clays than for coals. This means that elemental analysis of "vitrinites" is more variable and scattered in clays than in coals. Anisotropy is also likely to be much stronger in clays, even at low maturity stage (flow anisotropy) because S.O.M. being a minor constituent, it will be more easily orientated by mechanical stresses at the contact of mineral surfaces. Also, clays being poor in S.O.M., the relative contribution of reworked material is likely to be relatively important.

Figure 10 shows a comparison of R_o of raw coals and clays from Gironville 101 well with R_o of the same samples after $CHCl_3$ extraction and demineralization by HCL and HF. Obviously chemical treatments have here no influence on R_o values (see also table II).

An example of coal measures where "vitrinites" of clays do not behave like vitrinites of coals was provided by Buiskool Toxopeus (1983). Figure 11 shows that in the studied well, vitrinites from clays (vitrinite 2) follow a maturation track which do not parallel that of vitrinites from coals (vitrinite 1). Coals may also contain vitrinite 2 (fig. 12). The difference between maturation tracks is here larger than for vitrinite A and B of Gironville coals. According to Buiskool Toxopeus, vitrinite 1 and vitrinite 2 correspond to telocollinite and desmocollinite, like vitrinite A and vitrinite B do in Gironville. However, the larger differences in the maturation, tracks of vitrinite 1 and vitrinite 2 indicate that the differences in H/C are larger between these vitrinites than between vitrinite A and vitrinite B of Gironville at a same maturation stage. Such large difference seems to great and might indicate that vitrinite 2 is not desmocollinite (bituminite ?).

We will conclude that, although reflectance studies in coals and clays containing SOM derived from higher plants provide a precious tool for the follow-up of maturation, calibration is not straightforward and needs further studies which can be made only by petrographers who have a clear conscience of the problems to be solved. In the present state of the art, there are rather large uncertainties on the determination of maturation stages, even when using only coals. The uncertainties may increase considerably when using dispersed organic matter and is all the greater as the organic content of the sediment is low.

Difficulties are increased, sometimes considerably, when only cuttings are available (and this is now the most frequent case), owing to casings and pollution by the drilling muds.

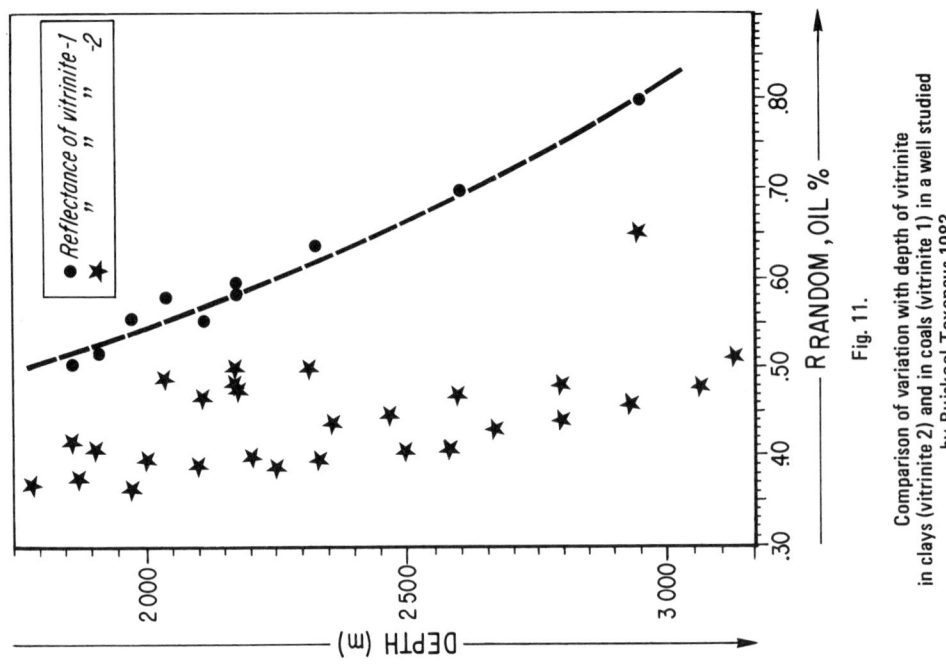

Fig. 11. Comparison of variation with depth of vitrinite in clays (vitrinite 2) and in coals (vitrinite 1) in a well studied by Buiskool Toxopeus 1983.

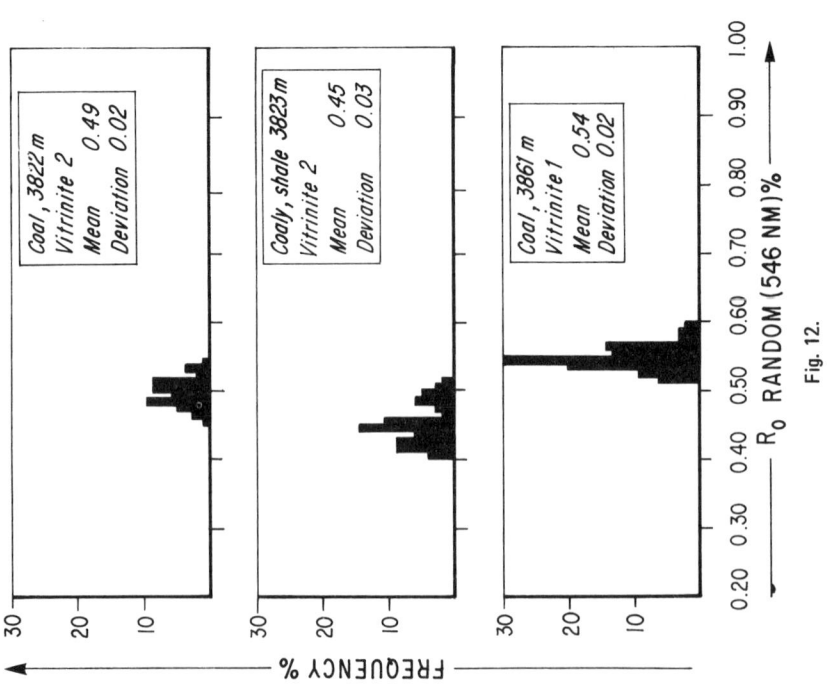

Fig. 12. Comparison at a same maturation level of the reflectance of "vitrinite" in a clay (vitrinite 2) and in coals (vitrinite 1 and vitrinite 2). After Buiskool Toxopeus 1983.

Table II

COALS AND CLAYS FROM GIRONVILLE 101 BOREHOLE.
INFLUENCE OF DEMINERALIZATION AND SOLVENT EXTRACTION ON REFLECTANCE OF VITRINITE.

1. INITIAL SAMPLE.
2. AFTER DEMINERALISATION BY HCL AND HF.
3. AFTER DEMINERALIZATION, THEN $CHCl_3$ EXTRACTION.

SAMPLE DEPTH (m)		1609	1738	2115	2240	3152	4262	5036	1738	3152	5675
1	R_m, %	0.73	0.79	0.94	1.0	1.3	2.4	3.1	0.79	1.53	3.9
	DISPERSION	0.55-0.9	0.6-1.05	0.7-1.1	0.8-1.2	0.95-1.65	1.95-2.85	2.6-3.7	0.55-1.05	1.2-1.85	2.75-5.25
2	R_m, %	0.72	0.71	0.98	0.93	1.6	2.45	3.8	0.8	1.66	4.2
	DISPERSION	0.5-0.9	0.55-0.9	0.7-1.2	0.7-1.2	1.3-1.85	1.9-3.05	3.1-4.6	0.5-1.1	1.25-2	3.55-4.85
3	R_m, %	0.77	0.82	0.89	1.04	1.5	2.5	3.7		1.6	4.3
	DISPERSION	0.5-1.0	0.55-1.1	0.7-1.05	0.85-1.3	1.2-1.75	1.95-3.05	2.75-4.6		1.15-1.85	3.6-4.95
		COALS							CLAYS		

459

4. "VITRINITES" OF SEDIMENTARY ORGANIC MATTER DERIVED FROM MARINE OR LACUSTRINE BIOMASSES

In shales, marls, carbonates, organic matter when present is generally derived from a planctonic biomass, aquatic or lacustrine. This is also the case of all kinds of preDevonian sediments, just because higher plants did not exist yet.

Such S.O.M. contain particles which under the microscope look more or less like vitrinites of coals. However, contrarily to coals, they make up only a minor part of S.O.M.

These "vitrinites" were at first identified to vitrinites of coals and therefore used for maturation studies. It might be in some cases that they are really derived from higher plants material and therefore are a small "terrestrial contribution" to S.O.M.

In most cases however these "vitrinites" are not derived from higher plants and have not the same initial composition as vitrinites of coals. Therefore they do not follow the same maturation track. There was for a long time a deep misunderstanding between petrographers and geochemists: geochemists did not understand that vitrinite meant mainly for a petrographer a general aspect of the particles and did not imply any definite initial chemistry and petrographers did not understand that "vitrinites" could not be used for a calibration of maturation if they had not all initially the same elemental composition.

"Vitrinites" of marine or lacustrine O.M. when they are not a terrestrial input are not true vitrinites and it would be preferable to call them bituminites, although they are not identical to bituminites of coals. They are generally richer in hydrogen than vitrinites of coals and therefore will have lower reflectances at the same maturation stage. Their reflectance will also increase very slowly down to great depths then increase abruptly after mid-catagenesis to join that of vitrinites of coals. In general, the richer in hydrogen is the organic matter as a whole, the richer in hydrogen will be its bituminite. Moreover, different populations may exist within the same sample.

Therefore there is a variety of possible maturation tracks, corresponding to the variety of bituminites whose corresponding calibration curves have not been established yet. If it had been done, it would be impossible anyhow to choose which one should be used for a given sample.

Another problem is the possible existence of true vitrinites impregnated by oily material; this results in a lowering of reflectance (Teichmüller and Ottenjahn 1977) at a same maturation stage.

REFLECTANCE OF VITRINITE AS A CONTROL OF THERMAL HISTORY OF SEDIMENTS

Because in any case the observable particles make such a small amount of the S.O.M., cavings, reworked material, pollution by the drilling muds, and existence of "solid bitumens", which are products of the thermal degradation of S.O.M. with maturation and resemble vitrinites but have a different chemistry, make often the search for an hypothetical "true" vitrinite even more difficult.

Some aspects of these problems were recently discussed by Price and Barker (1985).

An example of marine S.O.M. is the Lower Toarcian Shales of the Paris Basin, which are organic-rich marls. Three populations R_1, R_2 and R_3 of particles are observed (fig. 13), whose reflectance do not increase significantly at depth, while other parameters indicate that mid-catagenesis is approached in the deepest samples. Reflectance of coaly vitrinite would increase in these conditions from 0.3 to 0.8% approximately.

A bituminite with a very low reflectance is found in the Green River Shales from the Uinta Basin (U.S.A.) (Deroo et al, 1978). Its reflectance is very much lower at a same maturation stage than that of vitrinite of the coal layers encountered in this basin.

A demonstration of the different chemistry of "vitrinites" and "bituminites" according to the sediment type is given on figure 14 (after Alpern et al, 1978).

Three immature samples of S.O.M., isolated from their mineral matrix, were heated in an inert atmosphere from room temperature to 600° C at 4° C/min, and reflectance of the available particles measured regularly. These samples were :

- a terrestrial S.O.M. (Logbaba), containing true vitrinite from higher plants;

- a marine S.O.M. (Lower Toarcian Shales of the Paris Basin);

- a lacustrine S.O.M. (Green River Shales from the Uinta basin).

Reflectance was also measured on the bulk of S.O.M., which in the case of Logbaba is a vitrinite like material.

For the L.T.S. sample, only evolution of bituminite population R_1 was followed.

Obviously vitrinites and bituminites behave differently with heating for the three samples. Bituminites of LTS and GRS have low reflectances (this indicates hydrogen richness) which do not evolve up to high temperatures. Then they increase abruptly to join the reflectance of the vitrinite of the Logbaba sample. Such patterns are also those encountered in natural samples.

461

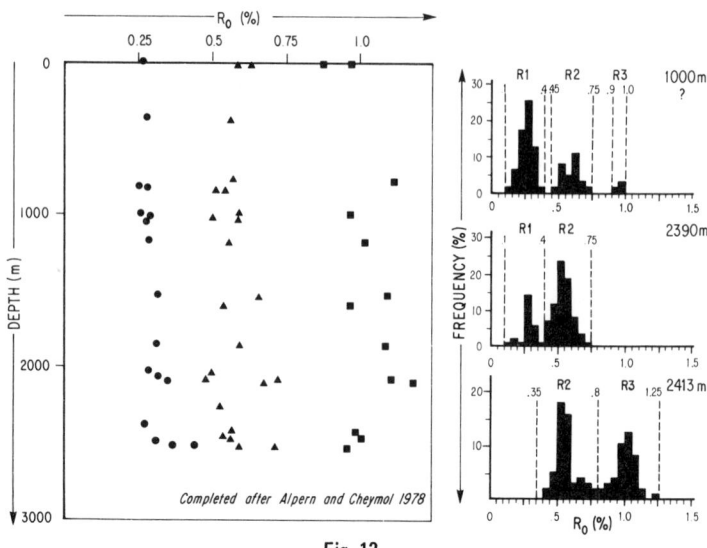

Fig. 13.

Different populations of "vitrinites" in the Lower Toarcian Shales of the Paris Basin and variation of their reflectance at depth. Completed after Alpern and Cheymol 1978.

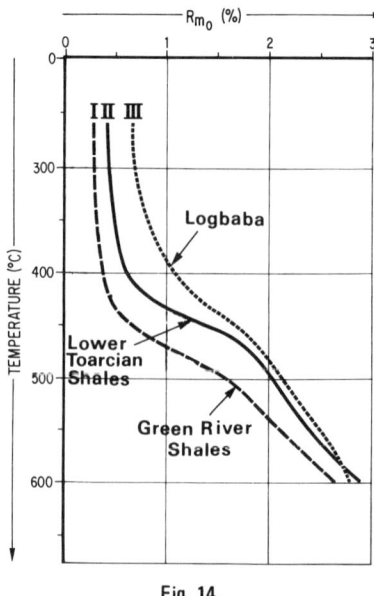

Fig. 14.

Evolution of the reflectance of "vitrinites" with heat treatment temperature on a heating experiment of immature samples of SOM from: Logbaba series, Lower Toarcian Shales of the Paris Basin, Green River Shales. After Alpern et al 1978.

REFLECTANCE OF VITRINITE AS A CONTROL OF THERMAL HISTORY OF SEDIMENTS

The conclusion is that "vitrinite" in marine or lacustrine sediments may be a suitable material for maturation and thermometric studies when it can be proven it is true vitrinite derived from higher plants that petrographers have the capacity to select in histograms such as the one of figure 1, which is typical of many of the series. In most cases it is not and the available particles are bituminites having a vary variable chemistry which makes them unreliable for a calibration of maturation. In any case, pre-Devonian sediments do not contain vitrinites.

The uncertainties in determining "vitrinite" reflectance in such series is illustrated by the very large scattering of "R_m" values produced by different laboratories when working on the same samples (Alpern 1980, in Kerogen, B. Durand ed., MOD ring analyses, p. 356 and Demkicki 1984). This scattering reflects primarily the choices made by the different laboratories in what they consider to be the true vitrinite in large histograms. Therefore it is highly advisable, for marine and lacustrine series, to rely not only on reflectance and to use other methods. Some of these methods are petrographical ones, such as the follow-up of fluorescence spectra of spores and algae (Teichmüller and Ottenjahn 1977).

5. VITRINITES REFLECTANCE AND THERMAL HISTORY OF SEDIMENTS

As explained above, the use of vitrinite reflectance as a control of thermal history of sedimentary basins is reliable only when based on vitrinites of coal beds and, but less safely, on vitrinites of terrestrial S.O.M. in clays having an organic carbon content higher than 0.5%. Even in these series, work is complicated when anisotropy becomes strong, i.e. over 2% R_o.

These conditions are rather restrictive and it is likely that a large part of the published work should be reevaluated, with a better confidence in work on coal measures, now that problems are better understood.

Recording vitrinite reflectances in terrestrial series provide an evidence of the variety of thermal regimes, as shown for instance by figure 15 where three coal series and a series of clays are compared. Although the vitrinites might be slightly different from one series to another, no doubt that we have here four clearly different thermal regimes.

Classical use of vitrinite reflectance for the study of thermal regimes are those by M and R Teichmüller on the Bramsche Massiv (fig. 16) and on the Rhine Graben (fig. 17 and 18).

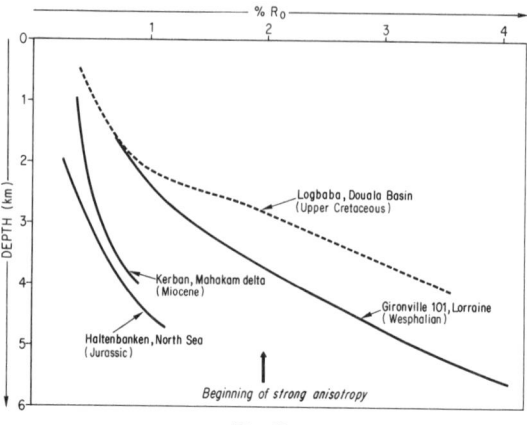

Fig. 15.

Vitrinite reflectance versus depth in four series containing terrestrial organic matter. In three cases (Kerbau, Haltenbanken, Gironville), vitrinite reflectance is recorded on coals. For the last one, Logbaba, it is recorded on SOM of clays, because there is no coal. After Alpern 1964 (Gironville), Durand and Espitalié 1976 (Logbaba), Pittion and Gouadin 1984 (Haltenbanken) and Pittion, unpublished results (Kerbau).

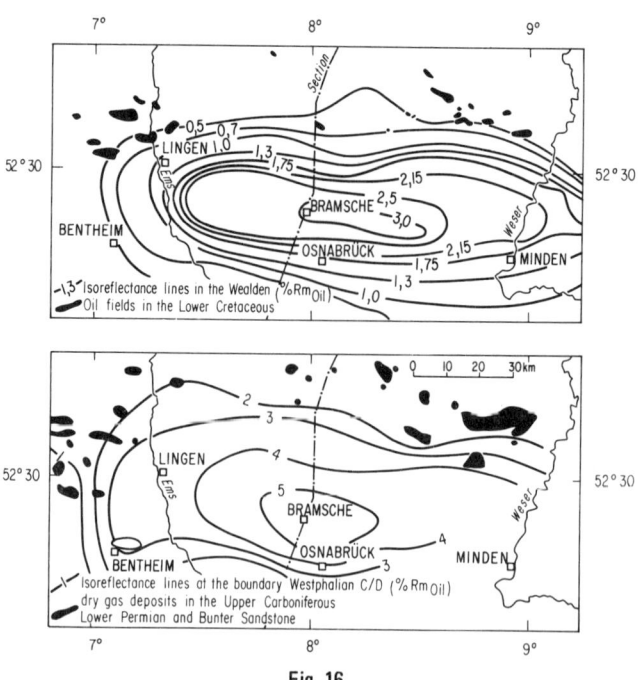

Fig. 16.

"Vitrinite" isoreflectance lines encircling the Bramsche Massiv., Western Germany. After M. and R. Teichmüller 1982 a).

464

REFLECTANCE OF VITRINITE AS A CONTROL OF THERMAL HISTORY OF SEDIMENTS

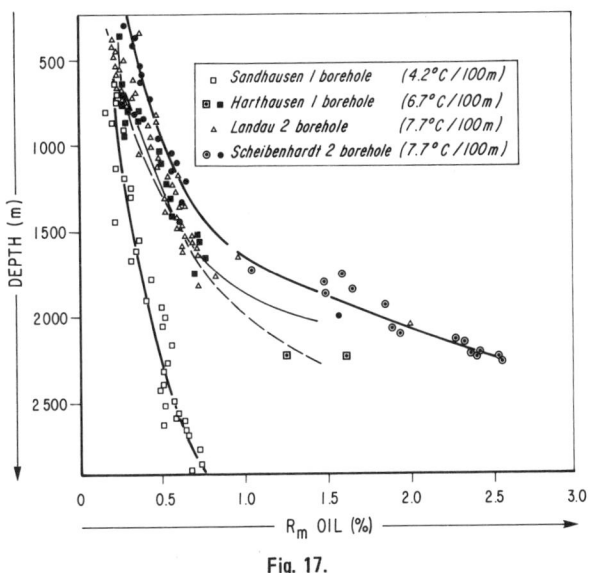

Fig. 17.

Variation of "vitrinite" reflectance with depth in some wells of the Rhine Graben with different present geothermal gradients. After M. and R. Teichmüller 1982 b.

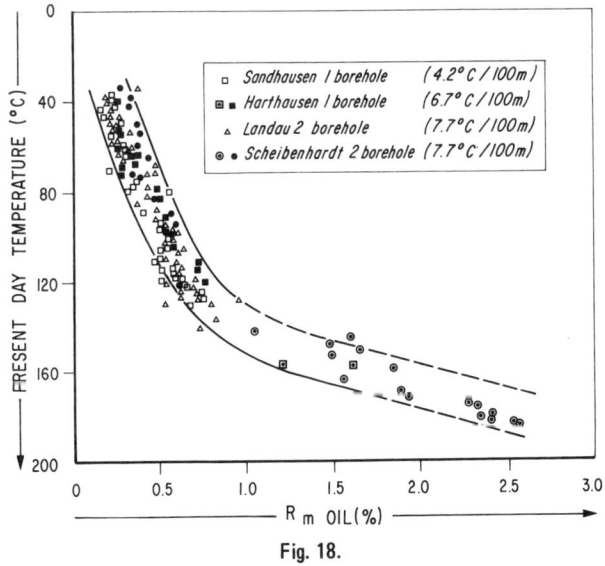

Fig. 18.

"Vitrinite" reflectance versus present day temperatures in wells of the Rhine Graben. After M. and R. Teichmüller 1982 b.

Thermal effects are clearly shown: isoreflectance lines encircling the Bransche Massiv, showing the effect of a deep igneous intrusion, variations at depth fitting the present geothermal gradients in the Rhine Graben. The material used in these two studies was mostly terrestrial clays and partly coals.

A considerable work using vitrinite reflectance, together with other petrographic methods, to reconstruct the thermal history of sedimentary basins is also that of P. Robert, who recently made a synthesis of his studies (P. Robert 1985).

When made on a suitable material, vitrinite reflectance studies allow to detect precise thermal phenomena, as shown by the study of Mahakam delta coals (fig.19 , 20, 21). In the Mahakam delta, the quality of the material permitted to establish precise iso-reflectance lines and to compare them with structural markers (Oudin and Picard, 1982). Therefore it was made possible to discuss the nature of thermal transfers in the delta and to suspect a heat transfer by fluid movement from the synclines toward the top of the structures.

Recording evidences of different thermal regimes is one thing, translating them in terms of a precise temperature history is another one.

Dow (1977) and Alpern (1980) assuming that increase of vitrinite reflectance at depth was exponential for a linear geothermal gradient, proposed to plot $\log R_o$ versus depth to evaluate the values of the gradients.

This pragmatical approach is disputable for a physicist, and do not integrate the effect of time, but is practical for a first approximation.

However a correct solution of the problem needs a model of the variation of vitrinite reflectance with temperature and time.

Such a model does not exist yet, because nobody was able to express reflectance mathematically as a function of temperature and time.

A possible way is the making of a tabulation of vitrinite reflectances versus transformation ratios of terrestrial S.O.M. given by a kinetic model of S.O.M. degradation. The first model of this kind was that of Karweil (1956), but the most known of these models are those of Lopatin (1971) and Tissot and Espitalié (1975).

Lopatin's tabulation was calibrated on Münsterland 1 borehole; he supposes that coals are degradated with a single activation energy corresponding to a doubling of the reaction rate every 10° C. This is a crude, but simple assumption. Transformation ratio of S.O.M. is expressed in time/temperature index (TTI).

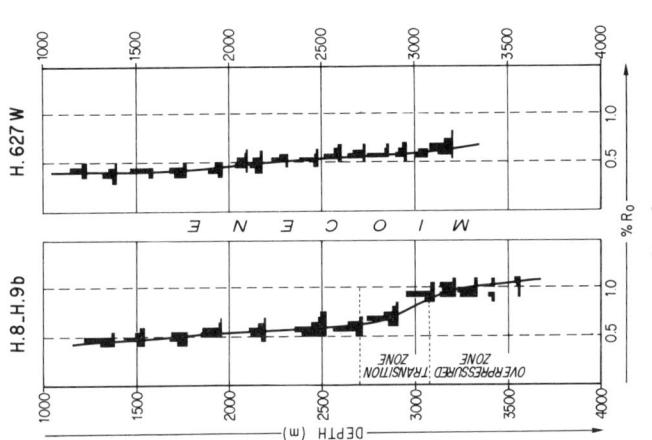

Fig. 19.

Differences in the evolution of coals vitrinite reflectance versus depth for a top well (H9 bis) mid-flank well (H6 27 W) of the Handil structure, Mahakam delta, Indonesia. After Oudin and Picard 1982.

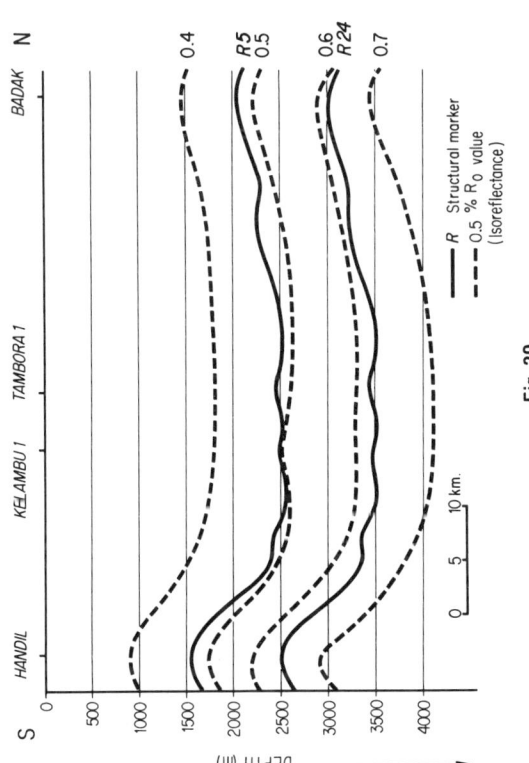

Fig. 20.

Coals vitrinite isoreflectance lines compared at a large scale to structural markers on a cross-section from top of Handil structure to top of Badak structure, i.e. perpendicular to delta progradation axis, Mahakam delta. After Oudin and Picard 1982.

REFLECTANCE OF VITRINITE AS A CONTROL OF THERMAL HISTORY OF SEDIMENTS

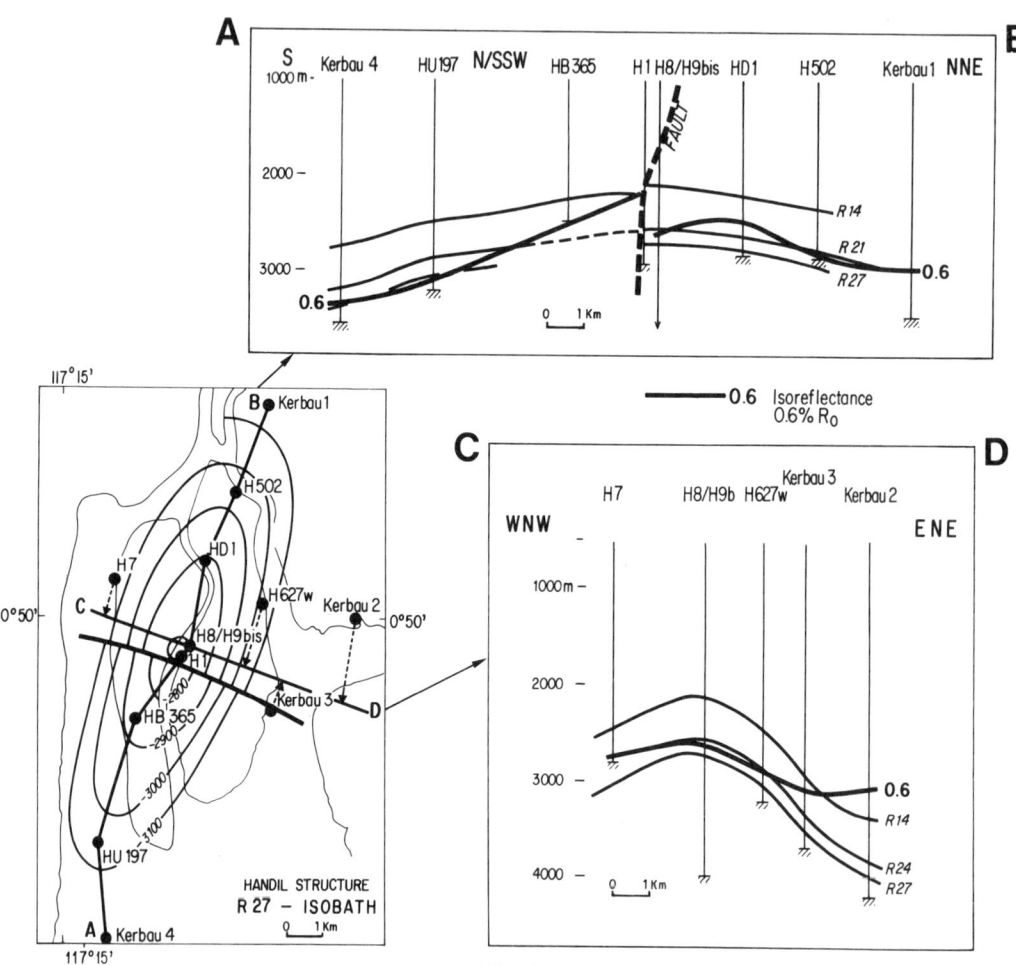

Fig. 21.

Coals vitrinite isoreflectance 0.6% Ro compared to structural markers at the scale of Handil structure, Mahakam delta.

Tissot and Espitalié's model for terrestrial S.O.M. was calibrated with the upper Cretaceous series of clays of Logbaba, Cameroon and on laboratory experiments. It supposes that S.O.M. is degraded into hydrocarbons by a set of six parallel reactions with different activation energies. This is more realistic than Lopatin's model. However this model cannot be used when degradation of kerogen is over, i.e. when reflectance is higher than 2 % R_o.

But the major problem with these models is they are in calibrated on series whose temperature history cannot be known with precision. Therefore the errors made on the assessment of temperature histories of this series are reflected in any further use of the models.

This difficulty will not be encountered when using the OPTIM model (Ungerer et al. 1985), which is a variation of the Tissot and Espitalié's model where the kinetic parameters are derived from laboratory experiments.

6. CONCLUSIONS

There are serious limitations to the use of vitrinite reflectance alone for a control of thermal history of sedimentary basins. They are of two kinds:

(1) Only series containing a fair amount of organic matter derived from higher plants are safely usable and not without discussion: coals are safer than dispersed organic matter. This excludes many series containing marine and lacustrine organic matter, and pre-Devonian series. In such series, other components, or other methods must be used whose some are of petrographical nature (fluorescence for instance). Sandstones, and more generally organic lean sediments must also be avoided, because then organic content is likely to be altered or reworked.

Even in terrestrial series, values over 2% R_o are not easy to interprete in terms of thermal history, because of the existence of anisotropy, but also because they are likely to depend not only on temperature and time, but also on pressure.

(2) In terrestrial series, a pragmatical comparison of thermal regimes is easy, but a direct translation of R_o values in terms of precise thermal history is not possible yet, because of the present absence of a mathematical model of the variation of R_o as a function of temperature and time and a bias must be taken by using kinetic models of terrestrial S.O.M. degradation.

However, vitrinite reflectance remains a precious tool in terrestrial series. Such limitations exist for any kind of geothermometer, organic ones and mineral ones as well. But their conditions of application are generally not seriously delimitated and their use is overstated. Such overstatements are dangerous because they prevent serious work. There cannot be any geothermometer of general use and a severe criticism of each one would be of considerable interest in helping the geologist to choose the best suited to the series he wants to study.

REFERENCES

ALPERN B., 1964
    Un exemple intéressant de houillification dans le bassin lorrain et ses prolongements.
    CERCHAR, document intérieur n° 1492.

ALPERN B. and CHEYMOL D., 1978
    Réflectance et fluorescence des organoclastes du Toarcien du Bassin de Paris en fonction de la profondeur et de la température. Rev. Inst. Franç. Pétrole, 33, 4, 515-535.

ALPERN B., DURAND B. and DURAND-SOURON C., 1978
    Propriétés optiques de résidus de la pyrolyse des kérogènes. Rev. Inst. Fr. Pét. 33, 867-890.

ALPERN B., 1980
    Pétrographie du kérogène.
    In Kerogen, insoluble organic matter from sedimentary rocks, B. DURAND ed.
    Editions Technip, Paris 339-384.

BERTRAND R., BERUBE J.C., HEROUX Y. et ACHAB A., 1985
    Pétrographie du kérogène dans le Paléozoïque inférieur : méthode de préparation et exemple d'application.
    Rev. Inst. Fr. Pét. 40, 2, 155-167.

BUISKOL TOXOPEUS J.M.A., 1983
    Selection criteria for the use of vitrinite reflectance as a maturity tool.
    In Petroleum geochemistry and Exploration of Europe, J. Brooks ed.
    Blackwell Scientific Publications, 295-307.

DEMBICKI H. Jr. 1984
    An interlaboratory comparison of source rock data.
    Geochim. et Cosmochim. Acta, 48, 2641-2649.

DOW W.G., 1977
    Kerogen studies and geological interpretations.
    Journ. Geochemical Exploration, 7, 79-99.

DURAND B. et ESPITALIE J., 1973
    Evolution de la matière organique au cours de l'enfouissement des sédiments.
    C.R. Acad. Sci. Ser. D. 276, 2253-2255.

DURAND B. and ESPITALIE J., 1976
   Geochemical studies on the organic matter from the Douala Basin (Cameroon). II - Evolution of kerogen.
   Geochim. Cosmochim. Acta 40, 801-808.

ERGUN (1967)
   Determination of longitudinal and transverse optical constants of absorbing uniaxial crystals. Optical anisotropy of graphite.
   Nature 6B, 14, 135-136.

HEVIA RODRIGUEZ V., 1977
   Le concept de la réflectivité moyenne statistique des substances anisotropes. Relations avec le rang des anthracites et de la matière organique dispersée dans les sédiments.
   In Advances in Organic Geochemistry 1975.
   Eds R. CAMPOS and J. GONI ENADIMSA, Madrid, 655-673.

KARWEIL J. (1956)
   Die Metamorphose des Kohlen vom Standpunkt der physikalischen Chemie. Z. dtsch. geol. ges. 107, 132-139.

LOPATIN N.V., 1971
   Temperature and geologic time factors in coalification.
   Izv. Akad. Nauk SSR ser. Geol. 3, 95-106.

OBERLIN A., BOULMIER J.L. and VILLEY M., 1980
   Electron microscopic study of kerogen microtexture. Selected criteria for determining the evolution path and evolution stage of kerogen.
   In Kerogen, insoluble organic matter from sedimentary rocks, B. DURAND ed. Editions Technip, Paris, 191-242.

OUDIN J.L. and PICARD P.F., 1982
   Genesis of hydrocarbons in the Mahakam delta and the relationship between their distribution and the overpressured zones.
   Paper at the XIth Annual Indonesian Petroleum Association Convention. June 8-9 1982.

OUDIN J.L., 1984
   Thermal maturation indices in organic geochemistry.
   In thermal phenomena in sedimentary basins.
   B. DURAND ed. Editions Technip Paris, 117-125.

PITTION J.L. and GOUADIN J., 1984
   Maturity studies of the Jurassic coal unit in three wells from the Haltenbanken area.
   Paper at the NPF Symposium on Organic Geochemistry, in exploration of the Norwegian Shelf, Stavanger, Oct. 22-24, 1984, in press.

PRICE L.C. and BARKER C.E., 1985
    Suppression of vitrinite reflectance in amorphous rich kerogen. A major unrecognized problem.
    Journal of Petroleum Geology 8, 1, 59-84.

ROBERT P., 1985
    Histoire géothermique et diagenèse organique.
    Bulletin des centres de recherches Exploration - Production Elf - Aquitaine. Mémoire 8, Pau 1985.

ROUZAUD J.N. and OBERLIN A., 1984
    Contribution of high resolution transmission electron microscopy (TEM) to organic materials characterization and interpretation of their reflectance.
    In thermal phenomena in sedimentary basins.
    B. DURAND Ed. Editions Technip, Paris, 127-134.

ROUZAUD J.N., 1984
    Relations entre la microtexture et les propriétés des matériaux carbonés. Application à la caractérisation des charbons.
    Thèse, Université d'Orléans.

SITTLER J.A., 1979
    Effects of source material on vitrinite reflectance.
    Master's thesis, West Virginia University, Morgantown, Virginia.

STACH E., MACHOWSKY M.Th., TEICHMULLER M., TAYLOR G.H., CHANDRA D., TEICHMULLER R. (1982)
    Textbook of coal petrology, 3rd edit., Borntraeger, Berlin-Stuttgart, 535 p.

TEICHMULLER M. and TEICHMULLER R. (1982 a)
    The importance of coal petrology in prospectivy for oil and natural gas. In Stach et al., Textbook of coal petrology, 3rd edit., Borntraeger, Berlin-Stuttgart, 399-412.

TEICHMULLER M. and TEICHMULLER R. (1982 b)
    Causes of coalification. In Stach et al., Textbook of coal petrology, 3rd edit., Borntraeger, Berlin-Stuttgart, 55-66.

TISSOT B., 1959
    Premières données sur les mécanismes et la cinétique de la formation du pétrole dans les sédiments. Simulation d'un schéma réactionnel sur ordinateur.
    Rev. Inst. Fr. Pét. 24, 470-501.

TISSOT B., DURAND B., ESPITALIE J. and COMBAZ A., 1974
    Influence of nature and diagenesis of organic matter in
    formation of petroleum.
    Am. Assoc. Pet. Geol. Bull., 58, 3, 499-506.

TISSOT B. et ESPITALIE J., 1975
    L'évolution thermique de la matière organique des sédiments :
    applications d'une simulation mathématique.
    Rev. Inst. Fr. Pet. 30, 743-777.

UNGERER Ph., ESPITALIE J., MARQUIS F. and DURAND B., 1986
    Use of kinetic models of organic matter evolution for the
    reconstruction of paleotemperatures. Application to the
    case of Gironville 101 well, France, in this volume.

J. ESPITALIÉ[1]

USE OF Tmax AS A MATURATION INDEX FOR DIFFERENT TYPES OF ORGANIC MATTER. COMPARISON WITH VITRINITE REFLECTANCE

1. INTRODUCTION

Tmax is a pyrolysis parameter that can be used for making a quick estimate of the degree of maturation of the organic matter in rocks. An increase is generally seen in Tmax values when this maturation increases. Tmax is defined as the pyrolysis temperature at which a maximum amount of hydrocarbon compounds coming from the thermal degradation of kerogen is released. This parameter must thus be linked directly to the cracking kinetics of organic matter, as shown by the observation of natural series and laboratory pyrolysis-test results.

The correspondence between Tmax and different oil ang gas formation zones thus varies with the type of organic matter, with Tmax sensitivity being greatest for determining the degree of maturation of Type III (coals or dispersed organic matter). This correspondence may be obtained by correlating the values of Tmax :

- either directly with the values of the reflectance of vitrinite in the case of Type III series or that of marine or lacustrine series containing interbedded coals (Green River Shales).

- or with the amounts of the chloroform extract in series in which vitrinite is not of continental origin, as in the Lias marine series in the Paris Basin.

This correspondence may be complicated by the variations of Tmax observed according to different operating conditions, heavy oil accumulations in source rocks and interactions between the mineral

---
(1) *Institut Français du Pétrole, Rueil-Malmaison, France.*

USE OF Tmax AS MATURATION INDEX FOR DIFFERENT TYPES OF ORGANIC MATTER. COMPARISON WITH VITRINITE REFLECTANCE

matrix of the rocks and their organic matter which occur during pyrolysis.

After making a critical analysis of the Tmax parameter (operating conditions, effect of mineral matrix, etc...), concrete cases are used to propose a scale of correspondence between Tmax and different stages of oil and gas formation for each of the three major types of organic matter.

2. EXPERIMENT

2.1. Rock-Eval Pyrolysis

The method consists (Espitalié et al. 1977, 1984) of the programmed-temperature heating (25°C/minute on the average) in an inert atmosphere (helium) of a small sample of rock (about 100 mg) to quantitatively and selectively determine : (i) the free hydrocarbons in the form of gas and oil contained in the rock sample (peak S1) volatilized at 300°C and (ii) the hydrocarbon compounds and the CO_2 (peaks S2 and S3) that are expelled during the cracking of the unextractable organic matter in the rock (kerogen) between 300 and 600°C. The temperature of the oven reached at the top of peak S2 is called Tmax and is expressed in degrees Celsius. The parameters S1 (oil + gas), S2 and S3 (HC and CO_2 issued from cracking) are expressed in mg HC or CO_2 per gram of rock.

In the most recent versions of the apparatus (Rock-Eval II and Rock-Eval III also called "OIL SHOW ANALYZER" ; Espitalié et al., 1985) the total organic carbon (TOC) of the sample can be determined automatically by oxidation in air of the organic matter remaining in the sample after pyrolysis (residual organic carbon ; S4 peak). The TOC is then determined by adding the residual organic carbon thus detected to the pyrolyzed organic carbon which in turn is measured from the hydrocarbon compounds issuing from pyrolysis (Espitalié et al. , 1984).

From these basic parameters the microprocessor of the device computes :

- The Hydrogen Index (HI) and the Oxygen Index (OI), as the values of S_2 ans S_3 expressed, respectively in mg of HC per gram of TOC and in mg of CO_2 per gram of TOC. This two parameters allow a quick determination of the organic matter (OM) types.

- The Production Index (PI), as the $S_1/(S_1+S_2)$ ratio, i.e. the proportion of free hydrocarbons in relation to the total amount of hydrocarbon compounds obtained by pyrolysis of the sample analyzed (Rock-Eval II).

USE OF Tmax AS MATURATION INDEX FOR DIFFERENT TYPES OF ORGANIC MATTER. COMPARISON WITH VITRINITE REFLECTANCE

With the Rock-Eval III "OIL SHOW ANALYZER" the two other parameters obtained, S_0 (gas) and S'_1 (oil), together with S_2 can be used to compute the gas production index : GPI = $S_0/(S_0 + S'_1 + S_2)$ and the oil production index : OPI = $S'_1(S_0 + S'_1 + S_2)$.

Under normal operating conditions, the reproducibility of the measurements made on the IFP reference rock (No 37133) is ±1°C for Tmax, ±5 % for total organic carbon values, ±8 % for the S_2 values, and ±10 % for S_3.

2.2. Geological series studied

For the study of the evolution of Tmax with the increasing maturation of the main types of organic matter, we used the following series :

- Green River Shales from the Uinta basin in Colorado (U.S.A.) for the analysis of OM of lacustrine origin (Type I) ;

- Lias from the Paris basin (France) for OM of marine origin (Type II) ;

- Carboniferous (Wesphalian) from the Paris basin for OM of continental origin (Type III).

We will bear in mind that the series investigated here probably do not represent all existing types of OM. Intermediate types (case of diatomites ?) or of mixtures of OM can complicate the results obtained. Likewise we do not fully understand the influence of the sedimentary environment (carbonates, shales, etc.) on the transformations of each type of OM during burial.

The vitrinite reflectance (R_0) is currently taken as a scale of maturity. Therefore we will refer to it for investigating the evolution of pyrolysis parameters. However we know that R_0 values have meaning only in series where the OM is of continental origin (Price and Barker, 1955). In other series comparisons of R_0 and pyrolysis parameters can be made when there is coal interbedding (case of the Green River Shales in Colorado). Such comparisons become highly problematic in formations not containing such interbedding (Lias of Paris basin).

3. CALIBRATION OF Tmax

In general, the values of S_0, S'_1, S_1, S_2 and Tmax are calibrated

USE OF Tmax AS MATURATION INDEX FOR DIFFERENT TYPES OF ORGANIC MATTER. COMPARISON WITH VITRINITE REFLECTANCE

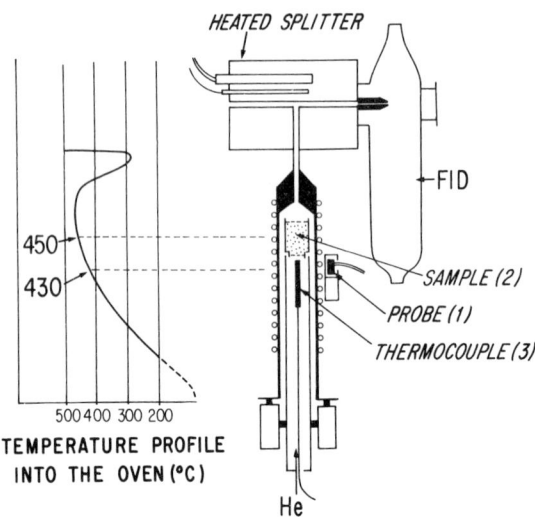

Fig. 1 - Variation of the temperature profile inside the pyrolysis furnace of the Rock-Eval device for a temperature of 430°C given by the probe (**1**).

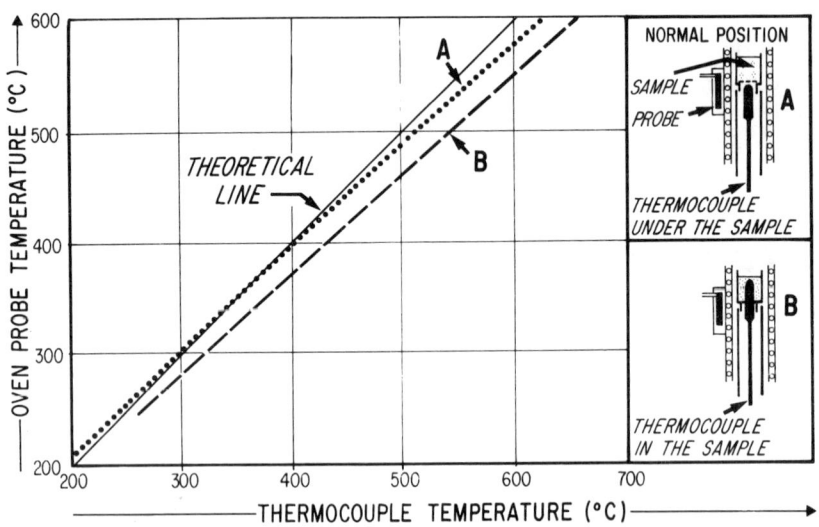

Fig. 2 - Variations of temperature during pyrolysis of the sample analyzed (line B), of the thermocouple (line A), and of the probe (Tmax ; theoretical line).

USE OF Tmax AS MATURATION INDEX FOR DIFFERENT TYPES OF ORGANIC MATTER. COMPARISON WITH VITRINITE REFLECTANCE

from peak S_2 for a reference rock such as sample N° 37133 which IFP proposes as a standard. Peak S_4 is calibrated from the residual organic carbon (remaining after pyrolysis) of the same rock.

Tmax is computed as being the average of 20 temperature measurements made on both sides of the top of peak S_2. These measurements are provided by the control probe of the pyrolysis oven. This probe is situated on the outer wall of this oven (|1|, Fig. 1). Tmax does not correspond to the temperature of the sample in the combustion boat (|2|, Fig. 1) but to the temperature of the probe (1).

The actual temperature of the sample in its combustion boat is really 30 to 50°C higher than the one given by Tmax (Fig. 1 and 2) because of a temperature gradient that increases toward the top of the oven (proximity of the heated splitter adjusted to 500°C). Tmax is thus a relative measurement of the pyrolysis temperature and can be used to compare samples among each other and thus to work out maturity scales.

Rock-Eval Pyrolysis can be used for the adjustment of the kinetic equations concerning the thermal degradation modelling of organic matters (Ungerer, 1984). In that case the increasing values of temperature into the analyzed sample will be determined using the correspondance between the curve B of the figure 2 and the temperature values recorded by means of the thermocouple situated into the piston, (curve A, fig. 2), or given by the probe (theoretical line, fig. 2) on the timer (Rock-Eval I) or on the microprocessor (Rock-Eval II and III).

At each calibration the Tmax value of the reference rock used is printed out. It is also compared by the microprocessor to its value given as a reference and which is placed in memory (429°C in IFP rock No 37133). If these two Tmax values do not coincide, the microprocessor computes their difference and corrects the Tmax values of the samples that are then analyzed. This process allows to compare all Tmax values among each other in the light of possible instrument drifts (improper positioning of the probe on the oven after reassembly, for example) and especially no matter what the rate of heating that is used (case of different analysis cycles).

4. VARIATIONS OF Tmax WITH OPERATING CONDITIONS

4.1. Rate of heating

Kerogen cracking is known to depend not only on the temperature but also on the time (Arrhenius' law). Therefore, peak S_2 (and hence Tmax) is formed at a pyrolysis temperature that is all the lower as the temperature programming of the oven is slow (Fig. 3). The Tmax of the reference rock follows the same law (constant deviation of

USE OF Tmax AS MATURATION INDEX FOR DIFFERENT TYPES OF ORGANIC MATTER. COMPARISON WITH VITRINITE REFLECTANCE

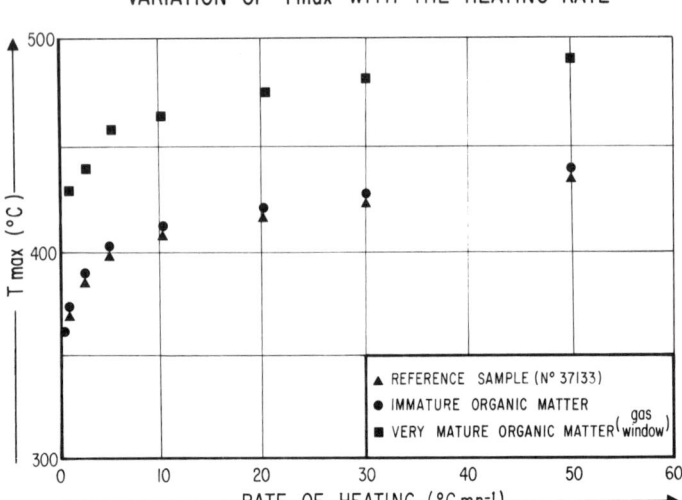

Fig. 3 - Variation of Tmax with the rate of heating

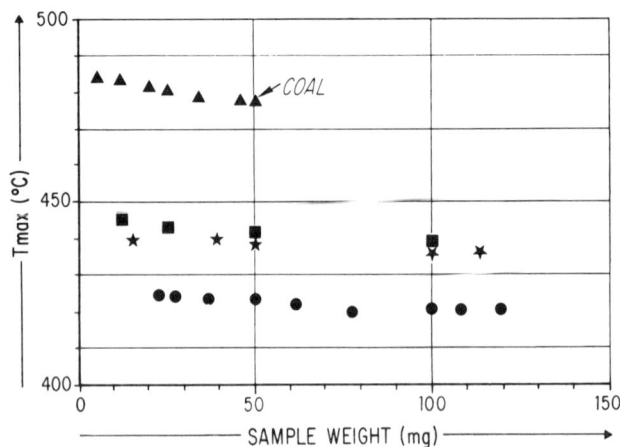

Fig. 4 - Variations of Tmax as a function of the weight of the sample analyzed.

USE OF Tmax AS MATURATION INDEX FOR DIFFERENT TYPES OF ORGANIC MATTER. COMPARISON WITH VITRINITE REFLECTANCE

values, Fig. 3), and the correction made by the microprocessor can thus cancel out the variations of Tmax due to the heating rates for all the samples. A calibration must thus be performed at the beginning of any new pyrolysis cycle.

4.2. Weight sample

The weight of the sample has an effect on Tmax values. Figure 4, compiled from various rock and coal samples, illustrates the variations in Tmax values for samples varying from 5 to 120 mg. The Tmax values can be seen to increase when the weight of the sample is less than 60 to 70 mg, with the maximum difference being about 6°C. This delay in the pyrolysis of the sample when the weight is small can be attributed to a slight cooling of the rock due to easier circulation of the carrier gas, helium, through the combustion boat. It can be also attributed to phenomena of retention and coking on the inside walls of the oven and divider. In that case the hydrocarbon compounds retained are relatively greater when the sample is small. This generally also causes poor linearity of the S_2 responses as a function of sample weight for crude-rock samples smaller than 20 mg.

4.3. Other effects

The grain size of the sample can also influence Tmax which increases of 3 to 4°C for unground samples (grain size of about 2 to 3 mm). It the temperature of the splitter (above the oven) is too high (>500°C) in some types of rock a more or less appreciable rise in the FID signal, caused by the ionization of mineral salts, may be observed just after the S_2 peak. S_2 value is then increased and Tmax can be perturbed (Espitalié et al., 1985).

5. VARIATION OF Tmax WITH HEAVY-OIL ACCUMULATIONS

The complications caused by the accumulation of more or less heavy oil in rocks were described for the first time by Clementz (1980). Peak S_2 is known not to be made solely of hydrocarbon compounds coming from kerogen cracking. It also contains hydrocarbon compounds coming either from the volatilization of heavy hydrocarbons (C_{35+}) or from the cracking of other hydrocarbon compounds such as asphaltenes or resins. In general this category of hydrocarbon compounds that is formed at the start of pyrolysis (300-400°C range, approximately) represents only a few percent of the S_2 total and affects neither the HI nor the Tmax values. On the other hand, when the rock has an accumulation of more or less heavy oil, Tmax may be more or less strongly lowered due to the pyrolysis of resins and asphaltenes.

USE OF Tmax AS MATURATION INDEX FOR DIFFERENT TYPES OF ORGANIC MATTER. COMPARISON WITH VITRINITE REFLECTANCE

Table 1

EFFECTS ON T_{max} OF INTERACTIONS BETWEEN THE MINERAL MATRIX AND THE ORGANIC MATTER, DURING PYROLYSIS

MINERAL OIL SHALE RATIO	TOC CONTENT IN MIXTURES (%)	T_{max} (°C) IN MIXTURES			
		ILLITE	CALCITE	SAND	ILLITE + SAND (50/50)
20	0.57	415	415	419	420
10	1.09	422	414	418	421
6.6	1.56	424	416	417	423
5	2.00	424	416	417	423
4	2.40	424	418	419	423
3.3	2.77	425	418	418	420
2.8	3.11	422	418	420	420
2.5	3.43	421			419
2.0	4.00	420			419
1.6	4.50	420			419
0	12.0			419	

USE OF Tmax AS MATURATION INDEX FOR DIFFERENT TYPES OF ORGANIC MATTER. COMPARISON WITH VITRINITE REFLECTANCE

Furthermore, heavy hydrocarbon compounds can possibly form solid pyrobitumens in rocks, and these may be confused in cuttings with pieces of coal. However, when they have the same degree of maturity their Tmax values are generally much lower (Vandenbroucke et al., 1981).

In the case of impregnated source rocks, the chloroform extraction of heavy products prior to pyrolysis eliminates the phenomenon and thus produces parameters that are representative of the kerogen (Clementz, 1980 ; Orr, 1983).

6. VARIATION OF Tmax WITH MINERAL MATRIX EFFECTS

The mineral matrix may interfere with pyrolysis parameters depending on the type or rock and the amount of organic matter it contains, by retaining certain hydrocarbon compounds coming from pyrolysis (heavy hydrocarbons, aromatics, heteropolar compounds). These compounds are then more or less completely coked on the mineral.

These effects were analyzed from the pyrolysis of kerogen mixtures (or of an OM-rich rock) with different minerals, either by Rock-Eval type pyrolysis or pyrochromatography (Horsfield and Douglas, 1980 ; Tarafa et al., 1983 ; Katz, 1983 ; Larter, 1984 ; Espitalié et al., 1980, 1985).

It was found that it is the heaviest hydrocarbon compounds (heteropolar compounds and C_{20+} hydrocarbons) making up peaks S_2 and S_1 that are the most easily retained on mineral matrices and "coked" during heating. This phenomenon is related to the specific surface area of the mineral matter (Espitalié et al., 1985).

This retention will have two main effects on the pyrolysis parameters :

- A decrease in peaks S_1 and S_2 : Therefore, the HI values are then underestimated (Katz, 1983 ; Larter, 1984), making the determination of types of OM from HI-OI and HI-Tmax diagrams more problematic.

- An increase in Tmax values : Pyrolysis tests performed with OM mixtures having different minerals (Table I) generally show that : (a) Tmax increases with the activity of the mineral matrix, minerals such as quartz and carbonates having little effect ; (ii) for a very active mineral such as illite, Tmax increases when the illite/OM ratio increases, although without going above an increase of 5 to 6°C for Type II and 10 to 12°C for Type III (coal), although for high values of the illite/OM

USE OF Tmax AS A MATURATION INDEX FOR DIFFERENT TYPES OF ORGANIC MATTER. COMPARISON WITH VITRINITE REFLECTANCE

ratio (higher than 10) Tmax again decreases (Table I) ; (iii) this effect of the mineral matrix on Tmax varies little with the degree of maturity of the O.M.

The pyrochromatography of hydrocarbons of the peak S_2 can be used to show that it is the hydrocarbon compounds formed at the start of the cracking of kerogen (heavy compounds) are the ones most retained on the minerals (mineral matrix not yet having any sites saturated by the hydrocarbons retained).Their retention thus changes the shape of peak S_2, causing a displacement of the top of this peak at higher temperatures.

All these phenomena (operating conditions, mineral matrix effects ...) are probably responsible of the fluctuations of Tmax values frequently observed during pyrolysis of samples taken in a same interval depth and having a same type of organic matter. An example is given by the figure 5 which shows the evolution of Tmax with depth for a great number of samples (cuttings) taken in a westphalian series of Paris basin. This series is composed of coal banks interbedded with levels containing a type III dispersed organic matter. The evolution of Tmax with depth is well marked but the dispersion of its values can estimated to ±5°C.

7. VARIATION OF Tmax WITH THE TYPE OF ORGANIC MATTER AND CORRELATION WITH VITRINITE REFLECTANCE

As a general rule, an increase is observed in the values of Tmax when the state of evolution of OM increases (Barker, 1974 ; Espitalié et al., 1977). Actually, the chemical bonds that persist in the most highly evolved kerogens are the one requiring the most energy during pyrolysis to be broken (Tissot and Espitalié, 1975). Artificial laboratory maturity tests (pyrolyses in open or closed systems) and the observation of natural evolutive series (Green River Shales, Lias and Carboniferous of Paris basin, Cretaceous of Douala basin ..) show that, for the same type of OM, the S_2 peaks corresponding to increasing states of maturity all fall within the area of peak S_2 for the least evolved kerogen of this type (Fig. 6). The geometric situation of the corresponding Tmax values outlines a trajectory situated in the vicinity of the drop-off of the kerogen peak, either parallel to or merged with it.

Tmax thus appears to be linked directly to the cracking kinetics of the OM. Types I and II are known to have a relatively simpler structure than Type III, thus implying a narrower distribution of cracking energy for the first two and hence a pyrolysis in a smaller temperature range. This means that, for the same heating law, the S_2 peaks recorded are all the wider and the movements of Tmax with evolution are all the more marked as the OM goes from Type I toward Type III (Fig. 6).

USE OF Tmax AS A MATURATION INDEX FOR DIFFERENT TYPES OF ORGANIC MATTER. COMPARISON WITH VITRINITE REFLECTANCE

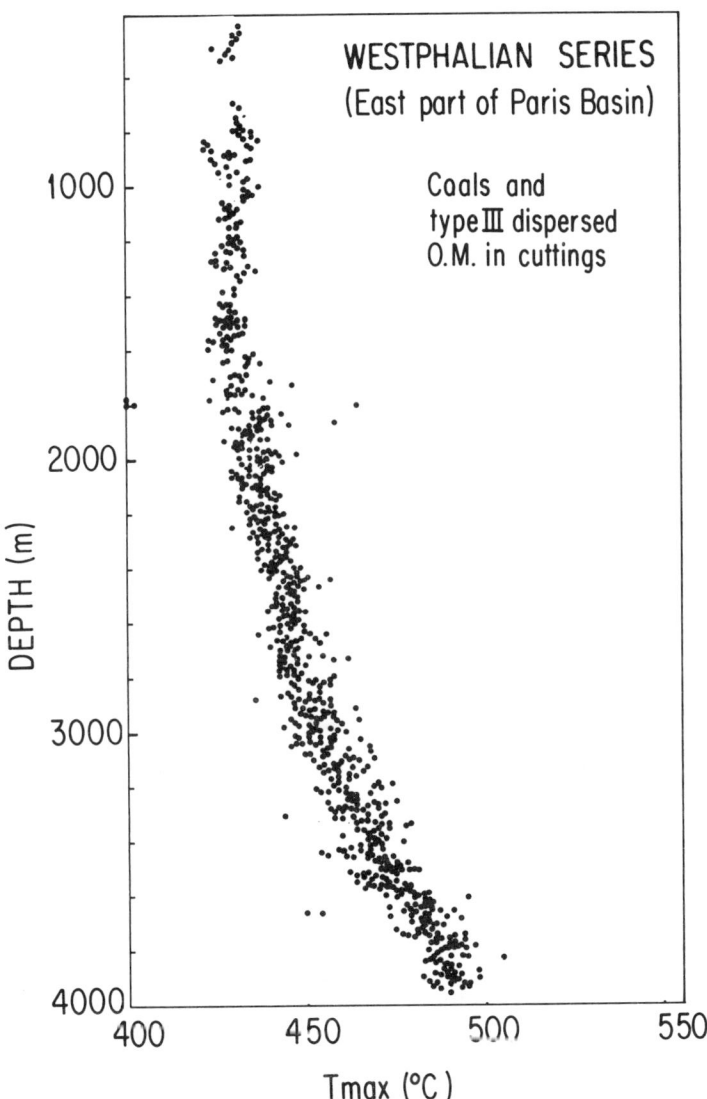

Fig. 5 - Evolution of Tmax with depth in a carboniferous series (Borehole of the East part of Paris basin). Pyrolyses are performed on cuttings more or less rich in coal grains.

USE OF Tmax AS MATURATION INDEX FOR DIFFERENT TYPES OF ORGANIC MATTER. COMPARISON WITH VITRINITE REFLECTANCE

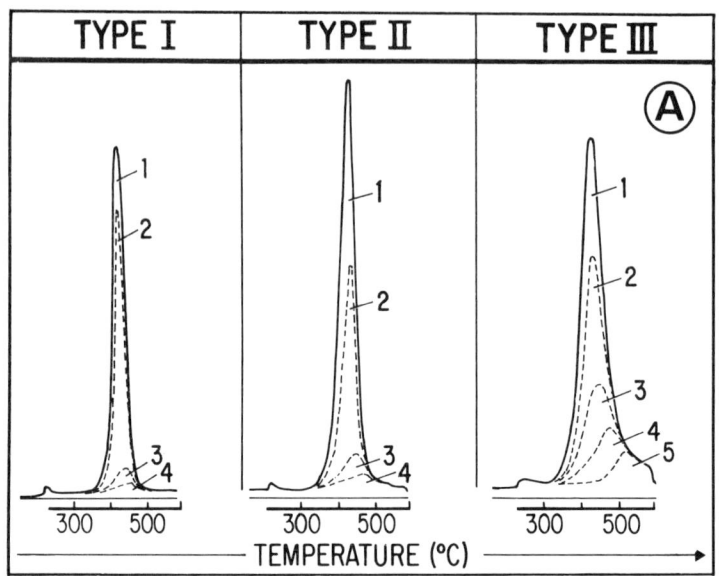

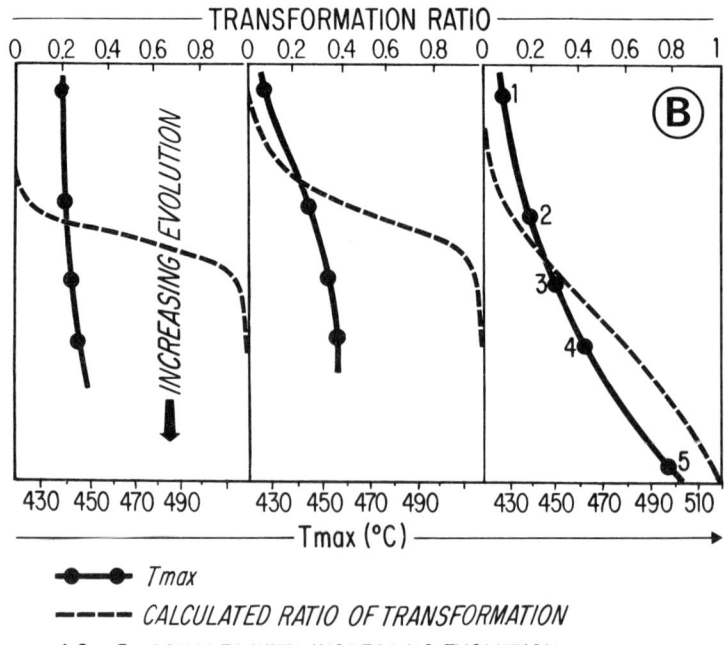

A) Variation in the shape of S_2 peak during maturation for the 3 main types of O M.
B) Evolution, as a function of maturation, of Tmax and the ratio of degradation computed from Hydrogen Index values.

Figure 6

USE OF Tmax AS MATURATION INDEX FOR DIFFERENT TYPES OF ORGANIC MATTER. COMPARISON WITH VITRINITE REFLECTANCE

It is in the case of Type III OM (coals, etc...) that the increase of Tmax during maturation is the greatest. At the beginning of evolution (peat) the values of Tmax are less than 400°C. They reach 600°C at the anthracite stage for a vitrinite reflectance value (R_o) of about 2.6. This value of 600°C, moreover, is a limit imposed by the Rock-Eval heating program. The pyrolysis of coals continues well beyond that, but the values of S_2 are then very low (less than 1 mg of HC per g of rock).

An example enabling the evolution of Tmax to be followed as a function of depth is observed in a borehole of the eastern part of the Paris basin which crosses through a series of Westphalian (Carboniferous) coal banks 2000 m thick. It was found that the Tmax of these coals varied from 435°C at 1800 m depth to 500°C on the average at 4000 m (Fig. 7). The evolution of the OM in the same depth interval also shows a great decrease in HI (250 to 40 mg of HC per g of TOC on the average).

In this borehole, the decreasing HI of coals can thus be used to compute the transformation ratio of the type III OM. Indeed, for a homogeneous evolutive series (same type of OM) the transformation ratio of the OM at depth (p) can be defined as the ratio of the decrease in HI resulting from the formation of oil and gas between depth (o) and depth (p), given by ($HI_o - HI_p$), to the mean initial HI at the start of the oil window (HI_o).

$$\text{Transformation ratio} = (HI_o - HI_p) / HI_o.$$

It has been shown (Pelet, 1985) that greater accuracy is obtained with the following equation:

$$\text{Transformation ratio} = \frac{1200 \, (HI_o - HI_p)}{HI_o \, (1200 - HI_p)}$$

in which 1200 = maximum amount of HC (in mg) corresponding to one gram of organic carbon pyrolyzed on the basis of a TOC content of 83 %

HI_o = mean initial Hydrogen Index

HI_p = mean Hydrogen Index at depth (p)

The variations in the Tmax values can also be compared to the variation in the corresponding R_o values of vitrinite reflectance (Teichmüller and Durand, 1983 ; Espitalié et al., 1984). The correlation obtained (Fig. 8) shows that the start of the oil zone, generally detected by an R_o value of 0.5 %, corresponds to a mean Tmax value of 430°C, and the start of the wet-gas zone (R_o of 1.35 %) corresponds to a mean value of 465°C. It also shows that the increase

USE OF Tmax AS MATURATION INDEX FOR DIFFERENT TYPES OF ORGANIC MATTER. COMPARISON WITH VITRINITE REFLECTANCE

in Tmax as a function of the reflectance is not constant. It is much less between 0.5 and 1.5 % R_O in the oil-formation zone than beyond 1.5 % in the gas-formation zone.

For <u>Type I OM</u> of aquatic origin and deposited in a confined environment (usually lacustrine), Tmax reaches a relatively high value of between 440 and 445°C beginning with the first stages of evolution, and it stays there during the entire phase of thermal degradation of the OM. For example, in the Paleocene Green River Shales formation in the Uinta basin (U.S.A.) (Fig. 9) the Tmax values vary almost not at all in the depth range from 6000 to 18 000 feet, and especially in the oil and gas formation zones situated between 13 000 and 18 500 feet, as indicated by the HC contents in the chloroform extract and by the R_O values of vitrinite of the interbedded coal levels (Tissot et al., 1978).

Tmax correlation of the Type I OM levels with the R_O values of the interbedded coals shows (Fig. 9) that, as opposed to Type III, Tmax evolves very little (440 to 445°C on the average) in the reflectance range from 0.40 to 1.50 %. Likewise, beyond 1.50 % R_O (about 20 000 feet depth) <u>there is no longer any Tmax</u> because of the disappearance of peak S_2.

<u>Type II OM</u> of marine origin has evolutive features that are intermediary between those of the other two types. For example, in the Lias in the oil wells in the Paris basin (Fig. 10) Tmax values go from 420°C at about 1000 m depth to 450°C at around 2800 m. The corresponding evolution of the OM is characterized by a steady decrease in the hydrogen index, with the transformation ratio computed reaching about 0.90 at 2800 m.

In this series of marine origin, the correlations with the R_O of vitrinite are very vague because the organic particles thought to be vitrinite are generally of doubtful certainty (Alpern and Cheymol, 1978). However, the Tmax values can be correlated with the hydrocarbons contained in the chloroform extract, with the computed transformation ratio and with the values observed for the production index. It is then found that the start of oil formation must be situated at about 1800 to 2000 m depth for Tmax values between about 430 and 435°C (Fig. 10).

As for Type I OM, the disappearance of peak S_2 with evolution imposes an extreme value on Tmax. Actually it is difficult to determine this extreme value because of the lack of sufficiently evolved marine reference series. However, Lower Toarcian samples from Germany have shown that Tmax can reach about 460° with an R_O of 1.30 %. In this same series there is no longer any peak S_2 for the most evolved samples (R_O = about 1.50 %).

USE OF Tmax AS MATURATION INDEX FOR DIFFERENT TYPES OF ORGANIC MATTER. COMPARISON WITH VITRINITE REFLECTANCE

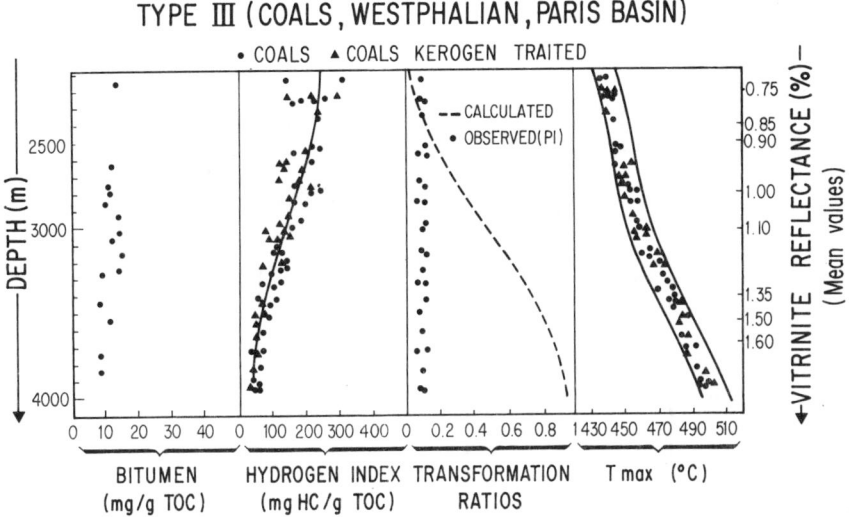

Fig. 7 - Analysis of a Paleozoic coal series (eastern Paris basin). Comparison, as a function of depth, of Tmax with other geochemical parameters.

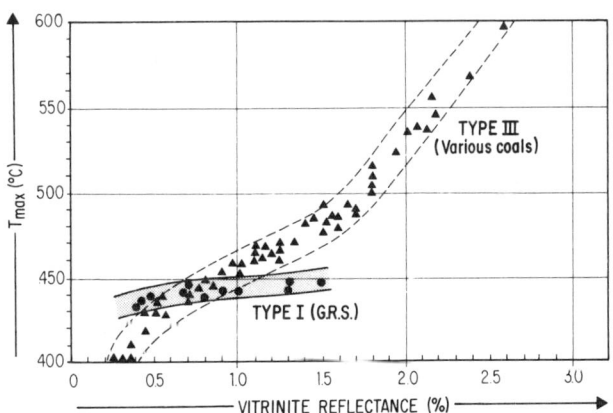

Fig. 8 - Correlation between Tmax and the vitrinite reflectance for types III and I.

USE OF Tmax AS MATURATION INDEX FOR DIFFERENT TYPES OF ORGANIC MATTER. COMPARISON WITH VITRINITE REFLECTANCE

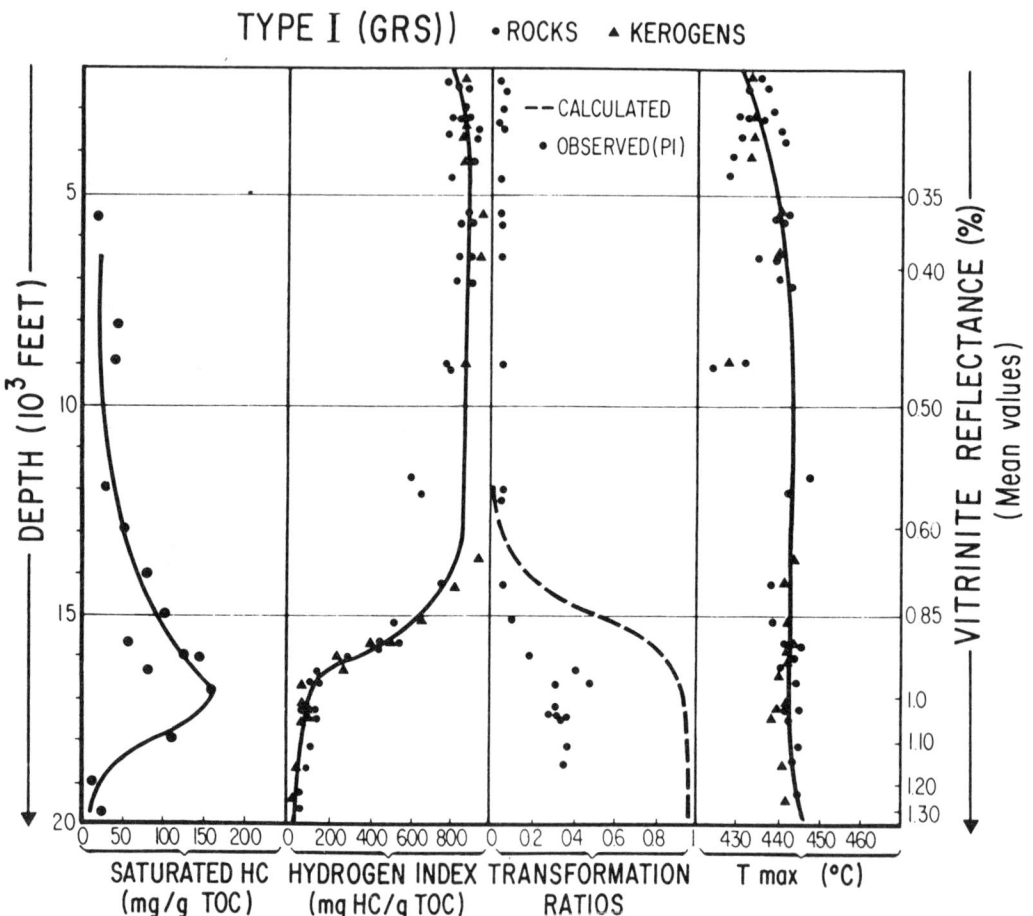

Fig. 9

Type I (GRS). Comparison, as a function of depth, of Tmax with other geochemical parameters.

USE OF Tmax AS MATURATION INDEX FOR DIFFERENT TYPES OF ORGANIC MATTER. COMPARISON WITH VITRINITE REFLECTANCE

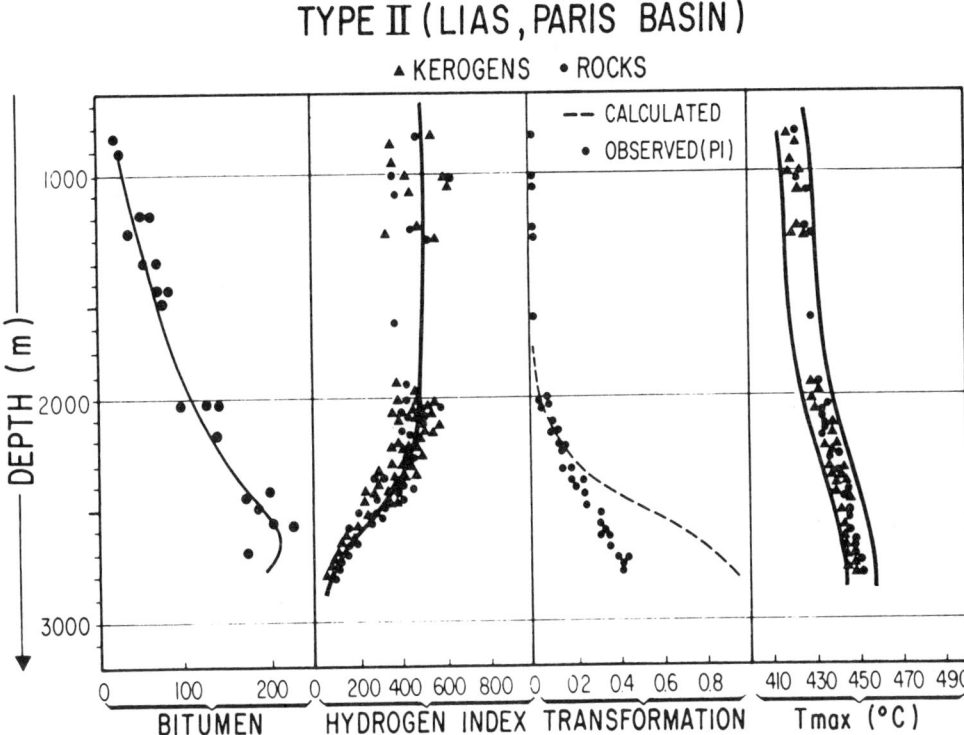

Fig. 10

Type II, Lias of Paris Basin. Comparison as a function of depth, of Tmax with other geochemical parameters.

USE OF Tmax AS MATURATION INDEX FOR DIFFERENT TYPES OF ORGANIC
MATTER.COMPARISON WITH VITRINITE REFLECTANCE

8. USE OF Tmax FOR DETERMINING OIL FORMATION ZONES

Figure 11 shows, for the three types of OM investigated, the evolution of the mean values of the computed transformation ratios and Tmax as a function of the Ro values of vitrinite. For Type II this calibration is difficult (doubtful vitrinite). It was done here on the basis of different series of marine origin, i.e. Lias from the Paris basin, Lower Toarcian from Germany, Tertiary from Angola and from pyrolysis tests of Types II and III kerogens in a closed cell. It can be seen that :

- <u>For Type I</u> (Green River Shales) oil formation begins at an R_O of about 0.70 % (interbedded coals) and at a mean Tmax of about 442°C. The thermal degradation of kerogen, which is abrupt, is almost over for an R_O of 1.00 % and almost the same Tmax. Judging from the HC contents in the chloroform extract (Tissot et al. , 1978), an R_O value of 1.30 % should correspond very closely to the beginning of oil cracking (wet-gas zone),with a mean Tmax corresponding to about 445°C. Tmax no longer exists for R_O values higher than 1.40 %. Its extreme value is approximately 450°C.

- <u>For Type II</u> the beginning of oil formation should be much earlier and should occur in an R_O range from 0.40 to 0.50 %, corresponding to Tmax values of between 430 and 435°C. Kerogen degradation should be almost ended, as in the case of Type I, at an R_O value of 1.00 % (Tmax around 450°C). The transition to the condensate zone (R_O = 1.30 %) should correspond to a Tmax of 455°C, and the extreme value of Tmax can be estimated at about 460°C.

- <u>For Type III</u> hydrocarbon formation begins in an R_O range from 0.50 to 0.60 % (Tmax between about 430 and 435°C), and the transition to the condensate zone (R_O = 1.30 %) is equal to a Tmax of 465°C. The end of Type III OM thermal degradation is very late since it is not completely over at an R_O of 1.60 %, i.e. at a Tmax of 600°C. In this case, moreover, the extreme value of Tmax is higher than 600°C. Figure 12 sums up the preceding results concerning the practical use of Tmax to determine oil and gas formation zones.

9. CONCLUSION

By way of conclusion, we will note that Tmax is not an evolution marker for Type I OM (Green River Shales). At the same time, Tmax is a more representative maturity criterion than the vitrinite reflectance for OM of marine origin (Type II). Indeed, as opposed to reflectance measurements determined from a tiny fraction of total OM (macerals of the vitrinite type, very doubtful diagnosis, generally not greatly represented in the marine environment, Price and Barker - 1985), measurements of Tmax are the result of the

USE OF Tmax AS MATURATION INDEX FOR DIFFERENT TYPES OF ORGANIC
MATTER. COMPARISON WITH VITRINITE REFLECTANCE

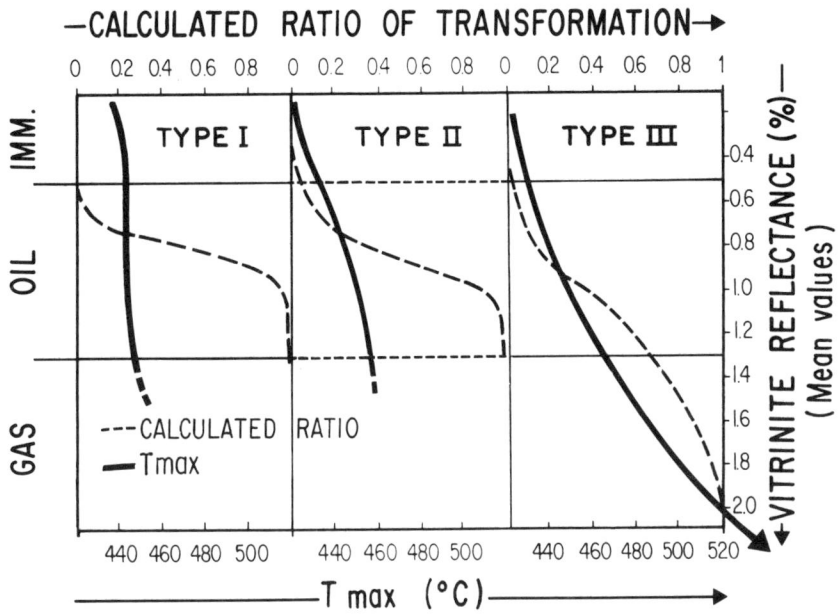

Fig. 11 - Comparative evolution of
Tmax and transformation ratios
for the three major types of O.M.

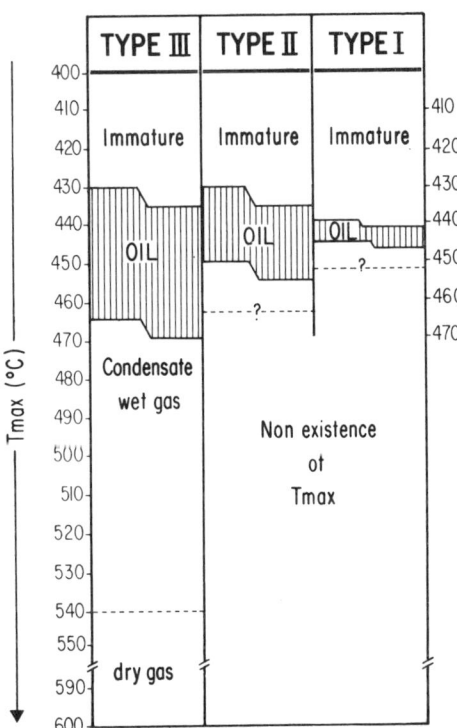

Fig. 12 - Use of Tmax to determine
the principal zones of oil and gas
formation.

493

pyrolysis of all the OM. Furthermore, such measurements obtained for a great many samples can be processed and interpreted statistically, with statistical interpretation and processing also being valid for Type III OM.

The interpretation of Tmax and its application to petroleum prospection will be facilitated by :

- a careful preparation of samples (pollutant elimination) ;
- a critical examination of the recordings obtained and the elimination of non-meaningful data (Espitalié et al. , 1986) ;
- the upgrading of results from geochemical logs so that a statistical analysis of data can also be made concerning the determination of accumulations or drainages and the characterization of source rocks, of their petroleum potential and of the type of OM they contain.

Acknowledgments

The author would like thank R. Pelet and B. Durand from the Institut Français du pétrole for their help and advice.

*
* *

REFERENCES

ALPERN B. et CHEYMOL D. (1978)
"Réflectance et fluorescence des organoclastes du Toarcien du bassin de Paris en fonction de la profondeur et de la température". Rev. Inst. Fr. Pétr., 33, 4, 515-535.

BARKER C. (1974)
"Pyrolysis techniques for source-rock evaluation".
Am. Assoc. Pet. Geol. Bull., 58, 2349-2361.

CLEMENTS D.M. (1980)
"Effect of oil and bitumen saturation on source-rock pyrolysis".
Am. Assoc. Petr. Geol. Bull., 64, 2227-2232.

ESPITALIE J., LAPORTE J.L., MADEC M., MARQUIS F., LEPLAT P. et PAULET J. (1977)
"Méthode rapide de caractérisation des roches-mères, de leur

potentiel pétrolier et de leur degré d'évolution".
Revue Inst. Fr. Pétr., 32, 23-43.

ESPITALIE J., MADEC M. and TISSOT B. (1980)
"Role of mineral matrix in kerogen pyrolysis ; Influence on petroleum generation and migration".
Am. Ass. Petr. Geol. Bull., 64, 59-66.

ESPITALIE J., MARQUIS F. and BARSONY I. (1984)
"Geochemical logging". Analytical Pyrolysis, ed. by K.J. Voorhees, Butterworth and Co (Publishers), 276-304.

ESPITALIE J., SENGA MAKADI K., TRICHET J. (1985)
"Role of the mineral matrix during kerogen pyrolysis".
Advances in Organic Geochemistry 1983, ed. by P.A. Schenck, Pergamon Press, Oxford.

ESPITALIE J., DEROO G. et MARQUIS F. (1985)
"La pyrolyse Rock-Eval et ses applications. Parties 1 et 2".
Revue Inst. Fr. Petr., 40, 5 et 6.
"Partie 3" (1986)-Revue Inst. Fr. Petr., 41, 1.

KATZ B.J. (1983)
"Limitations of Rock-Eval pyrolysis for typing organic matter".
Org. Geochem., 4, 3/4, 195-199.

LARTER S.R. (1984)
"Application of analytical pyrolysis techniques to kerogen characterization and fossil fuel exploration/exploitation".
Analytical Pyrolysis, ed. by K.J. Voorhees, Butterworth and Co. (Publishers), 212-275.

ORR W.L. (1983)
"Comments on pyrolytic hydrocarbon yields in source-rock evaluation". Advances in organic geochemistry 1981, ed. by M. Bjoroy et al., John Wiley and Sons Ltd., 1983, 775-787.

PELET R. (1985)
"Evaluation quantitative des produits formés lors de l'évolution géochimique de la matière organique".
Revue Inst. Fr. Pétr., 40, 5.

PRICE L.C. and BARKER C.E. (1985)
"Suppression of vitrinite reflectance in amorphous rich kerogen - A major unrecognized problem". Journal of Petroleum Geology, 8, 1, 59-84.

TARAFA M.E., HUNT J.M. and ERICSSON I. (1983)
"Effect of hydrocarbon volatility and adsorption on source-rock pyrolysis". Journal of Geoch. Explor., 18, 75-85.

TEICHMULLER M. and DURAND B. (1983)
"Fluorescence microscopal rank studies on liptinites and vitrinite in peat and coals and comparison with result of the Rock-Eval pyrolysis". International Journal of Coal Geology, 2, 197-230.

TISSOT B., DEROO G. and HOOD A. (1978)
"Geochemical study of the Unita Basin ; formation of petroleum from the Green River Formation". Geochim. Cosmochim. Acta, 42, 1469-1485.

TISSOT B. et ESPITALIE J. (1975)
"L'évolution thermique de la matière organique des sédiments : application d'une simulation mathématique". Rev. Inst. Fr. Pétr. 30, 743-777.

UNGERER P. (1984)
"Models of petroleum formation : how to take into account geology and chemical kinetics" in Thermal Phenomena in Sedimentary Basins, ed. by B. Durand, Editions Technip, Paris.

VANDENBROUCKE M., DURAND B. and OUDIN J.L. (1981)
"Detecting migration phenomena in a geological series by means of C1-C35 hydrocarbon amount and distribution".
Adv. Org. Geoch. 1981, ed. by M. Bjoroy et al., John Wiley and Sons Ltd, 1983, 147-155.

E. BROSSE, G. DEROO, J. ROUCACHÉ[1],
T. A. BOTNEVA[2]

ORGANIC GEOCHEMISTRY AS A TEST OF VALIDITY FOR THE RESULTS OF A MODELISATION IN THE PRIPIAT BASIN (BIELORUSSIA)

ABSTRACT

The Pripiat Basin (Bielorussia) functionned as an intracontinental rift during Late Devonian. Then were deposited :

- two carbonate series, which are at once reservoirs and source-rocks,

- and two evaporitic series, which act as seals for the traps.

A sequence of numerical models is applied to this basin in order to reconstruct its evolution : the subsidence history of the substratum is reached by a "backstripping" model ; the subsidence is linked to a thermal history through a geodynamic model ; then the temperature of the source-rocks may be calculated at any moment using a model of heat transfer in the sediments ; and last, the timing of oil formation is determined with a kinetic model of hydrocarbon generation.

Three classes of data are inputted :

- chronostratigraphic determinations ;

- porosity/depth and conductivity/depth curves ;

- a value for the radiogenic heat-flow.

The results computed by the models, in terms of maturation degree reached by three types (A, B & C) of organic matter, are compared with the geochemical evolution indexes provided by the

---

(1) *Institut Français du Pétrole, Rueil-Malmaison, France.*
(2) *VNIGNI, Moscow, USSR.*

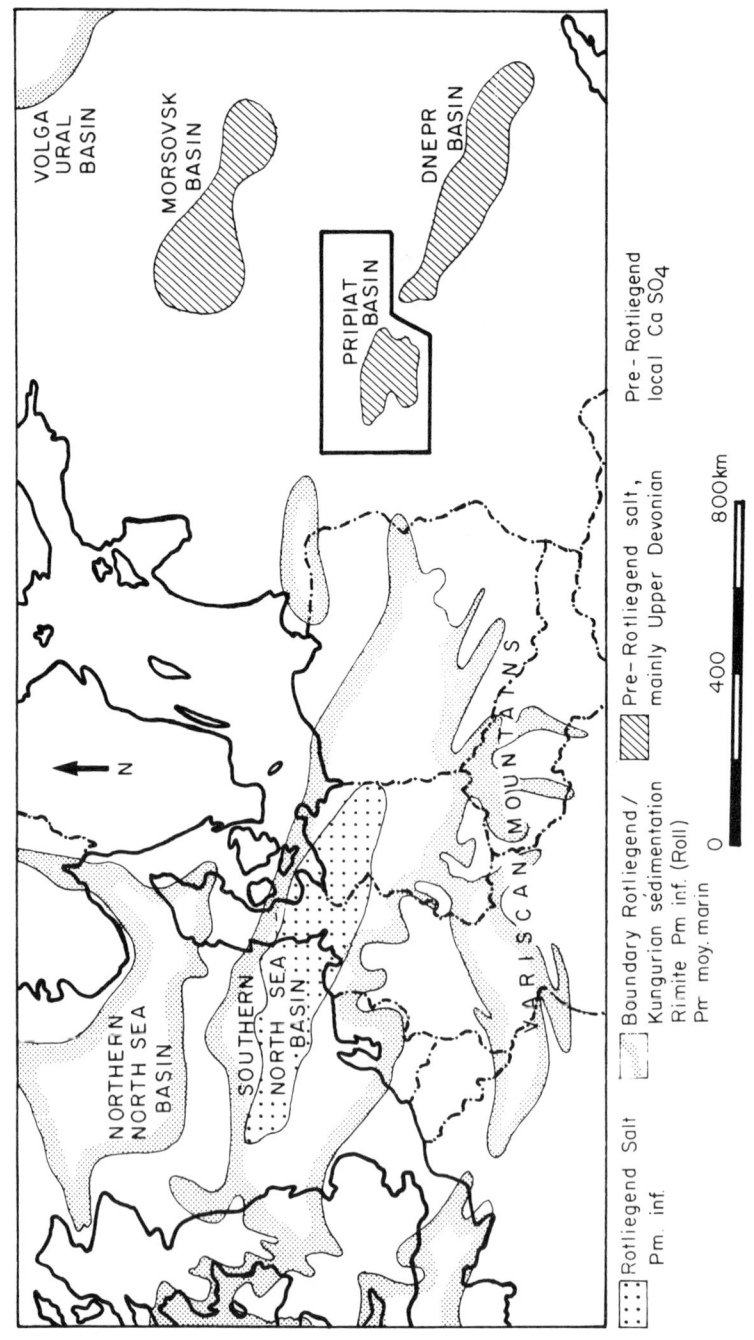

Fig. 1

The graben system, or aulacogen, Pripiat-Dnepr-Donetz, in a map of Europe.

ORGANIC GEOCHEMISTRY AND MODELS IN THE PRIPIAT BASIN

ROCK EVAL pyrolysis. Several groups of hypothesis, concerning especially the subsidence history (possible erosions during the Carboniferous) and thermal regimes, are discussed, together with the accuracy obtained for the adjustment.

In a basin as the Pripiat Basin, where the geology, the thermal environment and the petrophysical properties are burdened with large uncertainties, and where, on the other hand, the application of models depends on a debatable choice, the fact that organic matter study can provide particularly dependable informations on the thermal history is emphasized.

INTRODUCTION

One of the greatest difficulties when making use of geological models, is to control their results with the help of observations and measures. Among the controlling parameters, several are provided by a usual method of organic geochemistry, the ROCK EVAL pyrolysis. Because of its minor cost, its simplicity and its possible employment in any type of organic matter, ROCK EVAL is perhaps today the most prolific supplier of data concerning the maturation level reached by oil source-rocks.

The actual cases requested by the promoters of models in order to ascertain their validity are spectacular examples, where the series is exceptionally thick and homogeneous, together with a rather simple and well-known geothermal history. At the same time, one of the main targets of modeling is to propose new tools for hydrocarbons prospection in any kind of sedimentary basins. It is then another approach, where controlling turns more on the results of geological hypothesis tested through models than on the reliability of the models themselves.

The Pripiat Basin is a very interesting example where exploration difficulties arise, which one can hope to clear up by the help of a modelisation.

1. GEOLOGICAL SITUATION OF THE PRIPIAT BASIN

a) Geographical frame and lithostratigraphic series

The Pripiat Basin is a part, together with the famous coal-bearing Dniepr and Donetz Basins, of a long graben system extending from NW to SE between the Russian Platform and the Ukrainian Shield (figure 1) and filled principally with a thick Paleozoic series.

In the Pripiat Basin, the main part of the sediments is of Upper Devonian age (figure 2).

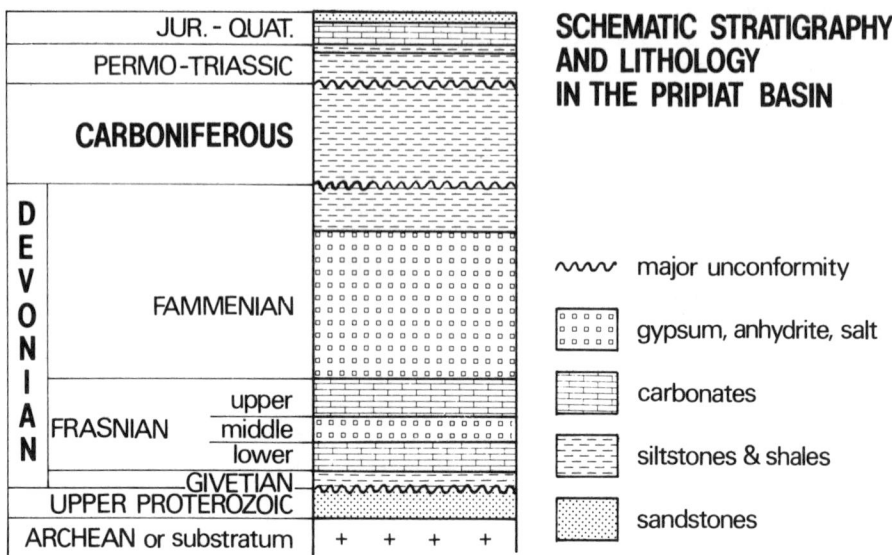

Fig. 2
The schematic litho-stratigraphic series of the Pripiat Basin

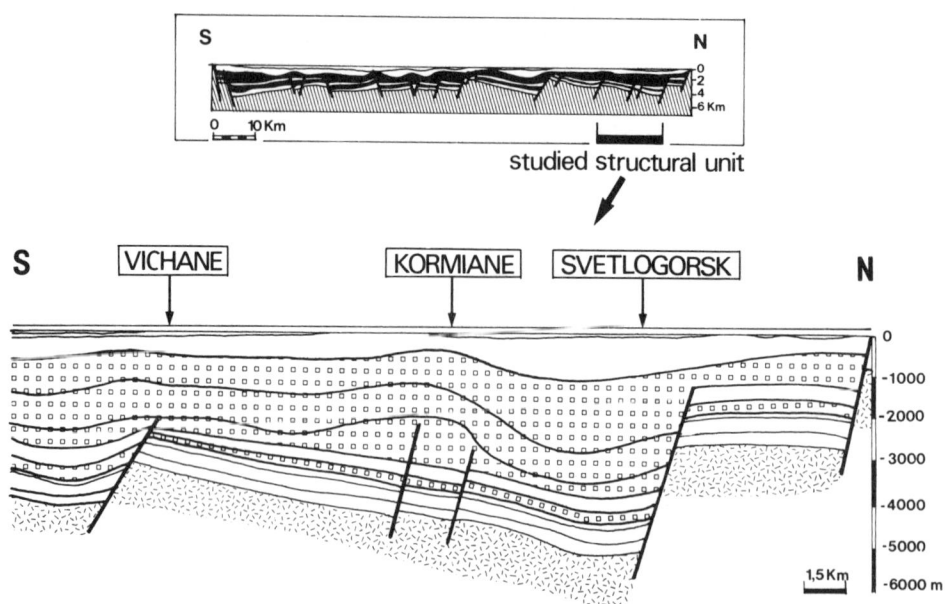

Fig. 3

The tectonic structure of the Pripiat Basin, from seismic profiles :
  a) general cross-section
  b) the tilted-block especially studied in the present paper.

b) The tectonic structure and the problem of the rifting age

The Pripiat Basin has the tectonic structure of an ancient intracontinental rift (figure 3). The substratum began to collapse at the end of the Givetian (BOTNEVA & MIZULINA, 1980). But the termination of the rifting phase, a key-point for the modelisation, is not known with a great accuracy.

The interpretation of seismic profiles (MACKHNACH et al, 1977 ; GARECKIJ, 1979) shows that :

- generally, the faults which delimit the tilted-blocks vanish in the Upper Salt. Nevertheless a few faults seem to stay working up to the Carboniferous.

- major unconformities do exist at the end of the Devonian and at the end of the Carboniferous, and the thicknesses of Latest Devonian, and of Carboniferous likewise, vary in the highest degree. However, these features could be explained by severe halokinesis episodes.

On the other hand, the sedimentation rates, in the wells where Latest Devonian and Carboniferous are present, remain very high, as during a rifting phase, up to the end of Tournaisian stage (KORNISCHCHEV, 1974 ; and our data).

We can conclude that :

- the rifting came to an end between the end of the Upper Salt ($\sim$ 361 MY) and the end of the Tournaisian (352 MY), which means an imprecision of about ten million years.

- deposition and erosion of Carboniferous terrains did happen, but with very different thicknesses from one point to another. Yet there is no place in the Basin where total thickness of Carboniferous exceeds 1300 metres (RADIONOVA & MAKSIMOV, 1981).

2. THE OIL SITUATION OF THE PRIPIAT BASIN

a) The position of the source-rocks in the series

In the Pripiat Basin, two lithologic units play at once the part of source rocks and reservoir rocks. They are the mainly carbonate formations of Infra-Salt and Inter-Salt, of Lower and Upper Frasnian age respectively (figure 2).

The Lower Salt formation, and above all the Upper Salt one, act as excellent seals for the oil traps.

b) The situation of production and exploration

Up to present, the Northern part of the Pripiat Depression has always revealed itself richer in oil-fields than the Central and Southern ones. Two kinds of arguments occur in order to explain that difference :

- firstly, arguments of structural nature, for instance :

. the faulting in tectonic units is much more pronounced in the South than in the North, and reverse faults are worthy of note (MAKSIMOV et al., 1977);

. the source-rocks has never been buried enough in the South, as tends to suggest the study of organic matter conducted by IFP in this area of the Basin;

- secondly, arguments of geothermal nature, for instance :

. the heat-flow at the base of the sediments has been higher in the North (TSYBULYA & ANPILOGOV, 1977);

. at the end of the Carboniferous, ultrabasic intrusions occurred in the North, which provided the energy required for the formation of hydrocarbons.

Anyway, exploited since 1965, the Pripiat Basin has made Bielorussia, during ten years, a major oil-bearing country. The Northern part has been intensively explored and is today fairly well known. In the seventies, production reached its highest level (8 Mt in 1975), and then the prospection has taken a renewed interest in the South region, but with poor results. Thus the hope to discover other fields in this area declines, without having found for that a final explanation.

In such a difficult context, the prospector needs the assistance of a predictive modelisation which, being able to grade the different phenomena, and being adjusted on the Northern zone, may help in better understanding what happened in the Southern zone.

3. THE ADOPTED MODELISATION

Our purpose is to reconstruct a geological, and especially geothermal, pathway which explains the available observations, and to examine what degree of accuracy one may look for to such a reconstruction.

In the Pripiat Basin, the observations are of :

ORGANIC GEOCHEMISTRY AND MODELS IN THE PRIPIAT BASIN

- stratigraphical and structural nature : the sedimentary column in several wells throughout the Graben ;

- thermal nature : some temperature gradient and heat-flow measurements ;

- geochemical nature : state of maturation of the organic matter.

The sequence of utilized models, and the survey of basic hypothesis

A first hypothesis is that during the Devonian and perhaps a part of the Carboniferous, the Pripiat Basin behaved as an intracontinental rift, to which we may apply the Mc KENZIE's geodynamic model (1978). The history of the thermal flux is thus linked to the subsidence history of the substratum.

All the models considered from this time onwards have been achieved in the Institut Français du Pétrole, and we shall utilize them in the following order :

- a "backstripping" method (BESSIS, 1983), applied here to well data, computes the tectonic subsidence of the substratum by decompaction of the sediments. The lithosphere response to charge movements is supposed to be an AIRY's one (local isostasy). In the Pripiat case, the porosity is introduced in terms of depth for six "pure" lithologies, and the sedimentary column is represented by twenty lithostratigraphic units, each one being a linear combination of these six lithologies.

The tectonic subsidence curve provides, by comparison to a chart, the value of β, that is to say the lithospheric stretching coefficient of the Mc KENZIE's model, and consequently a law for the decreasing of heat-flow since the end of the rifting phase.

- a thermal model (CHENET, 1984), applied alike to well data, computes at any moment the temperature of chosen horizons in the series. It considers a priori that the thermal transfers in the sediments result of conduction plus natural convection of connate waters during the compaction which accompanies the burial. But in general, the first phenomenon is to a large extent the leading one. Before the rifting phase, the heat-flow is arbitrarily taken as 33 mW/m^2, and after it, it decreases according to an exponential law. During the rifting phase, it increases in a linear manner between 33 and βx 33. In this model, the lithology is taken as homogeneous, which means a unique curve porosity/depth and unique conductivity data for the mineral matrix. It is a very rough approximation in the Pripiat Basin, where two

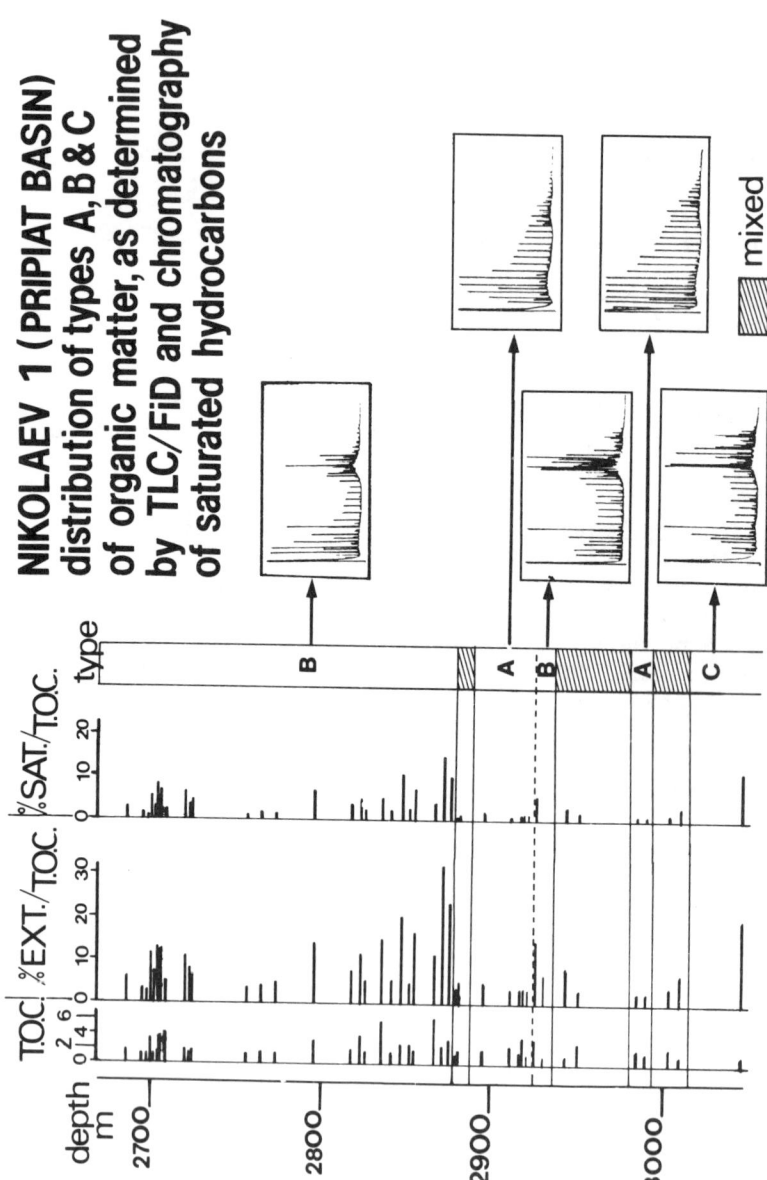

Fig. 4 The three families of organic matter as studied in Nikolaev 1 well.

thick salt formations are present, which have certainly petrophysical properties rather different from those of other rocks.

- a kinetic model for the evolution of organic matter (UNGERER, 1984, and this volume), based upon the ARRHENIUS'law. Each type of organic matter being characterized by a set of kinetic parameters determined as a preliminary (see hereafter), its theoretical state of evolution may be computed for a given thermal history, through three different parameters :

. the "transformation ratio", τ :

$$\tau = \frac{\text{part of the kerogen already transformed in hydroc.}}{\text{initial petroleum potential of the kerogen}}$$

. the residual petroleum potential, IH *

. the theoretical maximum temperature of pyrolysis, TMAX *.

These calculated parameters are then compared to their measured equivalents.

4. CHARACTERIZATION OF THE ORGANIC MATTERS AND MEASUREMENT OF THEIR MATURATION STATE

a) The methods

We have, deliberately, chosen simple, rapid and cheap methods to study the organic matter, as can do the prospector faced with an actual situation.

- the ROCK EVAL
In the first place, we access through a very great number of samples to quantities of carbon, to petroleum potentials, and to several indices which allow an estimation of the evolution rank.

Three parameters supply an evolution index, throughout the trend they present in terms of depth, providing that there are enough samples in the so-called "major phase of formation of hydrocarbons" (ESPITALIE et al., 1977) :

. TMAX, that is to say the temperature of maximum release of cracked hydrocarbons during pyrolysis (S_2 peak) ;

. IH = S_2/TOC, the residual petroleum potential of the present kerogen ;

. IP = $S_1/(S_1 + S_2)$, where S_1 is the quantity of free hydrocarbons in the rock.

REFERENCE SAMPLE	TYPE	E_1 X_{01}	E_2 X_{02}	E_3 X_{03}	E_4 X_{04}	E_5 X_{05}	E_6 X_{06}	E_7 X_{07}	E_8 X_{08}	A_i
49151	A	46	48	50	52	54	56	58		0.53×10^{14}
		8.7	0.0	43.3	203.7	63.	3.4	4.3		
49129	B	48	50	52	54	56	58	60	62	0.602×10^{15}
		7.8	10.4	23.5	140.2	169.5	26.0	9.4	3.6	
49172	C	46	48	50	52	54	56	58		0.133×10^{15}
		12.4	25.6	43.4	187.4	186.6	27.2	8.0		

Table 1 - Parameters of the kinetic model for the three types of organic matter observed in the Pripiat Basin

The primary cracking of kerogen into hydrocarbons is represented by several reactions in parallel obeying the ARRHENIUS' law and characterized by an activation energy E_i (Kcal/mole), a related part of the initial petroleum potential X_{oi} (mg/g T.O.C.), and a kinetic constant A_i (s^{-1}). These parameters are adjusted for each type of organic matter from ROCK EVAL data. The secondary cracking, into gas, is simulated by one single kinetic reaction (E = 52,6 Kcal/mole and A = 0,414 x 10^{15} s^{-1}) (UNGERER, 1984 and this volume).

In the second place, we may, with this apparatus, calibrate the kinetic parameters which control the transformations of each type of kerogen into hydrocarbons under the effect of time and temperature.

- the characterization of organic extract by IATROSCAN TH-10 TLC/FID analyser (BERRUT & JONATHAN, 1984) (1) plus the gas chromatography of saturated hydrocarbons

This method permits to distinguish, with fairly good sharpness, the type of organic matter, and especially slight differences among the three classical types of kerogens as defined by TISSOT & WELTE (1978).

b) The results in the Pripiat Basin

The organic matter of Devonian carbonate source-rocks, in the Pripiat Basin, has an algal origin, and is connected to the type II family of kerogens (ibid). But the previous method allowed us to discern three subclasses, named A, B & C, the latter having properties which bring it near to type I family (figure 4).
This has been conducted on a Southern well, Nikolaev 1, where the samples are numerous enough, immature, and rich in organic carbon. Every type, A, B & C, is recurrent in the series, the transition terms being mixed types. This is a prerequisite observation to ensure you can't proceed from one type to another through an evolution pathway. The existence of these three families of organic matter has then been controlled in a well of the Northern area, Vichane 18.

The kinetic parameters of the ARRHENIUS' law have been computed for each one of the three types, A, B & C, according to UNGERER's method (this volume). The results are presented in table 1.

Almost one hundred samples, from fifteen wells in the Northern region, provide sufficient organic carbon and sufficient S2 peak during ROCK EVAL pyrolysis, to be utilizable for adjustments. Generally speaking, we have not enough samples in the deepest parts of the wells, that is to say below 4000 metres, right in the oil-window. By another way, there is a very important hypothesis underlying the diagrams we will examine hereafter (figures 5 and 6) : to report the maturation data of different wells on the same diagram in terms of depth, an absolute prerequisite is that these wells were in the same relative structural situation when the principal phase of kerogen

---

(1) The organic extract is deposed on a silica rod (CHROMAROD) covered by an adsorbing film of silicagel. The rods are developed in the normal manner by solvent elution. The separations (saturated, aromatics, resins and asphaltenes) are performed as on conventional TLC plate. The direct and quantitative detection is applicable to almost all organic substances, with an automated system based on the classical GC/ Flame Ionisation (FID) principle.

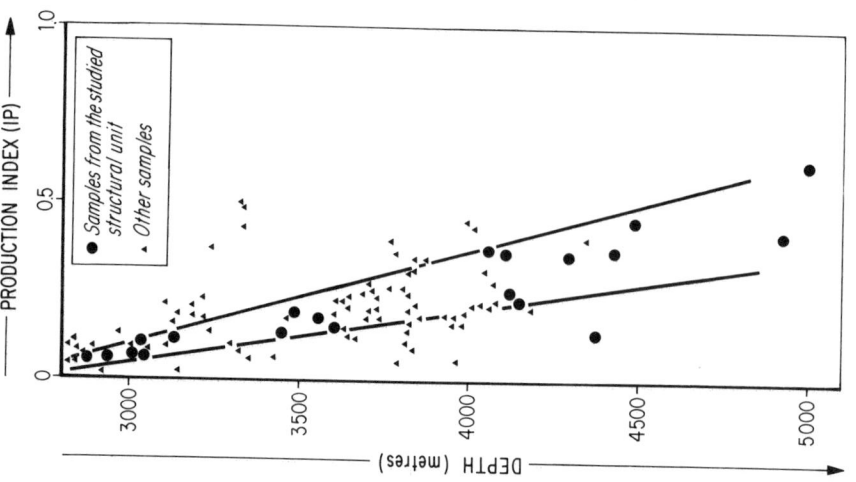

Fig. 6

The production index IP (= S1/S1+S2) in terms of depth for samples of the Pripiat Basin.

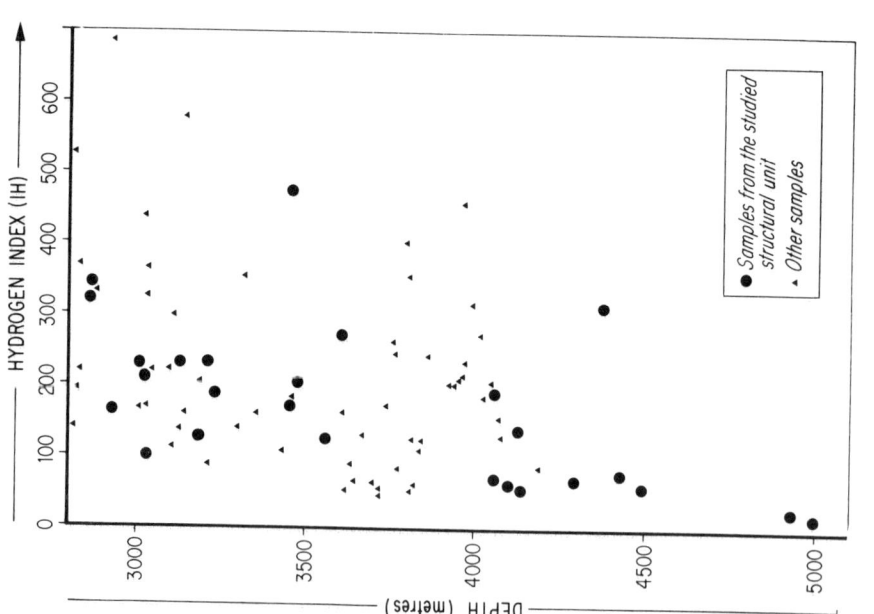

Fig. 5

The hydrogen index IH (= S2/TOC) in terms of depth for samples of the Pripiat Basin.

evolution took place, and that they have undergone thermal histories which are not very different. This is a disputable assumption in a rift context, but if it is not accepted, the work has to be made well by well, and there will never be enough data to conclude. In the subclasses of organic matter we are studying here, intermediate between types I and II, TMAX is not practically varying with increasing maturation (ESPITALIE, this volume), and so it is not a useful evolution criterion, neither a comparison tool for controlling the results of a modelisation. From a theoretical point of view, IH will be preferred (ibid.). Nevertheless, the figure 5 shows that its utilization in the present case is not very convincing. The very large dispersion of data may be ascribed to :

- the variety of kerogen types, intermediate values being mixed terms ;

- early alteration phenomena of the organic matter, perhaps before or during the sedimentation, as shown by representations like (IO, IH) diagram ;

- specific mineral matrix effects, particularly badly known in carbonate rocks.

So it is rather IP which will serve as a reference here. Its dispersion is large (figure 6), for the same reasons as already mentioned, but yet it is possible to draw a general trend zone, because its variations are progressive enough. Nevertheless, its a priori limitations must be kept in mind :

- IP is greatly disturbed by all the phenomena linked with the migration of hydrocarbons ;

- it is generally an under-estimation of τ, the "transformation ratio" (calculated by the models), which corresponds to effectively produced quantities, because you have never access to the whole products formed by cracking, the more as proportion of volatile fraction is greater (with increasing maturation). Practically, IP is useful until about 0.5.

Remark : the organic matter studied here doesn't contain vitrinite. However an equivalent of the vitrinite reflectance, named ER, has been measured by P. ROBERT. This index provides an additional and valuable information in order to locate, roughly, the oil-window. But it is not calibrated here with enough accuracy to be useful in controlling a modelisation. Moreover, it would be necessary to compute a theoretical ER, and this is a debatable process because, as for vitrinite reflectance, it is not based upon a physical law, whose mathematical expression is known and subject to experimental provings.

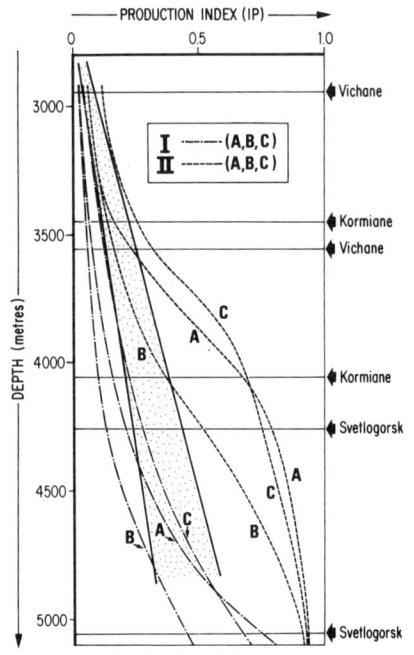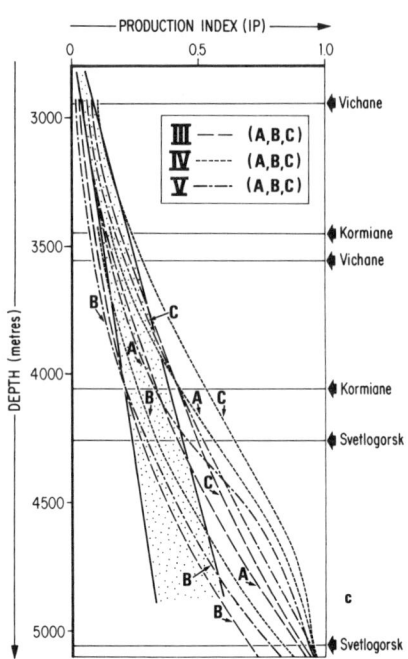

Fig. 7

Comparison between IP as measured into the tilted-block of Vichane-Kormiane-Svetlogorsk (see fig. 6) and the "transformation ratio" τ computed by the models for several geological hypothesis. Each hypothesis is represented here by a taper of three curves obtained for the three types (A, B, C) of organic matter. The adjustment can be made keeping in mind that IP is an underestimation of τ, above all for values greater than 0.5, that is to say below about 4500 metres.

I - No Carboniferous, $\beta = 1.5$, end of the rifting = 360 MY.
II - 1000 m Tournaisian (360-352 MY), eroded immediately at the end of Tournaisian age (352-347 MY), $\beta = 1.7$, end of the rifting = 352 MY.
III - 1000 m Mississipian (360-320 MY), eroded during the remaining of the Carboniferous (320-286 MY), $\beta = 1.5$, end of the rifting = 360 MY.
IV - 500 m Tournaisian, eroded during the remaining of the Carboniferous, $\beta = 1.7$, end of the rifting = 352 MY.
V - 500 m Tournaisian, eroded just after Tournaisian age, $\beta = 1.7$, end of the rifting = 352 MY.
For all these hypothesis, the radiogenic heat-flow is taken as 15 mw/m2.

5. THE RESULTS OF THE MODELISATION. COMPARISON WITH ORGANIC GEOCHEMISTRY MEASUREMENTS. DISCUSSION.

The preceding modelisation has been applied to three wells, drilled in the same tilted-block, Vichane 18, Kormiane 1, and Svetlogorsk 1 (figure 3).

The "back-stripping" model, whatever burial history tested (deposition and erosion of 0,500 or 1000 m of Carboniferous, during Tournaisian or during Mississipian), gives for β values between 1.2 and 1.7. In absence of Carboniferous, or with a slow deposition of this one (Mississipian), we must prefer β = 1.5, and a rifting phase ending around 360 MY. On the contrary, with a deposition concentrated during the Tournaisian, we will take the figures of 1.7 and 352 MY.

In order to adjust the thermal data available on the Pripiat Basin (TSYBULYA & ANPILOGOV, KUTAS et al., 1979 ; CERMAK, 1979 ; PINTCHUK et al., 1983), we had to take a rather weak radiogenic heat-flow : 15 mW/m^2.

The figure 7 shows the comparison between τ , computed by the models for different geological histories, and the measured IP presented above (figure 6).

a) the lack of deposition during Carboniferous doesn't allow to reach present levels of maturation. That's especially true at Svetlogorsk 1, which is the most buried well among those studied at IFP.

b) so it is necessary to have sedimentation during the Carboniferous

- A 1000 metres Tournaisian (360-352 MY) deposit, eroded at any moment of the remaining Carboniferous, is a too strong load.

- In return, the adjustment tool used (τ/IP) doesn't enable to choose between three other hypothesis :

. a 1000 metres Mississipian (360-320 MY) deposit, eroded during the remaining of Carboniferous (320-286 MY) ;

. a 500 metres Tournaisian deposit, eroded during the remaining of Carboniferous (352-286 MY) ;

. a 500 metres Tournaisian deposit, eroded just after the Tournaisian age (352-347 MY).

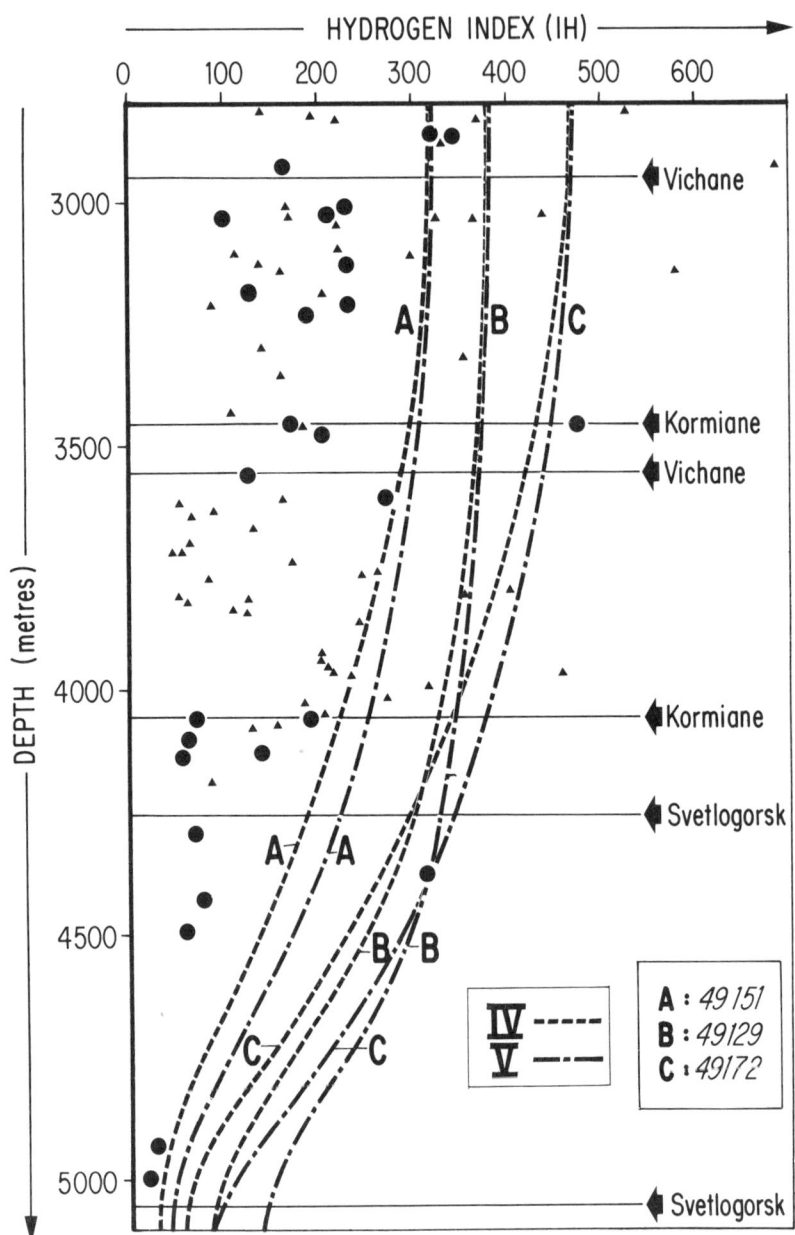

Fig. 8

Comparison between IH as measured into the studied tilted-block (see fig. 5) and IH* computed by the models for the hypothesis IV and V (see figure 7 caption).

ORGANIC GEOCHEMISTRY AND MODELS IN THE PRIPIAT BASIN

As a trial measure, in spite of the bad quality of the (IH, depth) diagram (figure 5), an attempt is proposed in the figure 8 to compare IH and IH* for the two last hypothesis, which are the maximum and the minimum of the satisfactory adjustments between $\mathcal{C}$ and IP. There is no contradiction between the two parameters, but as expected this comparison doesn't permit to choose in the present case. Nevertheless, it is easily understood that this parameter is theoretically a very sensitive one in the main zone of hydrocarbons formation.

CONCLUSION

In a moderately known basin, the usual criterions of organic matter evolution, as provided by ROCK EVAL pyrolysis, are likely to control to some extent the validity of geological hypothesis tested with the help of a modelisation. Thus, in the case of a structural unit (a tilted-block of a Devonian rift) considered in the Pripiat Basin, for which the Carboniferous history is rather badly known, we can conclude that sediments had to be deposited during this epoch, but no more than 1000 metres if the deposition rate is low (Mississipian) or 500 metres if it is high (Tournaisian). Such an assessment, established on the most studied parts of the downwarp, may constitute a useful guide for exploration in the Southern region.

It would be of interest to examine the efficiency of other evolution criterions, organic or inorganic, in such a geological setting. These criterions often require sophisticated techniques, which consequently can't be conducted on a large amount of samples, and it is sure that the parameters we have discussed here are unbeatable in terms of price and easy use.

Substantial improvements could be realized in the adopted modelisation, for instance a better way to take into account varied lithologies in the thermal model, and especially salt layers. This is a work in progress at IFP today. On the other hand, the calibration of the kinetic model, realized on rocks after extraction, could be re-examined using kerogens, and the adjustment done for these kerogens, a method which eliminates the possible matrix effects.

The most conspicuous conclusion concerns the sampling quality. Organic geochemistry measurements and methods offer currently to the petroleum explorationist very performing tools. But the only way to turn the models into account is to achieve a representative sampling of the region under study, especially without large gaps in the main zone of oil formation.

REFERENCES

BESSIS, F. (1983)
Thèse de 3è cycle, Paris.

BERRUT, J.B., JONATHAN, D. (1984)
Application du système CCM-DIF à l'analyse quantitative des constituants lourds du pétrole.
Symposium International sur la caractérisation des huiles lourdes et des résidus pétroliers.
Lyon, 25-27 juin 1984.

BOTNEVA, T.A., MIZULINA, N.N. (1980)
Short geological characterization of the Northern part of Pripiat downwarp (french transl.)
Appendix of a report of the Sovietic Institute for Scientific Research about petroleum exploration.

CERMAK, V. (1979)
Heat-flow map of Europe.
Terrestrial heat-flow in Europe, Springer-Verlag, Berlin, Heidelberg, New-York, pp. 3-40.

CHENET, P.Y. (1984)
Thermal transfer in sedimentary basins paleotemperature reconstruction and maturation studies in the Gulf of Lion Margin.
Thermal phenomena in sedimentary basins, Int. Coll., Bordeaux, 7-10 june 1983, ed. by B. DURAND, Technip, pp. 235-246.

ESPITALIE, J., LAPORTE, J.L., MADEC, M., MARQUIS, F., LEPLAT, P., PAULET, J., BOUTEFEU, A. (1977)
Méthode rapide de caractérisation des roches mères, de leur potentiel pétrolier et de leur degré d'évolution.
Rev. Inst. Fr. du Pétrole, vol. 32, n°1, pp. 23-42.

GARECKIJ, R.G. (1979), ed.
Pripiat Basin tectonics (in russian).
MINSK, "Nauka i tekhnika".

KORNISHCHEV, V.S. (1974)
Carboniferous evolution of regional faults in the Pripiat Basin (in russian).
Doklady Akadimi Nauk BSSR, 18, n°11, pp. 1032-1034.

Mc KENZIE, D. (1978)
Some remarks on the development of sedimentary basins.
Earth Planet. Sc. Letters, 40, pp. 25-37.

MAKHNACH, A.S., SVERZHINSKIJ, A.I., DIKHTIEVSKIJ, V.V. (1977)
USSR Geology. III. SSR of Bielorussia (in russian).
Moscou, "Nedra".

MAKSIMOV, S.P., ANTSUPOV, P.V., GONCHARENKO, B.D. (1977)
Ways for increasing effectiveness of geological exploration for
oil and gas in the Dnieper-Donets depression.
Geologiya Nefti i Gaza, n°4, pp. 9-14.

PINTCHUK, A.P., GERASIMOVA, J.A., IVANOV, U.P., KOROVKIN, B.A.
(1983)
Régimes thermiques dans la Dépression du Pripiat (in russian).
english transl. in Sov. Geol., n°7.

RADIONOVA, K.F., MAKSIMOV, S.P. (1981)
Organic matter geochemistry of Devonian sediments of the Pripiat
Basin (in russian).
Geokhimija organicheskogo Veshchestva i neftematerinskie porody
fanerozoja, pp. 169-181.

TISSOT, B.P., WELTE, D.H. (1978)
Petroleum Formation and Occurrence.
Springer Verlag, 540 p.

TSYBULYA, L.A., ANPILOGOV, A.P. (1977)
Question of inhomogeneity of the thermal field in the Pripiat
downwarp.
Doklady Akad. Nauk BSSR, vol. 21, n° 4, pp. 339-341.

UNGERER, P. (1984)
Models of petroleum formation : how to take into account
geology and geochemical kinetics.
Thermal phenomena in sedimentary basins, Int. coll., Bordeaux,
7-10 june 1983, ed. by B. DURAND, Technip, pp. 235-246.

ACKNOWLEDGEMENTS

We are grateful to MM. CAILLET, MARQUIS and PICHAUD for the
analytical part of this work ; to Mme DEROZE and Mme FEUILLAS for
the figures ; to Mme GRANDIN for the dactylography ; and to MM.
CHENET, ESPITALIE, PELET and UNGERER for useful discussions and
suggestions.

E. BROSSE, A.Y. HUC[1]

ORGANIC PARAMETERS AS INDICATORS OF THERMAL EVOLUTION IN THE VIKING GRABEN

ABSTRACT

The purpose of the present paper is to evaluate if the current evolution indicators provided by the organic geochemistry (coloration of palynomorph species, vitrinite reflectance and TMAX of the Rock-Eval pyrolysis) can help in controlling models of thermal history, in a complex geological basin such as the Viking Graben (North Sea).

I. INTRODUCTION

A set of modelling tools is currently available to the explorationist in order to reconstruct the geological history of a sedimentary basin. As far as hydrocarbon generation is concerned the main aspect of this reconstruction is the assessment of the paleotemperature history of the basin. The adjustment of such a thermal model is performed in order to account for, at least, the present day thermal pattern of the basin. However, in most of the cases, such a test is inadequate: informations on the current thermal pattern of a basin are not available to a sufficient extent, measurements are not always reliable and moreover the data are not unambiguous evidences of the paleotemperature history (Espitalié, 1984).

So, other adjustment procedures are needed. They are based upon indices, measuring a chemical or physical property of the sediments, which is known to change as a function of the amount of thermal energy received by the sediment during its geological history.

---

(1) *Institut Français du Pétrole, Rueil-Malmaison, France.*

Two kinds of thermal indicators can be used: those relying on mineral transformations, such as the clay minerals recrystallisations (Welde, 1984) and those relying on organic matter transformations (Oudin, 1984).

Three organic thermal indicators will be considered in this paper: the coloration of palynomorph species, the vitrinite reflectance and the temperature of S_2 peak generated by Rock-Eval pyrolysis.

Practically, the adjustment is performed by testing successive hypothesis in order to achieve a fit between the results of a model and actual data obtained from a series of samples. The key point of such a procedure is the selection of a suitable set of samples for which the measured organic parameters are solely a reflection of the considered thermal evolution.

The selection is easy when sampling can be done on a single well in which the burial history was continuous and in which organic bearing sediments are homogeneous and thick enough to monitor a large maturation range.

Such a situation occurs for instance in the Mahakam delta (Ungerer, 1984).

In the Viking Graben area (North Sea) the situation is much more complex. The moderate thickness of the source rocks in the basin leads to a selection of samples related to wells located in different structural settings. Such wells might have undergone different thermal and burial histories, moreover in such a geological framework the lateral changes of the properties of the sedimentary organic matter might be important. Nevertheless, the present depth stays generally the only objective ranking parameter which can be used in order to design a reference set of samples in a basin. The aim of the present paper is to evaluate if the current thermal indicators provided by the organic geochemistry can help in controlling models in such a complex geological situation.

A large amount of organic geochemical data generated from Jurassic samples by SNEA(P), CFP, IFP, IKU and Robertson Research International have been available to us, and have been included in a data base of over 6 000 samples.

II. ORGANIC MATTER IN THE JURASSIC OF THE VIKING GRABEN AREA

During the Jurassic the structural evolution of the North Sea have been dominated by the development of fault controlled rifts of which is the Viking Graben. In this complex tectonic framework the sedimentation was highly variable and organic matter reflects these changes of depositional environment (Cornford, 1984).

(1) <u>Lower Jurassic/Statfjord formation</u>
The Statfjord formation occurs as fluvio-deltaic sediments including coals and minor organic lean shales: Total Organic Carbon (TOC) < 1%.

(2) <u>Lower Jurassic/Dunlin group</u>
The Dunlin group is an admixture of silts, shales and marine sandstones deposited during a transgressive period reaching its maximum during Toarcian/Aalenian time.
Few coals occurences are recognized and shales containing up to 3% TOC are reported in the Northern part of the Viking Graben.

(3) <u>Middle Jurassic/Brent group</u>
Middle Jurassic time is characterized by the deposition of fluvio-deltaic sediments including mainly sandstones and coals interfingered by few coaly shales.

(4) <u>Upper Jurassic/Heather formation</u>
A widespread marine transgression occurs during the late Jurassic. During this period of time, organic containing mudstones are laid down. In the Heather formation TOC averages 2-2.5%.

(5) <u>Upper Jurassic/Draupne formation</u>
Marine transgression is in progress, Draupne formation is mainly developed as an organic rich mudstone and exhibits high TOC values (up to 10%). However the organic content of the Draupne formation varies considerably in the lateral and vertical dimensions (Demaison et al. 1983, Huc et al. 1984).

In the two latter formations (Draupne and Heather) belonging to the Viking group, the organic material basically derives from marine organisms, but according to its lateral and vertical quality variations (Demaison et al. 1983, Huc et al. 1984) it can be mixed with variable amount of organic matter of terrigenous origin or can be altered, to various extent, during sedimentation processes. Nevertheless organic matter in the Viking group can be considered as marine derived when compared to organic matter associated with Brent deposits where terrestrial plants are the main precursors.

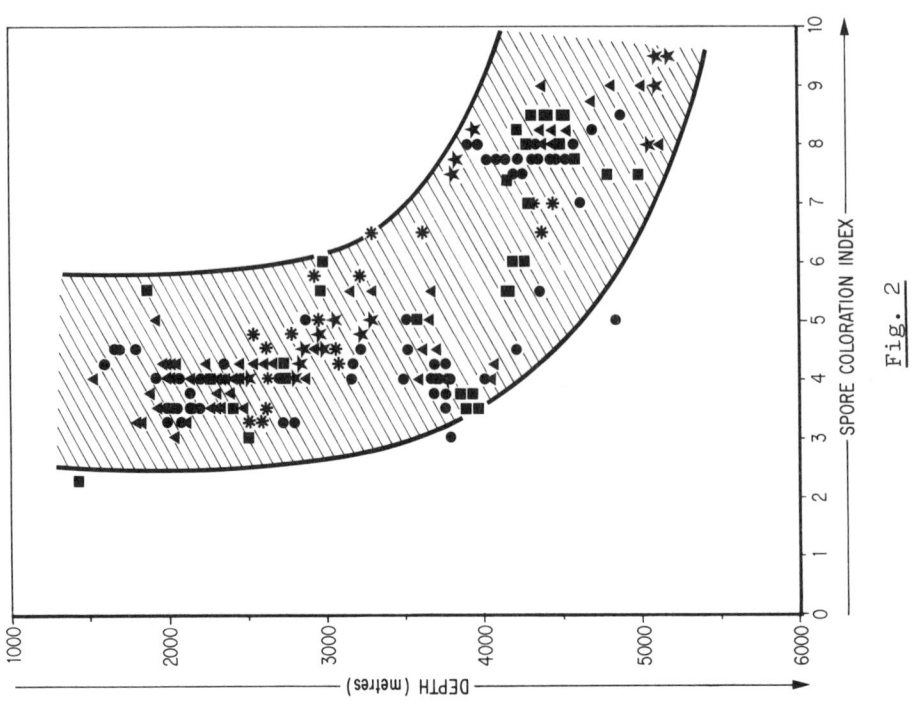

Fig. 2
Spore coloration index versus depth. Symbols represent data from different formations.

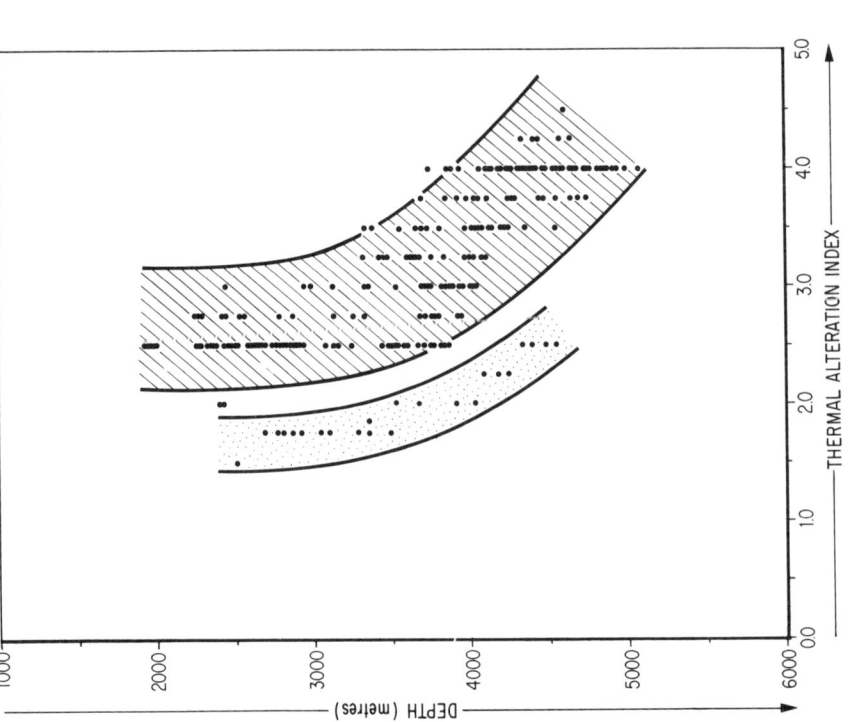

Fig. 1
Thermal alteration index versus depth. Comparison of data from two laboratories.

ORGANIC PARAMETERS IN THE VIKING GRABEN

III. ORGANIC MATURATION INDICATORS

1. Microscopic examination in transmitted light

By studying microfossils and translucent organic particles, one can observe the gradual darkening and ultimately the complete opacity of the organic matter as thermal evolution proceeds. Evaluation of the degree of evolution is carried out by matching palynomorph species coloration against a reference scale. The result is expressed as a numerical index ranging between 1 and 5 for the thermal maturation index (TAI) and between 1 and 10 for the spore coloration index (SCI). Some advantages of this technique are:

a) It can be used over a wide range of evolution stages ;

b) It can be applied on stratigraphical microfossils, avoiding mistakes due to allochtonous material occurrence ;

c) It is a low cost analysis when linked to stratigraphical studies.

But the limitations are mainly due to the fact that this is a visual comparison with a coloration scale. This scale is a set of slides which are not strictly identical between laboratories. Moreover data are dependent of the palynomorph species under consideration.

This situation results in discrepancies between laboratories. Figure 1 examples this laboratory dependence for TAI measurements in the Viking Graben area.

The technique by itself is not very precise and plotting generated by such a method are rather scattered (figures 1 and 2).

2. Vitrinite reflectance (R_o)

The reflectance of a coal maceral, the vitrinite, measured in reflected light is a widespread and well calibrated method which has long been used by coal petrographers to evaluate coal ranks (Van Krevelen, 1961). Further on, reflectance measurements have been extended to particles of vitrinite occurring in organic matter dispersed in sediments (Vassoevich et al., 1969 ; Lopatin, 1971 ; Alpern, 1980). The precision of vitrinite reflectance (R_o) determination for coals is held to be 0.1% when R_o < 1.5% (Alpern, 1980). However when dispersed organic matter is considered, the precision of data decreases down to 0.5% (Alpern, 1980). Reflectance measurements must be made only on vitrinite since the others particles follow different law of increasing reflectivity. In other words

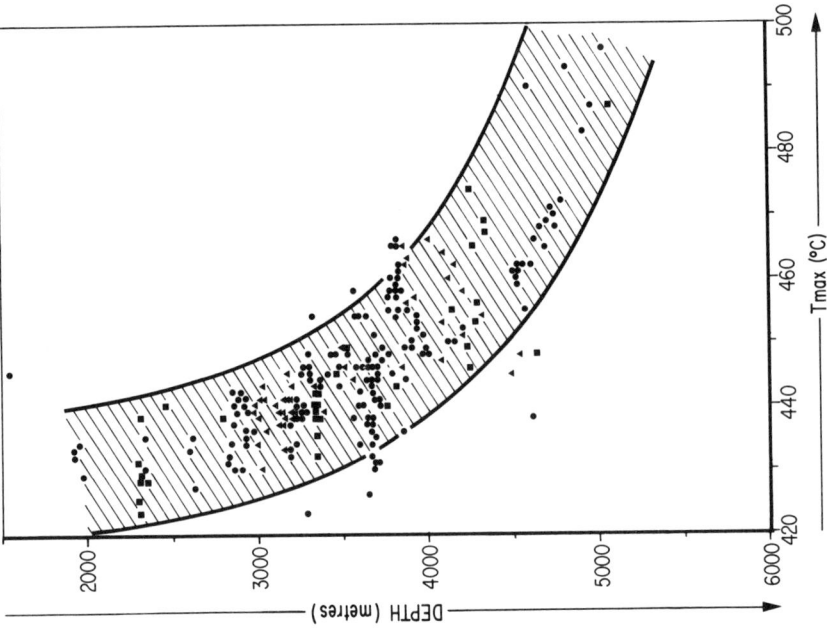

Fig. 4
TMAX versus depth for Brent/Vestland, Dunlin and Statfjord coals.

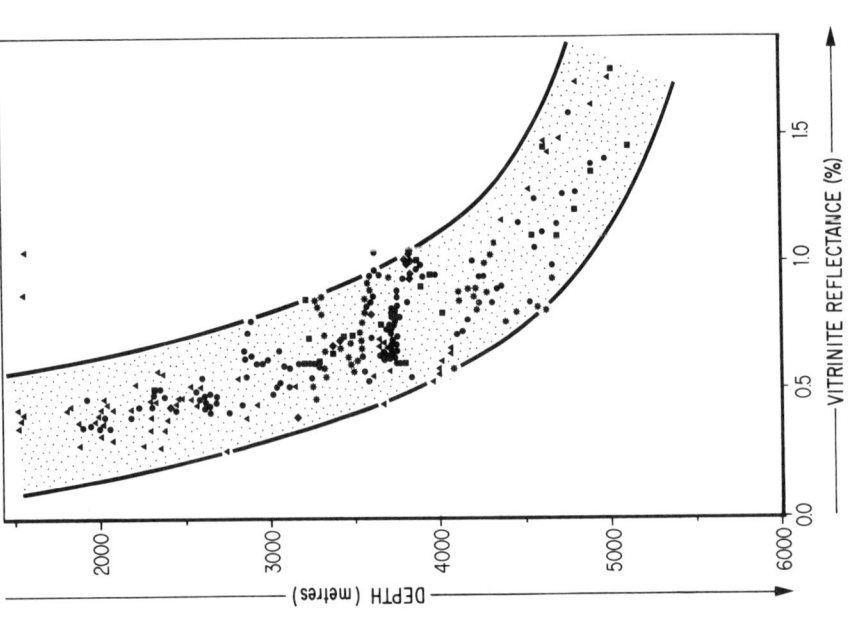

Fig. 3
Vitrinite reflectance versus depth, for Brent and Vestland samples. Symbols represent data from different laboratories.

sediments must contain vitrinite particles in sufficient amount (on a statistical point of view) to be considered. A purely marine derived organic matter will be devoid of autochtonous vitrinite particles and should be discarded for any direct measurement of reflectance. Moreover precise vitrinite identification is a key point because it is well known that even among the "vitrinite family" variation in the microstructure will affect the reflectance value.

According to these remarks only samples from the Brent group have been considered in figure 3. In this set of data it should be pointed out that vitrinite reflectance determinations are very consistent between the different laboratories. In very few wells discrepancies arise between laboratories, data from these peculiar wells have been discarded from our compilation.

The resulting R_O versus depth profile(figure 3) shows a regular increase of R_O as a function of depth.

The spread of reflectance values is rather large when compared with literature data (Dow, 1977 ; Alpern, 1980 ; Robert, 1980 ; Oudin, 1984 ; Teichmuller & Durand, 1983).

As previously mentionned interlaboratories discrepancies cannot be put forward to justify this pattern, moreover the spread related to each laboratory is as large as the general spread. This scattering of reflectance values might be explain by a diversity of materials from which the vitrinite is formed, or more certainly by the regional heterogeneity of the thermal regime in this area, as reflected by the present day thermal pattern of the basin (Oxburgh et al. 1981 ; Eggen, 1984), together with regional differences in burial history of sediments.

3. Rock-Eval TMAX

The measurement of the temperature at the maximum of S_2 peak generated by organic matter cracking during Rock-Eval pyrolysis is a well developed method to monitor the state of maturation of sedimentary organic material. It is a very rapid technique which allows the processing of numerous samples. Moreover this parameter can be measured on any kind of samples.

Limitations arise from samples which exhibit a shape and a size of the S_2 peak preventing any precise determination of TMAX: low organic carbon content (< 0.5%), low S_2 generation (< 0.3 mg/g TOC) or for bitumen rich samples for which high molecular weight hydrocarbons might interfer with the S_2 peak and cause a TMAX shift.

All these samples (TOC < 0.5%, S_2 < 0.3 mg/g TOC, and S_1/S_1+S_2 > 0.5) have been discarded from our compilation.

An other limitation, so called "Mineral Matrix effect" results from a retention effect of mineral matrix which might occur in some samples. When acting this phenomenon alters the information by delaying the released cracked hydrocarbons (S_2) and then results in an overestimation of TMAX (Espitalié et al., 1980). Providing that coals are mostly pure organic material no "mineral matrix effect" are expected to alter their TMAX values.

The TMAX values versus depth for Jurassic coals are plotted in figure 4. The TMAX curve shows the same regular increase with depth that the R_o curve (figure 3). As previously stated for R_o, the spread of TMAX values is rather large. The same explanations involving organic precursors variations or more probably the heterogeneity of the thermal regime can be put forward to account for this observation.

The relationship between TMAX and R_o for coals samples for which both parameters have been measured fit perfectly, in terms of location and dispersion, with the curve published for coals by Teichmuller et al., 1983 (figure 5).

This observation comforts our assumption that the spread of dots on TMAX versus depth and R_o versus depth does not reflect an analytical quality problem or an interlaboratories discrepancy, but rather a response to a complex geological situation.

The Jurassic sequence of the Viking Graben area provides a unique opportunity to compare the TMAX variation of two different kinds of organic matter. The Brent, Statfjord and Dunlin coals which derive from continental higher plants and the Viking group matter which derives mainly from marine organisms.

The TMAX versus depth curve of Viking sediments can be seen in figure 6, in which it is compared to the Jurassic coals profile.

Two remarks can be outlined:

First, the spread of TMAX values is larger for the Viking samples than for the Jurassic coals.

Second, the two profiles are not surimposed: at a given depth TMAX of Viking sediments are lower than TMAX of Jurassic coals, moreover the shift of TMAX between the two groups of samples seems to increase as a function of depth.

Such a behavior cannot be explained by a "mineral matrix effect" since the temperature alteration should have been inverse that the one observed. Moreover a comparison of TMAX between whole sediments and kerogens isolated from the same rocks (by dissolving the inorganic minerals with HCl and HF) as displayed in figure 7 shows that no

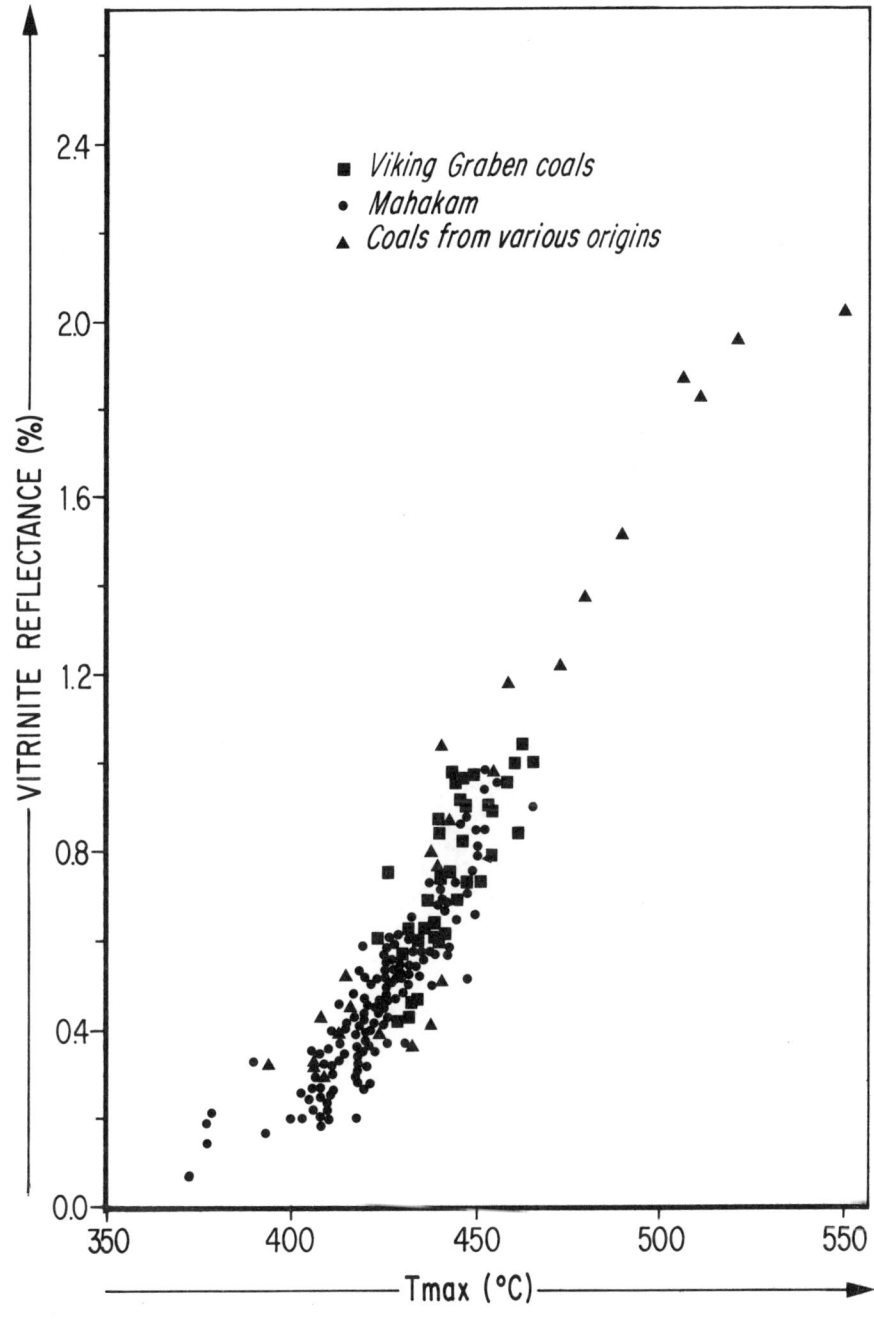

Fig. 5

TMAX versus Vitrinite Reflectance.
Comparison between Jurassic coals from the Viking Graben and coals from various origins (TEICHMULLER & DURAND, 1983).

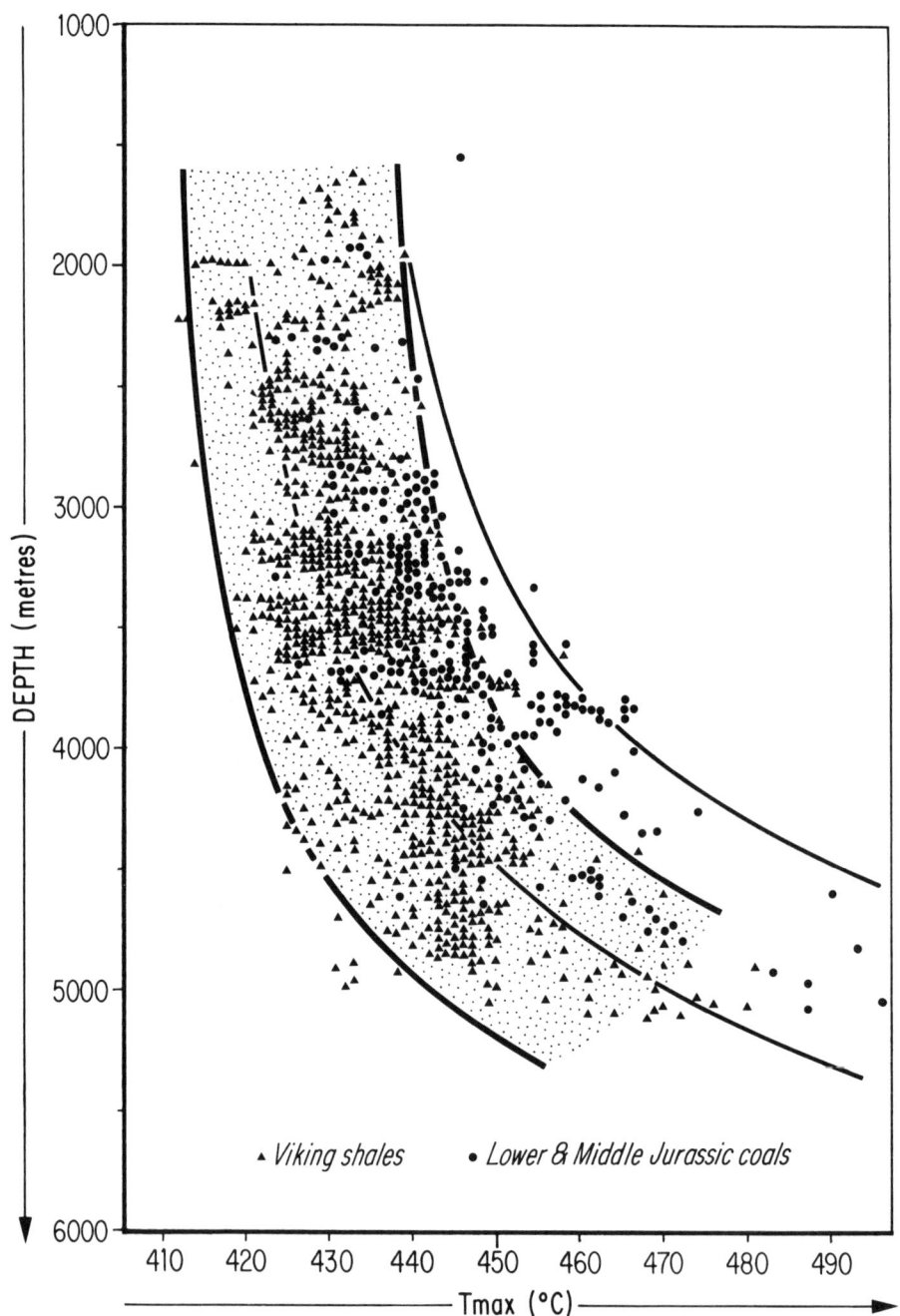

Fig. 6
TMAX versus depth. Comparison between Viking shales and Lower and Middle Jurassic coals.

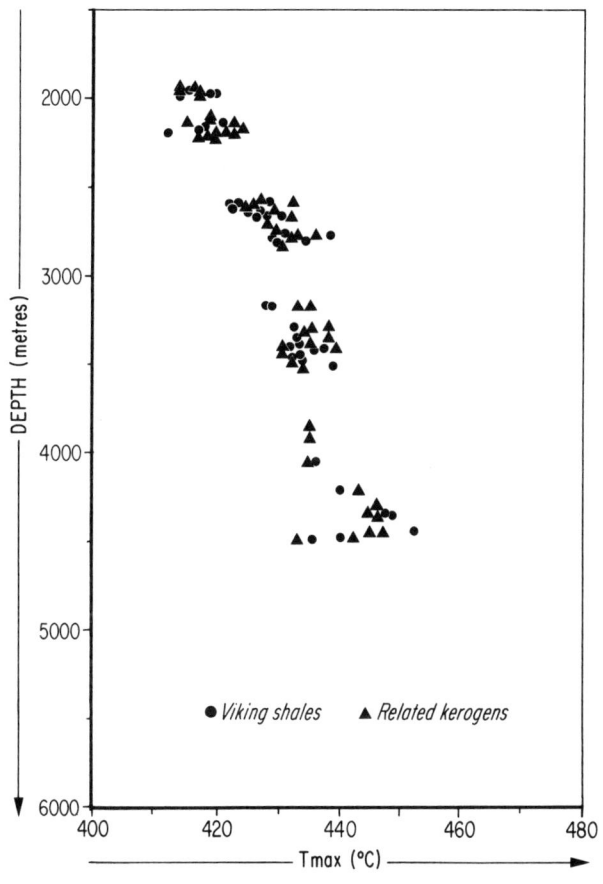

Fig. 7

TMAX versus depth. Comparison between whole sediments and related Kerogens.

systematic difference can be pointed out and then suggests that "Mineral matrix effect" is a subordinate factor in TMAX determination for Viking group sediments. The difference is, thus, related to the very properties of the two kinds of organic matter.

As a consequence the large dispersion of TMAX values of the Viking sediments would be interpreted as a reflection of the variation of organic matter nature inside the Viking group itself.

IV. CONCLUSION

This study of organic maturation indicators in the Jurassic of the Viking Graben area suggests that:

1) The coloration of palynomorph species is a valuable, unexpensive, rough indicator of the degree of thermal evolution, but its laboratory dependency and its low degree of precision does not justify its use for testing model ajustment ;

2) R_o and TMAX stay the best choice to quantify the stage of maturation of organic matter ;

3) TMAX is affected by the type of organic matter under study. For instance a significant difference occurs between the Jurassic coals and the Viking organic material. Changes of organic material nature inside the Viking group itself has been inferred to explain variability of TMAX in this sedimentological interval ;

4) Regional heterogeneity of the thermal regime and of the burial history of sediments have also been inferred to explain the rather large scatter of R_o and TMAX.

To sum up, the best interval to be used to calibrate the thermal evolution of organic matter in the Viking Graben area is the Brent group (or its southern equivalent: the Vestland group) in which the organic matter is the most homogeneous. Moreover the occurrence of coals provides well suited samples for vitrinite reflectance determination and TMAX measurement without any "mineral matrix effect" concern.

However special care should be taken in selecting wells on similar structural settings in order to minimize heterogeneity of the thermal regime. These considerations should be kept in mind when a series of samples has to be selected in order to test adjustment procedures of thermal models in this basin.

REFERENCES

1. ALPERN, B. (1980). - Pétrographie du kérogène, in Kerogen, ed. by B. DURAND, Technip, pp. 339-383.

2. CORNFORD, C. (1984). - Source Rocks and Hydrocarbons of the North Sea, in Introduction to the Petroleum Geology of the North Sea, ed. by K.W. GLENNIE, Blackwell Scientific Publications, pp. 171-204.

3. DEMAISON, G., HOLCK, A.J.J., JONES, R.W., MOORE, G.T. (1983). - Predictive Source Bed Stratigraphy ; a Guide to regional Petroleum Occurrence, in Origin, Migration and Accumulation of Hydrocarbons, 11th World Petroleum Congress, London, pp. 1-13.

4. DOW, W.G. (1977). - Kerogen studies and geological interpretations in J. of Geochem. Explor., 7, p. 79-99.

5. EGGEN, S. (1984). - Modelling of subsidence hydrocarbon generation and heat transport in the Norwegian North Sea, in Thermal phenomena in sedimentary basins, int. coll., Bordeaux, 7-10 June 1983, ed. by B. DURAND, Technip, pp. 271-283.

6. ESPITALIE, J., MADEC, M., TISSOT, B. (1980). - Role of Mineral Matrix in Kerogen Pyrolysis : Influence on Petroleum Generation and Migration, in Am. Ass. of Pet. Geol. Bull., vol. 64/1, pp. 59-66.

7. ESPITALIE, J. (1984). - Tentative reconstruction of geothermal paleogradients in some wells of the Rhine Graben, in Thermal phenomena in sedimentary basins, int. coll., Bordeaux, 7-10 June 1983, Technip, pp. 147-159.

8. HUC, A.Y., IRWIN, H., SCHOELL, M. (1984). - Organic Matter Quality Changes in an upper Jurassic Shale Sequence from the Viking Graben, in Organic Geochemistry in Exploration of the Norwegian Shelf, Stavanger, 22-24 Oct. 1984 (in press).

9. LOPATIN, N.V. (1971). - Temperature and geological time as factors of carbonification, in Akad. Nauk SSSR, Izv. Ser. Geol., n° 3, pp. 95-106.

10. OUDIN, J.L. (1984). - Thermal maturation indices in organic geochemistry, in Thermal phenomena in sedimentary basins, int. coll., Bordeaux, 7-10 June 1983, Technip, pp. 117-125.

11. OXBURGH, E.R., ANDREWS-SPEED, C.P. (1981). - Temperature, thermal gradients and heat-flow in the south-west North Sea, in ILLING & HOBSON, ed., Petroleum geology of the Continental Shelf of north-west Europe, pp. 114-151.

12. ROBERT, P. (1980). - The optical evolution of kerogen and geothermal histories applied to oil and gas exploration in B. DURAND, Kerogen, Technip, pp. 385-414.

13. TEICHMÜLLER, M., DURAND, B. (1983). - Fluorescence microscopical rank studies of liptinites and vitrinites in peat and coals, and comparison with results of the Rock-Eval pyrolysis in Int. Journal of Coal Geology, 2, pp. 197-230.

14. UNGERER, P. (1984). - Models of Petroleum Formation: how to take into account geology and chemical kinetics, in Thermal phenomena in Sedimentary Basins, int. coll., Bordeaux, June 7-10, 1983, ed. by B. DURAND, Technip, pp. 235-246.

15. VAN KREVELEN, D.W. (1961). - Coal, rééd. 1981, Elsevier, 514 p.

16. VASSOYEVITCH, N.B., KORCHAGINA, J.I., LOPATIN, N.V., TCHERNITCHEV, V.V. (1969). - Principal phase of oil formations Westnik Mosk. Univ., 6, pp. 3-27 (in russian). Engl. trans. Intern. Geol. Rev., 1970, 12, pp. 1276-1296.

17. VELDE, B. (1984). - Transformations of clay minerals, in Thermal Phenomena in Sedimentary Basins, int. coll., Bordeaux, June 7-10, 1983, ed. by B. DURAND, pp. 111-116.

ACKNOWLEDGEMENTS

This research has been conducted as a part of a project including Elf Aquitaine Norge, Total Marine Norsk, IKU and IFP participations. We thank EAN and TMN for the permission to publish this paper, and we are grateful to Robertson Research International for the authorization of using their data.

P. UNGERER, J. ESPITALIÉ[1],
F. MARQUIS, B. DURAND[1]

USE OF KINETIC MODELS OF ORGANIC MATTER EVOLUTION FOR THE RECONSTRUCTION OF PALEOTEMPERATURES.

APPLICATION TO THE CASE OF THE GIRONVILLE WELL (FRANCE)

INTRODUCTION

The kinetic models of organic matter (OM) evolution are most useful to quantify the processes of oil and gas formation in the undrilled areas of sedimentary basins. Additionnally, they may help to estimate the paleotemperatures encountered by the sediments, provided geochemical data are available.

Several kinds of kinetic models have been exposed in the literature, the majority of them being based on the classical formulation for the thermal cracking of petroleum products: first-order kinetics and Arrhenius law. Among these models, two approaches may be pointed out:

- 1° The first one (Tissot and Espitalié, 1975, Ungerer, 1983) considers maturation data documented in sedimentary basins and from pyrolysis experiments. The pyrolysis, that is a heating of the OM in an inert atmosphere, is known to be a simulation of the natural evolution from a qualitative and approximate quantitative standpoint, which advocates such a calibration of the kinetic parameters. The method needs however to reconstruct the paleotemperatures for the geochemical series used as reference, which may bring noticeable uncertainties.

- 2° The second approach examplified by Pitt (1961), Juntgen & Klein (1975), Akihisa (1978) and Lewan (1985) among others, considers only pyrolysis data from immature samples. This method does not rely upon any paleotemperature reconstruction and may be applied even to uncomplete geological series. Beside these important

(1) *Institut Français du Pétrole, Rueil-Malmaison, France.*

advantages, it needs a very careful calibration in order to keep a reasonable accuracy when extrapolating from the pyrolysis to the sedimentary basins.

The calibration presented here is a development of the former one (Ungerer, 1983), which allows both approaches mentioned above, depending on the data available. It is deviced to use standard Rock-Eval pyrolysis results and thus permits to derive easily the kinetic parameters specific for the type of OM considered.

The main use of a calibrated set of kinetic parameters is to predict the localization and amount of oil formation in the undrilled areas of a basin in the course of exploration. This topic is treated by other models which have to take into account the thermal history of the basins, such as the THEMIS model (Doligez, this conference). It will not be discussed here, since we will focus on the use of geochemical models for paleotemperature reconstruction.

The calibration on pyrolysis presents a serious advantage for this purpose, compared to former calibrations. Indeed, it does not rely on the paleotemperature reconstruction of any reference series. In the example presented here, that is a well of the Eastern part of the Paris Basin, the maximum depth of burial of the Paleozoic series is unknown, because of the Hercynian tectonic phase which caused uplift and erosion. The estimation of paleotemperature is thus a help for the reconstruction of burial.

I. CALIBRATION OF THE KINETIC MODEL

1. Kinetic scheme

The kinetics of OM evolution is represented in the same way as described by Tissot (1969), Tissot & Espitalié (1975) or Ungerer (1983): the formation of oil is described by several parallel first order reactions obeying the Arrhenius law:

$$\frac{dx_i}{dt} = -A_i \exp(-\frac{E_i}{RT}) x_i$$

with: E_i activation energy of reaction i (kcal/mole)
      A_i Arrhenius constant (s^{-1})
      R   2 cal/mole.K

KINETIC MODELS OF ORGANIC MATTER EVOLUTION

T temperature (K)

x_i residual petroleum potential relative to reaction i (mg/g org.C)

x_{io} initial petroleum potential relative to reaction i (mg/g org.C)

t time (s)

The formation of oil is quantified by:

$$q = \sum_i (x_{io} - x_i)$$

with: q amount of oil generated (mg/g org. C)

In this model, the process of organic diagenesis, that is the loss of oxygenated functions, is not accounted for. As a consequence, the denomination "initial" refers to the end of diagenesis. For instance, the initial petroleum potentials x_{io} are expressed in proportion to the amount of organic carbon at the end of the diagenesis. This amount may be different from the organic carbon at the beginning of diagenesis because CO_2 may be lost during diagenesis in significant amounts (Boudou, 1981).

The process of secondary degradation of oil into gas by thermal cracking is neither accounted for in this study. However, the parameters that are characteristic of the kerogen, such as the residual petroleum potential, are independant from secondary cracking and thus are well reproduced by the model, even in the metagenesis zone where oil is transformed into gas.

From one case to another, the distribution of the various chemical bonds of the kerogen may vary considerably and cause severe changes as well in kinetic behavior as in petroleum potential. As a consequence, the kinetic parameters (E_i, A_i, x_{io}) are specific for the organic matter type considered. The number of parallel reactions involved in the model must be great enough, so that the distribution of actual activation energies E_i and Arrhenius constants A_i can be reproduced with a good approximation. In order to describe correctly the shape of pyrolysis peaks, a spacing of activation energies of about 2 kcal/mole is necessary, at least in the range of the major part of the distribution (50 to 60 kcal/mole generally). The possible activation energies range from about 40 kcal/mole (Ungerer, 1983) to 80 kcal/mole that corresponds very roughly to the most stable carbon-carbon bonds. A unique value of A_i is assigned to each of the 18 activation energies E_i selected. As a consequence, there are 19 free parameters for the kinetic model: the initial petroleum potentials x_{io} and the common Arrhenius constant A. This limitation of the degrees of freedom makes the calibration easier and more reproducible.

KINETIC MODELS OF ORGANIC MATTER EVOLUTION

2. Calibration method

A computer program, called OPTIM, has been deviced to calibrate the kinetic parameters x_{io} and A_i for a given type of organic matter, on the basis of geochemical data that may be:

- Hydrogen index (HI) measured by the Rock-Eval method on a homogeneous geochemical series covering all the oil formation zone.

- Temperature history of the levels corresponding to HI values.

- Digitalized pyrolysis curves (peak S_2 registered by the Rock-Eval method) for various heating rates, obtained from samples representative of the OM type.

The calibration may be achieved either on a geochemical series (HI, temperature histories) and one pyrolysis curve (which is similar to Ungerer 1983) or on several pyrolysis curves of an immature sample with heating rates different by a factor greater than 100.

The determination of the kinetic parameters is made by an optimisation technique. Initial values of x_{io} and A_i are set and theoretical values of residual potentials and reaction rates are computed from the model. Assimilating the overall residual potential (Σx_i) to HI values corrected for organic carbon loss and the oil formation rate ($-\Sigma \frac{dx_i}{dt}$) to the hydrocarbon formation rate along the pyrolysis, an error function $F(A_i, x_{io})$ is then computed by summation of the quadratic errors. The search for the minimum of $F(A_i, x_{io})$ is achieved by a gradient method with variable step, and it is repeated to ascertain reproducibility.

3. Calibration on pyrolysis only

Changing the heating rate in Rock-Eval pyrolysis from 50° C/min to 0.2° C/min results in a shift of the peak towards lower temperatures by about 100° C - this value depending on the type of OM considered -. This shift keeps the general shape of the peak unchanged. For instance, coals exhibit a tailing peak, significant of high energy bonds for both high and low heating rates. On the opposite, marine origin OM present a sharply decreasing peak whatever the heating rate.
The area of the peak may vary with the heating rate but these variations of the oil yield have been neglected in this study by a suitable correction.

The use of pyrolysis curves for the calibration of the kinetic parameters needs that the pyrolysis temperature is well controlled. For this purpose, the temperature has been measured by a thermocouple in contact with the OM along several blank pyrolysis, which leads to correct the displayed temperatures by a linear relationship to obtain the actual temperatures (Espitalié, this congress).

4. Test of the method (Mahakam Delta)

The Mahakam Delta, of Tertiary age, is located in Indonesia. The sedimentary organic matter, which is homogeneous all over the basin, is of continental origin (Boudou, 1981). Coal beds and shales contribute comparably to the overall organic carbon in the basin. A coal sample from the end of the diagenesis zone (T_{max} = 431° C) has been selected and submitted to pyrolysis at 55.7° C/min and 0.557° C/min, which resulted in a shift of the S_2 peak as mentioned above (fig. 1). After the calibration by the OPTIM program, the general shape of both peaks is well reproduced (fig. 1) and the distribution of the activation energies (fig. 2) is found centered on 58 kcal/mole. This distribution is much alike the one determined by calibration on a reference series of the Mahakam Delta coals (Ungerer, 1983).

The set of kinetic parameters has been tested by comparison between the observed evolution of HI and T_{max} at Handil and the predicted evolution from the model. For this purpose, the temperature history of the various levels have been reconstructed on the basis of a burial reconstruction and of a constant geothermal gradient of 37° C/km. This value, somewhat higher than the available present-day temperature data, is yet consistent if the uncertainty on temperatures is considered. Besides, the temperatures may have been higher in the past, since the fluid convection at depth (Perrin, 1983) causes unstable temperature regimes.

The decrease of HI values in the oil formation zone at Handil is well reproduced by the model (fig. 3), which advocates the method. Although the very large extrapolation on the time scale (from about one day to several million years, that is 9 orders of magnitude), the temperature at which oil formation is predicted seems valid with no more than ± 15° C uncertainty. Another test of the method consists in comparing T_{max} values at Handil with computed ones. Assuming the same temperature history as above for a given level, it is possible to determine the theoretical distribution of activation energies at the corresponding depth (which is of course not the same as the initial one in figure 2). It is then easy to determine the theoretical S_2 peak at this depth and then the peak temperature T_{max}. The comparison (fig. 4) shows a difference between computed and observed T_{max}, the computed evolution occuring about 500 m deeper than the natural one. This difference is probably related to the uncertainty of the method. It is not surprising that the uncertainty on T_{max}, which is related to a reaction rate, is more important than on HI which is related to a reaction yield. Additionnally, it is possible that the sample used for calibration has already been submitted to some limited evolution, which prevented the first stages of evolution to be correctly described by the model.

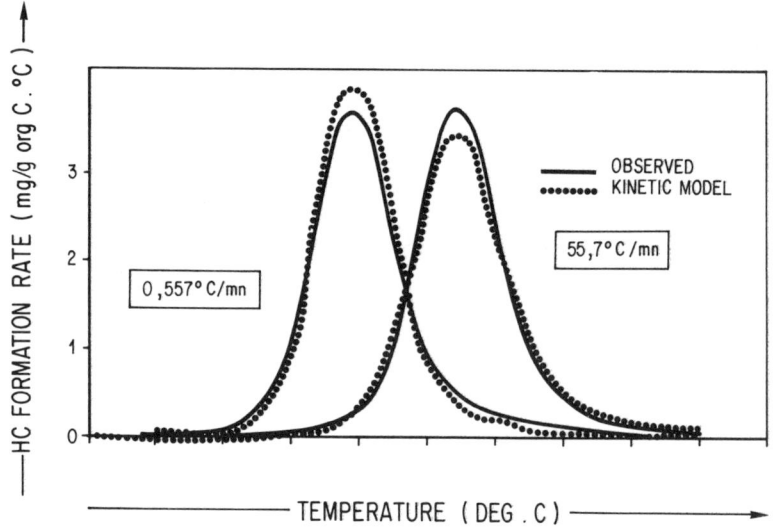

Fig. 1
COMPARISON OF PYROLYSIS CURVES AS OBSERVED AND AS COMPUTED BY THE KINETIC MODEL OF MAHAKAM DELTA COALS

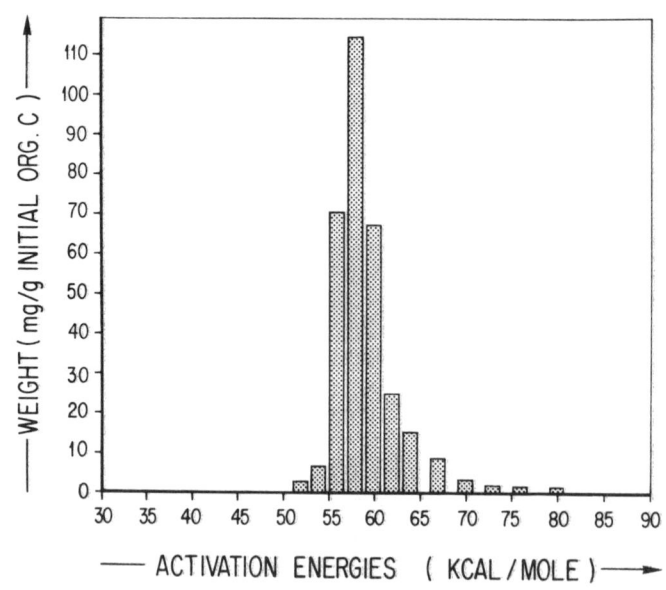

Fig. 2
DISTRIBUTION OF THE ACTIVATION ENERGIES INVOLVED IN THE KINETIC MODEL OF MAHAKAM DELTA COALS

KINETIC MODELS OF ORGANIC MATTER EVOLUTION

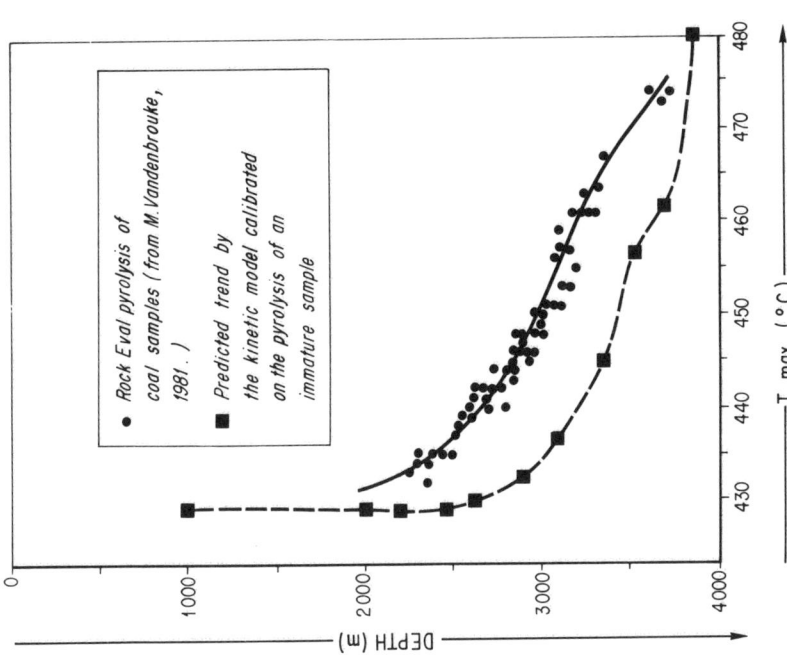

Fig. 4

EVOLUTION OF T MAX AT HANDIL
H8 AND H9 BIS WELLS (MAHAKAM DELTA)

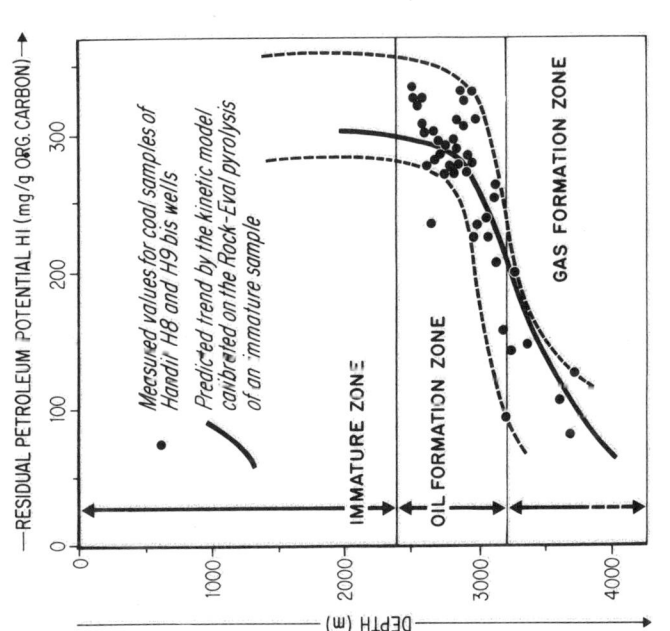

Fig. 3

LOCATION OF THE OIL FORMATION ZONE AT HANDIL
Mahakam Delta, Indonesia

537

II. PALEOTEMPERATURE RECONSTRUCTION FOR THE WELL GIRONVILLE 101

1. The Gironville paleozoic coals

As described by Durand (this congress) the paleozoic series in the eastern part of the Paris Basin is made of sand, shale and coal beds intercalations which deposited in a great deltaic system through Wesphalian times. The origin of the organic matter is uniformly continental. The progressive decrease of HI and the increase of T_{max} versus depth (fig. 5 and 6) show that the oil formation zone (T_{max} = 435° C) starts at 1 200 m approximately, and that the whole wet gas and dry gas generation zone ($T_{max} \geqslant$ 470° C) is documented between 2 500 and 5 700 m. These evolution stages cannot be explained by the present burial depths, unless very high geothermal gradients (more than 50° C/km) are assumed. In fact, the present-day geothermal gradients in the Paris Basin range between 30 and 40° C/km.

The burial reconstruction of the Paleozoic series is difficult. Sedimentation has been probably continuous throughout the Westphalian and Stephanian times as known regionally. During Permian times, the Hercynian tectonic phase resulted in the erosion of the whole Stephanian and the uppermost part of the Westphalien at the location of the Gironville 101 well. From the thickness of the Stephanian in preserved locations, the amount of erosion is at least 1 000 m. Sedimentation occured again through Jurassic and Cretaceous times, and was followed by a slight uplift from the beginning of Tertiary to present time. The resulting erosion was probably limited to a few hundred meters. Since the Mesozoic cover is about 1 100 m thick presently, its maximum thickness was at most 1 500 m. These informations cannot help to determine whether the maximum burial occured in the Paleozoic or at the end of the Mesozoic. As we will see, our results may help to know it.

2. Calibration of the kinetic parameters

In the same way as described above for the Mahakam Delta, an immature sample was selected at the top of the series, at a depth of 1 170 m, and submitted to pyrolysis at 55.7° C/min and 0.223° C/min (fig. 7). The T_{max} of this sample is 434° C, which corresponds to the limit between the diagenesis zone and the oil formation zone. The distribution of activation energies, as found by the OPTIM program, shows a dysymetric shape with a maximum at 54 kcal/mole (fig. 8). This shape is possibly due to the disappearance of the low activation energies during the very first evolution stage. As it was the case for the Mahatam Delta calibration, the Arrhenius constant found by the program lies in the expected range (10^{13} to 10^{16} s^{-1}) which is given by theoretical considerations (Frost & Pearson, 1961) and experimental determinations (Albright et al., 1983)

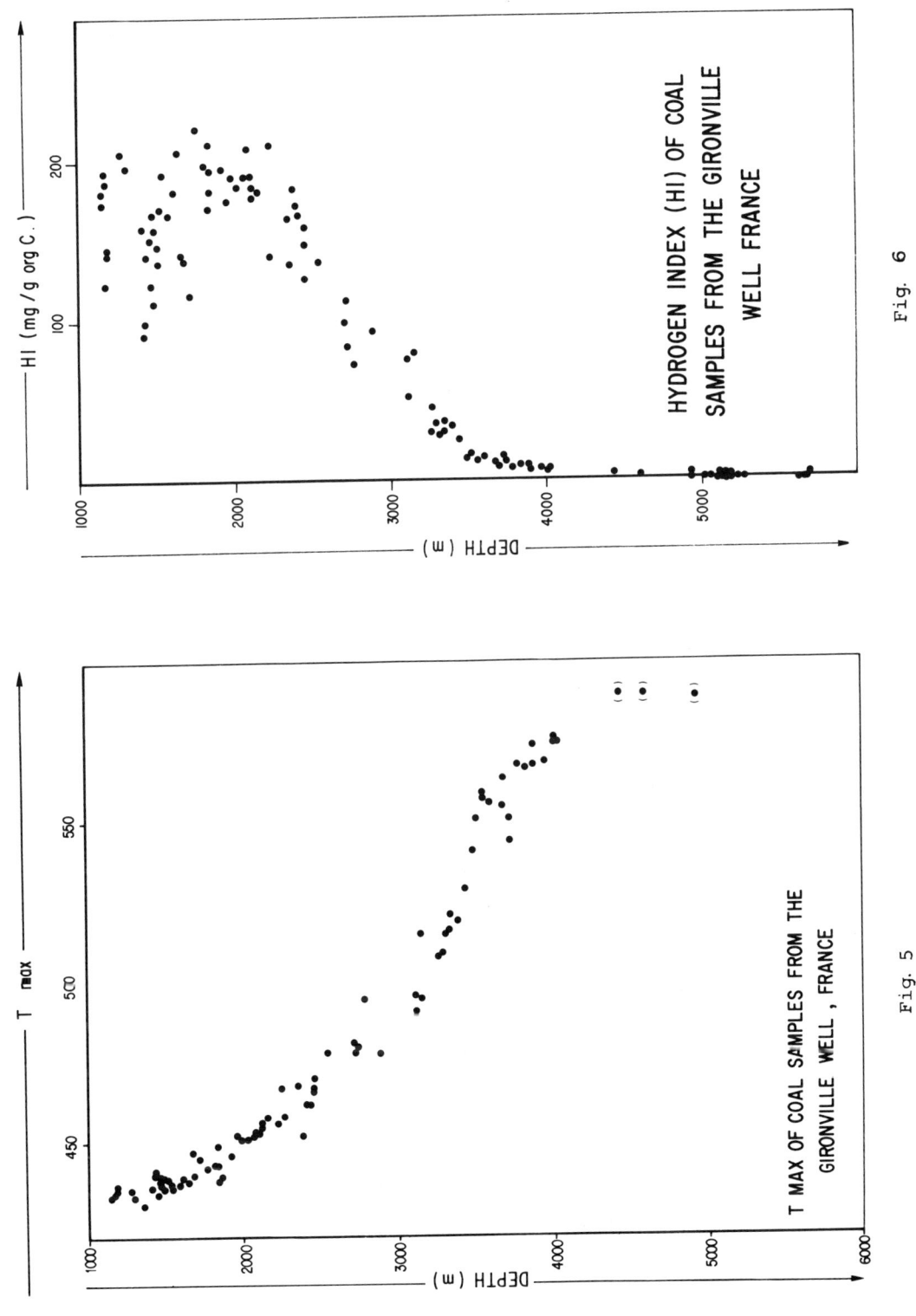

Fig. 6

Fig. 5

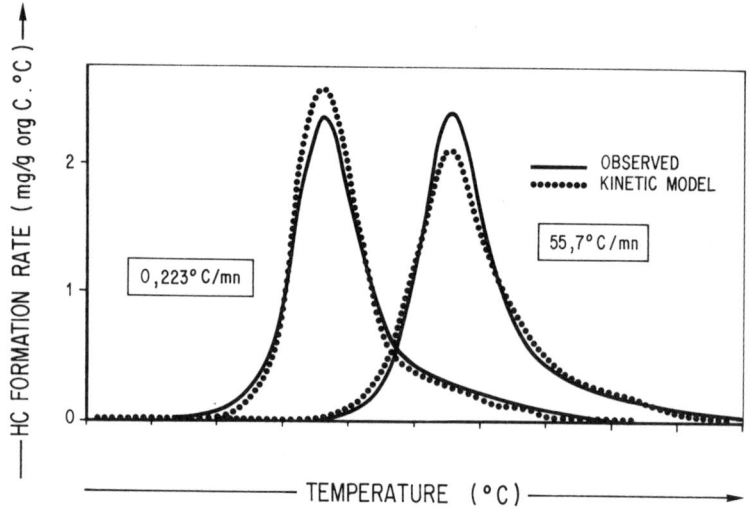

Fig. 7
COMPARISON OF PYROLYSIS CURVES AS OBSERVED AND AS COMPUTED BY THE KINETIC MODEL OF GIRONVILLE COALS

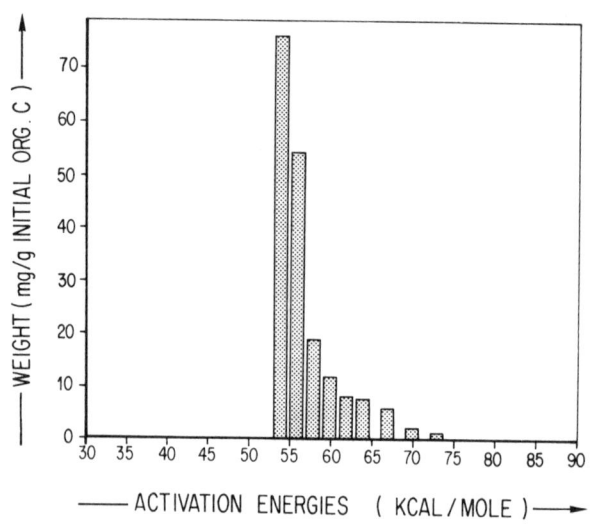

Fig. 8
DISTRIBUTION OF THE ACTIVATION ENERGIES INVOLVED IN THE KINETIC MODEL OF GIRONVILLE COALS

for carbon-carbon bonds. This by the way justifies the calibration method.

3. Sensitivity of geochmical maturation to heating

At first, the set of kinetic parameters was used to evidence that the organic matter is essentially sensitive to the maximum paleotemperature reached. As shown by the tableau I, the maturation

Tableau I

Comparison of the kinetic model results for six thermal histories (the histories 2, 4 and 6 summarize the histories 1, 3 and 5 by a constant temperature interval)

Thermal history	HI	T_{max}
1. From 25° C to 115° C in 90 MY and then from 115° C to 120° C in 10 MY	178.9	433
2. 120° C for 10 MY	179.7	433
3. From 25° C to 145° C in 120 MY and then from 145° C to 150° C in 10 MY	74.5	468
4. 150° C for 10 MY	75.9	466
5. From 25° C to 175° C in 150 MY and then from 175° C to 180° C in 10 MY	31.7	530
6. 180° C for 10 MY	31.7	530

computed by the model is approximately the same when the whole temperature history is replaced by a steady interval at the maximum paleotemperature. The duration of this interval is the time spent within 5° C from the maximum paleotemperature in the whole temperature history. As a consequence, the maturation parameters T_{max} and HI are little sensitive to paleotemperatures lower than the maximum. They cannot be used to reconstruct the whole temperature history. Since time and temperature may compensate each other in the maturation process, the paleotemperature cannot be assessed exactly if the time is unknown.

In fact, the effect of time is slight compared to temperature. From the model, doubling the reaction time is equivalent to an increase of temperature of 3 to 5° C. This means that an important uncertainty on the time does not result in great errors on the maximum paleotemperature.

This aspect is evidenced by figure 9 which represents the influence of geologic time for a constant geothermal gradient (30° C/km). For each depth was computed from the model, assuming a heating at a temperature T = 10 + 30 Z with various times from 0.3 to 100 MY. The shape of the HI versus depth curve keeps the same when times varies, translating about 250 m for each time increase by a factor of 3.

The effect of the geothermal gradient is larger, as demonstrated by figure 10. Although extreme values such as 20° C/km or 60° C/km have not been figured, the maturation curve is much steeper and translated more than 1 500 m when the geothermal gradient changes from 30 to 50° C/km.

4. Paleotemperature reconstruction

Comparing figure 10 curves with the observed values of figure 6 leads to select 40° C/km as the best fit, although the model does not predict a decrease to zero HI as it is observed in the Gironville well at 4 000 m. This is caused by the temperature program of the standard pyrolysis method (Rock-Eval II) which ends at 550° C. This final temperature is not sufficient to measure the tail of the pyrolysis peak. This tail is accounted for by the model, because it is calibrated on special pyrolysis experiments (up to 650° C). The hypothesis of a paleogradient around 40° C/km is also supported by the computation of T_{max} (fig. 11). For the same reason as in the case of the Mahakam Delta coals, the model suffers probably a noticeable inaccuracy in the first stages of maturation because the sample used for calibration is nearly in the oil generation zone. Nevertheless, the average slope of the observed trend between T_{max} = 450° C and T_{max} = 550° C is best reproduced with the hypothesis of 40° C/km. Of course, the depths corresponding to a given maturation level are not the same from the 40° C/km computation and from the observed trend, the latter being about 800 m shallower. This means that be apparent erosion since maximum burial is 800 m. It suffers however an important uncertainty because it depends on the determination of the geothermal gradient, on the surface temperature selected and the duration of heating assumed.

The above estimation assumed 30 million years were spent at the highest paleotemperature within ± 5° C, for the purpose of computation. In our case, this duration cannot be assessed and only general considerations may be used to set boundary values. From the general evolution path, which is consistent with a gentle temperature profile, it can be stated that the maximum heating lasted at least one million years. Otherwise, a gentle temperature profile would not have established, considering that the time constant of heat conduction within a sedimentary column is about one million years. On the other hand, a duration of more than 50 million years for maximum burial is

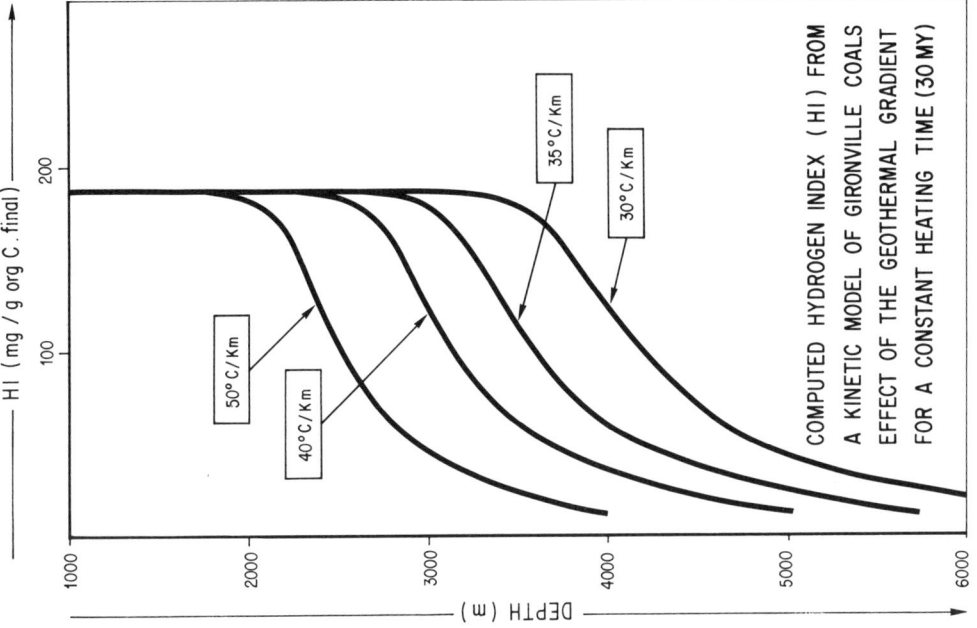

Fig. 10

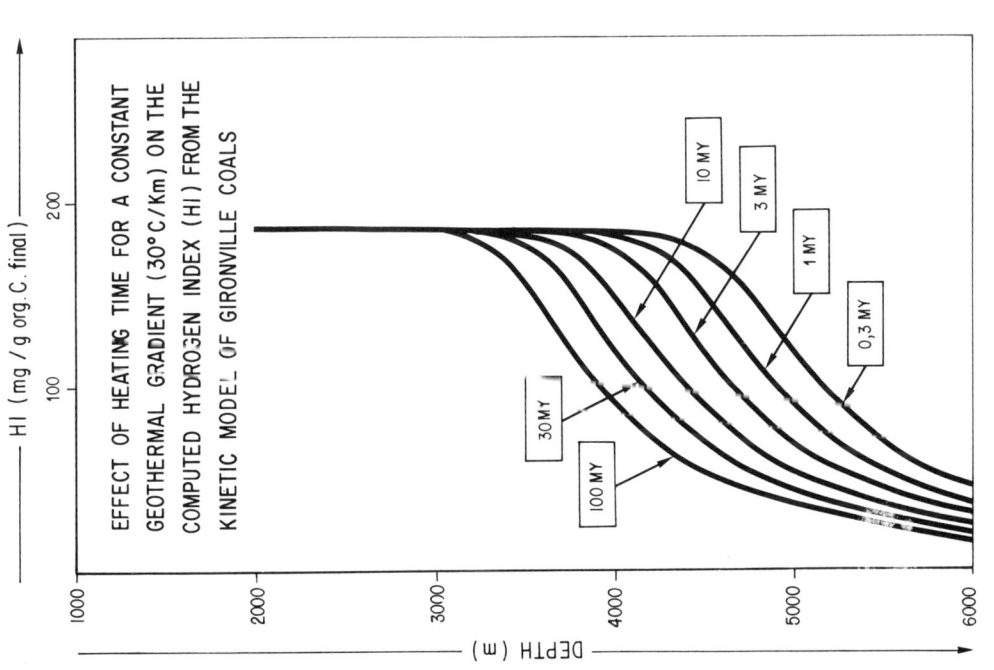

Fig. 9

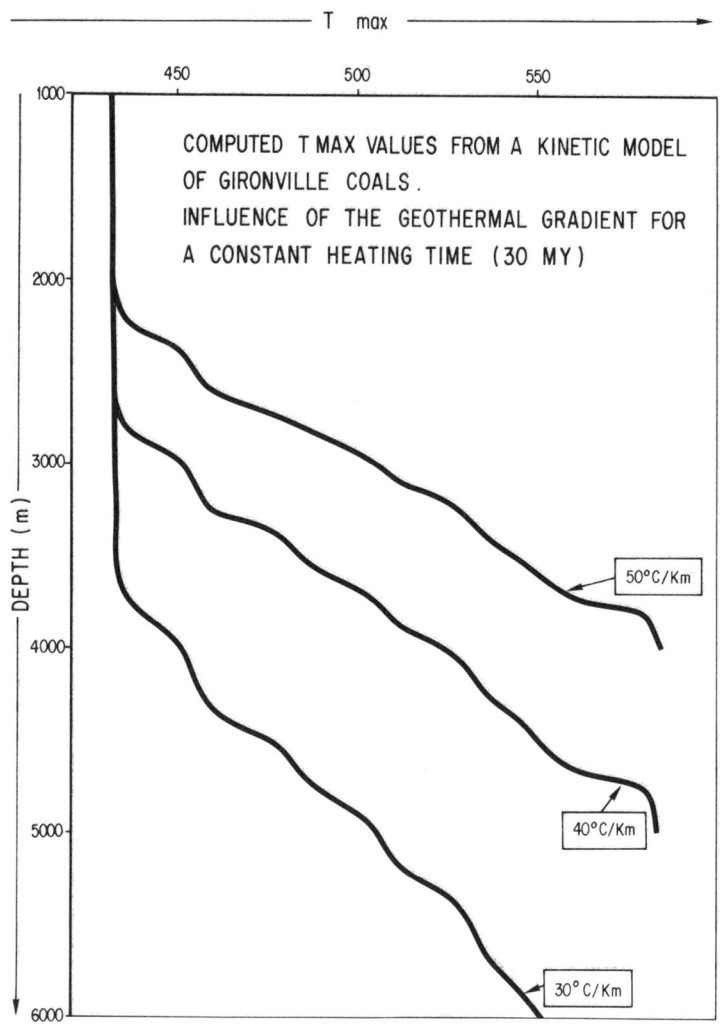

Fig. 11

hard to imagine. These extreme values may influence the paleo-temperature reconstruction by 20° C, which corresponds to 500 m for burial reconstruction.

Conclusion

The calibration program developed on the same principles as the model of Tissot & Espitalié (1975) has proved efficient to calibrate the kinetic parameters on the basis of pyrolysis data. Such calibrations are most useful for the oil exploration, since they need only one immature sample and they account for the specific petroleum potential and kinetic behavior of the organic matter type considered. The test of the method on the series of the Mahakam Delta coals was satisfactory for the prediction of the residual petroleum potential HI and to a lesser extent for the prediction of the peak temperature T_{max}. Applied to the case of the Paleozoic coals of the Gironville well*, in the Eastern part of the Paris Basin, the program was used to calibrate specific kinetic parameters for this series, which in turn made possible to determine which paleotemperatures could explain the observed trends of HI and T_{max} versus depth. The most probable hypothesis is that 800 m sediments have been eroded since the maximum burial and that the average geothermal gradient was then of 40° C/km. Since the Tertiary erosion does not exceed a few hundred meters, this maximum burial occured probably during the Paleozoic, before a 1 500 m or 2 000 m erosion took place (Hercynian phase).

These results suffer however great uncertainties. At first, the calibration itself leads to an uncertainty of about 15° C when applied to geological situations. Second, the scatter of the observed geochemical values and the poor knowledge of the time of burial brings additional uncertainty. If a restricted range is selected for burial time (10 to 30 MY) the uncertainty is about ± 20° C. Further progress in the calibration method should enhance its accuracy, especially by a better control of pyrolysis conditions.

REFERENCES

- AKIHISA K. (1978): Etude cinétique des roches mères de pétrole (rapport n° 4). Formation de produits pétroliers par pyrolyse du kérogène à basse température. Journal of the Japanese Assoc. of Petr. Technologists, vol. 44, n° 2, pp. 26-33.

- ALBRIGHT L.F., CRYNES B.F., CORCORAN W.H. (1983): Pyrolysis: Theory and industrial practice. Academic Press, 482 p.

- BOUDOU J.P. (1981): Diagenèse des sédiments deltaïques (Delta de la Mahakam, Indonésie). Thesis Univ. Orléans, France.

- FROST A.A., PEARSON R.G. (1961): Kinetics and Mechanism, 2nd edition, Wiley, 405 p.

- JUNTGEN H., KLEIN J. (1975): Entstehung von Erdgas aus Kohligen Sedimenten. Erdoel und Kohle-Erdgas-Petrochemie vereinigt mit Brennstoff-Chemie. Bd 28, Heft 2, Februar 1975, pp. 65-73.

- LEWAN M.D. (1985): Evaluation of petroleum generation by hydrous pyrolysis experimentation. A paraître dans : Philosophical Transactions of the Royal Society of London, series A. Feb. 1985.

- PERRIN (1983): Modélisation du champ thermique dans les bassins sédimentaires. Application au bassin de la Mahakam, Indonésie. Univ. of Bordeaux, Doctoral Thesis.

- PITT G.J. (1961): The kinetics of the evolution of volatile products from coal. 4th Int. Conf. on Coal Science, Le Touquet, May 30 - June 2, 1961.

- TISSOT B.P., ESPITALIE J. (1975): L'évolution de la matière organique des sédiments : application d'une simulation mathématique. Revue de l'IFP, vol. 30, pp. 743-777.

- UNGERER P. (1983): Models of petroleum formation: how to take into account geology and chemical kinetics. In: Thermal phenomena in sedimentary basins, Technip, 1984.

J. J. SWEENEY, A. K. BURNHAM, R. L. BRAUN[1]

A MODEL OF HYDROCARBON MATURATION IN THE UINTA BASIN, UTAH, U.S.A.*

I. INTRODUCTION

The Uinta Basin in northeastern Utah provides an ideal setting to study the evolution of kerogen to petroleum. Oil shale rocks containing Type I (lacustrine) kerogen of the Eocene-age Green River Formation outcrop extensively at the southern edge of the synclinal basin. The same rocks are also found at depths of 12000 ft (3650 m) in the deepest part of the syncline, where mature petroleum is presently being recovered. Thus we know that kerogen has been converted to petroleum within the basin.

Oil shale from the Green River Formation has been studied in pyrolysis experiments for many years at Lawrence Livermore National Laboratory (LLNL). The purpose of these experiments has been to better understand the process of extraction and to calculate rates and amounts of oil formation for a variety of pyrolysis conditions. Recently we have developed a detailed computer model that can accurately predict the rates and amounts of chemical products (or species) developed at heating rates, temperatures, and pressures prevalent in different kinds of oil shale retort processes (Burnham and Braun, 1985).

We tested the fundamental accuracy of this model by extending the range of heating rates and temperatures from laboratory conditions to those prevalent in a geologic basin. In this paper geophysical and geologic data are used to develop a time-temperature history of the kerogen-rich lithologies of the Uinta Basin. The geologic model is then used as the data set to characterize conversion of the kerogen to various hydrocarbon products through time. The end result is a prediction of the

---

* Work performed for the U.S. Department of Energy by the Lawrence Livermore National Laboratory under contract number W-7405-ENG-48.

(1) *Lawrence Livermore National Laboratory, Livermore, California, USA.*

A MODEL OF HYDROCARBON MATURATION IN THE UINTA BASIN

present-day kerogen maturity level for various depths and locations within the Altamont-Bluebell and Redwash oil fields of the Uinta Basin. We then compare these results with maturity data from material extracted from the oil fields.

II. GEOLOGIC SETTING AND HISTORY

The Uinta Basin is a structural and topographic depression in the northeastern corner of Utah. During the early Tertiary as much as 20,000 ft (6000 m) of lacustrine and alluvial sediments were deposited. These sediments represent a central core of open lacustrine claystone and carbonate mudstones; a marginal lacustrine facies with sandstone, claystone, and carbonate; and a peripheral alluvial facies of conglomerates, claystones, and carbonaceous shales (Fouch, 1975). The lacustrine period of deposition waned in the late Eocene and later deposition was primarily alluvial in character. The bulk of the Type I kerogen contained in post-Cretaceous rocks of the Uinta Basin is found in the Green River Formation between the Mahogany shale and the base of the Eocene.

A series of correlation markers, established by Fouch (1975), are used to establish a basin-wide stratigraphy. Other details of the Cenozoic geologic history are obtained from Anderson and Picard (1974) and Hansen (1984). Potassium-argon dating of biotite in tuff by Mauger (1977) provides age control for the stratigraphic units. The time-stratigraphic markers chosen for the geologic thermal history model are listed below:

Event	Time Period (Mya)
Renewed uplift	10 - present
Period of stability and peneplaination	30 - 10
Deposition of Duchesne River formation	41 - 30
Deposition of Uinta formation	44 - 41
Deposition of Mahogany oil-shale	45
Paleocene-Eocene boundary	57.8

III. BASIN ANALYSIS AND DEVELOPMENT OF THE BURIAL HISTORY

The complete burial history of a stratigraphic unit is determined from the maximum depth of burial and the thickness of the unit through time. The time-thickness relation can be determined from the density or porosity of the unit as a function of burial depth and from the burial history using a method known as

A MODEL OF HYDROCARBON MATURATION IN THE UINTA BASIN

backstripping (Sclater and Christie, 1980 and Steckler and Watts, 1978). With knowledge or estimates of past geothermal gradients, we can then calculate the time-temperature relation.

Geologic evidence indicates that as much as 5988 ft (1825 m) of the stratigraphic section may have been eroded from the Uinta Basin. A method that uses borehole interval velocity data, demonstrated by Magara (1978), was used to estimate the amount of overburden removed at locations of wells in the Redwash and Altamont-Bluebell oil fields. The procedure involves determining a normal compaction trend of seismic wave interval transit time for shale units from an area with little or no erosion and then fitting that curve to data from an area nearby where erosion has taken place. From $\Delta t_0'$ (the zero-depth intercept value of interval transit time) for a particular well the thickness of the removed overburden can be calculated. The problem with applying this method to the Uinta Basin is that erosion has taken place everywhere. Thus a normal compaction curve can only be determined by averaging the slopes of a number of interval velocity-depth logs.

In this study the electric log (E-log) was used to identify shale-rich zones. These zones were then matched with depths on the down-hole acoustic log (DHAL) to obtain interval transit times. We selected well locations where detailed lithologic logs were available. We had to exercise caution when using the interval transit time (Δt) data because of the possibility of present or past overpressures at depth which result in undercompaction of shale and a corresponding roll-off of interval transit time (Magara, 1978). This is seen in most logs from the Uinta Basin. Figure 1 is a typical plot of Δt versus depth (z) for the Shell Tennaco-Brotherson well in the Altamont-Bluebell field. Below depths of 5000 ft (1500 m) Δt begins to increase, due to overpressuring, and then varies up and down in a cyclic fashion.

We estimated a normal compaction trend for a number of wells by averaging the slope of the Δt-z curve at depths above the overpressured zone. The average value for the slope of the interval velocity compaction curve is .000096 ft^{-1}, which is within the .000085 to .000147 range that Magara (1978) obtained for Cretaceous shales in western Canada. We used this average slope to fit data like that of Figure 1 for other wells in the basin and to calculate values of Z_{OB}, the amount of overburden removed. Calculated values of Z_{OB} for wells in the Altamont-Bluebell field of the Uinta Basin are shown in Figure 2.

Uncertainty in the estimate of Z_{OB} is about ±1000 ft (±300 m). Values of Z_{OB} generally range from 6000 to 6600 ft with values about 1000 ft less for wells in the Redwash field. This agrees with estimates used by Tissot et al. (1978).

A MODEL OF HYDROCARBON MATURATION IN THE UINTA BASIN

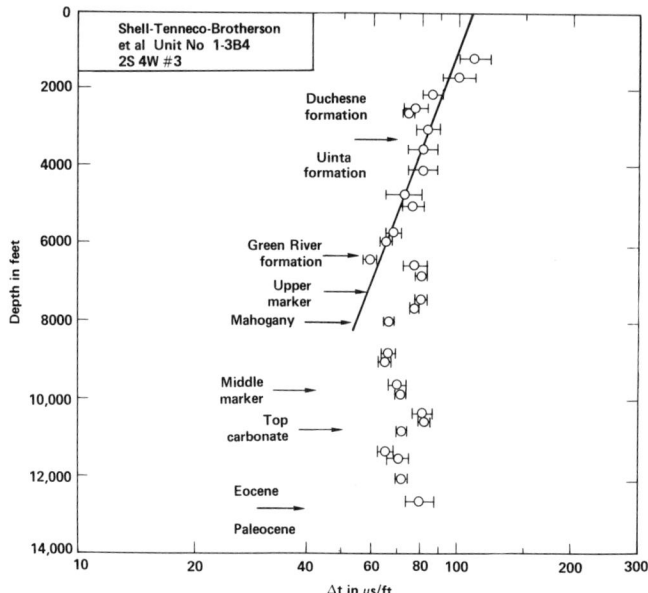

Figure 1: A typical interval transit time-depth plot for the Altamont-Bluebell field.

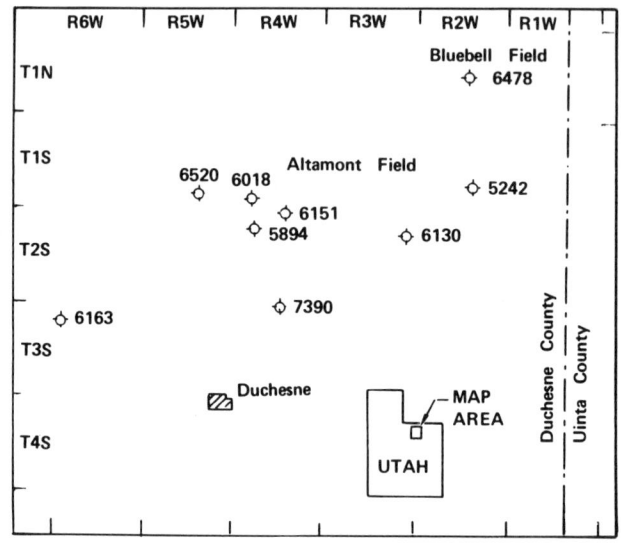

Figure 2: Locations of wells in the Altamont-Bluebell field used in this study. The calculated amount of removed overburden (in feet) is listed next to each well.

A MODEL OF HYDROCARBON MATURATION IN THE UINTA BASIN

For the backstripping analysis we chose to use a simple porosity curve ($\phi = 0.35e^{-0.7z}$). Detailed analysis of temperature data by Chapman et al. (1984) revealed the present day geothermal gradient in the Uinta Basin to be 25°C/km. We assume that the geothermal gradient throughout the Tertiary has been constant and ignore localized effects such as overpressuring and lithologic variation. We chose a value of 10°C for the long-term average surface temperature. From these assumptions and the time-burial depth data corrected for compaction, we have constructed time-temperature-depth plots for well locations in the Altamont-Bluebell and Redwash oil fields. A representative plot is shown in Figure 3, along with values of heating rates for the base of the Eocene during the separate time intervals. Note that temperatures above 110°C are reached after 40 Mya only by units below the base of the Uinta formation.

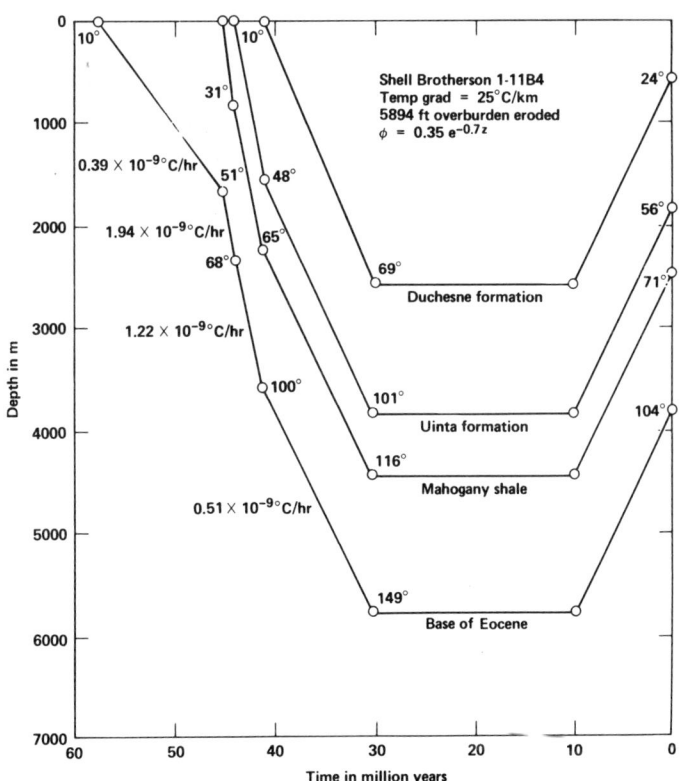

Figure 3: Burial history and temperature curve, corrected for compaction, for the Shell Brotherson 1-11B4 well.

A MODEL OF HYDROCARBON MATURATION IN THE UINTA BASIN

A change of one standard deviation in the value of Z_{OB} results in a 6°C to 8°C change in T_{max} (the maximum temperature attained) and a change of about 0.1×10^{-9} °C/hr in the heating rate. Uncertainty in the porosity function has no effect on T_{max}, but leads to an uncertainty in the heating rate of 0.1×10^{-9} °C/hr. The geothermal gradient varies throughout the basin (Chapman et al., 1984) and the present-day gradient may be different from past gradients, thus its uncertainty is difficult to assess. For a ±5°C/km uncertainty in geothermal gradient and a ±300 m uncertainty in Z_{OB}, the net uncertainty in T_{max} is about ±35°C. This range of temperatures would have a profound effect on hydrocarbon maturity, which can be evaluated by comparing the geologic model with hydrocarbon maturation data. We show later that the low uncertainty in the heating rate has an insignificant effect on hydrocarbon maturity calculations when compared to the uncertainty in T_{max}.

The most deeply-buried Tertiary strata in the basin are in the Altamont-Bluebell field. Kerogen-bearing units have been heated to maximum temperatures of 149°C to 175°C. Maximum heating rates occurred between 41 and 30 Mya and range from 0.51 to 0.74×10^{-9} °C/hr. The maximum temperature attained by the base of the Eocene in a well from the Redwash field was much lower (111°C) with a lower heating rate between 41 and 30 Mya of 0.16×10^{-9} °C/hr.

IV. ESTIMATION OF KEROGEN TRANSFORMATION

Although it has been known for many years that pyrolysis of kerogen yields liquid hydrocarbons, there has been some doubt about whether petroleum formation could be explained by a pyrolysis process. In the 1960s research established firmly that petroleum is formed predominantly by thermal transformation of kerogen (e.g., Philippi, 1965). Later, others tried to establish the time-temperature relationship for petroleum formation (Tissot and Espitalie, 1975; Ishiwatari et al., 1976; Waples, 1978). However, there is a persistent conflict between the 15 to 20 kcal/mole activation energies for kerogen conversion reactions determined from geological studies and the 40 to 60 kcal/mole activation energies determined from laboratory pyrolysis.

The rate of reaction in a sample undergoing pyrolysis at a constant heating rate of H_r is given by Van Heek and Juntgen (1968):

$$\frac{dV}{dt} = AV_\infty \exp\left[-\frac{E}{RT} - \frac{A}{H_r}\int_0^T \exp\left(-\frac{E}{RT}\right) dT\right] \qquad (1a)$$

A MODEL OF HYDROCARBON MATURATION IN THE UINTA BASIN

$$\simeq AV_\infty \exp\left[-\frac{E}{RT} - \frac{ART^2}{H_r E}\exp(-\frac{E}{RT})\right]. \tag{1b}$$

Campbell et al. (1978) used Eq. (1b) to determine an accurate rate expression for the generation of oil from Green River oil shale. They determined A to be 2.8×10^{13} and E/R to be 26390 (°K). The E/R term corresponds to an activation energy of 52.4 kcal/mole.

To generate the curves of Figure 4, we used this nonisothermal method to simulate the macroscopic conversion of Type I kerogen to petroleum at various heating rates. Heating rates of the same order of magnitude as the geologic rates estimated for the Uinta Basin were used in the calculations. For a heating rate of 10^{-8}°C/hr T_p, the temperature of the peak rate of oil production, is 170°C, while T_p is only 139°C for a heating rate of 10^{-10}°C/hr. A 10% change in heating rate causes about a 1 to 2° change in T_p. The values of T_p shown are close to the values estimated for the maximum temperatures attained in the Uinta Basin, indicating that a potential for oil production is present and that values used for Z_{OB} and the geothermal gradient are approximately correct.

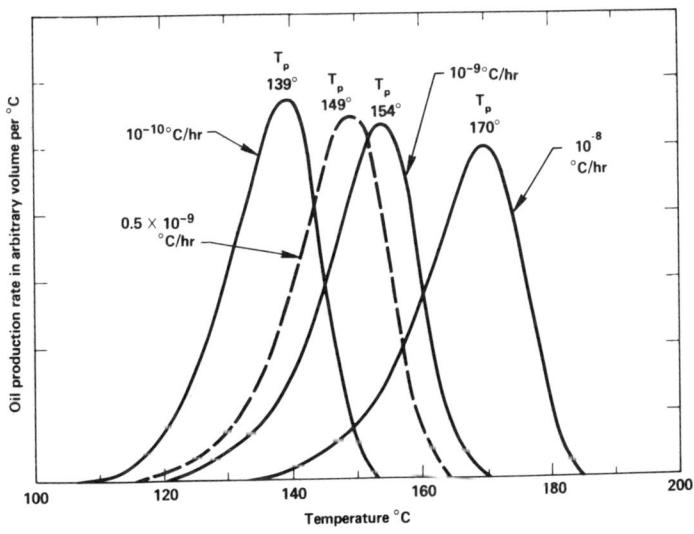

Figure 4: Oil production rate curves for the conversion of Type I kerogen using the results of Campbell et al. (1978).

A MODEL OF HYDROCARBON MATURATION IN THE UINTA BASIN

These preliminary results are encouraging. They indicate that the geologic model is reasonable and consistent with extrapolations of simple laboratory retort models of kerogen evolution. Next we employ a more detailed geochemical model and predict hydrocarbon compositions as well as production rates. We then compare the modeled composition and production rate data with published data for the oil fields.

V. FORMULATION OF A DETAILED GEOCHEMICAL MODEL

While the rate of kerogen decomposition is central to understanding the origin of petroleum, many other reactions are also important. Continued heating of oil can convert the oil to gas, provided it does not migrate to a cooler reservoir. Secondary reactions are also important in oil shale processing, and we have determined rate expressions from laboratory experiments. We combined our understanding of the organic pyrolysis reactions into a detailed chemical model (Burnham and Braun, 1985). The model accurately calculates the amounts and rates of oil and gas formed under a wide range of pyrolysis conditions. It consists of 67 first-order, nonlinear, ordinary differential equations which are solved by numerical integration. The equations specify the rate of change each of gas, liquid, and solid components in terms of the vaporization or chemical reactions. The oil is divided into 50°C boiling point intervals, allowing (for laboratory experiments) a direct calculation of liquid-phase residence time prior to evaporation. For the geological case oil evaporation is not included (to conserve computer time), but this feature allows us to calculate how distillation characteristics of the oil vary with thermal history.

We can not at present vary pressure with time in the model calculations. The main effect of pressure is on the rate expression, which affects rates and composition of the products of cracking (Burnham and Braun, 1985). Higher pressure inhibits cracking and results in less gas formation for a given amount of cracking. Pressures representative of averages during the high temperature intervals (maximum burial portion) were estimated to range from 40 to 90 MPa and kept as a constant throughout each run.

Oil can be destroyed by both coking and cracking. Oil coking is the polymerization and condensation of oil components, usually aromatic heterocycles, to form a predominately solid residue (coke). It leads to an increase in aliphatic content and a decrease in heteratom content. In contrast, oil cracking is the fission of aliphatic structures to smaller molecules, ultimately methane. It leads to a concentration of aromatic content. There is, of course, overlap between the types of reactions, but this separation has proved useful.

A MODEL OF HYDROCARBON MATURATION IN THE UINTA BASIN

The model includes rate expressions for gas formation during both the original kerogen decomposition and during secondary pyrolysis of the carbonaceous residue. While the model does not calculate oil composition explicitly, it does calculate qualitative oil composition indicators based on the amount of each kerogen type remaining and their current rate of decomposition. These indicators are used to obtain estimates of the degree of hydrocarbon maturation.

VI. APPLICATION OF THE DETAILED GEOCHEMICAL MODEL

We ran the model with heating rate and temperature data from several wells to calculate oil formation and degradation at various depths. The conversion of kerogen ranges from 7 to 100%. The elemental composition of the remaining solid, plotted in Figure 5, indicates that the material follows, as expected, the maturation curve of Type I kerogen shown by Tissot et al. (1978). The oil becomes noticeably lighter with increasing maturity. This is due to three effects: the oil generated from kerogen is lighter than the original bitumen, coking tends to remove heavier oil components, and cracking converts heavy components to light ones. One interesting observation is that essentially all the generated cokable oil (30% of the total) has coked. This accounts, in part, for the difference in aliphatic content of petroleum from most laboratory pyrolysates. Another observation is that although very little of the oil has cracked to gas, very few high boiling point components remain in two of the oil samples. For the other oil samples cracking is probably small enough that biomarker/normal alkane ratios can be related to the extent of kerogen conversion. This provides an additional basis for comparing the geological and chemical model calculations.

Three published studies (Reed and Henderson, 1971; Tissot et al., 1978; and Anders and Gerrild, 1984) give analyses of hydrocarbons at various depths from different lithologic facies in the Uinta Basin. Gas chromatograms of the hydrocarbon fractions are given by Reed and Henderson and Tissot et al. In order to equate these data to our calculations of predicted maturity (in the form of percent of oil generated) we derived cumulative and instantaneous maturity indicators from laboratory pyrolysis gas chromatographs. Indicators used were the pristane/($C_{17} + C_{18}$) ratio, phytane/($C_{17} + C_{18}$) ratio, and $C_{17}/(C_{16} + C_{18})$ ratio.

Data from the wells modeled can be used to make another type of calibration curve. In each model, the percent of oil ungenerated, percent of liquid oil created, and the atomic H/C ratio are calculated. From this, we can determine the transformation ratio

A MODEL OF HYDROCARBON MATURATION IN THE UINTA BASIN

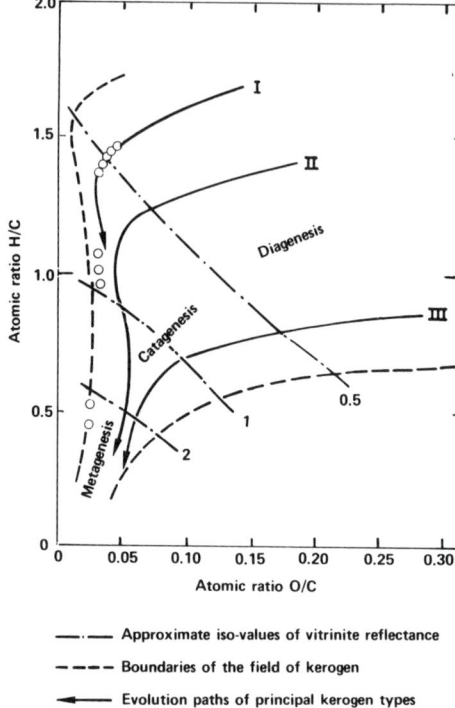

Figure 5. Van Krevelen diagram showing calculated changes in elemental composition of kerogen from the wells modeled.

(or production index) $[S_1/(S_1 + S_2)]$. In the Rock-Eval pyrolysis method (commonly used in the oil industry) S_1 corresponds to the free hydrocarbons that are released between 90°C and 300°C in flowing helium. Hydrocarbons that are generated during kerogen pyrolysis between 300°C and 600°C correspond to S_2. In our models liquid oil produced is equivalent to S_1 and the oil ungenerated is equivalent to S_2. Transformation ratios for various depths in the wells in the Altamont-Bluebell field have been determined by Anders and Gerrild (1984). Figure 6, which compares the amount of kerogen converted with the transformation ratio and H/C ratio for a range of T_{max}, was prepared from model runs and is the basis for comparison of our geologic-geochemical models with published data. In each case, we assumed that the value of Z_{OB} (which determines T_{max}) is the determining variable for hydrocarbon maturation and that small changes in heating rate (due to different burial histories and different values of T_{max}) are negligible.

In Table I, the data of Tissot et al. (1978) are compared with our model. Maturity levels for the published oil compositions are

A MODEL OF HYDROCARBON MATURATION IN THE UINTA BASIN

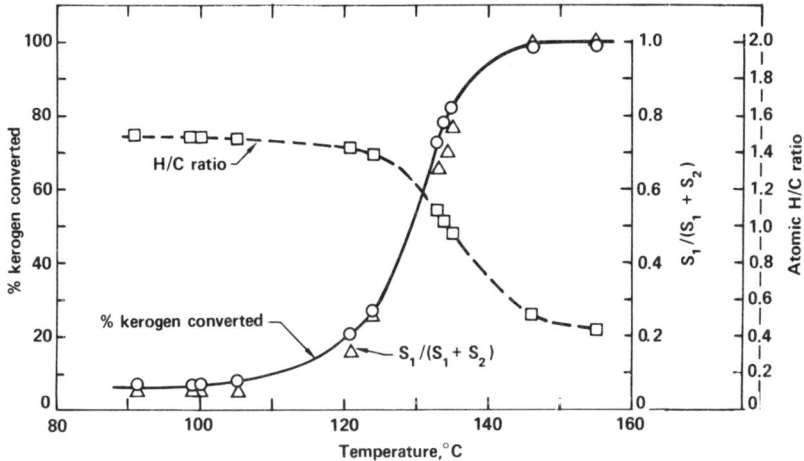

Figure 6. Calibration of transformation ratio $[S_1/(S_1 + S_2)]$ and H/C ratio with the percent of kerogen converted and maximum temperature, T_{max}, attained by the kerogen. Data points are from the model runs.

determined from the indicators discussed above. In this case, the ratios (such as $C_{17}/(C_{16} + C_{18})$) were estimated using the gas chromatograms in the Tissot et al. article.

The Shell Brotherson 1-11B4 well, which we modeled, is located near the Shell Brotherson 1-14B4 and 1-23B4 wells used by Tissot et al. Using 5894 ft for Z_{OB}, our results agree well with Tissot et al. for all four depths sampled. A range of values of Z_{OB} were used to compare maturity levels at a depth of 9127 ft. Our estimate of maturity from Tissot et al. is 30 to 60% for the Shell Brotherson 1-23B4 well. The low value of Z_{OB}, 5074 ft, results in a 19% maturity, while the high value of Z_{OB}, 7029 ft, results in a 73% calculated maturity. Both of these calculated values are out of the range by roughly equal amounts, suggesting that the value of 5894 ft for Z_{OB} is a good determination for this area.

In Section B of Table I, the Shell Murdock well data of Tissot et al. is compared to our calculations for the Shell Brotherson 1-11B4 (next section east) and Shell Christensen (next section north) wells at different depths. In each case, the calculated values (56 and 78% and 99 and 100%) are consistent with the (>40% and >60%) values obtained from the oil sampled. In this case, it is difficult to make any conclusions, except that the Z_{OB} of 6520 ft from the Christensen well may be slightly high.

Table I. Comparison of hydrocarbon maturity of wells studied by Tissot et al. (1978) with model calculations.

Wells from Tissot et al. (1978)				Models from this study for comparison				
Well Name	Depth(ft)	Estimated maturity level (%oil produced)		Model Well	Depth(ft)	Z_{OB}(ft)	T_{max}	Estimated maturity level (%oil produced)
A. Shell Brotherson 1-14B4 (2S,4W)	7540	0-25%	A.	Shell Brotherson 1-11B4 (2S,4W)	7540	5894	112°	11%
Shell Brotherson 1-23B4 (2S,4W)	7250	0-20%		Shell Brotherson 1-11B4	7250	5894	110°	9%
Shell Brotherson 1-23B4	8389	15-35%		Shell Brotherson 1-11B4	8389	5894	119°	20%
Shell Brotherson 1-23B4	9127	30-60%		Shell Brotherson 1-11B4	9127	5894	125°	36%
				Shell Brotherson 1-11B4	9127	7029	133°	73%
				Shell Brotherson 1-11B4	9127	5074	118°	19%
B. Shell Murdock 1-11B4 (2S,5,W)	9782	> 40%	B.	Shell Brotherson	9782	5894	129°	56%
				Shell Christensen (1S,5W)	9782	6520	134°	78%
Shell Murdock 1-11B4	11655	> 60%		Shell Brotherson	11655	5894	144°	99%
				Shell Christensen	11655	6520	149°	100%
C. Standard of Calif. Redwash 32	8692	30-60%	C.	Chevron Redwash (7S,24E)	8692	4625	111°	10%
				Broadhurst #13 (7S,23E)	8692	5914	121°	24%
Redwash 164	5572	0-20%		Chevron Redwash	5572	4625	88°	6%
				Broadhurst #13	5572	5914	98°	7%

Wells from the Redwash area, to the east of the Altamont-Bluebell field, are compared in Section C of Table I. Tissot et al. did not give the exact locations of the wells in their paper, so we compared their data with a range of values for Z_{OB} in the Redwash area. For the shallower wells, the 6 to 7% calculated maturity is within the range of 0 to 20% obtained for the samples. Values of maturity from our calculations at the deeper levels are smaller than those that come from actual samples from the area. The highest value of Z_{OB} used, 5914 ft, results in a maturity level of 24% at 8692 ft, below the 30 to 60% maturity indicated by Tissot et al. This discrepancy would be explained if the oil sampled from the field had migrated from a more mature source area. Recovered oil samples analyzed by Reed and Henderson were also compared with similar results.

A MODEL OF HYDROCARBON MATURATION IN THE UINTA BASIN

Higher values for the geothermal gradient would produce higher calculated maturity levels. In the Redwash area, at the 5500 ft level with an assumed value for Z_{OB} of 5500 ft, a geothermal gradient of 34°C/km would produce a T_{max} of 124°C. The corresponding maturity level calculated with our model is about 30% (Fig. 6) and would agree well with measured values of 20 to 40%. However, the lower values of geothermal gradient work best for the Shell Brotherson wells in the deepest part of the basin where migration is less likely. There is no geologic evidence (such as magma intrusions at depth) for large changes in geothermal gradient over such a small area, and it is unlikely that the magnitude of change required could be accounted for by hydrologic circulation. Furthermore, the nature of fracturing in the basin is conducive to the migration of hydrocarbons. Thus, we conclude that migration, rather than large changes in Z_{OB} or geothermal gradient, is the most likely explanation for the difference between modeled predictions and measurements of the maturity levels indicated in Table I.

Anders and Gerrild (1984) sampled five wells from the Uinta Basin and compared various maturity indicators with total organic carbon (TOC) determinations and parameters such as stratigraphic facies. We compared our determinations of the transformation ratio with those determined from material sampled by Anders and Gerrild. The results show good agreement between predicted and measured maturity levels.

VI. CONCLUSIONS

Predicted maturity levels of the evolved hydrocarbons in the basin agree well with measured maturity levels. We feel that this agreement gives us confidence in the general applicability of the pyrolysis model. We emphasize the following conclusions:

- The value of activation energy (52.4 kcal/mole) used in the pyrolysis model results in good prediction of amounts and compositions of evolved hydrocarbons in the deepest part of the basin.

- Discrepancies between predicted and measured maturity levels in the Redwash Field are probably related to migration.

- Estimated values for removed overburden of approximately 5900 ft in the Shell Brotherson 1-11B4 well result in very good agreement between predicted and measured maturity levels, and this value of Z_{OB} is constrained by about ± 500 ft by the model (assuming a constant geothermal gradient)

A MODEL OF HYDROCARBON MATURATION IN THE UINTA BASIN

By incorporating the ability to vary pressure in the pyrolysis model (in future work) and adding calculations of pore volumes and pressures, the true potential for geological application of our model can be realized. Utilization of a refined pyrolysis model with more detailed geologic data will enable us to study the phenomena of migration and overpressuring in detail. Refinement of these techniques could eventually lead to the use of the pyrolysis model in quantitative evaluations of the thermal history of sedimentary basins and the associated hydrocarbon resource potential.

REFERENCES

ANDERS, D. E. and GERRILD, P. M., 1984, "Hydrocarbon Generation in Lacustrine Rocks of Tertiary Age, Uinta Basin, Utah - Organic Carbon, Pyrolysis Yield, and Light Hydrocarbons," Hydrocarbon Source Rocks of the Greater Rocky Mountain Region, J. Woodward, F. F. Meissner, and J. L. Clayton, eds., Rocky Mountain Association of Geologists, Denver, pp. 513-529.

ANDERSON, D. W., and PICARD, M. D., 1974, "Evolution of Synorogenic Clastic Deposits in the Intermontane Uinta Basin of Utah," Tectonics and Sedimentation, W. R. Dickinson, ed., SEPM Sp. Publ. 22, pp. 167-189.

BURNHAM, A. K. and BRAUN, R. L., 1985, "General Kinetic Model of Oil Shale Pyrolysis," In Situ, Vol. 9, No. 1, pp. 1-23.

CAMPBELL, J. H., KOSKINAS, G. H., and STONE, N. D., 1978, "Kinetics of Oil Generation from Colorado Oil Shale," Fuel, Vol. 57, June, pp. 372-376.

CHAPMAN, D. S., KEHO, T. H., BAUER, M. S., and PICARD, M. D., 1984, "Heat Flow in the Uinta Basin Determined from Bottom Hole Temperature (BHT) Data," Geophysics, Vol. 49, No. 4, pp. 455-466.

FOUCH, T. D., 1975, "Lithofacies and Related Hydrocarbon Accumulations in Tertiary Strata of the Western and Central Uinta Basin," Symposium on Deep Drilling Frontiers in the Central Rocky Mountains, D. W. Bolyard, ed., Rocky Mountain Asso. Geol. Spec. Publ. pp. 163-173.

HANSEN, W. R., 1984, "Post-Laramide Tectonic History of the Eastern Uinta Mountains, Utah, Colorado, and Wyoming," The Mountain Geologist, Vol. 21, No. 1, pp. 5-29.

ISHIWATARI, R., ISIIWATARI, M., KAPLAN, I. R., and ROHRBACK. B. G., 1976, "Thermal Alteration of Young Kerogen in Relation to Petroleum Genesis," Nature, Vol. 264, pp. 347-49.

A MODEL OF HYDROCARBON MATURATION IN THE UINTA BASIN

MAGARA, K., 1978, Compaction and Fluid Migration - Practical Petroleum Geology, Elsevier, New York, pp. 11-25.

MAUGER, R. L., 1977, "K-Ar Ages of Biotites from Tuffs in Eocene Rocks of the Green River, Washakie, and Uinta Basins, Utah, Wyoming, and Colorado," Contributions to Geology, University of Wyoming, Vol. 15, No. 1, pp. 17-41.

PHILIPPI, G. T., 1965, "On the Depth, Time, and Mechanisms of Petroleum Generation," Geochimica et Gosmochimica Acta, Vol. 25, pp. 1021-1049.

REED, W. E. and HENDERSON, W., 1971, "Proposed Stratigraphic Controls on the Composition of Crude Oils Reservoired in the Green River Formation, Uinta Basin, Utah," Advances in Geochemistry, Pergamon Press, Oxford, pp. 499-515.

SCLATER, J. G. and CHRISTIE, P. A. F., 1980, "Continental Stretching: An Explanation of the Post Mid-Cretaceous Subsidence of the Central North Sea Basin," Journ. Geophys. Res., Vol. 85, No. 87, pp. 3711-3739.

STECKLER, M. S. and WATTS, A. B., 1978, "Subsidence of the Atlantic-type Continental Margin Off New York," Earth and Planet. Sci. Ltrs., Vol. 41, pp. 1-13.

TISSOT, B., DEROO, G., and HOOD, A., 1978, "Geochemical Study of the Uinta Basin: Formation of Petroleum from the Green River Formation," Geochim. Cosmochim. Acta, Vol. 42, pp. 1469-1485.

TISSOT, B. and ESPITALIE, J., 1965, "L'evolution Thermique de la Matiere Organique des Sediments; Applications d'une Simulation Mathimatique Potential Petrolier des Bassins Sedimentaires et Reconstitution de l'histore Thermique des Sediments," Revue de l'Institut Francais de Petrole, Vol. 30, pp. 743-777.

VAN HEEK, K. H. and JUNTGEN, H. 1968, "Determination of Reaction-kinetic Parameters from Nonisothermal Measurements," Ber. Bensenges. Physik. Chem., Vol. 72, p. 1223.

WAPLES, D. W., 1980, "Time and Temperature in Petroleum Formation - Application of Lopatin's Method to Petroleum Exploration," AAPG Bull, Vol. 64, pp. 916-926.

PART B

MINERAL METHODS

M. PAGEL[1], F. WALGENWITZ[2], J. DUBESSY[1]

FLUID INCLUSIONS IN OIL AND GAS-BEARING SEDIMENTARY FORMATIONS

In the study of the evolution, through time, of a thermal sedimentary basin regime, fluid inclusions represent one of the most powerful tools. In the case of a chemical regime, they provide important salinity and gas content data. Additionally, interpretation of data obtained by fluid inclusions studies may be enlarged if it is combined with stable and radiogenic isotopes or fission track data. Also, it may benefit by being associated with organic geochemistry however, at present, little published material exists (HASZELDINE et al., 1984, HORSFIELD and McLIMANS, 1984). During the past few years fluid inclusions studies in oil and gas-bearing sedimentary formations have significantly increased data.

First, a review of the interpretation of fluid inclusions data is presented, with application to petroleum formations, and then certain specific applications, which take into account unpublished data and examples from the literature are discussed. General statements on fluid inclusions applications to sedimentary environments may be found in the proceedings of the First International Meeting on Thermal Phenomena in Sedimentary Basins (PAGEL and POTY, 1984) and in ROEDDER (1979, 1984).

1. BASIC PRINCIPLES FOR STUDYING FLUID INCLUSIONS IN THE FIELD OF PETROLEUM GEOLOGY

To begin with it is important to remember that fluid inclusions might have been re-equilibrated by various processes such as necking down, stretching, leaking, decrepitation, hydraulic fracturation,

---

(1) *Centre de Recherche sur la Géologie de l'Uranium et Groupement Scientifique, Centre National de la Recherche Scientifique-CREGU, Vandoeuvre-les-Nancy, France.*
(2) *Centre de Recherches, Société Nationale Elf Aquitaine (Production), Pau, France.*

recrystallization.... These alterations occur in the geological environment but some also may result from dressing, heating, or cooling of the samples. A precise and detailed microscopic observation recognizes these alterations. The study of fluid inclusions under the microscope must be carried out with great care. In addition, such examinations assist considerably in establishing the relative chronology which exists between the various generations of fluid inclusions.

Microthermometry, which is the basic method for fluid inclusions studies, consists of observing, under the microscope, the phase transitions during slow and controlled temperature variations. The following assumptions must be made in order to interpret these observations :

1) that, since trapping, the cavity volume remains unchanged ;
2) that, since entrappment, the fluid chemistry and density remain unchanged, so are representative of fluid present in the surroundings.

If the fluid present in the inclusions is trapped as a single homogeneous phase, the fluid inclusion contains, at room temperature, a liquid and a small vapour bubble. The temperature which occurs at the beginning of melting, after a metastable solidification of the liquid which has been frozen and then slowly heated, may provide valuable information concerning the chemical composition of the inclusions (CRAWFORD, 1981). This is especially so if a significant quantity of divalent cations, in the inclusions, is associated with Na+. However, this is never easy to determine accurately. The final melting temperature (T_m) is interpreted in terms of salinity, expressed generally as equivalent weight % NaCl. Upon heating, the gas phase dissolves in the liquid which gives the homogenization temperature (T_h). This temperature is the minimum trapping temperature and, combined with salinity and thermodynamic data, provides the fluid density. As the density remains constant in the inclusion upon heating, temperature increase also induces a pressure increase. When pressure is greater than the host-mineral resistance, the inclusion decrepitates. For quartz, the decrepitation temperature (T_d) corresponds to a pressure of about 850 bars (LEROY, 1979). The pressure depends upon the hardness of the mineral (TUGARINOV and NAUMOV, 1970).

In oil reservoirs, fluid inclusions, containing hydrocarbons with no visible aqueous solution, coexist in the same crystal with

aqueous fluid inclusions where no visible trace of hydrocarbons is observed. Inclusions containing an aqueous solution and hydrocarbon droplets generally are scarce. Usually, in the first case the following questions arise :
  a) do these inclusions represent trapping of the immiscible hydrocarbon and aqueous phases ?
  b) what information on the P-T conditions may be derived from the microthermometric measurements ?

1.1. Immiscibility in the H_2O-CH_4 system

General rules concerning immiscibility with application to fluid inclusions have been described by PICHAVANT et al., (1982) and RAMBOZ et al., (1982). The first condition, is the thermic and pressure equilibrium between the two phases. The second condition, of equilibrium, deals with the fugacities :

$$f^L_{CH_4} = f^V_{CH_4} \text{ and } f^L_{H_2O} = f^V_{H_2O}$$

The two following equations may be written :

$$x^L_{H_2O} \times \gamma^L_{H_2O} = x^V_{H_2O} \times \gamma^V_{H_2O} \text{ and}$$
$$x^L_{CH_4} \times \gamma^L_{CH_4} = x^V_{CH_4} \times \gamma^V_{CH_4}$$

Consequently, this enables us to draw the conclusion that the fluid inclusions representing the aqueous part must contain a significant amount of CH_4. Methane generally may be identified by clathrate melting or low temperature liquid-vapour equilibrium during microthermometric studies, or by micro Raman spectrometry.

For each fluid, there is an isopleth which delineates the P-T plane into two areas, with the low temperature side corresponding to the coexistence of the two phases (Fig. 1).

For a system with two components, the liquid-vapour equilibrium degree of freedom is $V = 4 - \Phi = 2$. If the composition is specified for each phase, there is only one immiscibility point in the P-T plane, that corresponds to the intersection of the two isopleths (Fig. 1).

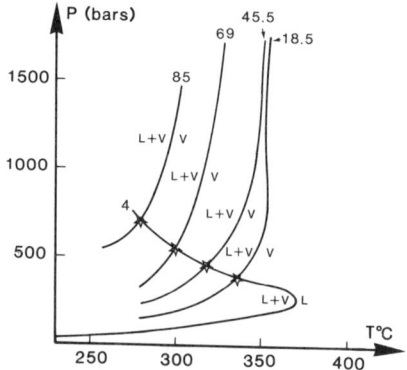

Figure 1 : Five isopleth projections of liquid-vapour equilibria in the CH_4-H_2O system. The composition is given in mole % CH_4. L+V, L, V, indicate the presence of liquid + vapour, liquid or vapour in the P-T plane relative to each isopleth. The isopleth 4 moles % CH_4 intersects the isopleths 85, 69, 45.5 and 18.5. The intersection points define the P-T conditions for which a fluid with 4 moles % CH_4 may represent the aqueous rich phase in an immiscibility equilibrium with CH_4 rich fluids of 85, 69, 45.5 and 18.5 moles % CH_4. (WELSCH, 1973).

This means that the homogenization temperature of the two immiscible phases, measured in the fluid inclusions, represents the immiscibility process temperature (PICHAVANT et al., 1982; RAMBOZ et al., 1981). MULLIS et al., (in prep.) illustrate such a process on fluid inclusions from alpine clefts (Fig. 2).

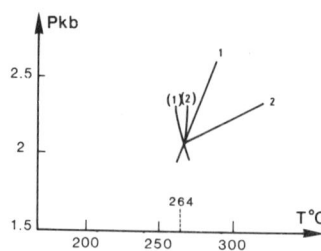

Figure 2 : P-T projection of two immiscible fluids in the H_2O-CH_4 system.
1,2 : isochores corresponding respectively to the aqueous rich fluid and the CH_4 rich fluid.
(1) and (2) : liquid-vapour isopleths for the corresponding compositions. 264 is the measured homogenization temperature of the aqueous-rich fluid inclusions. Natural example from an alpine cleft (MULLIS et al., (in prep.) Thermodynamic data from WELSCH, 1973).

1.2. Immiscibility in the H_2O - Hydrocarbon system

The immiscibility theory, illustrated above, on the H_2O-CH_4 system may be applied to the H_2O-Hydrocarbon system. However, according to PRICE (1976), the petroleum solubility in water is generally quite low, around 100ppm. This suggests that the PVT properties of the aqueous fluid inclusions, with the invisible oil droplet, may roughly be considered equivalent to those of an aqueous solution. Thus, as a first approximation, the immiscibility process occurs along the isochore of this aqueous fluid. In the same way, the immiscibility process P-T conditions are along the hydrocarbon-rich fluid isochore (Fig. 3). An estimation of the immiscibility P-T conditions is derived from the intersection of these two isochores.

In figure 3, the isochore for aqueous inclusions is based upon POTTER'S data (1977). The 60 MPa decrepitation pressure taken for feldspar (TUGARINOV and NAUMOV, 1970) is consistent with the average pressure determined from the aqueous fluid inclusion in this sample.

Isochores for hydrocarbon inclusions are based upon the homogenization temperatures on the one hand and the average decrepitation temperature on the other hand. The most difficult problem in such an interpretation is that of determining the hydrocarbon bubble point curve.

The hydrocarbon bubble point curve is based upon PVT data obtained from the production tests at the same depth, in the same drill-hole, where the fluid inclusions were sampled. The actual P-T conditions of the reservoir fluid are shown by circle 1. If the hydrocarbon fluid inclusions densities and compositions are the same as the present hydrocarbon at the sample depth, the expected homogenization temperature and pressure will be along the hydrocarbon liquid-vapour curve.

For the isochore, defined by points (1) and (2), a 60 MPa pressure corresponds to a 195°C temperature. This temperature is well below the decrepitation temperatures measured on hydrocarbon fluid inclusions. This shows that the PVTX properties of the present hydrocarbon are not consistent with those of the hydrocarbons trapped in fluid inclusions. Considering that the presence of CH_4 and hydrocarbons increases the pressure of the bubble point curve for a given T, this different behaviour suggests that the present hydrocarbons contain lighter molecules. Consequently, the maximum pressure of the hydrocarbon fluid inclusions at Th corresponds to the pressure at Th, along the bubble point curve, of present hydrocarbons (3). Thus, in the P-T plane, the hydrocarbon fluid inclusion isochore is located between the upper isochore, defined by point (3) and the decrepitation point, and the lower isochore defined by point (4) and the decrepitation point.

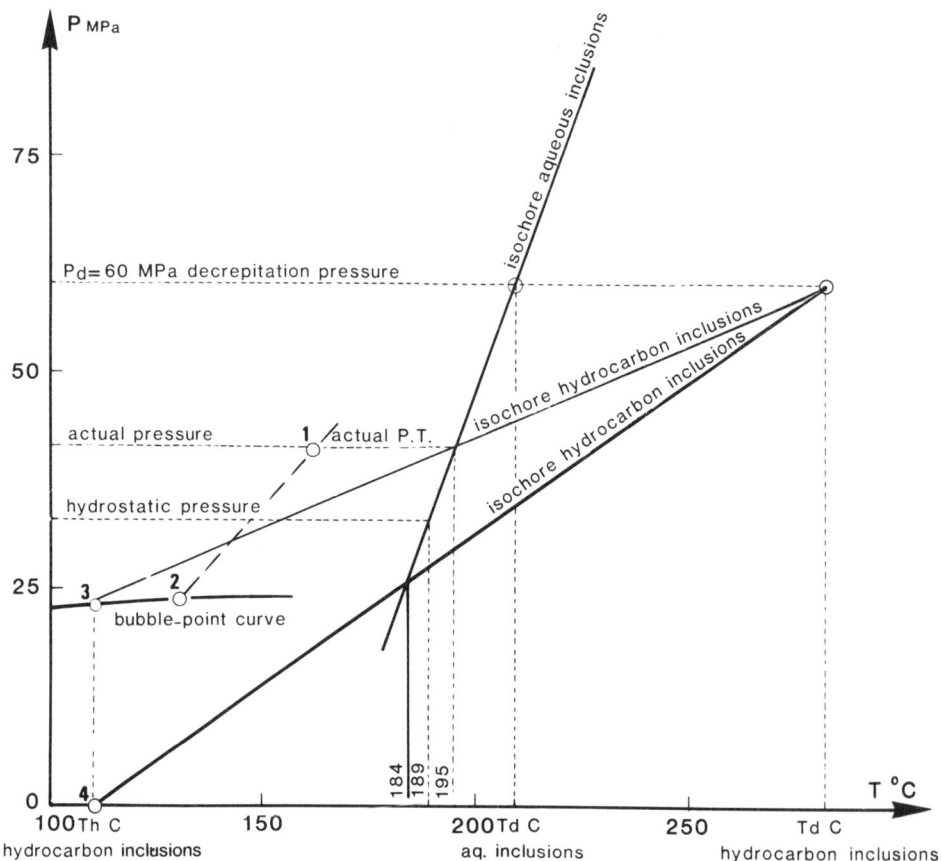

Figure 3 : Microthermometric data interpretation on coexisting aqueous two-phases inclusions and hydrocarbon two-phases inclusions on a natural example.

T_h°C : homogenization temperature.

T_d°C : decrepitation temperature.

Pd : decrepitation pressure.

Trapping temperature might be bracketed in the following way : if hydrostatic pressure is considered, the entrappment temperature is 189°C. If actual overpressure is taken into account, entrappment temperature is 195°C. If the intersection of isochores are considered, the entrappment temperature is between 184 and 195°C.

It is obvious that the actual P-T conditions are different from the fluid inclusions formation ones.

Consequently, it is dangerous to assume that PVT hydrocarbon properties, within the inclusions, are identical to those of the crude oil as postulated by NARR and BURRUSS (1984). For, at the same time, they demonstrate that methane concentration has decreased since fracture initiation.

An independent geothermometer may be used in order to derive the fluid pressure from aqueous fluid inclusions. To define the P-T entrappment conditions in the Beatrice Oilfield (North Sea), HASZELDINE et al., (1984) used the normoretane and norhopane ratio calibrated against temperature in the sediments and the isochore derived for aqueous inclusions.

1.3. Methane solubility in brines

The PVT properties of aqueous fluid inclusions are modified significantly when small quantities of methane are present (HANOR, 1980), under the detection limit of non-destructive analytical methods for fluid inclusions. This presently uncontrollable problem is probably the origin of the most important errors for determining fluid inclusion trapping pressures and temperatures.

2. APPLICATIONS OF FLUID INCLUSIONS STUDIES IN THE FIELD OF PETROLEUM GEOLOGY

2.1. Chemical and density fluid variations during the diagenetic history

Data obtained for one diagenetic event from fluid inclusions such as salinites and densities (interpreted in favorable cases, as temperatures and pressures) contribute significantly to understanding the cementation and dissolution processes in a reservoir. For example, in the Viking Graben, (North Sea), homogenization temperature of aqueous inclusions in quartz overgrowths are around 100-110°C. If only conduction is invoked, corrected temperature for Cretaceous ages imply high geothermal gradients. It is, however, more realistic to consider that this thermal gradient results from a large hot water circulation. This may also account for the important silica deposition.

Also, fluid inclusions are an important aid in deciphering the relationship between authigenic minerals. For example, in the pre-salt sandstone from the Angola coastal basin it is shown that silicification and carbonate development occurred from higher salinity solutions, but from lower homogenization temperatures than the authigenic adularia and albite development (Fig. 4).

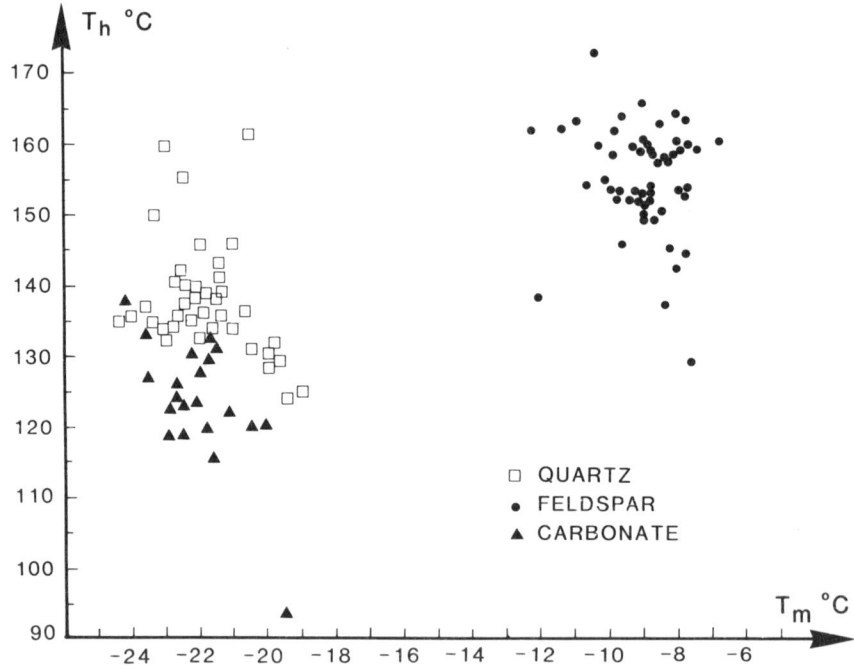

Figure 4 : Relationships between final ice melting temperature (T_m°C) and homogenization temperature in the liquid phase (T_h°C) for two-phases aqueous inclusions in quartz-feldspar and carbonate : pre-salt sandstones from Angola basin.

The mean homogenization temperatures for carbonate, quartz and feldspars are respectively 120°C, 135°C and 155°C. When combined with observations on texture, the temperatures show that the relative chronology of minerals is quartz - carbonate - feldspars. It should be noted that the aqueous fluid salinity, during the stage of feldspars growth, is close to the actual salinity solution in the reservoir. The fluid salinity decreases during evolution of sediment which implies an input of a more dilute fluid in the formation. The absence of inclusions with intermediate salinity indicates, that quartz and calcite on the one hand, and feldspars on the other hand, were formed during two distinct events and probably after a long time interval.

The significance of homogenization temperatures variations in feldspars may be questioned and illustrated by the following study of 156 aqueous two-phases inclusions in authigenic albite zones arising from a single detritic plagioclase grain. (1) In the same zone, there is a homogenization temperature zoning (Fig. 5) which is verified as not resulting from the measurement procedure.

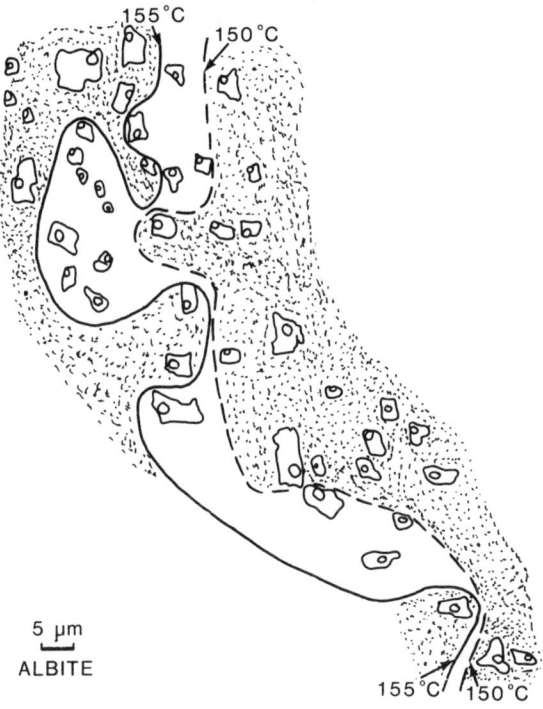

Figure 5 : Homogenization temperatures zoning for two-phases aqueous inclusions in albite (Angola).

(2) There are significant differences between the various zones (Fig. 6).

As the entrapped fluid salinity remains constant, it is clear that albite development takes place during a slight variation of the fluid density. No evidence has been found to attribute these variations to differences in the CH_4 content.

In conclusion, as the evolution of porosity and permeability in a sedimentary basin depends partly upon minerals deposition or dissolution, which are controlled by the fluid phase, fluid inclusions are useful as a tool in this research area.

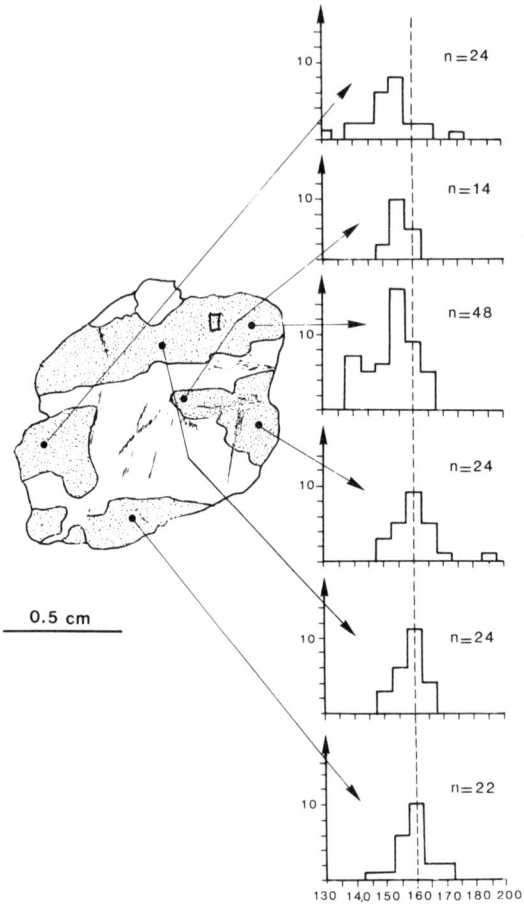

Figure 6 : Homogenization temperatures histograms in several areas of a single albite grain (Angola).

2.2. RELATIVE TIME OF HYDROCARBON MIGRATION

The microscopic location of hydrocarbon inclusions in minerals might be characterized by non-destructive methods such as microthermometry, U-V-fluorescence or infrared spectra under microscope. The latter new technique will be described elsewhere (BURNEAU et al., in prep.). Fluid inclusions provide invaluable data concerning the hydrocarbon migration age relative to diagenetic processes, i.e.

FLUID INCLUSIONS IN OIL AND GAS-BEARING SEDIMENTARY FORMATIONS

authigenic mineral depositions. Furthermore, of particular interest are the comparisons between the chemical composition of hydrocarbons present in the fluid inclusions and hydrocarbons actually found in the reservoir. Several hydrocarbon generations corresponding to different migration epochs might be observed very simply by using U-V fluorescence (BURRUSS, 1981). In addition, fluid inclusions have enclosed fresh, non- or low-altered oil, subsequent to trapping in contrast to the present reservoir oil.

The use of fluid inclusions to constrain the relative hydrocarbon migration time in sedimentary formations is illustrated by the two following cases : The Jurassic sandstones from the Viking Graben (North Sea) and the albo-aptian feldspathic sandstones from Angola. In the first case, hydrocarbon two-phases inclusions are mainly located at the boundary between the detritic-quartz grain and its quartz overgrowth (Fig. 7). It should also be noted that, in the same rock, aqueous two-phases inclusions are observed in the same position. Fluid inclusions occur in this context (PAGEL, 1975) due to the irregularity and dusty surface of the detritic-quartz grain which induces an irregular quartz growth and fluid inclusion genesis.

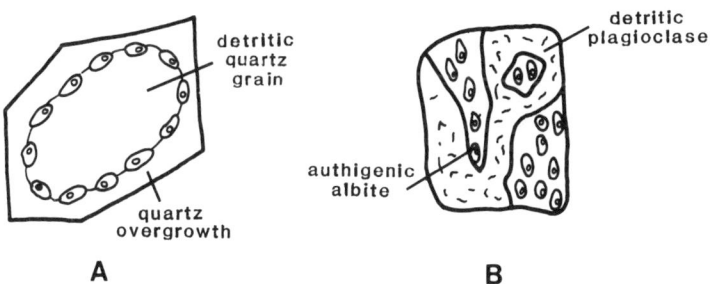

Figure 7 : Schematic location of hydrocarbon two-phases inclusions in authigenic minerals.
A. Jurassic sandstones from the Viking Graben (North Sea).
B. Albo-Aptian feldspathic sandstones from Angola.

In the case of the Viking Graben, this means that hydrocarbon migrations occur before, or, at the same time as the silicification begins. Other recent data obtained in different North Sea areas (HASZELDINE et al., 1984 ; MOGE, 1985 ; MALLEY et al., 1985) show that this observation is common in all the samples studied.

In the albo-aptian feldspathic sandstones from Angola, primary two-phases hydrocarbon inclusions are found, especially in authigenic albite or adularia, and are associated with aqueous two-phases inclusions. In quartz only secondary hydrocarbon inclusions occur. The presence of hydrocarbon inclusions is limited to the present oil zones. This could be considered representative of the timing of the hydrocarbon appearance present in the reservoir. Furthermore, the scarcity of aqueous inclusions and the lack of visible water in the hydrocarbon inclusions show that they were trapped from an oil dominated system.

By contrast to the diagenetic minerals, the primary minerals (halite, anhydrite, carbonate) of chemical sediments may record several fluid generations throughout the burial history. In halite layers from Gabon, one stage of primary inclusions is observed along growing faces ("en chevron structure"). In addition to these primary inclusions, two-phases aqueous inclusions and several generations of hydrocarbon inclusions are present.

As a result of the technical approach presently available, fluid inclusions might give <u>directly</u> the relative timing of hydrocarbons migration in sedimentary formations and, <u>indirectly</u> if combined with other isotopic (K/Ar, Rb/Sr, $^{39}Ar/^{40}Ar$) or fission track data, the absolute time of hydrocarbon migrations. Isotopic studies provide with two ways to date an event during which fluid inclusions are formed. Directly, as in the case of Angola where host mineral datation, i.e. feldspar, is possible, or indirectly, for quartz when it is associated with a K-bearing mineral which has grown simultaneously (see paragraph on local hydrothermalism from Gabon).

Other constraints, which may be envisaged, have been illustrated by HORSFIELD and McLIMANS (1984) for the Fatch Field, Dubai. Combination of burial history and fluid inclusion temperature data shows that petroleum migration occurs during the late-Oligocene to mid-Miocene. Furthermore, because the oil droplets were trapped from an H_2O dominated system, the calcite deposition occurs during early petroleum generation and migration, rather than at the peak generation and migration (HORSFIELD and McLIMANS, 1984).

2.3 - EVIDENCE FOR UNMIXING BETWEEN HYDROCARBON AND AQUEOUS INCLUSIONS

In numerous reservoirs, the coexistence of hydrocarbon inclusions and aqueous inclusions in the same mineral is interpreted as unmixing.

FLUID INCLUSIONS IN OIL AND GAS-BEARING SEDIMENTARY FORMATIONS

In albo-aptian feldspathic sandstones from Angola, microthermometric measurements on fluid inclusions present in albite and adularia show two different behaviour patterns during heating and freezing. For the hydrocarbon inclusions, an increase of the gas bubble volume is noted during cooling, and the homogenization in the liquid phase occurs around 110 ± 20°C (Fig. 8). These hydrocarbon inclusions also are easily distinguished by a yellow fluorescence under ultraviolet light.

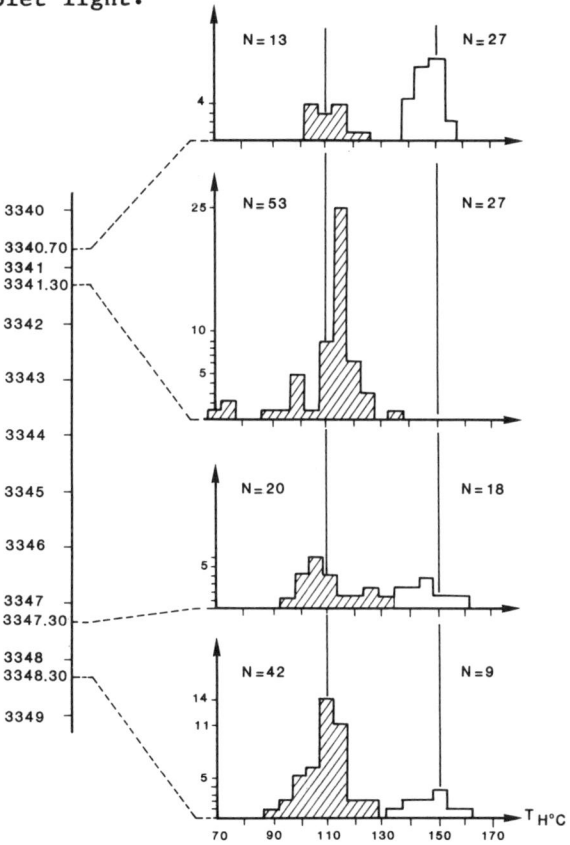

Figure 8 : Histograms of homogenization temperatures for hydrocarbon (hached) and aqueous (white) inclusions in samples taken at different levels in albo-aptian feldspathic sandstones from Angola.

For the aqueous inclusions, the gas bubble volume decreases slightly during cooling and there is solidification of the liquid phase. Slow heating after cooling shows final melting of the solid around -10°C, and homogenization in the liquid phase at 150 ± 15°C.

The fluid inclusions distribution supports the fact that they were trapped at the same time from a heterogeneous medium, despite the fact that no fluid inclusions containing both water and hydrocarbons are invisible under optical microscopy. Laser Mass Microspectrometry (LAMMA) destructive microanalysis of these inclusions are in agreement with unmixing : small quantities of water were detected in hydrocarbon inclusions, whereas small quantities of hydrocarbon were detected in aqueous inclusions. Evidence, for water in hydrocarbon inclusions, invisible under microscope because it forms a thin layer all around the cavity, is satisfactory for the feldspar crystallization. These data support an immiscibility process and a separate trapping of the two end members.

The homogenization temperatures of hydrocarbon inclusions are lower, by about 40°C, than the homogenization temperatures of aqueous inclusions. This fact is in accordance with the fluids unmixing PVT properties in the water hydrocarbon system.

Hydrocarbon and aqueous inclusions coexistence in the same occurrence already have been published (NARR and BURRUS, 1984 ; HASZEDLINE et al., 1984). In figure 9, two different cases are shown. In North Dakota, the inclusions are localized in quartz associated with thin, vertical, planar fractures (NARR and BURRUSS, 1984). The presence of some three-phases inclusions : an aqueous liquid, a hydrocarbon liquid and gas provide substantial evidence for unmixing. The difference between the homogenization temperatures of the hydrocarbon inclusion and of the aqueous inclusion is less than 20°C. HASZELDINE et al., 1984 show a case where homogenization temperature of an hydrocarbon inclusion is close to aqueous inclusions homogenization temperatures. However, current studies, in other areas from the North Sea, show that this is not always so.

The difference between homogenization temperatures of hydrocarbon and aqueous inclusions increases with the hydrocarbons maturity. The lighter hydrocarbons having the lowest critical point, their homogenization temperature is usually lower.

As previously shown, hydrocarbon and aqueous inclusions might be used to constrain the P-T trapping conditions.

Unfortunately, little data on petroleum source rocks exist since fluid inclusions may prove to be a valuable tool in the discussion of the oil transport in one- or two-phases. VISSER (1982) shows that in La Luna petroleum source rock (Venezuela) secondary aqueous and oil inclusions occur in calcite and quartz. This case indicates that oil migrates as a separate phase.

FLUID INCLUSIONS IN OIL AND GAS-BEARING SEDIMENTARY FORMATIONS

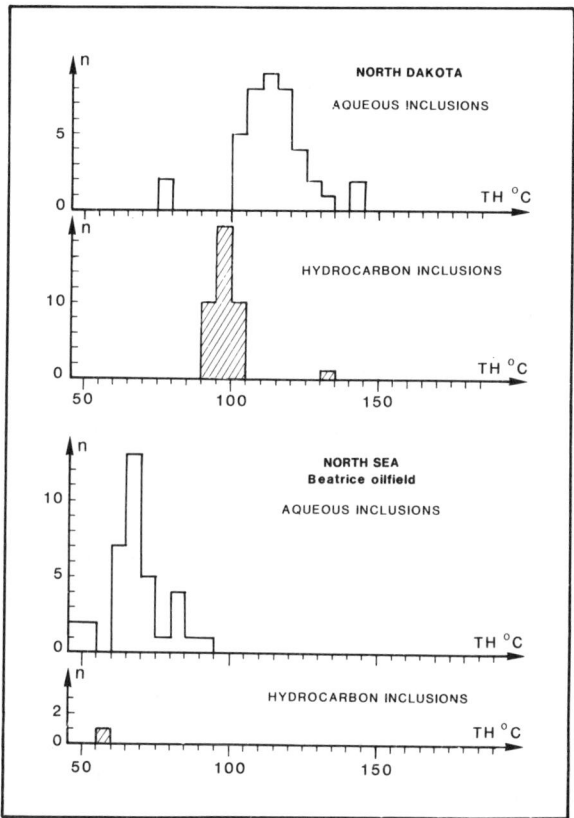

Figure 9 : Histograms of homogenization temperatures for coexisting hydrocarbon and aqueous inclusions in quartz associated with North Dakota fractures (NARR and BURRUSS, 1984) and in quartz overgrowths from North Sea Jurassic Sandstones (HASZELDINE et al., 1984).

2.4 REGIONAL VARIATIONS OF CHEMICAL AND THERMAL GRADIENTS IN SPACE AND TIME

Such reconstructions imply that samples must be studied at various depths, from different areas and using authigenic minerals in a petroleum reservoir, or in a source rock. Such an approach is not yet published. However, similar data has been obtained from sandstone-filled basins where uranium deposits occur at the bottom of the formation (PAGEL and POTY, 1984). It is interesting to note that recent fission track data (KOUL et al., 1984) on the Kombolgie sandstones (Australia) confirm the maximum temperature derived from fluid inclusions (DURAK et al., 1983). A combination of precise data on deposition and erosion history of sediments on the one hand and timing of diagenetic events on the other hand may be used with fluid inclusions data to obtain the thermal variation gradients in space and time.

2.5. LOCAL HYDROTHERMALISM IN A BASIN

In the southern part of the Gabon coastal basin, the basal sandstones (Hauterivian) show quartz veinlets containing aqueous inclusions with homogenization temperatures ranging from 250 to 310°C (average 283°C ± 12°C). The pressure correction is no more than 20°C. As no intense erosion has been noted in this area, this might be due to a hydrothermal fluid circulation related probably to deep-seated fracturing (as infered also by STORZER and SELO (1984)). K-Ar and Rb-Sr dating of the associated authigenic illite in the same sample gives an age of 120 ± 2 M.a..

2.6. OTHER APPLICATIONS OF FLUID INCLUSIONS AND FUTURE DEVELOPMENTS

Fluid inclusions might also be able to resolve other problems such as the fluid flow patterns and hydrocarbon content in aqueous solutions, the consequence of magmatic intrusions in a sedimentary sequence, the relationship between tectonics and hydrocarbon migration (NARR and BURRUSS, 1984), the alteration of the basement by diagenetic fluids underneath a deep basin and the nature of fluid in halite-bearing strata in oil and gas provinces (ROEDDER, 1984).

Acknowledgements : Many thanks to Professor J.F. MULLER from Metz University for giving us the opportunity to carry out LAMMA analyses and to Arnaud MEYER who worked with us during his University holidays.

REFERENCES

BURRUSS, R.C., (1981). Hydrocarbon fluid inclusions in studies of sedimentary diagenesis. In L.S. HOLLISTER and M.L. CRAWFORD, eds : Short course in fluid inclusions, applications to petrology. Mineralogical Association of Canada, v. 6, pp 138-156.

BURRUSS, R.C., CERCONE, K.R., and HARRIS, P.M., (1983). Fluid inclusion petrography and tectonic-burial history of the Al Ali No 2 well : Evidence for the timing of diagenesis and oil migration, northern Oman Foredeep. Geology, v. 11, pp 567-570.

CRAWFORD, M.L., (1981). Phase equilibria in aqueous fluid inclusions. In L.S. HOLLISTER and M.L. CRAWFORD, eds. : Short Course in fluid inclusions : applications to petrology. Mineralogical Association of Canada, v. 6, pp. 75-100.

CURRIE, J.B., and NWACHUKWU, S.O., (1974). Evidences on incipient fracture porosity in reservoir rocks at depth. Bull. Canadian Petroleum Geology, v. 22, pp 42-58.

DURAK, B., PAGEL, M., et POTY, B., (1983). Températures et salinités des fluides au cours des silifications d'une formation gréseuse surmontant un gisement d'uranium du socle : l'exemple des grès Kombolgie (Australie). C.R. Acad. Sc. Paris, t. 296, série II, pp 571-574.

HANOR, J.S., (1980). Dissolved methane in sedimentary brines ; potential effect on the PVT properties of fluid inclusions - Economic Geology, v. 75, pp 603-617.

HASZELDINE, R.S., SAMSON, I.M., and CORNFORD, C., (1984). Dating diagenesis in a petroleum basin, a new fluid inclusion method. Nature, v. 307, pp 354-357.

HORSFIELD, B., and MC LIMANS, R.K., (1984). Geothermometry and geochemistry of aqueous and oil-bearing fluid inclusions from Fatch Field, Dubai : Org. Geochem., v. 6, pp 733-740.

KOUL, S., WALL, V.J., and JOHNSTON, J.D., (1984). Fission-track studies of the East Alligator River uranium field (Northern Territory, Australia) and their implications. Abstr. Geol. Soc. Aust. 12, pp 311-313.

LEROY, J., (1979). Contribution à l'étalonnage de la pression interne des inclusions fluides lors de leur décrépitation. Bull. Mineral. 102, pp 584-593.

MALLEY, P., JOURDAN, A., and WEBER, F., (1985). Fluid inclusions in silicified sandstones from the Brent Formation, North Sea (Abstr.). E.U.G. Strasbourg. Terra Cognita, v.5, n°2-3, p. 296.

MOGE, M., (1985). Evolutions diagénétiques et caractéristiques pétrophysiques de formations gréseuses à porosité secondaire. Thèse Université Nancy I, (To be published in Sciences de la Terre, Nancy).

NARR, W., and BURRUSS, R.C., (1984). Origin of reservoir fractures in Little Knife Field, North Dakota. The American Association of Petroleum Geologists Bulletin, v. 68, n°9, pp. 1087-1100

PAGEL, M., (1975). Détermination des conditions physico-chimiques de la silicification diagénétique des grès Athabasca (Canada) au moyen des inclusions fluides. C.R. Acad. Sc. Paris. v. 280, série D pp 2301-2304.

PAGEL, M., and POTY, B., (1984). The evolution of composition, temperature and pressure of sedimentary fluids over time : a fluid inclusion reconstruction In : Thermal phenomena in sedimentary basins (Ed B. DURAND), Technip., pp 71-88.

PICHAVANT, M., RAMBOZ, C., and WEISBROD, A., (1982). Fluid immiscibility in natural processes. Use and misuse of fluid inclusion data I. Phase equilibria analysis - A theoretical and geometrical approach. Chem. Geol. 37, pp 1-28.

POTTER, R.W., (1977). Pressure corrections for fluid inclusion homogenization temperatures based on the volumetric properties of the system $NaCl-H_2O$ - J. Res. U.S. Geol. Surv., 5, pp. 603-607.

PRICE, L.C., (1976). Aqueous solubility of petroleum as applied to its origin and primary migration - AAPG, 60, pp 213-244.

RAMBOZ, C., PICHAVANT, M., and WEISBROD, A., (1982). Fluid immiscibility in natural processes. Use and misuse of fluid inclusion data. II : Interpretation of fluid inclusion data in terms of immiscibility. Chem. Geol. 37, pp 29-48.

ROEDDER, E., (1979). Fluid inclusion evidence on the environments of sedimentary diagenesis, a review. SEPM Special Publication n° 26, pp 86-107.

ROEDDER, E., (1984). Fluid inclusions. Reviews in Mineralogy, v. 12, 644 p.

STORZER, D., and SELO, M., (1984). Toward a new tool in hydrocarbon resource evaluation : the potential of the apatite fission track chrono-thermometer. In : Thermal Phenomena in sedimentary basins (Ed. B. DURAND), Technip. pp 89-110.

TUGARINOV, A.I., and NAUMOV, V.B., (1970). Dependence of the decrepitation temperature of minerals on the composition of their gas-liquid inclusions and hardness. Dokl. Acad. Sci. URSS, 195, pp 112-114.

VISSER, W., (1982). Maximum diagenetic temperature in a petroleum source-rock from Venezuela by fluid inclusion geothermometry. Chemical Geology, 37, pp 95-101.

WELSCH, H., (1973). Die System Xenon-Wasser and Methan-Wasser bei Drücken and Temperaturen - PhD Thesis, Karlsruhe.

D. W. PHELPS[1], T. M. HARRISON[2]

APPLICATION OF $^{40}Ar/^{39}Ar$ THERMOCHRONOLOGY ON DETRITAL POTASSIUM FELDSPARS TO THE STUDY OF SEDIMENTARY BASINS

I. ABSTRACT

$^{40}Ar/^{39}Ar$ age spectra of detrital potassium feldspars can be used to reconstruct the thermal history of sedimentary basins and to date the provenance of sandstones. Analyses of potassium feldspars from sandstones from the southern Viking Graben and Sole Pit Basin yield plateau ages that are consistent with the age of surrounding basement rocks. On the basis of the amount of recent $^{40}A_{rad}$ loss in the Viking Graben samples, a temperature history was constructed that agrees well with that derived independently by standard backstripping techniques.

II. INTRODUCTION

One of the outstanding problems in the study of sedimentary basins is the accurate determination of thermal history. Thermal history has received so much attention because it is directly related to such fundamental topics as the mechanism of basin formation, and the degree of lithospheric attenuation, if any, that a basin has experienced (McKenzie, 1978, Sclater and Christie, 1980; and many others).

Reconstruction of the thermal history of a basin generally utilizes a combination of two approaches. One approach involves constructing a thermal history by applying theoretical models of basin formation in conjunction with stratigraphic analysis such as

---
(1) *Exxon Production Research Company, Houston, Texas, USA.*
(2) *Department of Geological Sciences, State University of New-York at Albany, New-York, USA.*

sedimentary backstripping (e.g. Sclater and Christie, 1980; Barton and Woods, 1984). The second approach involves modifying the calculated thermal history to match the level of organic maturation observed in fine-grained sediments of a basin. The major limitation of this second aspect is that the kenetics of the organic reactions are not well known (Tissot and Welte, 1978; McKenzie et al., 1983), thus the resulting thermal history is somewhat imprecise.

Recently, however, inorganic reactions have also been used to constrain thermal history. These include fission track annealing and track length studies (Gleadow et al., 1983) and the application of $^{40}Ar/^{39}Ar$ incremental heating techniques to detrital potassium feldspars (Harrison and Be, 1983). In this paper we present the results of applying $^{40}Ar/^{39}Ar$ incremental heating experiments to potassium feldspars from sandstones of the southern Viking Graben and the Sole Pit Basin. The results have allowed us to accurately model the late Cenozoic thermal history of the southern Viking Graben and to make heretofore unobtainable observations on the age of the provenance of the sandstones.

III. GEOLOGIC SETTING

The North Sea is underlain by several related but separate sedimentary basins (Fig. 1). In this paper we are concerned with two of these basins; the Viking Graben and the Sole Pit basin.

The Viking Graben is a Mesozoic rift basin overlain by a thick sequence of nearly undeformed Cenozoic sediments. Although there is evidence of Permian extension in the Viking Graben, the phase of extension that generated the present-day structural trends began in the Triassic, culminated in the Late Jurassic and finally terminated near the end of the Early Cretaceous (Ziegler, 1981). Extension and normal faulting was followed by thermally controlled subsidence that continues into the Recent (Sclater and Christie, 1980; Barton and Woods, 1984). Thus, sediments in the Viking Graben are currently at their maximum depth of burial and maximum temperature.

In Sole Pit, Permian normal faulting occurred during deposition of the Rotliegende sandstones and Zechstein evaporites. Mesozoic extension which so strongly affected the Viking and Central Grabens, had relatively little effect on the Sole Pit basin (Glennie and Boegner, 1981). However, the Sole Pit basin underwent a period of structural inversion during the Late Cretaceous that caused up to 1500 meters of uplift and created the present-day Sole Pit structural axis (Glennie and Boegner, 1981).

IV. $^{40}Ar/^{39}Ar$ TECHNIQUE

The $^{40}Ar/^{39}Ar$ technique is a variation of the K/Ar dating method in which the sample is irradiated to generate ^{39}Ar from the normally stable ^{39}K. Thus, the ratio of ^{39}Ar to ^{40}Ar is implicitly the parent to daughter ratio which is a measure of the age of the sample. The age spectrum technique involves incremental heating of samples to progressively higher temperatures and measurement of the $^{40}Ar/^{39}Ar$ ratio in gas expelled during each temperature step. Any partial ^{40}Ar loss caused by slow cooling of the host intrusion or by subsequent episodic loss is expressed as anomalously young ages in the low temperature gas fractions. Details of the technique can be found in Mitchell (1968) and Turner (1970).

Harrison and McDougall (1982) demonstrated that the diffusion parameters for Ar in potassium feldspar derived from laboratory stepwise heating experiments can be extrapolated to geological conditions. Using these parameters, Harrison and McDougall (1982) were able to show that measurable loss of $^{40}Ar_{rad}$ occurs at temperatures as low as 110°C over geologically significant periods of time. Harrison and Be (1983) have applied the stepwise heating technique to detrital potassium feldspars collected from sandstones to model the thermal history of sedimentary rocks. They demonstrated that the time necessary to account for the observed $^{40}Ar_{rad}$ loss (as determined from the age spectrum) can be determined using the diffusion parameters obtained from the stepwise heating experiments along with knowledge of the present-day temperature and burial history.

V. ANALYTICAL PROCEDURES

High purity K-feldspar separates, sized to between 0.25 and 0.125 mm in diameter were irradiated for 80 hours in a fast neutron flux in the core of the Ford reactor, Phoenix Memorial Laboratory, University of Michigan, along with the Australian National University inter-laboratory standard GA1550 biotite. This mica has a K-Ar age of 97.9 Ma using the decay constants and isotopic abundances recommended by Steiger and Jager (1977). The fast neutron flux in this position is about 1.5×10^{13} n-cm^{-2} sec^{-1} at 2 MW power and the fast to thermal flux ratio is about unity. The samples were placed in flat-bottomed, evacuated, 100 mm long quartz tubes which were filled to about a 50 mm height with alternating unknowns and monitors. The quartz vials were positioned in the reactor in such a way as to ensure that the middle region on the samples was coincident with the maximum flux. This serves to minimize the flux gradient across the 50 mm length of the samples to about 2%. Correction factors used for interfering isotopes produced by nuclear reactions on K and Ca are $(^{40}Ar/^{39}Ar)_K = 3.82 \times 10^{-2}$, $(^{36}Ar/^{37}Ar)_{Ca} = 2.86 \times 10^{-4}$ and $(^{39}Ar/^{37}Ar)_{Ca} = 7.51 \times 10^{-4}$. Argon extractions were performed in a resistance heated, double vacuum extraction furnace, a much modified version of the system described by Staudacher and others (1978). Integral temperature control by a solid state controller and thyrmistor device allows precision in temperature monitoring of $\pm 0.5°C$, more than sufficient for kinetic studies (e.g. Harrison and McDougall, 1982). The blank imparted by this furnace and the clean up system, consisting of an ion-pump and 50 l s^{-1} SEAS getter pump averages about 5×10^{-15} mol ^{40}Ar at temperatures below 1200°C. The evolved gas at each temperature step was isotopically analyzed using a Nuclide 4.5-60RSS instrument with automated, digital data acquisition. Uncertainties in the derived $^{40}Ar/^{39}Ar$ ages include the precision of the isotope ratio measurements in both the sample and flux monitor, but does not include systematic uncertainty of the age of the flux monitor of about 0.5%.

VI. RESULTS

We analyzed potassium feldspar separates from five sandstone cores from three wells in the southern Viking Graben, from two sandstone cores from two wells in the Sole Pit Basin and from an outcrop of Rotliegende sandstone near Durham, England (Fig. 1). Stratigraphic age, depth below the sea bottom, present-day estimated temperature and age spectra for all 8 samples are shown in Figures 2 to 6.

We analyzed the outcrop sample of the Rotliegende sandstone as an example of a sedimentary rock that has not been heated to temperatures greater than about 50°C. The sample (DUR-1) comes from the Crime Riggs sand pit near Durham, England. At this locality the Rotliegende consists of unconsolidated sands (Glennie and Buller, 1983) that most likely have not been buried by more than about one kilometer of additional sediments. The age spectrum generated from the potassium feldspar separate (Fig. 2) is consistent with shallow burial. The spectrum shows a well developed plateau age of 360 $\pm$ 1 Ma that incorporates greater than 95% of the total gas released with only slightly younger ages characterizing the low temperature steps. The shape of the age spectrum exhibited by DUR-1 is typical of that of thermally undisturbed potassium feldspars derived from plutonic rocks (Harrison and Be, 1983).

The two core samples of the Rotliegende from the Sole Pit Basin exhibit much more complex age spectra (Fig. 3a and 3b). Both samples

$^{40}Ar/^{39}Ar$ THERMOCHRONOLOGY TO SEDIMENTARY BASINS

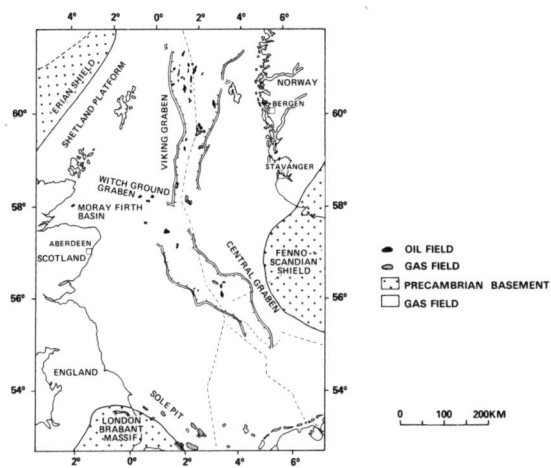

FIGURE 1. Map of the North Sea showing major Mesozoic structural trends and the distribution of basement rock types (after Ziegler, 1981).

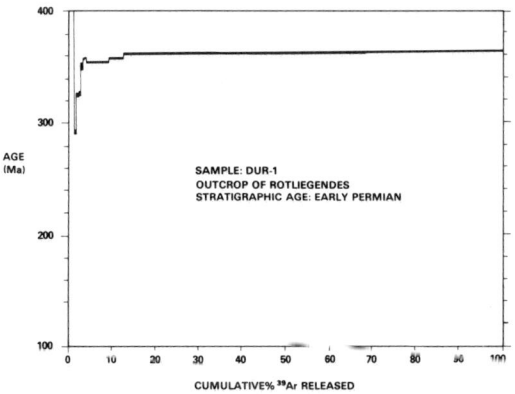

FIGURE 2. $^{40}Ar/^{39}Ar$ age spectrum of Sample DUR-1.

589

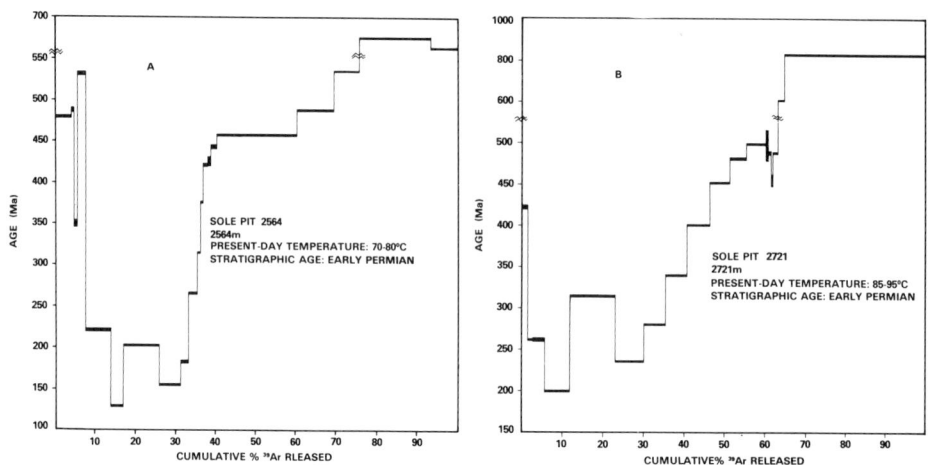

FIGURE 3. Spectra of Samples 2564 and 2721, Sole Pit Basin.

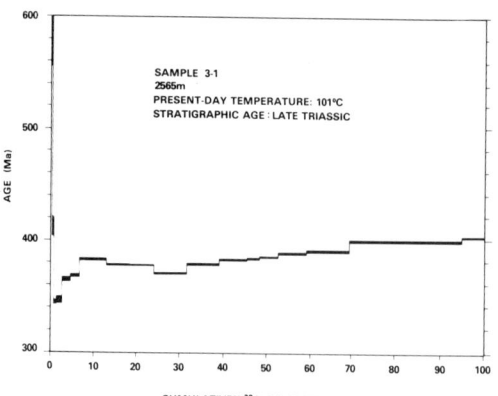

FIGURE 4. $^{40}Ar/^{39}Ar$ age spectrum of Sample 3-1, southern Viking Graben.

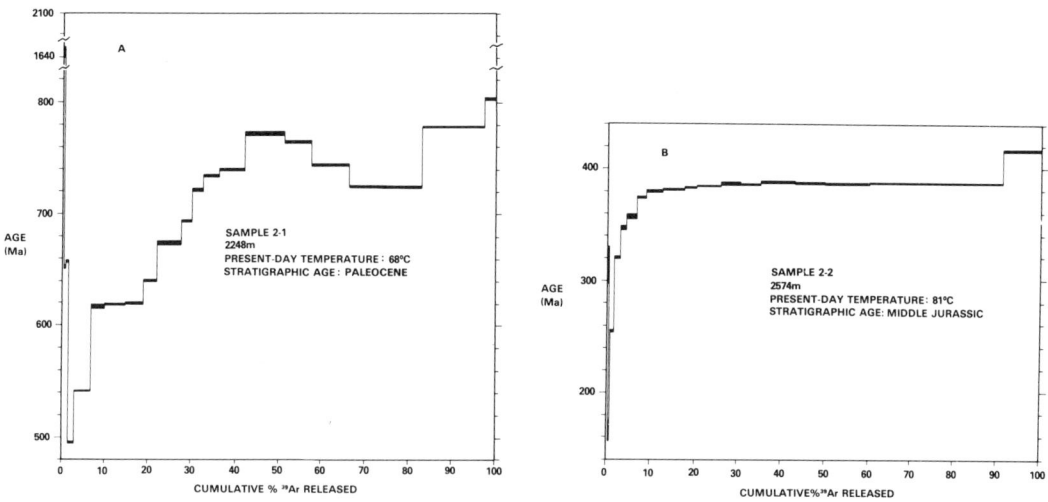

FIGURE 5. $^{40}Ar/^{39}Ar$ age spectra of Samples 2-1 and 2-2, southern Viking Graben.

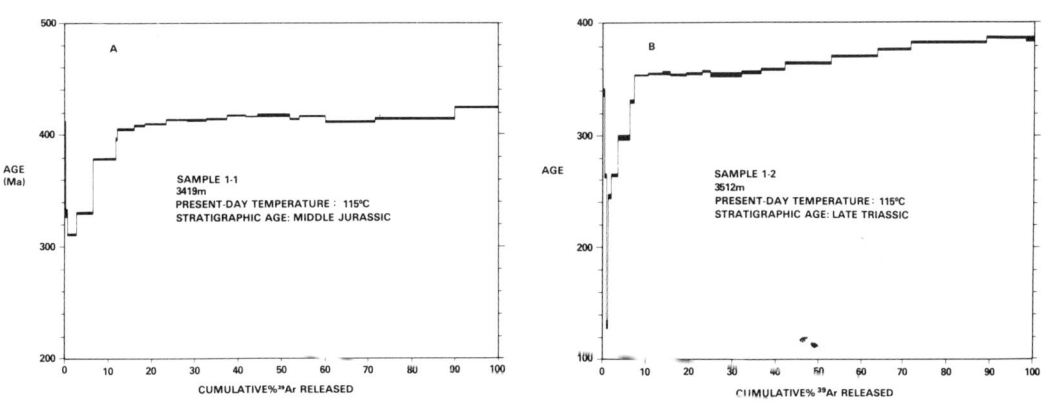

FIGURE 6. $^{40}Ar/^{39}Ar$ age spectra of Samples 1-1 and 1-2, southern Viking Graben.

are probably mixtures of feldspars of at least two ages. In addition the ages obtained from the first 30% of the gas released from sample 2564 are younger than the stratigraphic age of the sample (Early Permian), suggesting partial degassing during a thermal event that occurred sometime after deposition. Excess ^{40}Ar present in the lowest temperature gas fractions from both 2564 and 2721 renders it impossible to precisely determine the age at which the samples were degassed.

Samples 3-1, 2-2, 1-2, and 1-1 from the southern Viking Graben all show plateau ages encompassing greater than 90% of the total gas released that cluster around 400 Ma or that rise monotonically from about 370 Ma to 400 Ma (Figs. 4, 5b, 6a and 6b). Sample 2-1 (Fig. 5a) has a much more complex spectra that probably results from a mixture of feldspars of different ages. All the southern Viking Graben samples show steadily decreasing ages in the lowest temperature gas fractions. However, only samples 2-2 and 1-2 have ages in the low temperature gas fractions that are as young or younger than their respective stratigraphic ages (Jurassic and Triassic), suggesting ^{40}Ar$_{rad}$ loss following deposition of the sedimentary rock.

VII. DISCUSSION

A. PROVENANCE STUDIES

Figure 7 is a plot of the stratigraphic age of each sample versus the plateau age or minimum age of the potassium feldspars as determined from the ^{40}Ar/^{39}Ar incremental heating experiments. Included in Figure 7 are results from Jurassic sandstones from the Tartan field in the Moray Firth area (Harrison, 1984). The ^{40}Ar/^{39}Ar age spectra of the potassium feldspars of all six Mesozoic sandstones indicate that these feldspars were derived from plutons that cooled below the closure temperature of Ar in potassium feldspar (about 130°C; Harrison and McDougall, 1982) between 360 and 400 Ma. None of these samples show any evidence of mixing with potassium feldspar from older terrains. In contrast, the Paleocene sandstone from the southern Viking Graben and the two Rotliegende core samples from the Sole Pit Basin contain a mixture of potassium feldspars, some of which are at least as old as Late Precambrian. These contrasting feldspar populations can be explained in terms of paleogeography and depositional environment.

The Triassic sandstones in the Viking Graben are interpreted as fluvial deposits laid down during active rifting, perhaps in fault bounded basins (Fisher, 1984). The Jurassic sediments are considered to be deltaic, deposited by a system of deltas that advanced

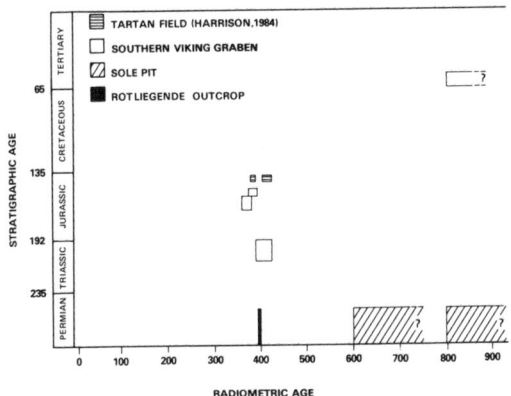

FIGURE 7. Radiometric age as determined by the $^{40}Ar/^{39}Ar$ age spectrum technique of potassium feldspar separates from sandstones from the North Sea area plotted against the stratigraphic age of each sample.

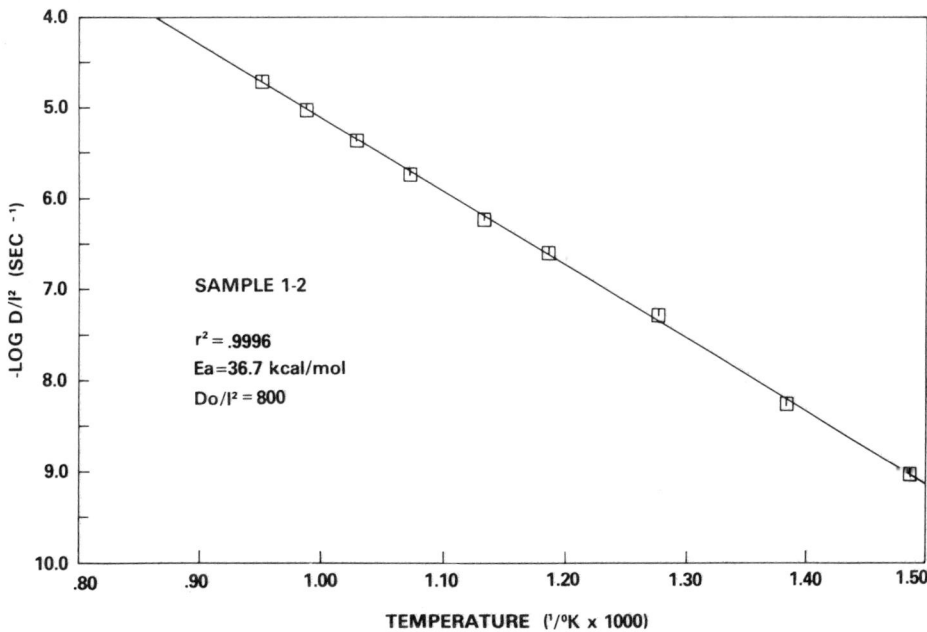

FIGURE 8. Arrhenius relationships for Sample 1-2.

from south to north, probably in response to Middle Jurassic uplift near the intersection of the Viking, Central, and Moray Firth Grabens (Eynon, 1981; Brown, 1984). The crystalline basement underlying the Viking graben is predominantly SiluroDevonian intrusive and metamorphic rocks of the Caledonian orogeny (Frost et al., 1981; Fig. 1). A single $^{40}Ar/^{39}Ar$ age spectrum determined on micas from a biotite microgranite from well 16/2-1 in the Viking Graben resulted in an estimated age of intrusion of 409 $\pm$ 6 Ma (Frost et al., 1981). This age correlates well with the cooling ages obtained on potassium feldspars from all sixx Mesozoic sandstones. We suggest that the potassium feldspars in the Mesozoic sandstones were derived directly from the surrounding basement rocks, were not transported great distances and received no significant detrital component from the Precambrian terrains that flank the North Sea.

The single sample of Paleocene sandstone from the southern Viking Graben (sample 2-1) contains potassium feldspars that are at least as old as Late Precambrian (Fig. 5a). These Paleocene sandstones are interpreted as submarine fan deposits derived from the Shetland platform area (Rochow, 1981). Significantly, the eastern part of the Shetland platform is underlain by Precambrian crystalline rocks (Ziegler, 1981; Fig. 1). Thus, our date provide independent support for the conclusions of Rochow (1981) that the Paleocene sandstones of the southern Viking graben were derived from the Precambrian Erian Shield of the Shetland Platform. The complex age spectrum of sample 2-1 may result from mixing with younger feldspars during transport from the Shetland Platform to the Viking Graben.

The feldspar separates from the Rotliegende cores from the Sole Pit Basin also have complex age spectra that we interpret as resulting from a mixture of potassium feldspars of Late Precambrian and Caledonian age (plus post-depositional $^{40}Ar_{rad}$ loss, see the following section). The Rotliegende consists predominantly of eolian sandstones (Glennie and Buller, 1983). One might expect, in this depositional environment, to encounter feldspars from a variety of source regions. The nearby LondonBrabant Massif is a likely source for the Precambrian feldspars (Fig. 1), and the surrounding Caledonian basement is a likely candidate for the younger Siluro-Devonian aged feldspars.

B. THERMAL HISTORY

1. Southern Viking Graben

The age spectra of the potassium feldspar separates from samples 3-1 and 1-1 are typical of that expected for thermally undisturbed

$^{40}Ar/^{39}Ar$ THERMOCHRONOLOGY TO SEDIMENTARY BASINS

potassium feldspars derived from plutonic rocks (Harrison and Be, 1983). Thus these two samples have not recorded the thermal history of the sedimentary rock. However, the age spectra of samples 1-2 and 2-2 both have ages in the low temperature gas fractions that are as young as or younger than their respective stratigraphic ages, suggesting loss of $^{40}Ar_{rad}$ sometime after incorporation into the sandstone.

Sample 2-2 is somewhat enigmatic. Compared to the virtually unaffected spectrum of DUR-1, sample 2-2 appears to have lost about 1% of its $^{40}Ar_{rad}$ at some time after 170 Ma. Burial history curves for wells from this part of the Viking Graben (see Fig. 9) suggest that the present-day burial depth, and hence temperature, are the maximum that the sample has experienced. The present-day estimated temperature of 81°C is too low to account for even 1% $^{40}Ar_{rad}$ loss from potassium feldspars with an activation energy of between 34 and 36 kcal. However, 1% $^{40}Ar_{rad}$ loss could be accounted for if the present-day temperature was about 90 to 95°C instead of 81°C, provided that these temperatures were maintained for the past 1 to 2 Ma. The temperature profile for well #2 was drawn on the basis of only two data points, and we estimate that an error of 10°C is possible. Supporting evidence for higher present-day temperatures also comes from fission track analyses of apatites from this same sample. Unpublished track length studies (A. Gleadow, written communication, 1984) suggests annealing appropriate to temperatures of 90 to 95°C. We conclude that the temperature estimated for sample 2-2 is too low and that the correct present-day temperature is between 90 and 95°C, thereby accounting for the small amount of observed $^{40}Ar_{rad}$ loss.

Sample 1-2 shows unequivocal evidence of post-depositional $^{40}Ar_{rad}$ loss. The youngest age obtained from the low temperature steps is about 130 Ma, significantly younger than the Triassic stratigraphic age of the sample. The amount of $^{40}Ar_{rad}$ lost can be estimated at 2 to 4% by comparison with the undisturbed age spectrum of sample DUR-1. Using methods outlined by Harrison and McDougall (1982), the calculated diffusion parameters for sample 1-2 are an activation energy (E_a) of about 36 kcal/mole and a combined frequency factor/grain size parameter ($D_o/1^2$) of 800 (Fig. 8). Solving equation 5 of Harrison and Be (1983), we calculate that the observed $^{40}Ar_{rad}$ loss of 2 to 4% in sample 1-2 should have occurred in less than 5 Ma at the present-day temperature of 115°C, assuming a square pulse temperature history.

However, a square pulse temperature history is not geologically realistic. In order to construct a more reasonable temperature history we carried out a geohistory analysis of well #2 using standard backstripping techniques similar to those of Sclater and Christie (1980). These results are shown in Figure 9. The subsidence curve, corrected for the effects of sea level fluctuations, water depth,

595

$^{40}Ar/^{39}Ar$ THERMOCHRONOLOGY TO SEDIMENTARY BASINS

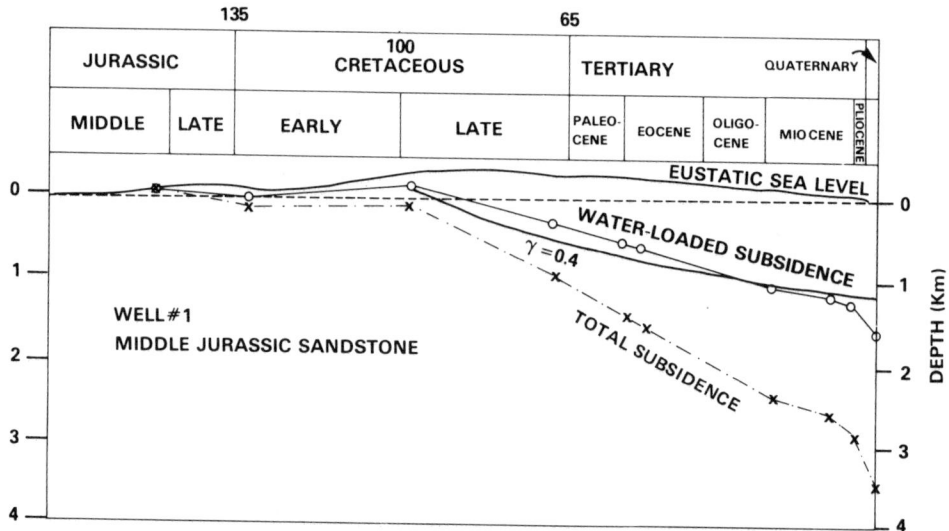

FIGURE 9. Geohistory analysis of well #1, southern Viking Graben. The analysis was done on the Middle Jurassic sandstone because of the lack of adequate age control on the older units. Theoretical subsidence curve assuming crustal thinning of 40% (curve labeled $\gamma = 0.4$) is shown for comparison.

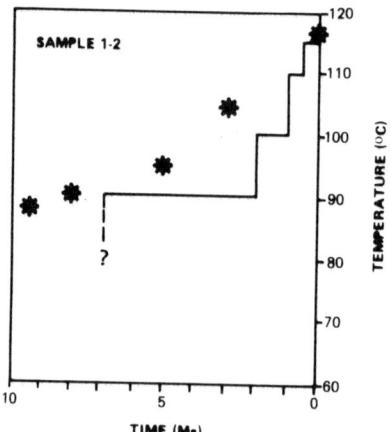

FIGURE 10. Temperature history for well #1, southern Viking Graben, calculated using the results of the $^{40}Ar/^{39}Ar$ data (solid line) assuming 3% $^{40}Ar_{rad}$ loss and geohistory analysis (stars).

sediment compaction, and sediment loading assuming Airy isostasy are consistent with crustal attenuation of about 40% in a rifting event that ended near the end of the Early Cretaceous (Fig. 9). Assuming a present-day heat flow of 63 mW/m^2 (Cermak and Hurtig, 1979), we reconstructed a heat flow history for well #1 on the basis of 40% crustal stretching at the end of the Early Cretaceous. We then solved the heat flow equation through time assuming one-dimensional conductive heat flow to generate a temperature history for sample 1-2. The results of these calculations (Fig. 10) indicate a rapid temperature rise over the past 5 Ma to the present-day temperature. This agrees qualitatively with the ^{40}Ar/^{39}Ar results which indicate that the present-day temperature has been maintained for less than 5 Ma.

We then used the theoretically derived temperature history as a guide and constructed a temperature history consisting of a number of small isothermal steps (Harrison and Be, 1983) that sum to a total of 2 to 4% ^{40}Ar$_{rad}$ loss for the feldspars in sample 1-2. The results (Fig. 10) show that our quota of 2 to 4% ^{40}Ar$_{rad}$ loss can be achieved with a temperature history that is nearly identical to that calculated using backstripping techniques. Both methods of calculating the temperature history show that the temperature of the Triassic sandstone, from which sample 1-2 was derived, has risen above 90°C only within the past 5 Ma.

C. SOLE PIT BASIN

The complexity of the age spectra from the Sole Pit Basin samples precludes a quantitative evaluation of their temperature history. However, the age spectra of both samples contain ages that are significantly younger than the stratigraphic age of the Rotliegende group (Early Permian), suggesting post-depositional ^{40}Ar$_{rad}$ loss. The amount of ^{40}Ar$_{rad}$ loss cannot be determined precisely because of the complex age spectra, but the fact that ages for sample 2564, over the first 30% of the gas released, are younger than the stratigraphic age suggests that the loss is on the order of 10 to 20% of the total ^{40}Ar$_{rad}$ present. On the basis of the age spectra alone, Ar loss can be constrained to have occurred during the last 120 Ma. Furthermore, the presentday temperature of sample 2564 is too low to account for significant Ar loss (even if our temperature estimates are 10°C too low), suggesting that the sample was exposed to higher temperatures in the past. This conclusion is supported by sonic velocity anomalies in the sedimentary rocks which indicate that as much as 1500 m of uplift occurred in the Sole Pit basin near the end of the Cretaceous (Glennie, 1981). In addition, our data demonstrate that the Rotliegende must have reached temperatures in excess of 110°C prior to the Late Cretaceous uplift. At present-day thermal gradients, a minimum of 1000 m of additional burial is required to raise the temperature to 110°C.

VIII. CONCLUSIONS

The $^{40}Ar/^{39}Ar$ age spectrum technique can be applied to detrital potassium feldspars from sandstones to derive valuable provenance and thermal history data. For the North Sea we found two distinct sediment source terrains that correlate well with paleogeography and depositional environment. In the southern Viking Graben we were able to use the $^{40}Ar/^{39}Ar$ age spectrum technique to document a rapid rise in temperature within the past 5 Ma. The temperature history derived using the $^{40}Ar/^{39}Ar$ technique correlates well with the burial history of significant Pliocene to Recent deposition. In the Sole Pit Basin the $^{40}Ar/^{39}Ar$ age spectrum technique confirmed the occurrence of a minimum of 1000 m of uplift sometime after 120 Ma ago. This finding is consistent with published estimates of 1500 m of uplift in the Late Cretaceous (Glennie, 1981).

IX. ACKNOWLEDGMENTS

The authors thank Mat Heizler for his help in analyzing the samples from Sole Pit. Esso Norway Inc., Esso Exploration and Production U. K. Ltd., and Exxon Production Research Company kindly granted us permission to publish these data. The paper was received by Dr. R. C. Vierbuchen whose helpful comments and encouragement throughout the entire project were appreciated.

X. REFERENCES CITED

BARTON, P. and WOOD, R., 1984, Tectonic evolution of the North Sea basin: crustal stretching and subsidence: Geophys. Jour. Royal Astr. Soc., v. 79, p. 987-1022.

BROWN, S., 1984, Jurassic, in Glennie, K. W. ed., Introduction to the Petroleum Geology of the North Sea: Oxford, Blackwell Scientific Publications, p. 103-131.

CORMAK, V. and HURTIG, E., 1979, Heat flow map of Europe, in Cermak, V. and Rybach, L., eds., Terrestrial Heat Flow in Europe: Berlin, Springer-Verlag, enclosure.

EYNON, G., 1981, Basin development and sedimentation in the Middle Jurassic of the northern North Sea, in Illing, L. V. and Hobson, G. D., eds., Petroleum geology of the continental shelf of northwest Europe. London, Institute of Petroleum, p. 196-204.

FISHER, M. J., 1984, Triassic, in Glennie, K. W., ed., Introduction to the Petroleum Geology of the North Sea: Oxford, Blackwell Scientific Publications, p. 85-101.

FROST, R.T.C., FITCH, F. J. and MILLER, J. A., 1981, The age and nature of the crystalline basement of the North Sea basin, in Illing, L. V. and Hobson, G. D., eds., Petroleum geology of the continental shelf of northwest Europe: London, Institute of Petroleum, p. 43-57.

GLEADOW, A.J.W., DUDDY, I. R. and LOVERING, J. F., 1983, Fission track analysis: a new tool for the evolution of thermal histories and hydrocarbon potential: Australian Petroleum Expl. Assoc. Jour., v. 23, p. 93-102.

GLENNIE, K. W. and BOEGNER, P.L.E., 1981, Sole Pit inversion tectonics, in Illing, L. V. and Hobson, G. D., eds., Petroleum geology of the continental shelf of northwest Europe: London, Institute of Petroleum, p. 110-120.

GLENNIE, K. W. and BULLER, A. T., 1983, The Permian Weissliegend of NW Europe: the partial deformation of aeolian dune sands caused by the Zechstein transgression: Sedimentary Geol., v. 35, p. 43-81.

HARRISON, T. M., 1984, Microcline thermochronology: an approach to determining the temperature history of sedimentary basins (abs.): Amer. Assoc. Pet. Geol. Bull., v. 68, p. 795.

HARRISON, T. M. and MCDOUGALL, I., 1982, The thermal significance of potassium feldspar K-Ar ages inferred from $^{40}Ar/^{39}Ar$ age spectrum results: Geochim. Cosmochim. Acta, v. 46, p. 1811-1820.

HARRISON, T. M. and BE, K., 1983, $^{40}Ar/^{39}Ar$ age spectrum analysis of detrital microclines from the southern San Joaquin Basin, California: an approach to determining the thermal evolution of sedimentary basins: Earth Planet. Sci. Lett., v. 64, p. 244-256.

MCKENZIE, D. P., 1978, Some remarks on the development of sedimentary basins: Earth Planet. Sci. Lett., v. 40, p. 25-32.

MCKENZIE, D., MACKENZIE, A. S., MAXWELL, J. R. and SAJGO, Cs., 1983, Isomerization and aromatization of hydrocarbons in stretched sedimentary basins: Nature, v. 301, p. 504-506.

MITCHELL, J. G., 1968, The argon-40/argon-39 method for potassium-argon age determination: Geochim. Cosmochim. Acta, v. 32, p. 781-790.

ROCHOW, K. A., 1981, Seismic stratigraph of the North Sea 'Palaeocene' deposits, in Illing, L. V. and Hobson, G. D., eds., Petroleum geology of the continental shelf of northwest Europe: London, Institute of Petroleum, p. 255-266.

SCLATER, J. G. and CHRISTIE, P.A.F., 1980, Continental stretching an explanation of the past mid-Cretaceous subsidence of the central North Sea Basin: Jour. Geophys. Res., v. 85, p. 3711-3739.

STAUDACHER, T.h., JESSBERGER, E. K., Dorflinger, D. and Kiko, J., 1978, A refined ultra-high vacuumed furnace for rare gas analysis: Jour. Phys. E.: Sci. Instrum., v. 11, p. 781-784.

STEIGER, R. H. and JAGER, E., 1977, Subcommission on geochronology: convention on the use of decay constants in geo- and cosmochronology: Earth Planet. Sci. Lett., v. 36, p. 359-362.

TISSOT, B. P. and WELTE, D. H., 1978, Petroleum Formation and Occurrence -A New Approach to Oil and Gas Exploration: Berlin, Springer-Verlag, 538 p.

TURNER, G., 1970, Thermal histories of meteorites, in Runcorn, S. K., ed., Paleogeophysics: London, Academic Press, p. 491-502.

ZIEGLER, P. A., 1981, Evolution of sedimentary basins in northwest Europe, in Illing, L. V. and Hobson, G. D., eds., Petroleum geology of the continental shelf of northwest Europe: London, Institute of Petroleum, p. 3-39.

Achevé d'imprimer
en juin 1986
par **CID** Éditions
44800 Saint-Herblain
N° d'éditeur : 726
Dépôt légal juin 1986,

Imprimé en France

COLLECTION COLLOQUES ET SÉMINAIRES

1. Les recherches et la production du pétrole en mer - Offshore. *(épuisé)*
2. Phénomènes d'auto-inflammation dans les moteurs alternatifs. *(épuisé)*
3. Le filtrage en sismique (Tome 1). *(épuisé)*
4. Techniques marines pour la recherche et l'exploitation du pétrole. *(épuisé)*
5. Le financement des investissements dans l'industrie du pétrole. *(épuisé)*
6. Planification et contrôle des constructions industrielles pour l'exploitation des gisements de pétrole. *(épuisé)*
7. Les ordinateurs en géologie pétrolière et dans les études de production.
8. Les méthodes de calcul sur ordinateurs appliquées au raffinage et à la pétroléochimie.
9. Les fluides de forage. *(épuisé)*
10. La fiabilité au service de l'entretien et de l'inspection du matériel. *(épuisé)*
11. Problèmes généraux de la distribution des produits pétroliers.
12. Les produits pétroliers et l'entretien.
13. Le forage aujourd'hui (3 tomes). 1. Méthodes de forage. 2. Matériels et techniques particulières. 3. Opérations spéciales.
14. La plongée profonde.
15. Pétrole et pétrochimie.
16. Connaissance de la houle, du vent, du courant pour le calcul des ouvrages pétroliers.
17. Têtes de forage sous-marines.
18. L'évolution récente des techniques et de l'économie de l'industrie pétrochimique.

18 *bis* Recent Technological and Economic Developments in the Petrochemical Industry.

19. Coûts de transport et approvisionnements en pétrole brut à long terme.
20. Les hauts polymères thermostables.
21. Méthodes rapides d'analyse des huiles usagées.
22. Histoire structurale du golfe de Gascogne (2 tomes).

23 Méthodes de développement en mer.

24 Les traitements des eaux dans l'industrie pétrolière.

25 Le positionnement en mer.

26 Le gaz naturel dans l'approvisionnement énergétique de l'Europe de l'Ouest.

27 Les lubrifiants industriels.

28 Dégradation microbienne des matériaux.

29 Les fluides de travail des métaux.

30 Economies d'énergie en raffinage et pétrochimie.

31 Tendances et perspectives de l'industrie des hydrocarbures. Actes du deuxième séminaire pétrolier international.

32 Perspectives énergétiques et impacts sur l'industrie des hydrocarbures. Actes du troisième séminaire pétrolier international.

33 Les lubrifiants moteur. Pertes par frottement et usure.

34 Climatologie de la mer. Sea Climatology.

35 La crise mondiale de l'énergie. Nécessité d'une coopération pétrolière internationale. Actes du quatrième séminaire pétrolier international.

36 Wave and Wind Directionality. Applications to the Design of Structures.

37 Symposium on the Pressuremeter and its Marine Applications.

38 Les polymères organiques utilisables à températures élevées. Applications. Durabilité. Environnement.

39 La lubrification industrielle (2 tomes). 1. Transmissions. Compresseurs. Turbines. 2. Travail des métaux. Graisses. Surveillance. Pollution.

40 Caractérisation des huiles lourdes et des résidus pétroliers. Characterization of heavy crude oils and petroleum residues.

41 Thermal Phenomena in Sedimentary Basins.

42 Interactions solide-liquide dans les milieux poreux. Solid-liquid interactions in porous media.

43 Numerical Methods in Offshore Piling. Méthodes numériques de calcul des pieux pour les ouvrages en mer.